Scientific Foundations of Audiology

Perspectives from Physics, Biology, Modeling, and Medicine

Editor-in-Chief for Audiology
Brad A. Stach, PhD

Scientific Foundations of Audiology

Perspectives from Physics, Biology, Modeling, and Medicine

Anthony T. Cacace, PhD
Emile de Kleine, PhD
Avril Genene Holt, PhD
Pim van Dijk, PhD

KH

5521 Ruffin Road
San Diego, CA 92123

e-mail: info@pluralpublishing.com
Website: http://www.pluralpublishing.com

Typeset in 10½/13 Palatino by Flanagan's Publishing Service, Inc.
Printed in the United States of America by McNaughton & Gunn, Inc.

The functional MRI figures on the front cover were provided courtesy of Dr. Chris Lanting (Lanting et al., 2014). The other graphs (Middle ear Power reflectance, DPOAEs, and Audiogram) were provided courtesy of Dr. A.T. Cacace (unpublished data).

Lanting, C. P., de Kleine, E., Langers, D. R. M., van Dijk, P. (2014). Unilateral Tinnitus: Changes in Connectivity and Response Lateralization Measured with fMRI. *PLoS 1, 9,* e110704, 1–14.

Library of Congress Cataloging-in-Publication Data

Names: Cacace, Anthony T., editor. | Kleine, Emile de, editor. | Holt, Avril Genene, editor. | Dijk, Pim van, 1960- , editor.
Title: Scientific foundations of audiology : perspectives from physics, biology, modeling, and medicine / [edited by] Anthony T. Cacace, Emile de Kleine, Avril Genene Holt, Pim van Dijk.
Description: San Diego, CA : Plural Publishing, Inc., [2016] | Includes bibliographical references and index.
Identifiers: LCCN 2015046757 | ISBN 9781597566520 (alk. paper) | ISBN 1597566527 (alk. paper)
Subjects: | MESH: Audiology—methods | Hearing Disorders—diagnosis | Hearing Disorders—therapy | Hearing—physiology
Classification: LCC RF290 | NLM WV 270 | DDC 617.8—dc23
LC record available at http://lccn.loc.gov/2015046757

9/23/16

CONTENTS

INTRODUCTION

This is not your typical textbook in audiology; rather, it represents a compendium of state-of-the-art chapters on unique topics dealing with hearing, vestibular, and brain science, the majority of which are not found in standard texts but are highly pertinent to the field. The underlying theme is that audiology is the primary "translational interface" between basic science and clinical concerns. Trained primarily as clinicians and clinical scientists, audiologists are situated in a unique position to implement breakthroughs in engineering, molecular biology, neuroimaging, genetics, medicine, nanobioscience, etc., and deliver them to the clinic. However, the underlying advancements require a fundamental understanding of advanced concepts and materials. Therefore, our intent is to provide a foundation for doctoral students in audiology, physics, neurobiology, and engineering and residents in various medical specialties (otolaryngology, neurology, pediatrics, and neurosurgery) with the background and concepts necessary to facilitate understanding in these different areas.

Of the "Current issues" subsumed within this book, we focus on topics that have practical, experimental, and theoretical value. The practical information is clearly apparent and is directly applicable to clinical situations. However, within this material, we also provide insight into basic areas of research where technical information is developing, where our understanding is incomplete, where theory has *not* been applied in a rigorous manner, and where experimental models can be improved upon to validate our concepts in complex areas. We hope that the end result will inspire new investigators to fill in the gaps and advance the field.

Moreover, it should be obvious that after viewing the table of contents, the topics being covered are expansive. They range from areas of basic science (anatomy, physiology, genetics, gene expression, molecular biology, neurochemistry) and clinical concerns (peripheral and central otopathology) to other relevant domains in assessment and treatment. They cover physical principles of middle ear and inner ear function (auditory, vestibular, balance), molecular and neural substrate underlying normal and pathologic activity in afferent and efferent pathways, implanted devices (cochlear and midbrain implants), mechanisms of speech perception associated with electrical stimulation, to the cortical processing of sound (normal and pathological) using noninvasive methods vis-à-vis magnetic resonance imaging (MRI).

We also consider "Future perspectives" in a similar context to those areas described above. However, these particular areas will no doubt be transformative in nature, where advancements are motivated by the ingenuity of the investigators and where the potential to produce large dividends (successful treatments and potential cures) is on the horizon. One area of interest concerns the combined use of manganese-enhanced MRI (MEMRI), gene expression, and functionalized nanoparticles to treat noise-induced tinnitus. Another very exciting domain concerns novel approaches for the protection and restoration of hearing. This highly fluid area is expected to have substantial impact on the field, where future developments remain extremely bright.

It is our hope that information derived from these topics expands one's knowledge base but also provides the incentive to improve the status quo. However, this is not an easy task. To succeed in this ambitious undertaking, we have assembled a stellar array of international world-class scientists, clinicians, and scholars to ensure that state-of-the-art technical information is explicated in an understandable, logical, and cohesive manner. The authors of these chapters have taken this task very seriously and share the common responsibility for giving an exposé on potential gaps in knowledge that currently exist in a thoughtful and unselfish manner. We are extremely grateful for their efforts and contributions.

To summarize, we believe that this book will have many beneficiaries. They will be independent of geographical boundaries but will have in common the desire to learn and apply new and advanced concepts to everyday situations. This includes a broad spectrum of individuals from multiple scientific disciplines, including medicine (otolaryngology, pediatrics, neurology, neurosurgery), engineering (biomedical, mechanical, electrical, chemical), basic science (neuro/molecular biology and neurochemistry), rehabilitation, physics, psychology, and of course audiology, where each group will have specific domains-of-interest and applications. We also believe that having a literary source in one book that contains a repository of diverse and highly technical information, presented in a coherent manner, should be extremely valuable to a wide range of individuals, but to our knowledge, such a document does *not* yet exist. Therefore, this book should fill an important void in the scientific literature as a combined reference text, research guide, and educational tool.

As science in this area evolves, the profession of audiology is in a unique position to integrate advanced technologies developed by clinicians, engineers, and basic scientists and apply them to the clinic. Consequently, audiologists and others in related fields like medicine and engineering represent the "translational interface" between basic science and current clinical concerns. It is a big responsibility to integrate new ideas and concepts into the clinic but it is one that encompasses the technical skills and educational background of those individuals already working in this field.

CONTRIBUTORS

Faith W. Akin, PhD
Vestibular/Balance Laboratory
Mountain Home VA Medical Center
Professor
Department of Audiology and Speech-Language Pathology
East Tennessee State University
Mountain Home, Tennessee
Chapter 4

Jont B. Allen, PhD
Professor
Department of Computer and Electrical Engineering
University of Illinois
Urbana, Illinois
Chapter 1

Deniz Baskent, PhD, MSc
Professor
Department of Otorhinolaryngology-Head and Neck Surgery
University of Groningen
University Medical Center Groningen
Research School of Behavioral and Cognitive Neurosciences
Groningen, The Netherlands
Chapter 12

Magnus Bergkvist, PhD
Assistant Professor of Nanobioscience
SUNY Polytechnic Institute
Colleges of Nanoscale Science and Engineering
Albany, New York
Chapter 6

Robert F. Burkard, PhD, CCC-A
Professor and Chair
Department of Rehabilitation Science
University at Buffalo
Buffalo, New York
Chapter 3

Anthony T. Cacace, PhD
Professor and Director of the Hearing Science Laboratory
Department of Communication Sciences & Disorders
Wayne State University
Detroit, Michigan
Chapters 3 and 13

Emile de Kleine, PhD
Medical Physicist-Audiologist
University of Groningen
University Medical Center Groningen
Groningen, The Netherlands
Chapter 15

Aniruddha K. Deshpande, PhD, CCC-A
Assistant Professor
Department of Speech-Language-Hearing Sciences
Hofstra University
Hempstead, New York
Chapter 8

Shruti Balvalli Deshpande, PhD, CCC-A
Visiting Assistant Professor
Postdoctoral Research Scholar
The University of Iowa
Iowa City, Iowa
Chapter 8

Angela R. Dixon, PhD
Postdoctoral Fellow
Department of Anatomy and Cell Biology
Molecular Anatomy of Central Auditory Related Systems
Wayne State University School of Medicine
Detroit, Michigan
Chapter 6

Camille Dunn, PhD
Research Assistant Professor
Department of Otolaryngology
University of Iowa
Iowa City, Iowa
Chapter 8

Bastian Epp, Dr. Rer. Nat.
Assistant Professor
Hearing Systems Group
Department of Electrical Engineering
Technical University of Denmark
Lyngby, Denmark
Chapter 2

Bruce Gantz, MD
Professor and Chair
Department of Otolaryngology
University of Iowa
Iowa City, Iowa
Chapter 8

Etienne Gaudrain, PhD, MSc
Senior Researcher
Lyon Neuroscience Research Center
Auditory Cognition and Psychoacoustics Team
Department of Otorhinolaryngology-Head and Neck Surgery
University of Groningen
University Medical Center Groningen
Research School of Behavioral and Cognitive Neurosciences
Groningen, The Netherlands
Chapter 12

Marlan Hansen, MD
Associate Professor
Department of Otolaryngology
University of Iowa
Iowa City, Iowa
Chapter 8

Avril Genene Holt, PhD
Associate Professor
Department of Anatomy and Cell Biology
Molecular Anatomy of Central Auditory Related Systems
Wayne State University School of Medicine
Health Science Specialist
John D. Dingell VA Medical Center
Detroit, Michigan
Chapters 5 and 6

Patricia S. Jeng, PhD
Mimosa Acoustics, Inc.
Mahomet, Illinois
Chapter 1

Paul R. Kileny, PhD
Professor of Otolaryngology
Director, Academic Program–Audiology
Department of Otolaryngology-Head and Neck Surgery
University of Michigan Health System
Ann Arbor, Michigan
Chapter 8

Dave R. M. Langers, PhD
Department of Otorhinolaryngology
University of Groningen
University Medical Center Groningen
Groningen, The Netherlands
Chapter 15

Min Young Lee, MD
Kresge Hearing Research Institute
Department of Otolaryngology-Head and Neck Surgery
University of Michigan Medical School
Ann Arbor, Michigan
Chapter 9

Thomas Lenarz, MD, PhD
Professor and Director
Department of Otolaryngology
Hannover Medical School
Hannover, Germany
Chapter 11

Harry Levitt, BSc, PhD
Professor Emeritus
The City University of New York
Director of Research
Advanced Hearing Concepts
Bodega Bay, California
Chapter 1

Hubert H. Lim, PhD
Assistant Professor
Biomedical Engineering and Otolaryngology
Institute for Translational Neuroscience Scholar
University of Minnesota, Twin Cities
Minneapolis, Minnesota
Chapter 11

Glenis Long, PhD
CUNY Graduate Center
Professor Emerita
Speech-Language-Hearing Science Program
New York, New York
Chapter 2

Lawrence R. Lustig, MD
Howard W. Smith Professor and Chair
Department of Otolaryngology-Head and Neck Surgery
Columbia University Medical Center
New York, New York
Chapter 7

Catherine A. Martin, BA
Kresge Hearing Research Institute
University of Michigan
Ann Arbor, Michigan
Chapter 6

Dennis J. McFarland, PhD
Research Scientist
National Center for Adaptive Neurotechnologies
Wadsworth Center
New York State Department of Health
Albany, New York
Chapter 13

Judi A. Lapsley Miller, PhD
Senior Scientist
Mimosa Acoustics, Inc.
Hearing Research Consultant
Wellington, New Zealand
Chapter 1

Antonela Muca
Wayne State University School of Medicine
Detroit, Michigan
Chapter 6

Owen D. Murnane, PhD
Vestibular/Balance Laboratory
Mountain Home VA Medical Center
Professor
Department of Audiology and Speech-Language Pathology
East Tennessee State University
Mountain Home, Tennessee
Chapter 4

Mirna Mustapha, PhD
Assistant Professor
Department of Otolaryngology-Head and Neck Surgery
Stanford University School of Medicine
Stanford, California
Chapter 5

James F. Patrick, AO, DEng, FTSE, FIE (AUST), CPEng (Biomed)
Chief Scientist, Senior Vice President Cochlear Limited
Adjunct Professor, Macquarie University
Associate Professor, University of Melbourne
Adjunct Professor, LaTrobe University
Sydney, Australia
Chapter 11

Roy D. Patterson, PhD
Professor
Department of Physiology, Development and Neuroscience
University of Cambridge
Cambridge, United Kingdom
Chapter 14

Yehoash Raphael, PhD
The R. Jamison and Betty Williams Professor of Otolaryngology-Head and Neck Surgery Kresge Hearing Research Institute
The University of Michigan
Ann Arbor, Michigan
Chapter 9

Kristal Mills Riska, AuD, PhD
Vestibular/Balance Laboratory
Mountain Home VA Medical Center
Assistant Professor
Department of Audiology and Speech-Language Pathology
East Tennessee State University
Mountain Home, Tennessee
Chapter 4

Sarah R. Robinson, MS
PhD Candidate
Department of Electrical and Computer Engineering
University of Illinois at Urbana-Champaign
Urbana, Illinois
Chapter 1

Terrin Nichole Tamati, PhD
Postdoctoral Researcher
Department of Otorhinolaryngology-Head and Neck Surgery
University of Groningen
University Medical Center Groningen
Groningen, The Netherlands
Chapter 12

Richard Tyler, PhD
Professor
Department of Otolaryngology
University of Iowa
Iowa City, Iowa
Chapter 8

Stefan Uppenkamp, PD Dr. Rer. Nat. Habil.
Physicist
Medical Physics Section
University of Oldenburg
Oldenburg, Germany
Chapter 14

Pim van Dijk, PhD
Medical Physicist and Audiologist
University of Groningen
University Medical Center Groningen
Groningen, The Netherlands

Douglas E. Vetter, PhD
Associate Professor
Department of Neurobiology and Anatomical Sciences
University of Mississippi
Jackson, Mississippi
Chapter 10

Anita Wagner, PhD, MA
Researcher
Department of Otorhinolaryngology-Head and Neck Surgery
University of Groningen
University Medical Center Groningen
Research School of Behavioral and Cognitive Neurosciences
Groningen, The Netherlands
Chapter 12

To my AuD and PhD students for their inspiration and interest in science and research, which in part motivated the need for such a book; To those students, scientists, and clinicians who will continue to advance the field; and To my wife Lydia, for her unwavering support.

—Anthony T. Cacace

To my wife Margreet and our girls Veerle and Céline.

—Emile de Kleine

To my laboratory team for their boundless energy and enthusiasm for science; To my colleagues at Wayne State University and The Kresge Hearing Research Institute, University of Michigan for thought-provoking and stimulating scientific conversations; To my Saline CoC family for helping me to stay grounded; To my parents for giving me a solid foundation and for their continuous encouragement; and To my husband Ron and our son Parker, who through their support, provide me with the opportunity to continue the work I love.

—Avril Genene Holt

To my wife Jacqueline and children Jop and Jet; and To my scientific colleagues at the University Medical Center Groningen, University of Groningen, University of Oldenburg, Graduate Center of the City University of New York, University of Tübingen, University of Cambridge and University of California, Los Angeles for great collaborations in the past and at present.

—Pim van Dijk

CHAPTER 1

Middle-Ear Reflectance: Concepts and Clinical Applications

Jont B. Allen, Sarah R. Robinson, Judi A. Lapsley Miller, Patricia S. Jeng, and Harry Levitt

The middle ear is a complex sound transmission system that converts airborne sound into cochlear fluid-born sound, in a relatively efficient way, over the bandwidth of hearing (about 0.1–15 kHz). The middle ear is the gateway to the auditory system, and it is involved in nearly every audiologic test. It is therefore critical to assess middle-ear status in any audiologic evaluation and, in the case of abnormal middle-ear function, pinpoint the source of pathology to enable an appropriate medical intervention. By the use of wideband acoustic measurements, the middle-ear structures can be noninvasively probed across the wide frequency range of hearing, allowing clinicians to make nuanced interpretations of hearing health. The term *wideband acoustic immittance* (WAI) has recently been coined as an umbrella term to identify a variety of acoustic quantities measured in the ear canal (Feeney et al., 2013). Here we focus primarily on wideband reflectance, from which other WAI quantities may be derived. The *reflectance* is defined as the ratio of reflected to forward pressure waves.

A middle-ear reflectance measurement involves inserting an acoustic measurement probe into the ear canal, fitted with an ear tip designed to create a sealed ear-canal cavity (Figure 1–1). A hearing aid loudspeaker in the probe transmits wideband sound into the ear canal. Any reflected sound, related to structures of the middle ear, is measured by the probe microphone. This probe is calibrated in such a way that the absorbed and reflected pressures in a cavity may be determined.

Reflectance measurements are clinically practical to make: The measurement takes less than a minute and the ear does not require pressurization. The

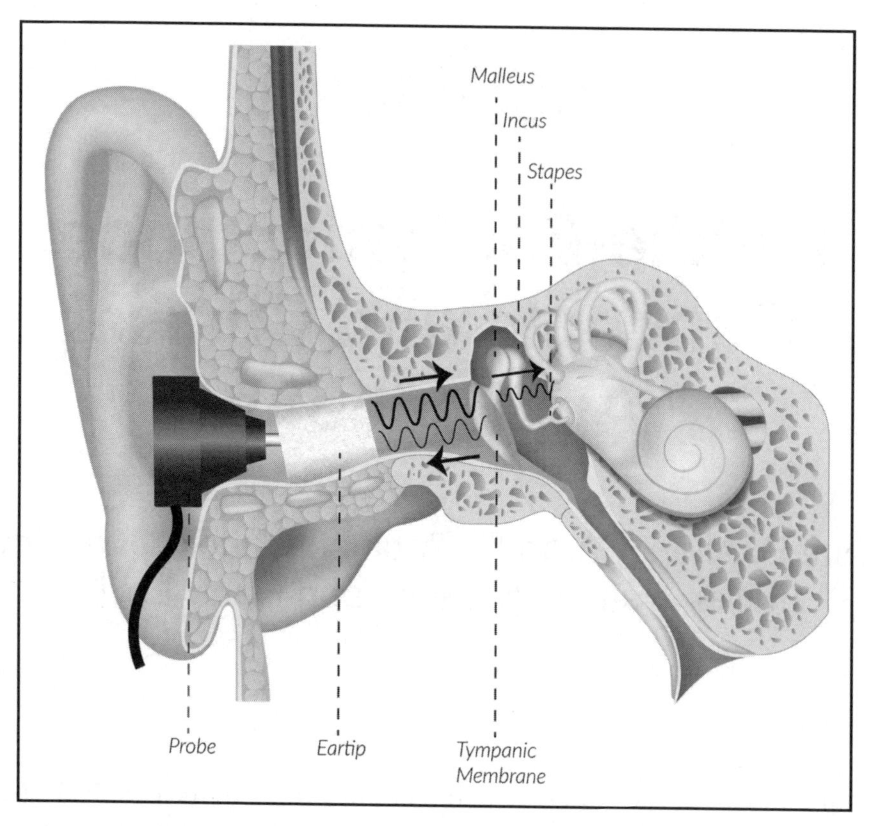

Figure 1–1. Probe configuration in the ear canal to measure middle-ear reflectance, showing the acoustic signal traveling down the ear canal until it reaches the TM. At the TM, the sound is partially reflected back into the ear canal and partially absorbed into the middle ear.

same probe can be used for other audiologic tests, such as otoacoustic emission (OAE) tests and pure-tone hearing threshold testing. Such testing, when a microphone is used in the ear canal, is known as *real-ear* testing. Given knowledge of the reflectance, it is possible to correct for troublesome ear canal standing waves, which can produce large artifacts in the real-ear calibrations. Alone, or together with other audiologic measurements, middle-ear reflectance measurements can help identify many abnormal conditions which may lead to conductive hearing loss (CHL), including degrees of otitis media, tympanic membrane (TM) perforations, otosclerosis, and ossicular disarticulation. The method is noninvasive, fast, and clinically available.

In this chapter, we cover the theoretical principles of middle-ear reflectance. We then move to clinical applications, showing how normal middle ears behave and how abnormal middle ears differ. We offer advice on how to make quality measurements and provide suggestions for future research.

Background to Middle-Ear Assessment

Noninvasive assessment of middle-ear status is of great importance in hearing health care. An early approach to middle-ear assessment is that of *tympanometry* (e.g., Feldman, 1976; Shanks, 1988), and it is still the clinical gold standard. The method relies on measurements at low frequencies (e.g., probe tones at 226 Hz and 1,000 Hz are commonly used) and provides no information on the status of the middle ear at higher frequencies relevant to speech perception (e.g., 0.2–8.0 kHz). The methods employed in tympanometry were developed prior to the introduction of digital technology, and these methods reflect the limitations of that era.

Reflectance of sound from the TM and the acoustic impedance of the middle ear are different facets of the same underlying mechanism. Historically, acoustic impedance of the ear was the first to be measured and studied (West, 1928). There is a substantial body of research on the acoustic impedance of the ear. Metz (1946) developed the first clinical instrument for measuring the acoustic impedance of the ear. This instrument was not easy to use and clinical measurement of acoustic impedance proceeded at a slow pace until more practical instruments were developed (Møller, 1960; Terkildsen & Nielsen, 1960; Zwislocki & Feldman, 1970). Tympanometry, the measurement of the middle-ear acoustic impedance as a function of static pressure in the ear canal, provided useful clinical data. Thus, practical instruments were developed for measurements of this type. The 1970s saw a rapid growth in the use of tympanometry, which is widely used today in audiologic evaluations (Jerger, 1970).

The introduction of small, inexpensive computers in the mid-1980s paved the way for a new generation of digital test equipment with capabilities well beyond that of conventional electronic instrumentation. It also facilitated new ways of thinking about audiologic measurement, resulting in the development of innovative wideband techniques. The evolution of wideband reflectance measurement allows for more detailed diagnostic assessment of the middle-ear status than the previous approach based on tympanometry. Early reflectance studies were conducted by Keefe, Ling, and Bulen (1992); Keefe, Bulen, Arehart, and Burns (1993); and Voss and Allen (1994).

The use of reflectance measurements in a computer-based system does not preclude the use of acoustic impedance data, where appropriate. Acoustic reflectance and acoustic impedance are both WAI quantities; different facets of the same underlying mechanism. If one is known, the other can be computed by means of a mathematical transformation. This mathematical transformation can be implemented conveniently in a computer-based instrument.

Acoustics of the Outer and Middle Ear

When a sound wave travels down the ear canal toward the TM, the acoustic power is continuous until it reaches an *impedance discontinuity,* such as the

Propagation of Sound: The Basics

Many of the concepts in WAI, including reflectance, are defined in mathematical or physics terms. This creates a problem for clinicians and others without the necessary background. Here we explain some acoustical concepts in lay terms.

The transmission of sound in the ear canal can be approximated quite well by a tube with a fixed diameter equal to that of the average adult ear canal. The tube is terminated at one end by a loudspeaker that delivers an acoustic signal in the frequency range up to at least 10,000 Hz. One may imagine that the air in the tube is partitioned into a very large number of infinitesimally thin discs (Beranek, 1949); each disc can be thought of as consisting of a layer of air particles. These discs of air are compressed or expanded by an applied force, such as a change in air pressure (air molecules will spread out from an area of high pressure to an area of lower pressure), and will return to their original volume once the applied force is removed.

Consider now what happens when the loudspeaker at one end of the tube generates an acoustic signal. When the speaker diaphragm moves inward, it displaces and compresses the adjacent discs of air, which then displace and compress the next layer of air, and so on. By this means, the in and out movements of the transducer diaphragm create a pressure wave that travels down the tube at the speed of sound, about 343 m/s at 20°C. The velocity of each disc of air about its quiescent position (the position of the disc at rest) multiplied by its cross-sectional area is known as the *volume velocity*, as the product of velocity and cross-sectional area encompasses a moving volume.

The air in the tube opposes being displaced and compressed by the transducer diaphragm. The force exerted by the transducer diaphragm is equal to the pressure times the area of the diaphragm. The *work* done by the force is equal to force times the displacement, and is stored as energy in the air as it travels along the tube. The *acoustic power*, $P(f)$ (the force times the volume velocity, often expressed in watts), inserted into the tube is equal to the rate of work done. The power propagated down the tube is transmitted without significant loss through the tube via the air.

TM. Impedance discontinuities result in frequency-dependent reflections of the sound wave, which we quantify using wideband reflectance.

The acoustic variables discussed in this section may be defined either in the time or frequency domain. It is important to always be aware of which domain is under consideration. In this chapter, we work almost exclusively in the frequency domain, where all variables are functions of frequency, f. These variables are also a function of location. For measurements in the ear canal, we define $x = 0$ as the measurement probe location and $x = L$ the TM location.

Pressure and Volume Velocity Waves

We denote the forward traveling pressure wave as $P_+(f,x)$ [Pa], using the plus sign subscript to signify the forward direction (toward the TM). This wave is a function of both frequency f (in Hz) and location and has units of Pascals. Similarly, the reflected, backward traveling *retrograde* pressure wave is denoted $P_-(f,x)$. At any location in the ear canal, the total pressure $P(f,x)$ is defined as

$$P(f,x) = P_+(f,x) + P_-(f,x). \quad (1)$$

The pressure is a scalar quantity (it has no direction). Any change in the pressure results in a force, which is a vector quantity (it has direction); this force leads to the motion (velocity) of air molecules in the direction of the force.

The corresponding acoustic *volume velocity* $U(f,x)$ may be decomposed into forward $U_+(f,x)$ and reverse $U_-(f,x)$ traveling portions, as

$$U(f,x) = U_+(f,x) - U_-(f,x). \quad (2)$$

The volume velocity is a vector quantity, which accounts for the change in sign of Equation 2 (here positive U_- values indicate propagation of the retrograde wave toward the probe, and positive U_+ values indicate propagation of the forward wave toward the TM).

The *complex acoustic reflectance*, which we represent using the uppercase Greek letter "Gamma," is defined as the ratio of retrograde to forward traveling pressure (or velocity) waves

$$\Gamma(f,x) = \frac{P_-(f,x)}{P_+(f,x)} = \frac{U_-(f,x)}{U_+(f,x)}. \quad (3)$$

Since $\Gamma(f,x)$ is complex, it may be expressed either as the sum of real and imaginary parts, or in terms of a magnitude and phase. The utility of the complex reflectance (as compared to other WAI quantities, such as impedance and admittance) is that the acoustic power is proportional to the square of the pressure. Thus, the squared magnitude of the reflectance describes the ratio of reflected to incident power (a value ranging between 0 and 1) as a function of frequency, while the reflectance phase codifies the latency of the reflected power (e.g., the depth at which the reflection occurs). Additionally, power absorbed by ear (potentially including the ear canal, middle ear, and inner ear) may be quantified as one minus the ratio of power reflected. The *power reflectance* at the probe may be defined as $|\Gamma(f,0)|^2$; thus, the power absorbed by the ear is $1 - |\Gamma(f,0)|^2$. These properties of reflectance are more intuitive than impedance for formulating diagnoses of middle-ear pathologies.

For reference, the *complex acoustic impedance* is defined as the total pressure over the total volume velocity

$$Z(f,x) = \frac{P(f,x)}{U(f,x)}. \quad (4)$$

The *complex acoustic admittance* is given by $Y(f,x) = \frac{1}{Z(f,x)}$ and various other WAI quantities may be calculated from $Z(f,x)$ and $Y(f,x)$, as outlined in Appendix 1–A. This variety of immittance quantities can be confusing, so it is important to remember that they may all be derived from the complex acoustic reflectance. Specifically, the complex impedance is related to the reflectance via

$$Z(f,x) = r_0\frac{1+\Gamma(f,x)}{1-\Gamma(f,x)}, \quad (5)$$

where the constant r_0 is called the *characteristic acoustic resistance* of the ear canal.

The characteristic resistance is defined as the ratio of pressure to volume velocity for a single wave propagating in the canal. Therefore, it applies to the forward and retrograde waves separately, as follows:

$$r_0 \equiv \frac{P_+(f,x)}{U_+(f,x)} = \frac{P_-(f,x)}{U_-(f,x)} = \frac{\rho c}{A}. \quad (6)$$

Here $\rho \approx 1.2$ kg/m^3 is the density of air, $c \approx 343$ m/s is the speed of sound in air, and A is the ear canal area. The average diameter of the adult ear canal is about 7.5 mm, with an average area of about 44.2×10^{-6} [m^2]. In a real ear, the canal area will vary with distance along the canal, and thus r_0 will also vary. Additionally, the area of the ear canal at the measurement location is typically unknown. Variation due to the use of an incorrect r_0 (Equation 5) has been shown to have a relatively small effect on reflectance and impedance compared to individual variation across ears (Keefe et al., 1992; Voss & Allen, 1994).

Other than for the simplest of cases (e.g., $|\Gamma=0|$), the transformation in Equation 5 can be mathematically challenging. Regardless of the mathematical complexity of the relationship, impedance and reflectance are mathematically equivalent (interchangeable). Since reflectance is more intuitive than other WAI quantities, we focus on reflectance.

Sound Transmission in the Middle Ear

The middle ear converts airborne sound into cochlear fluid-borne sound by the mechanical action of the TM and ossicles. Vibration of the TM drives the ossicles. The ossicles then transmit the signal to the annular ligament of the oval window, which in turn transmits the signal to the fluid-filled cochlea.

Sounds are transmitted efficiently from the canal to the cochlea in the normal middle ear, with little loss in acoustic power (Parent & Allen, 2010; Puria & Allen, 1998). The middle ear may be modeled as a cascade of transmission lines (mathematical models for the flow of acoustic power) that are approximately matched, having relatively little loss and few reflections along the pathway from the ear canal to the inner ear (Allen, 1986; Lüscher & Zwislocki, 1947; Møller, 1983; Puria & Allen, 1991, 1998). If the middle ear becomes unbalanced, abnormal reflections occur, transmission is impaired, and this translates to poorer hearing. Little power is lost for reverse traveling signals (Allen & Fahey, 1992), giving us the opportunity to study both the middle ear and cochlea using acoustic measurements made in the ear canal.

Small vibrations generated in the cochlea by nonlinear motions of the outer hair cells can be measured in the ear canal (Kemp, 1978). These are known as otoacoustic emissions (OAEs), and they are an important tool for studying the function of the inner ear. It is important to know the status of the middle ear when interpreting OAE measurements, since the OAE-evoking acoustic signal must first travel through the middle ear into the cochlea, then the evoked retrograde OAEs must travel back out through the middle ear to the ear canal. An abnormal middle ear will disrupt sound propagation both to and from the cochlea.

The auditory system is exquisitely sensitive. The reference level 0 dB SPL is defined by the threshold of hearing at 1 kHz and corresponds to pressure vibrations of about 2×10^{-5} Pa (for reference, ambient atmospheric pressure is on the order of 10^5 Pa). The middle ear is also remarkably robust, especially compared to the inner ear. The middle ear is not damaged by sounds of extremely high intensity (on the order of 120 dB SPL). The inner ear, in contrast, is subject to substantial damage from sounds of this intensity. The inner ear is protected by at least two efferent systems. The acoustic reflex from the stapedius muscle (Feeney & Keefe, 1999, 2001; Møller, 1983) helps protect the cochlea from intense low-frequency vibrations. The tensor tympani, which connects to the long process of the malleus, may be activated during speaking and chewing (Aron, Floyd, & Bance, 2015; Bance et al., 2013).

The middle ear may be represented as an acoustic transmission line, consisting of compliance (spring), resistance, and mass elements, as shown in Figure 1–2. Each of these elements has an impedance associated with it, which causes sound pressure waves to be reflected in a frequency-dependent manner; the combination of these elements determines the reflectance at the TM. The uniform tubes represent time delay. Figure 1–2 depicts the outer ear (pinna and concha) and ear canal as a series of tubes of varying area. The TM is also a transmission line, with approximately 36 μs of delay (Puria & Allen, 1998). Following are the ossicles, consisting of the masses of the malleus,

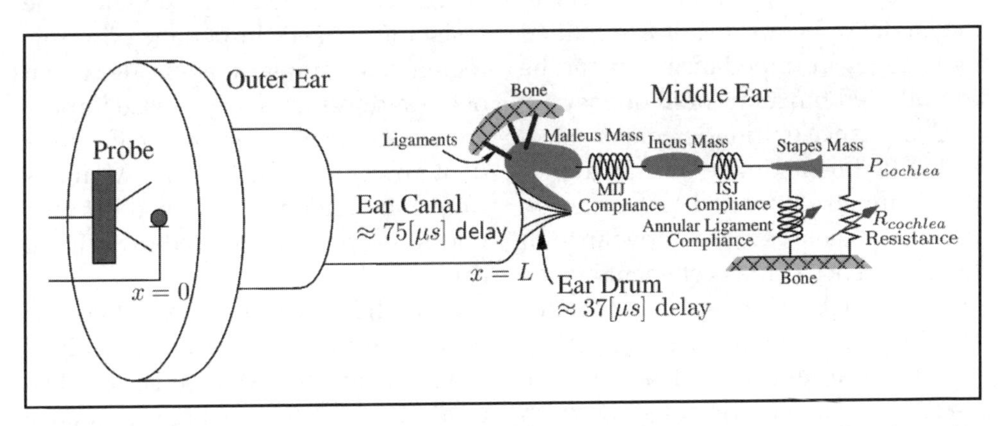

Figure 1–2. A transmission line model of the middle ear is shown, including the outer ear (pinna and concha), the ear canal (represented as a tube), and the TM, which is also a transmission line with approximately 36 μs of delay (Puria & Allen, 1998). Following are the ossicles, consisting of the masses of the malleus, incus, and stapes. Between each of these masses is the joint ligament, which is represented as a spring (MIJ: malleoincudal; ISJ: incudostapedial). The annular ligament holds the stapes footplate in the oval window; this spring is nonlinear since it changes its compliance when force is applied via the stapedius muscle. Finally, the cochlear impedance is represented as a nonlinear resistance. The middle ear, TM, and ear-canal transmission lines are all well matched, such that sound power is delivered efficiently to the ear over a broad range of frequencies (Allen, 1986; Lüscher & Zwislocki, 1947; Møller, 1983; Puria & Allen, 1991, 1998).

incus, and stapes. Between each of these masses is a spring, which represents the compliant ligament between each pair of ossicles. The annular ligament holds the stapes footplate in the oval window; this spring is nonlinear, as its compliance changes when force is applied via the stapedius muscle. Finally, the cochlear impedance is represented as a nonlinear resistance. In general, a nonlinear model element has an impedance (and reflectance) that depends on the stimulus.

Sound Reflections in the Outer and Middle Ear

Impedance discontinuities lead to reflected pressure waves, which contribute to the reflectance measured at the probe. When a sound pressure wave propagates through the middle ear in the model of Figure 1–2, it is modified by changes in impedance that can be complicated functions of frequency.

Compliance (spring) elements, such as the ligaments, have an impedance that is inversely related to frequency; thus they have a high impedance at low frequencies. Mass elements (such as the ossicles) have an impedance that increases linearly with frequency. Finally, ideal resistance elements have impedances that are constant with frequency. A combination of these elements results in a more complicated frequency dependence of pressure reflections.

Reflectance measured at the probe microphone $\Gamma(f,0)$ varies as a function of frequency and depends on how the acoustic impedance of the TM varies with frequency. At frequencies below 1 kHz, the impedance of the TM is due mostly to the compliance (stiffness) of the annular ligament (Allen, 1986; Lynch, Nedzelnitsky, & Peake, 1982; Lynch, Peake, & Rosowski, 1994) and other middle-ear structures. When pressure waves at these low frequencies reach the stapes, almost all of their power is briefly stored as potential energy in the stretched ligament (spring) and then reflected back to the ear canal as a retrograde pressure wave (Allen, Jeng, & Levitt, 2005). At even lower frequencies, below about 800 Hz, only a small fraction of the incident power is absorbed into the middle ear and cochlea (Parent & Allen, 2010; Puria & Allen, 1998).

In a normal ear, in the mid-frequency region between 1 and 4 kHz (or higher), the stiffness- and mass-based impedance effects of the middle ear largely cancel each other. Additionally, the resistance-based impedance in this region has a similar magnitude to that of the stiffness- and mass-based impedances. As a result, much of the incident power that reaches the TM in this frequency region is absorbed into the middle ear and transmitted to the inner ear.

At high frequencies above 6 kHz, the mass-based impedance of the ossicles can dominate the TM impedance (Allen et al., 2005). When a high-frequency pressure wave reaches the TM and mass-based impedance is substantial, most of the power in the incident pressure wave is momentarily stored as kinetic energy, primarily in the ossicles, and then reflected back to the ear canal as a retrograde pressure wave.

The ideal ear canal may be modeled as a rigid-walled (lossless) cylinder of constant area. Under this assumption, there are no reflections along the ear canal except at the TM. In this case, the reflectance at the probe microphone $\Gamma_m(f) = \Gamma(f,0)$ is related to the TM reflectance $\Gamma_{tm}(f) = \Gamma(f,L)$ by a pure delay, expressed as

$$\Gamma_m(f) = \Gamma_{tm}(f)e^{-j2\pi f2L/c}. \quad (7)$$

There is a round trip delay of $2L/c$ between the measurement point and the TM (Voss & Allen, 1994). Thus, when the ear canal is lossless,

$$|\Gamma_m(f)| = |\Gamma_{tm}(f)|. \quad (8)$$

This has been shown to be a good approximation in adult ears (Voss, Horton, Woodbury, & Sheffield, 2008), though it may be affected by the depth and quality of the probe insertion, and any compliance-related loss associated with the ear canal walls (Abur, Horton, & Voss, 2014).

Given the measured $\Gamma(f,0)$ and a known canal length L, one may estimate $\Gamma_{tm}(f,L)$, and thus the complex TM reflectance. Voss and Allen (1994) used the complex reflectance to estimate the acoustic properties of the TM, by removing pure delay from the reflectance phase. When the area of the canal depends on position (as in a real ear canal), the effective length L is a function of frequency, and removing the effects of the ear canal from the complex reflectance is nontrivial. A number of methods to estimate the ear-canal length and remove ear-canal effects have been proposed (including Keefe, 2007; Lewis & Neely, 2015; Rasetshwane & Neely, 2011; Robinson, Nguyen, & Allen, 2013).

Measurements and Procedures

Applications of Reflectance

Next we discuss some of the main applications of complex reflectance measurements. Many clinical studies to date have considered the reflectance magnitude, in the form of power reflectance or absorbance level, which we describe in the following sections. Additionally, current results indicate that *forward pressure level* (FPL) calibrations allow for accurate individualized calibration of stimuli for assessment of the inner ear (Scheperle, Neely, Kopun, & Gorga, 2008; Withnell, Jeng, Waldvogel, Morgenstein, & Allen, 2009). FPL is derived from pressure reflectance measurements in the ear canal to account for the effect of standing waves in the ear canal, which have large and highly variable frequency effects, dependent on the ear and the probe-insertion depth. Finally, methods to analyze the complex reflectance are presented. These include time- and frequency-domain methods to study the complex reflectance, with a particular focus on how to account for the variable phase effects of the *residual ear canal* between the probe and the TM. For clinical diagnoses, the TM reflectance is the quantity of interest.

Power Reflectance and Absorbance Level

As previously noted, the magnitude reflectance measured at the probe microphone location in the ear canal $|\Gamma_m(f)| = |\Gamma(f,0)|$ may be assumed to be approximately equal to the magnitude reflectance at the TM $|\Gamma_{tm}(f)| = |\Gamma(f,L)|$, while its phase is highly varying across ears. For this reason, clinical studies have focused on the *power reflectance*, $|\Gamma_m(f)|^2$. Because this value ranges between 0 and 1, expressing the ratio of power reflected from the middle ear, it is often expressed as a percentage.

A related quantity, the *power absorbance*, $1 - |\Gamma_m(f)|^2$, is a measure of middle-ear energy transmission, indicating approximately how much power is conveyed to the middle ear and cochlea (Allen et al., 2005; Rosowski et al., 2012). With some middle-ear abnormalities, power can be absorbed by the middle ear and does not reach the cochlea. The power absorbance expressed in decibels with reference to the total absorbance, $10log_{10}(1 - |\Gamma m\ (f)|^2)$, is referred to as the *power absorbance level*, and has a distinctive shape for normal ears, similar to the middle-ear transfer function. This quantity has also been referred to as *transmittance* (Allen et al., 2005).

Figure 1–3 shows example power reflectance and absorbance level measurements from normal ears (Voss & Allen, 1994), an ear simulator (the Brüel & Kjær 4157), and a rigid cylindrical cavity. The cavity has a power reflectance close to 100% across all frequencies, as expected, with some small losses due to viscous and thermal effects of airflow along the cylinder walls (Keefe, 1984). For normal ears, the absorbance level has a distinctive shape, with a rising slope below about 1 kHz, a relatively flat region with very little attenuation (e.g., −3 dB) between about 1 and 4 kHz (Rosowski et al., 2012), and a falling slope at high frequencies. Note that in the mid-frequency range (e.g., 1–4 kHz) large individual variations of the power reflectance (e.g., a 40% range) corresponds to a small range of decibel variation (e.g., 3 dB) of the absorbance level.

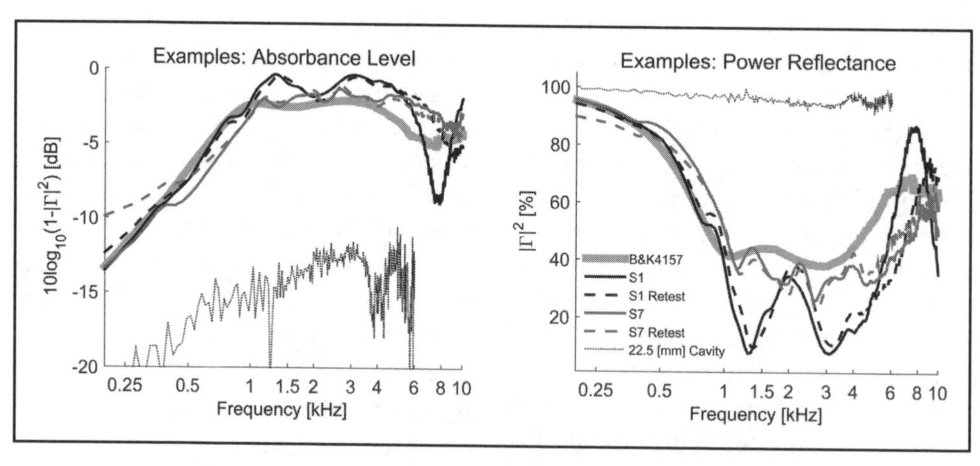

Figure 1–3. Absorbance level (*left*) and power reflectance (*right*) for two normal ears (with retest measurements) from Voss and Allen (1994). Additionally, a measurement of the Brüel & Kjær 4157 ear simulator from that study is presented, where the ear simulator is intended to mimic the response characteristics of the average adult ear. Finally, a measurement of a rigid cylindrical cavity (22.5 mm long) is shown.

Forward Pressure Level (FPL)

Many acoustic assessments of the inner ear, including OAE and hearing thresholds (e.g., the audiogram), rely on the transmission of sound stimuli to the cochlea via the outer and middle ear. Therefore, proper calibration of such stimuli requires an understanding of the magnitude and phase effects introduced by stimulus propagation through the ear canal and middle ear.

Standard practice to account for middle-ear effects is to calibrate stimuli using a middle-ear simulator, often referred to as *reference equivalent threshold sound pressure level* (RETSPL) calibration (ISO, 1997; Souza, Dhar, Neely, & Siegel, 2014). However, ear simulators do not account for the significant variation of middle-ear properties across normal individuals. More important, they do not account for probe-insertion depth, which varies across ears and probe insertions. Probe-insertion depth is of particular importance because acoustic standing waves occur in the ear canal between the measurement probe and the TM (Siegel, 1994).

A standing wave is created when the forward and retrograde pressures in the ear canal are out of phase, nearly canceling each other and creating a deep minimum (e.g., −20 dB) in the total pressure magnitude (Equation 1). The frequency at which this cancellation occurs is dependent upon the round-trip delay from the probe source ($x = 0$) to the TM ($x = L$) and varies considerably across probe insertions and across normal ears. To understand the standing wave effect, consider a rigid cylindrical cavity of uniform area and length L. The retrograde pressure at the microphone is related to the forward pressure by $P_{-}(f,0) = P_{+}(f,0)e^{-j2\pi f\tau}$, where the round trip delay is $\tau = 2L/c$. Therefore, the total pressure at the microphone is

$$P_m(f) = P(f,0) = P_{+}(f,0)(1+e^{-j2\pi f\tau}), \quad (9)$$

which goes to zero for any frequency where $e^{-j2\pi f\tau} = -1$ (which occurs when the quantity $2\pi f\tau$ is equal to π, plus or minus any integer multiple of 2π). In the case of a real-ear measurement, delay from the TM and middle ear will also contribute to the frequency location of the pressure null.

It is now recognized that the *forward pressure level* (FPL) should be used for such stimulus calibrations, to account for standing wave effects on the stimulus magnitude in individual ears (Scheperle et al., 2008; Souza et al., 2014; Withnell et al., 2009). Thus we present an ear-dependent variation of the RETSPL method here, which may be called *reference equivalent forward pressure level* (RETFPL) calibration. At the time of this writing, RETFPL calibration is available to hearing researchers, and will soon be available to clinicians.

The FPL is defined as the forward component of the total pressure wave, $P_{+}(f,x)$, as previously described. One may determine the forward pressure at the microphone from the total pressure as follows:

$$P_m(f) = P = P_{+}(f,0) + P_{-}(f,0) =$$

$$P_{+}(f,0)\left(1 + \frac{P_{-}(f,0)}{P_{+}(f,0)}\right) = P_{+}(f,0)\,(1 + \Gamma(f,0)).$$

Solving for $P_{+}(f,0)$ gives

$$P_{+}(f,0) = \frac{P_m(f)}{1+\Gamma_m(f)}. \quad (10)$$

Given the microphone pressure $P_m(f)$ and the measured complex reflectance $\Gamma_m(f)$, which may both be determined by the reflectance measurement system, the forward pressure at the microphone $P_+(f,0)$ can be estimated (Withnell et al., 2009). To perform the RETFPL stimulus correction for inner ear assessment, the loudspeaker voltage can be varied so that $P_+(f,0)$ is constant at the desired level. The reflectance measured at the probe characterizes individual magnitude and phase properties of both ear canal and middle-ear transmission; without it, the standing wave cannot be precisely removed.

Figure 1–4 shows the magnitude of the normalization factor $(1 + \Gamma_m(f))/2$ in decibels for 10 human ears and 2 ear simulators, computed from the measurements of Voss and Allen (1994). The extra factor of 2 is to compensate for the fact that $\Gamma_m(f)$ goes to 1 as the frequency goes to zero. The use of this factor of 2 (6 dB) is optional, depending on what magnitude correction is desired at low frequencies.

Considering Figure 1–4, at low frequencies, the forward and retrograde pressures are approximately in phase, requiring little correction. Above 3 kHz, the phase of $\Gamma_m(f)$ plays a very important role, as it results in a deep null in the correction factor, due to the ear canal standing wave. The frequency location of the correction factor null

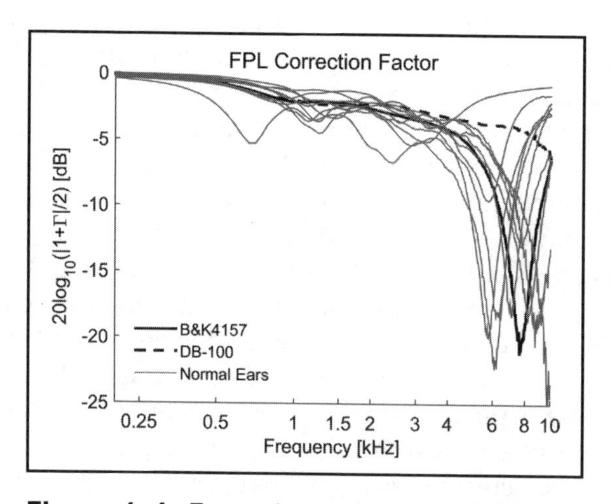

Figure 1–4. Forward pressure level (FPL) normalization factor $|1 + \Gamma_m|/2$, which corrects for the ear canal standing wave, for 10 normal ears and two ear simulators from Voss and Allen (1994). At the frequency of the null, the phase of the complex reflectance is approximately 180 degrees (Withnell et al., 2009). The frequency of the null critically depends on the round-trip delay between the probe tip and the TM (Equation 9). This delay is different for each ear, as it depends on the insertion depth of the probe and the geometry of the ear canal and TM. As the ear-canal delay decreases, the standing wave null shifts upward in frequency.

increases as the distance between the probe and TM decreases. To understand the effect of the TM reflectance on the standing wave null frequency, consider a distance of 1.5 cm between the probe and TM. This gives a round-trip delay of about 87 μs, and thus the standing wave null might be estimated to occur at 5.7 kHz. However, including additional delay from the TM (about 36 μs) and ossicles, the actual null frequency will be lower.

Based on the deep nulls in the FPL correction factor (due to ear canal standing waves) observed in Figure 1–4, it would not be reasonable to normalize the microphone pressure $P_m(f)$ to be constant when delivering stimuli to the cochlea. Such a normalization would boost the level at the standing wave null frequency by as much as 25 dB (in Equation 10, if $P_m(f)$ is held constant there is a peak in the forward pressure corresponding to the minimum of $|1 + \Gamma_m(f)|$). Calibration using an ear simulator will typically not be effective either, even if an artificial ear canal of similar length is included in the measurement. Consider the correction factors for the two ear simulators shown in Figure 1–4. The DB-100 has a very short ear canal, such that the correction factor is nearly constant. In the case of a longer simulated canal, the Brüel & Kjær 4157, it is unlikely that the length of the simulator canal will precisely equal the distance between the probe microphone and TM for the real-ear measurement. Because the correction factor minima are so narrow and deep, this length must be precise to avoid introducing a deep attenuation at the false null frequency, in addition to boosting the FPL much too high at the true null frequency.

Complex Frequency-Domain Reflectance

For middle-ear diagnostics, the quantity of interest is the reflectance of the TM, $\Gamma_{tm}(f) = \Gamma(f,L)$, as a function of frequency. Given $\Gamma_{tm}(f)$, other WAI quantities may be determined at the TM (e.g., Equation 5). As previously discussed, we are not able to measure the reflectance directly at the TM, and the residual ear canal length is unknown (it must be estimated). Therefore, clinical studies to date have considered only the magnitude reflectance (Equation 8), in the form of power reflectance or power absorbance level. However, taking the magnitude of the complex reflectance eliminates phase information from the TM, which may provide additional diagnostic information, when separated from the ear canal phase. Thus, a topic of recent interest is modeling the ear canal contribution to reflectance, with the goal of making the complex TM reflectance available for clinical investigation. Researchers and modelers have approached this with a variety of time- and frequency-domain methods (Keefe, 2007; Lewis & Neely, 2015; Rasetshwane & Neely, 2011; Robinson et al., 2013).

The measured reflectance phase has two additive components: phase from the middle-ear structures (e.g., the TM and ossicles) and ear-canal phase, due to the residual ear canal delay between the probe tip and the TM. The reflectance magnitude has two multiplicative components associated with energy loss and absorbance by the middle ear and ear canal. For the ideal case, where the ear canal is assumed to be a lossless uniform tube, this is described by Equation 7. More generally, we can represent the complex reflectance at the microphone as

$$\Gamma(f) = \Gamma_{tm}(f)\Gamma_{ec}(f). \qquad (11)$$

Here $\Gamma_{tm}(f)$ is the complex reflectance at the TM, which is the quantity of interest for middle-ear diagnostics, while the residual ear canal factor, $\Gamma_{ec}(f)$, accounts for round-trip sound propagation in the nonideal ear canal.

Due to nonideal properties of the human ear canal, it is nontrivial to estimate its contribution to the complex reflectance measured at the probe. In a real ear, the phase of $\Gamma_{ec}(f)$ will have a frequency dependence that is related to both the ear canal length and area variation between the probe and TM. Deviation of the magnitude of $\Gamma_{ec}(f)$ from 1 is due to acoustic losses, related to the compliance of the ear canal walls. These losses are assumed to be small in normal adult ears but increase with probe distance from the TM.

Robinson et al. (2013) factored the frequency-domain reflectance such that the ear canal component, $\Gamma_{ec}(f)$, has a magnitude of 1 for all frequencies, assuming the ear canal has no acoustic losses. The method accommodates any lossless delay from an ear canal of varying area (e.g., in Equation 7, the length parameter L may be considered to be a function of frequency). This factorization is unique when performed on a parameterized version of the complex frequency-domain reflectance. Though the ear canal must have at least small losses, this method is consistent with the common assumption used in reflectance analysis (i.e., $|\Gamma(f)| = |\Gamma_{tm}(f)|$), while also providing phase information associated with the TM factor, $\Gamma_{tm}(f)$. In this procedure, any lossless delay associated with the middle ear will also be attributed to the ear canal.

Other studies have considered frequency-domain transmission line models of the ear canal and middle ear (e.g., Figure 1–2), including ear-canal elements of variable area (Lewis & Neely, 2015). It is difficult to verify either of these methods using real-ear measurements, as measurements are highly affected by probe placement within a few mm of the TM, due to the complicated sound field close to the TM.

Time-Domain Reflectance

Time-domain methods have been proposed to analyze the reflectance phase. Due to the variation in distance between the probe and middle-ear structures (e.g., the malleus is closer to the probe than the stapes, along the sound pathway), reflections from these structures may be distinguished based on peaks in the time-domain reflectance (Neely, Stenfelt, & Schairer, 2013).

Using time-domain reflectance, Rasetshwane and Neely (2011) described a method to determine the ear-canal area as a function of distance along the canal. Estimates of the ear-canal area function may be used to model the complex TM reflectance from the reflectance measured at the microphone. For instance, one may approximate the ear canal as a series of concatenated tubes or conic sections of varying area and calculate a transmission line model result, in both the time and frequency domains.

A good time-domain (or frequency-domain) estimate of the TM reflectance contains delay information associated with various middle-ear structures, which may, in addition to the magnitude, be systematically altered in the presence of a middle-ear abnormality.

Wideband Tympanometry

The primary focus of this chapter is reflectance measured in the ear canal at ambient static atmospheric pressure. A clinical alternative to this technique involves pressurizing the ear canal as in tympanometry, to measure the reflectance as a function of frequency *and* ear-canal static pressure. This is typically represented as a three-dimensional magnitude plot (Margolis, Saly, & Keefe, 1999), where the canal pressure is swept from about +200 to −400 daPa. An advantage to this technique is that it combines wideband power absorbance with tympanometry, the standard clinical technology. Disadvantages of this measure include the need to pressurize the ear canal, and the effects pressurization can have on subsequent measurements, due to preconditioning of the TM (Burdiek & Sun, 2014). Wideband tympanometry generates more data to analyze than WAI, which must be interpreted for clinical decision-making, a subject of current research (Keefe, Hunter, Patrick Feeney, & Fitzpatrick, 2015). Clinical efficacy is still being established; for example, Keefe, Sanford, Ellison, Fitzpatrick, and Gorga (2012) did not find an advantage to adding pressurization for detecting CHL in children.

Measurement Technique

System and Calibration

Typically, a reflectance measurement is made by playing a broadband sound stimulus in the ear canal, such as a chirp, and measuring the sound pressure at the probe microphone. The sound source must be calibrated in order to correctly interpret the ear-canal sound pressure response. Most techniques for measuring reflectance use multiple cylindrical cavities of known lengths to calibrate the probe; these approaches differ primarily in terms of the size, length, and the number of calibration cavities used (Keefe et al., 1992; Møller, 1960; Neely & Gorga, 1998). For instance, Mimosa Acoustics' *HearID*® and *OtoStat*® systems (Champaign, IL) uses a four-cavity method described by Allen (1986). In this method, the pressure responses of four cylindrical cavities are compared to their theoretical values, in order to determine the *Thévenin equivalent* source parameters, describing the acoustic behavior of the probe loudspeaker(s).

Reflectance can also be derived using other calibration methods such as the two-microphone method or the standing-wave tube method (Beranek, 1949; Shaw, 1980). The two-microphone method requires a precise calibration of the microphones and precise knowledge of their relative placement; the standing wave tube method requires manual manipulation of the microphone placement with reference to the end of a tube. Many of these early methods have been shown to be relatively inaccurate due to their sensitivity to precise placement of the microphone(s). The four-cavity method is suitable for clinical use because it is less sensitive to precise placement of the microphone.

Any calibration may be verified by measuring a system of known impedance. The simplest method is to measure the reflectance of a syringe or hard-walled lossless cavity. In this case, we expect $|\Gamma_m(f)| \approx 1$ across the frequency range of measurement for an approximately lossless load impedance. Larger errors can occur above about 5 kHz due to more complicated patterns in the ear canal pressure at higher frequencies. Since an ear is not a rigid cavity, perhaps a more easily interpreted test is to measure an artificial ear. However, such a coupler and its expected frequency response may not be readily available, making this approach less practical.

Challenges and Sources of Variability

For investigators working with reflectance in the lab or clinic, it is important to be aware of sources of measurement variability that can be controlled (including calibration, noise, probe ear-tip insertion, and acoustic leaks) and those that cannot (such as normal variation across ears).

Acoustic probe calibration. Acoustic measurements can be very sensitive to probe calibration. In the case of a foam-tipped probe, parameters characterizing the sound source may drift over time with expansion and compression of the foam (ear-tip geometry), and be sensitive to differences across replaceable tips, such as small variations in the tip length. Environmental noise and vibration can affect calibrations and be propagated through to measurements. It is important to always carefully follow the manufacturer's guidelines when calibrating acoustic equipment. Manufacturers are working on systems that can be factory calibrated, removing the need for daily field calibrations.

Noise. Acoustic measurements can be sensitive to environmental noise. In the hospital or clinic, such noise is often unavoidable. If this is the case, some ways to obtain better data include increasing averaging times for data acquisition (if this parameter is under user control) or making repeated measurements. A set of measurements may, in some cases (e.g., the power reflectance), be averaged, or the measurement with lowest noise may be selected during analysis. Often this can be accomplished automatically, via artifact-rejection software included with the measurement device.

In addition to noise in the environment, acoustic noise from vocalizations and movement by the patient may also affect measurements. Data from infants often have the most measurement noise because they are collected in a busy hospital, and the infant cannot be asked to sit still and be quiet during measurement. Noise can also be propagated through the probe's electrical cable if it is draped over the patient. Normative data as well as individual measurements may be degraded by such noise.

Probe insertion. The depth of the probe in the ear canal and the probe seal may both have effects on the measured reflectance, though the effect of probe depth is typically assumed to be small. Power reflectance at low frequencies has been shown to decrease (absorbance level increases) with the probe distance from the eardrum (Lewis & Neely, 2015; Voss et al., 2008; Voss, Stenfelt, Neely, & Rosowski, 2013). This effect is relatively small in adult ears (infant ears have larger ear-canal losses) when the probe is situated in or close to the bony portion of the ear canal, because this region is most similar to a rigid-walled cavity and thus has smaller acoustic losses. Outside of the bony region, the cartilaginous section of the ear canal has much greater wall losses. Therefore, it is best if the probe tip is inserted deeply enough (e.g., 1 cm) to reach the bony part of the ear canal. Probe location in the ear canal may also vary due to different probe-tip types across measurement systems. For instance, "umbrella" tips, which may be used for pressurized wideband tympanometry measurements, are placed against the opening to the ear canal. By construction, measurements made with these tips will not extend as far into the ear canal as an insert ear tip, leading to a much lower ear-canal standing wave frequency and more ear-canal wall losses.

Finally, investigators should have a basic knowledge regarding the effects of acoustic leaks on reflectance, so that a leaky probe insertion may be corrected during the measurement sitting. The measurement device may include methods to detect these leaks, but they may not be effective 100% of the time (e.g., for small leaks). Most leaks occur because the probe tip does not properly seal in the ear canal, or may shift in the ear canal over the course of multiple measurements. Sometimes, using a smaller tip more deeply inserted will decrease issues with leaks.

An example of an acoustic leak is given in Figure 1–5. These measurements are from a normal ear, where the probe insertion drifted over the course of multiple tests made over a period of a few minutes (reflectance data from

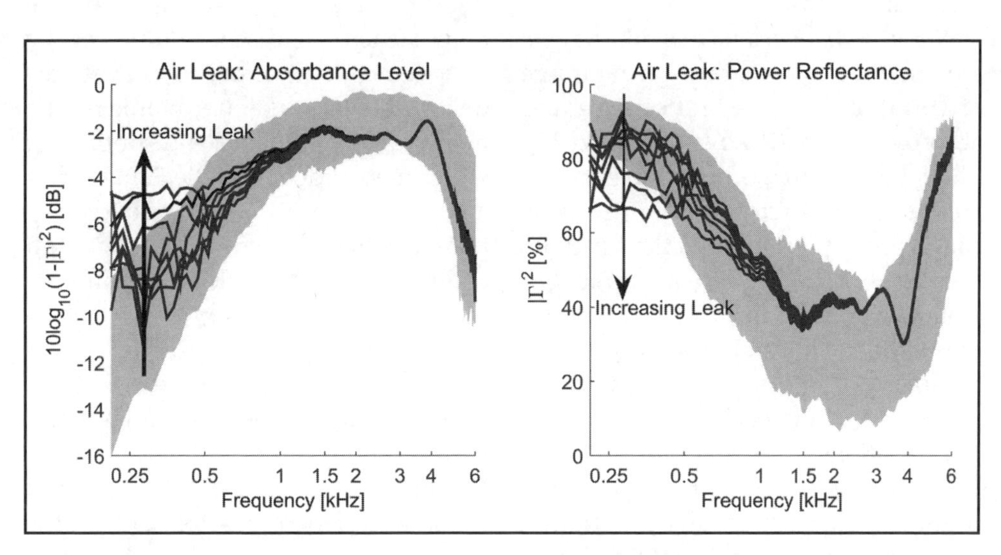

Figure 1–5. Example of the effects of a leak in probe insertion on the absorbance level and power reflectance in a normal ear, plotted against the 10th to 90th percentile (*gray region*) for normal ears from Rosowski et al. (2012). As the size of the acoustic leak increases (e.g., the probe insertion slowly loosens), there is an increase in the low-frequency absorbance level and a decrease in the low-frequency power reflectance. The effect propagates upward in frequency as the leak increases.

Thompson, 2013). As the leak in the probe seal increases in size, the low-frequency absorbance level increases, and the corresponding power reflectance curve decreases (Groon, Rasetshwane, Kopun, Gorga, & Neely, 2015). This increased absorbance is not from the middle ear absorbing the acoustic power but from the power dissipating through the leak around the probe tip. For most middle-ear conditions, unless the TM is abnormally compliant or perforated, the absorbance level should be relatively low (power reflectance close to one, or 100%) below 1 kHz. Groon et al. (2015) recommend that when the frequency range of interest extends as low as 0.1 kHz, low-frequency absorbance should be ≤0.20 and low-frequency admittance phase ≥61 degrees; for frequency ranges as low as 0.2 kHz, low-frequency absorbance should be ≤0.29 and low-frequency admittance phase ≥44 degrees.

Normal variation. Normal middle ears are known to have a fairly wide range of variation in power reflectance and absorbance level (Rosowski et al., 2012). The largest source of *intra*subject variability (i.e., test-retest within the same ear) is probe placement in the ear canal (Voss et al., 2013), though this variability is small compared to variability across a population of normal ears (Rosowski et al., 2012). Middle-ear pressure within a normal range may also cause intrasubject variations (Shaver & Sun, 2013).

When the probe is properly placed in the ear canal, *inter*subject variation (i.e., across ears) is due to differences in middle-ear physiology. Voss et al. (2008) found, based on manipulations in cadaveric ears, that variations in the volume of the middle-ear space produced larger variability in power reflectance measurements than variation in probe insertions. The middle-ear cavity can affect the power reflectance and absorbance level over a broad frequency range and may play a role in variability across normal subjects. Variability of the absorbance level (see Figure 1–3) between 1 to 4 kHz is small (e.g., ±3 dB) in the normal ear, and it is likely related to the acoustics of the TM and ossicles, in addition to middle-ear space (Rosowski et al., 2012; Stepp & Voss, 2005).

Clinical Applications

The middle ear is a complex mechanism with many components. It follows that there are many possible disorders of the middle ear. Some disorders include fluid or infection in the middle-ear space, ossification of the bony structures, discontinuities of the ossicular chain, perforation of the eardrum, and various abnormalities of the membranes, ligaments, and supporting structures. Wideband reflectance offers a novel approach to describe and diagnose middle-ear dysfunction. Although complex pressure reflectance offers a more complete picture, to date, clinical researchers have focused on power reflectance and power absorbance. These quantities contain similar information, and they are both widely used. We provide both power reflectance and absorbance where possible to help the reader navigate the literature. Arguably, the power absorbance or absorbance level is the preferable format, due to its close relationship with the middle-ear response when plotted in decibels.

There are many approaches to establish diagnostic criteria. Statistical methods are used to establish the normal range and to identify criteria for what is abnormal. Mathematical models are used to simulate and describe the underlining physics of normal variability and abnormal changes in the middle ear (Parent & Allen, 2010; Voss, Merchant, & Horton, 2012). Combining these methods allows us to derive useful diagnostic criteria. In the discussions below, different studies demonstrate different approaches to establish such criteria. In some cases, the diagnostic criteria are built into the measurement software, and in other cases, the clinician or researcher may need to derive the result from exported data. The clinical utility of WAI is still undergoing intensive research, and clinical applications are rapidly advancing.

Quantities Used

A single WAI measurement produces a wealth of information. The measurement technique provides frequency resolution on the order of 20 Hz over at least a 0.2- to 6-kHz range. Typically, this is too much information to work with statistically for the purpose of clinical decision making. Thus, clinical researchers use various tactics to reduce the number of variables and extract meaningful quantities to assist diagnostic decision making. Current approaches include:

1. *Looking for patterns.* Small-*N* and case studies are used to get an idea of the general pattern of normal and abnormal results, particularly when characterizing relatively unstudied pathologies. This can help focus attention onto specific frequency ranges in larger studies and improve detection based on physical modeling.
2. *Band averaging.* Band-averaging power absorbance level and power reflectance across frequency can describe frequency-dependent behavior using a smaller set of parameters (e.g., the reflectance area index (RAI) defined in Hunter, Feeney, Lapsley Miller, Jeng, & Bohning, 2010). One-third, one-half, and whole-octave bands are frequently used and fit nicely with other audiologic tests, while still capturing the shape of the power absorbance curve. It is not typically useful to take the average across the entire curve, because frequency-dependent behavior is obscured.
3. *Comparison to norms.* Comparing abnormal results to a norm can be done qualitatively and quantitatively. For instance, the Absorbance Level Difference (ALD), defined by Rosowski et al. (2012), is the absorbance level relative to a normal ear average over a specific frequency range. They used the ALD to quantify notches seen in abnormal absorbance curves.
4. *Parameterization.* The WAI response may be characterized with a small number of parameters, such as a three-line approximation to the absorbance level curve (Rosowski et al., 2012).
5. *Multivariate approaches.* Methods such as discriminant function analysis and multiple regression can help the researcher narrow down variables that provide the most unique information. These

approaches can also allow information to be combined across test types and across measurements with different units, providing a powerful basis for clinical decision making. Multiple parameters are combined to produce one number, which is then used for making a decision.

6. *Physical models.* Parameters are extracted using a physics-based model for the middle ear. For example, Robinson, Thompson, and Allen (in press) and Lewis and Neely (2015) extract parameters from simple transmission line models of the middle ear to characterize the condition of the middle ear.

Norms

For clinical use, we must first establish the normal range of immittance quantities, broken down by key demographics such as sex, ear, age, and ethnicity. With information about how normal ears behave, ears with middle-ear dysfunction may be identified. A norm is a statistically-defined range for a given quantity, derived from highly screened normal ears from a specific demographic group. The type and degree of screening for defining "normal" varies across studies, but includes audiologic history and audiologic tests such as auditory brainstem response (ABR), OAE, tympanometry, surgical discovery, and pneumatic otoscopy. Different screening tests and criteria can lead to differences between norms across studies. Norms can be expressed in various ways, including percentile ranges and means with standard deviations.

As summarized by Shahnaz, Feeney, and Schairer (2013), the demographic that has the greatest effect on the normal middle ear is age, followed to a much lesser extent by ethnicity. Sex and ear differences are much smaller and typically not statistically significant. Figure 1–6 shows norms for four key age groups, replotted from their original studies in consistent units (median and 10th–90th percentile range) to aid comparison. Here we focus on the age demographic and describe some key normative studies, with additional discussion about secondary demographic differences. Across many studies, the general consensus is that larger-*N* normative studies are needed using highly screened normal ears. Many existing norms are based on small *N*, especially by the time they are divided into various demographic groupings. Notable gaps include older children and teenagers, and the elderly. There has been a focus on infants and young children due to their high prevalence of middle-ear disorders.

Norms may obscure individual patterns across frequency and can also obscure the noise and variation seen in WAI measurements taken in real-world clinical settings where noise levels may be difficult to control. This is why we have overlaid the norms in Figure 1–6 with five randomly selected examples of normal ears from the same data sets.

Newborn and Infant Norms

The human ear undergoes significant maturation in the first 12 months of life (as summarized in Kei et al., 2013). They found that the largest changes in absorbance and reflectance norms occur between birth and 6 months of age,

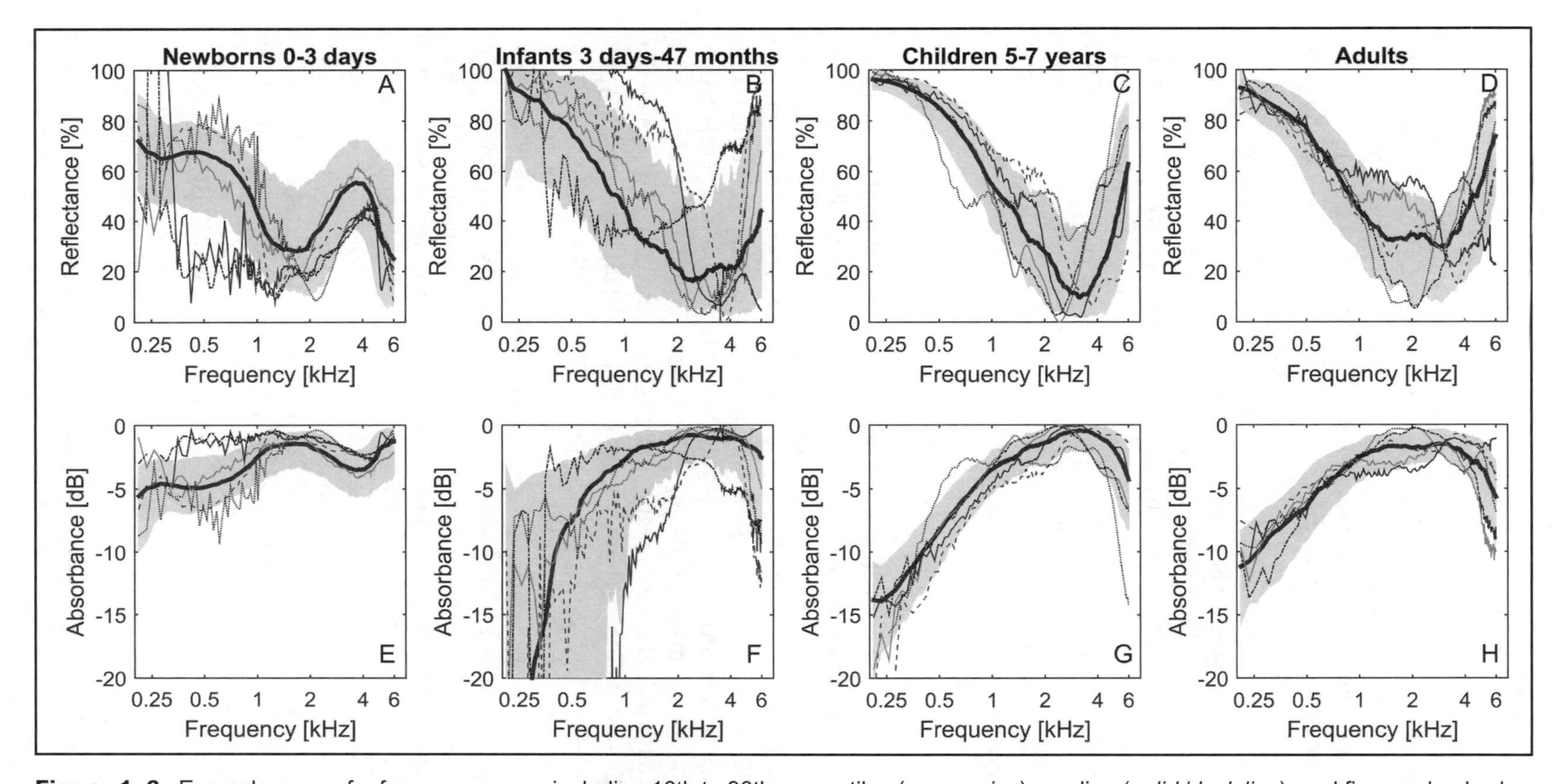

Figure 1–6. Example norms for four age groups, including 10th to 90th percentiles (*gray region*), median (*solid black line*), and five randomly chosen examples (*dotted lines*) for each group. The top row shows power reflectance (%) and the bottom row shows the same data plotted as power absorbance level (dB re 100% Absorbance). These norms were recalculated and replotted from data published elsewhere: Newborn norms are from Hunter et al. (2008); infant norms are from Hunter, Tubaugh, et al. (2008); child norms are from Beers et al. (2010); adult norms are from Rosowski et al. (2012).

indicating that this is the period of most rapid change in the outer and middle ear. Aithal, Kei, and Driscoll (2014b) showed that developmental changes in the outer and middle ear over the first 6 months of life cause a decrease in absorbance for low to mid-frequencies and an increase in absorbance at higher frequencies (>2.5 kHz). The absorbance at low frequencies is dominated by compliance characteristics of the ear canal and middle ear. With maturation, ossification of the inner two thirds of the ear canal causes the ear canal to be less compliant, corresponding to a lower middle-ear absorbance at low frequencies (Kei et al., 2013). Additionally, changes in ossicle bone density with age, along with the loss of mesenchyme and other middle-ear fluids, lead to decreased mass in the middle-ear system (Aithal et al., 2014b; Kei et al., 2013). This leads to an increase in the middle-ear absorbance at high frequencies, as the lower mass causes less sound to be reflected. This rapid maturation implies norms are potentially needed for many age ranges.

Norms from newborn babies in the first hours to first days of life are of particular interest due to large-scale newborn hearing screening programs that test babies soon after birth. Newborn absorbance level norms typically show unreliable results below 1 kHz due to environmental noise and ear-tip leaks, and high absorbance at 1 to 2 kHz (higher than in older ears), decreasing at 3 to 4 kHz, and rising again at 6 kHz (which is also not seen in norms for older ears). Figure 1–6 shows norms from the Hunter et al. (2010) study, recalculated to show the entire frequency range and to show absorbance. Similar norms were also shown by others (Aithal, Kei, Driscoll, & Khan, 2013; Sanford et al., 2009; Shahnaz, 2008), with differences mainly due to screening criteria for "normal," demographics, and equipment.

Aithal et al. (2013) pointed out that using just distortion product otoacoustic emission (DPOAE) pass/refer results (as the earlier studies did) was not sufficient as a gold standard for normal ears, because ears with strong DPOAEs can overcome middle-ear dysfunction; however, they found similar norms with a smaller, more highly screened group. For newborns, there is ambiguity in defining the "normal" condition, because it is natural for healthy newborns to have some fluid in their external and middle ears, which affects middle-ear measurements. Whether this is an issue or not depends on the purpose. If the aim is to understand the normal infant middle ear, it is important. But if the purpose is to assess infant inner-ear status (as in universal newborn hearing screening [UNHS] programs), any temporary middle-ear condition that affects sound propagation is of concern regardless of whether it is "normal" or not.

Sex and ear differences are typically not observed or are not clinically significant; however, differences in ethnicity were found by Aithal, Kei, and Driscoll (2014a) where Australian Aboriginal infants had lower wideband absorbance than Australian Caucasian infants. This may be of clinical importance due to the high otitis media with effusion (OME) prevalence among Aboriginal children.

For slightly older infants, Hunter, Tubaugh, Jackson, and Propes (2008) produced norms for infants 3 days to 47 months old. They did not find significant age or sex effects, although their age bands had only around 10 ears per band.

Children

The study with the largest number of normative subjects for young children is Beers, Shahnaz, Westerberg, and Kozak (2010). They tested wideband reflectance in 78 children (144 ears) ages 5 to 7 with normal middle-ear function for comparison to those with OME. They compared Caucasian and Chinese children's ears, and found significant differences at 2 and 6 kHz; 2 kHz falls within an important frequency range for detecting conditions that increase middle-ear stiffness, like OME, so this may be of clinical significance. It remains to be studied if body size is a better predictor of variation in absorbance and reflectance than ethnicity.

Adults

Rosowski et al. (2012) established norms on a medium-sized group of highly screened otologically normal adults (29 adults/58 ears, up to age 64). They found small sex and ear differences, and their overall average power reflectance curve was similar to previous studies of power absorbance and reflectance (see Figure 1–6). Of interest is the parameterization of the absorbance curve. In log-log coordinates, the curve can be modeled with three straight lines (Allen et al., 2005; Rosowski et al., 2012). Below 1 kHz, average absorbance increases by about 15 dB per decade. Above 4 kHz, absorbance decreases by 23 dB per decade. Between 1 and 4 kHz, absorbance is essentially constant at around −2.5 dB. Extracting the key features of the absorbance curves and deriving parametric values to characterize them could aid in clinical decision making. For instance, changes in slopes, frequency of intercepts or large deviation from a straight line may be indicative of abnormal middle-ear performance.

Middle-Ear Dysfunction

A number of small-*N* and case studies have suggested where power absorbance and reflectance might be useful for detecting middle-ear dysfunction, and they provide a descriptive patterns of the conditions relative to norms (e.g., Allen et al., 2005; Feeney, Grant, & Marryott, 2003; Sanford & Brockett, 2014). These studies suggest areas where larger studies should look. However, we need larger-*N* studies to understand how middle-ear power absorbance behaves statistically over a population, for each pathology and demographic. These results must then be combined and reduced to specific criteria for decision making. As well as larger-N studies for specific pathologies, broad studies across a range of confusable pathologies are needed to establish the bases for differential diagnoses. Fortunately, a number of such studies are in progress, and we will summarize some here.

Wideband Reflectance in Universal Newborn Hearing Screening (UNHS) Programs

The goal of UNHS programs is to detect babies who have sensorineural hearing loss so they can benefit from early intervention (e.g., cochlear implants). UNHS programs provide a pass or refer result from either OAE or ABR tests. These screening tests are not diagnostic but are used to determine referrals for more extensive diagnostic follow-ups.

It has long been best practice in UNHS programs to rescreen babies who get a refer result to reduce false positives for diagnostic referrals. This rescreening is usually done after a delay, because testing within 24 hours of birth is much more likely to produce a refer result than testing after 24 hours (and preferably 36 hours). The majority of these false-positive referrals are from transient middle-ear dysfunction from the birth process (e.g., amniotic fluid, mesenchyme, and meconium in the middle-ear space), which clears within the first few days of life. Figure 1–7A–D shows examples of five newborn ears from Hunter et al. (2010) that did not pass DPOAE testing. In all cases, their middle-ear reflectance was higher than normal (absorbance was lower than normal), indicating their DPOAE refer result was possibly a false positive. Hunter et al. (2010) showed why rescreening OAEs after a delay was often successful—middle-ear absorbance tends to increase over time, presumably as the middle ear clears, allowing more sound to propagate into the inner ear and back.

This transient middle-ear dysfunction is not reliably picked up with tympanometry in newborns. Hunter et al. (2010) and Sanford et al. (2009) both showed that wideband power absorbance and reflectance are superior to tympanometry in predicting which ears show low OAE levels due to middle-ear dysfunction. Adding absorbance to OAE screening can potentially identify those babies most in need of diagnostic follow-up, help determine the best time for repeat screening, and reduce false-alarm referrals.

WAI can be used to interpret OAE results in older infants, children, and adults similarly to newborns. In these older age groups, however, low absorbance is cause for follow-up for middle-ear dysfunction like otitis media.

WAI in UNHS

When using WAI with OAEs in testing newborns, a pass/refer result can be assigned to each test, giving four possible outcomes. Each outcome is illustrated in Figure 1–8 with real examples from real babies in the Hunter et al. (2010) study. This study showed that the reflectance or absorbance around 2 kHz was the best predictor of a DPOAE test pass (DPOAE levels at 3 or 4 out of 4 frequencies are normal) or refer (DPOAEs at 2 or more out of 4 frequencies are abnormally low). Absorbance below 1 kHz is not plotted because in babies, this region is often noisy (Hunter et al., 2010).

Also plotted are two normative regions in gray. For the absorbance plot, the gray norm represents an

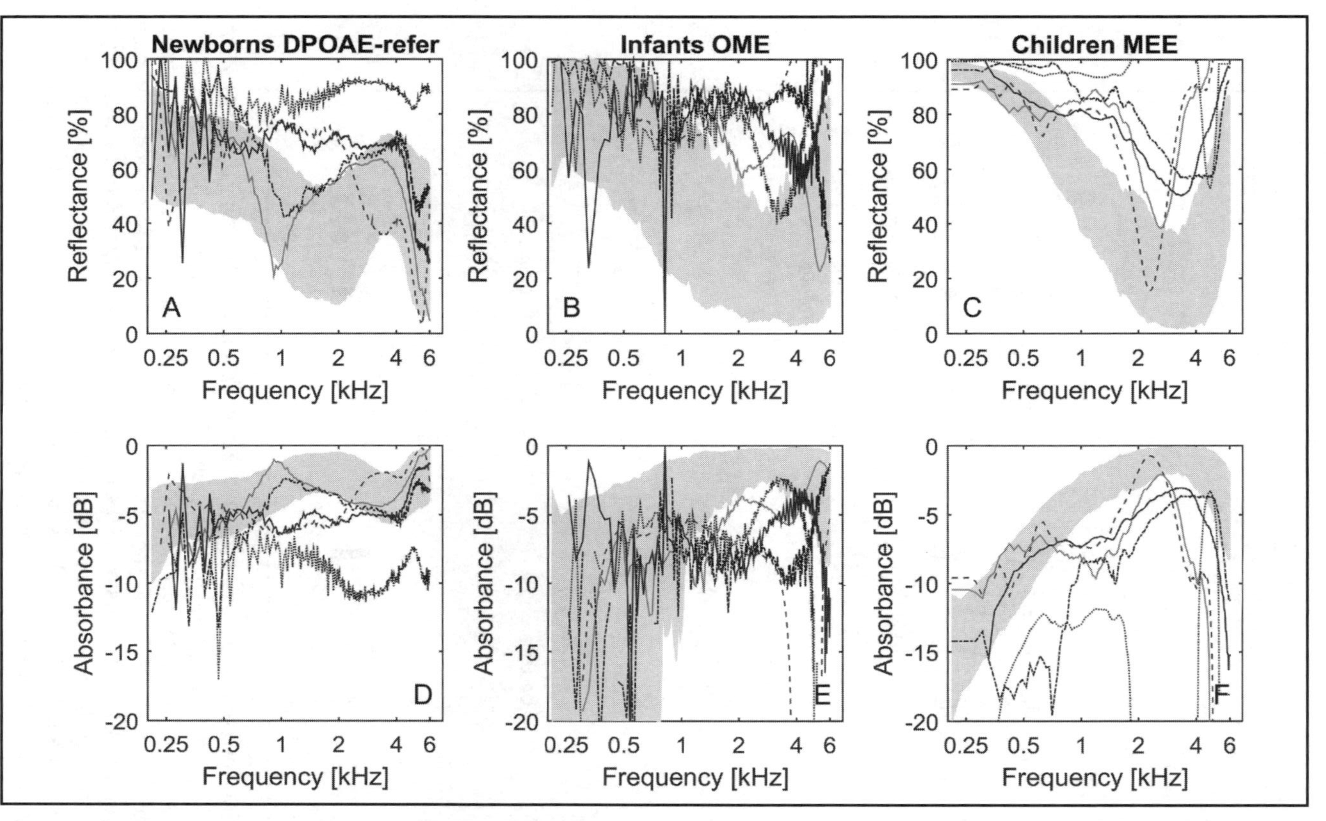

Figure 1–7. Five examples of middle-ear dysfunction in children, overlaying norms from Figure 1–6 (10th–90th percentiles, *gray region*). The top row shows power reflectance (%) and the bottom row shows the same data plotted as power absorbance level (dB re 100% Absorbance). **A–D:** newborns who received a DPOAE refer result (Hunter et al., 2010). **B–E:** infants diagnosed with otitis media with effusion (OME) (Hunter, Bagger-Sjoback, & Lundberg, 2008). **C–F:** children diagnosed with middle-ear effusion (MEE) (Beers et al., 2010).

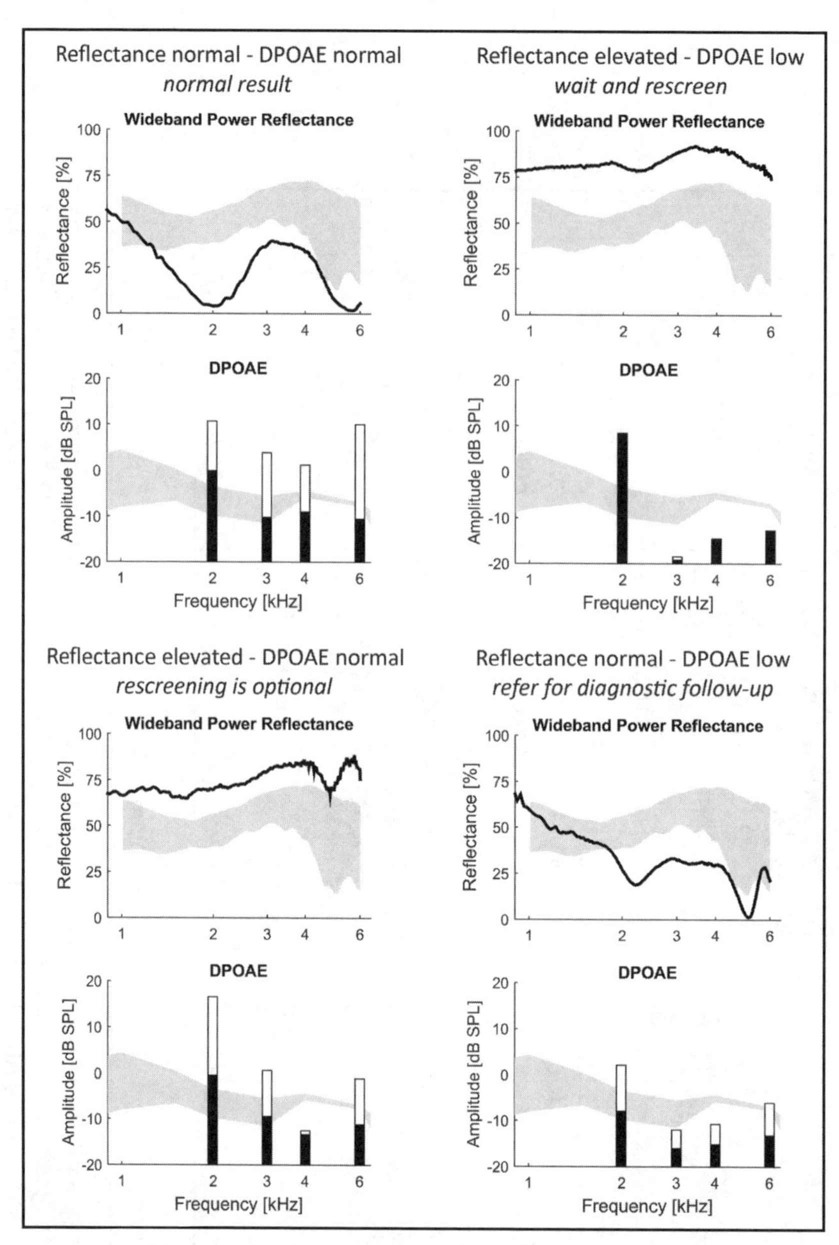

Figure 1–8. Four outcomes are possible when using wideband absorbance at 2 kHz with DPOAEs in a newborn hearing screening program. The absorbance plots show power absorbance (*black line*, dB re 100% absorbed) and the normative ambiguous region (*gray region*, dB re 100% absorbed). The DPOAE plots show DPOAE amplitude (*white bar*, dB SPL), the noise floor (*black bar*, dB SPL), and the Boystown 90% ambiguous region (*gray region*, dB SPL), where DPOAEs below the region are considered refer results. For the screening protocol used in this study, if 3 or 4 out of 4 DPOAE frequencies get a pass result, the overall result is a pass (*left plots*). If 2 or more out of 4 DPOAE frequencies get a refer result or are noisy, the overall result is a DPOAE refer (*right plots*). Referrals can be reduced by considering absorbance. If absorbance is low, the DPOAE refer is probably due to middle-ear dysfunction (*top right*). However, the possibility of underlying sensorineural hearing loss cannot be excluded, so repeat screening is needed. If absorbance is normal, the DPOAE refer needs diagnostic follow-up for possible sensorineural hearing loss (*bottom right*). Sometimes absorbance will be low but the DPOAEs are so strong, they are able to overcome the reduction in middle-ear transmission (*bottom left*).

ambiguous region. Absorbance *above* this region (especially at 2 kHz) was associated with DPOAE pass results. Absorbance *below* this region (especially at 2 kHz) was associated with DPOAE refer results. The ambiguous region describes where the pass and refer regions overlap (defined by the 10th and 90th percentiles). Similarly, for the DPOAE plot, the gray norm also represents an ambiguous region (from the Boystown norms Gorga et al., 1997). DPOAEs *above* this region are associated with normal hearing. DPOAEs *below* this region are associated with abnormal hearing.

How could WAI be used in UNHS programs? Specific guidelines are still in development. Potentially, WAI may be used to enable better timing for rescreening and follow-ups. The examples in Figure 1–8 suggest the following courses of action:

1. Normal absorbance—normal DPOAEs across all frequencies (top left). Screening passed and no rescreening or follow-up is needed.
2. Low absorbance—low DPOAEs: possible middle-ear fluid (top right). Wait a few hours and rescreen to see if absorbance is higher and DPOAEs pass. Refer for diagnostic follow-up if DPOAEs do not pass on rescreening. Chances are absorbance will increase as the middle ear clears and the true DPOAE status will be more clearly revealed. Low absorbance and low DPOAEs are commonly seen in newborn hearing screening programs and cause undue worry for parents and an increased workload due to unnecessary diagnostic follow-ups. With WAI + DPOAEs, testers can immediately see if there is middle-ear dysfunction and can reassure parents that this is common and not of concern.
3. Normal absorbance—low DPOAEs (bottom right). This ear is a priority for diagnostic follow-up because there may be permanent sensorineural hearing loss. Rescreening is optional because the usual reason for DPOAE false alarms—low middle-ear absorbance from transient middle-ear dysfunction—has been eliminated. Any rescreening can occur immediately because the WAI results show the middle ear is not impeding sound propagation into the inner ear.
4. Low absorbance—normal DPOAEs (bottom left). The DPOAEs are strong enough to overcome what is possibly a probe blockage or transient middle-ear dysfunction. In this situation, the tester should check for probe or ear canal blockage, or a collapsed ear canal, and then retest WAI. Since DPOAEs passed, rescreening is optional because an outer or middle-ear condition is not typically a reason for referral.

Identifying Conductive Hearing Loss (CHL) in Infants and Children

Identifying CHL in young infants can be difficult with tympanometry, and there is no standard interpretation. Prieve, Vander Werff, Preston, and Georgantas (2013) evaluated tympanometry variations along with wideband reflectance and showed the latter was just as effective as tympanometry in identifying CHL in infants less than 6 months old (3–26 weeks) who had been referred in an infant hearing screening program. The babies received both air and bone conducted ABR tests along with tympanometry and WAI (43 ears had normal hearing and 17 ears had CHL, determined from the air- and bone-conducted ABR thresholds). Prieve et al. (2013) found that wideband reflectance between 800 and 3000 Hz was higher in CHL ears compared with normal ears (absorbance was lower than normal). Prieve et al. found that a criterion for power reflectance greater than 69% (power absorbance less than 31%) in the one-third octave band around 1,600 Hz produced the highest likelihood ratio for CHL, compared to other frequency bands and compared to various quantities derived from multifrequency tympanometry (at 226, 678, and 1,000 Hz). These results indicate the frequency range most sensitive to CHL in infants and show that power absorbance is a suitable replacement for tympanometry in this age group. A larger study will be needed to determine statistically meaningful criteria, and investigate if further age-specific criteria are needed.

In children, otitis media is the most common reason for CHL. Otitis media can produce middle-ear effusion (MEE) and negative middle-ear pressure (NMEP), which both tend to stiffen the middle ear and thereby decrease the amount of power absorbed for low- to mid- frequencies below 2 to 4 kHz, compared to normal ears (Robinson et al., in press). The extent to which this stiffness affects the mid-frequency range (i.e., between 1 and 4 kHz) may depend on the severity of the condition. For MEE, there can be an additional decrease in absorbance at mid- to high frequencies due to the mass of the fluid. These decreases in absorbance can be used to identify MEE. Hunter, Tubaugh, et al. (2008) point out that tympanometry is unreliable in very young infants since it can produce normal results in the presence of MEE. They found that absorbance was decreased between 1 and 3 kHz in ears with suspected MEE. Figure 1–7 shows five examples of OME from this study, compared to the norms. Below 1 kHz, the data are noisy due to the difficulty in getting quiet measurements on this population. Above 1 kHz there is a decrease in absorbance, consistent with stiffening of the middle ear from OME. Noise rejection has been improved in clinical equipment since this study was conducted.

Beers et al. (2010) tested 78 children (144 ears) with normal middle ears and 64 children with abnormal middle ears (21 ears with suspected MEE, 21 ears with confirmed MEE, and 54 ears with abnormal NMEP). The children were aged 5 to 7 years. They found that reflectance in the frequency region around 1.25 kHz best separated the normal ears from those with MEE. Using the 90th percentile from the normal group as a criterion, all ears with MEE had higher reflectance (lower absorbance) than the criterion at 1.25 kHz (hit rate 100%), for a false-alarm rate of 10%. Figure 1–7 shows five examples of children with MEE from this study, overlaid on the norms, showing greatly increased reflectance and reduced absorbance. They also showed that WAI was much more sensitive than 226-Hz tympanometry in detecting MEE. On average, ears with abnormal NMEP had higher reflectance (lower absorbance) than normal but not as high as those ears with MEE. Ellison et al. (2012) also investigated MEE in children and found similar results to Beers et al. (2010), with decreased absorbance between 1.5 and 3 kHz in" ears with surgically verified MEE.

Middle-Ear Pathology in Adults

In infants and children, the main interest is hearing screening and otitis media. In adults, a range of other middle-ear conditions are common. Figure 1–9 shows five examples of four common conditions, compared to adult norms.

Ossicular Disarticulation/Discontinuity

The disruption or near-disruption in an ossicular joint causes a significant peak in the absorbance, typically below 1 kHz (Panels A, E in Figure 1–9). This peak is likely due to a resonance of the ossicle mass and TM stiffness. Although the absorbance is elevated in this narrow frequency band, this power is *not* transmitted to the inner ear but is dissipated in the resonant joint.

Otosclerosis

Stapes fixation (due to otosclerosis) typically reduces the absorbance at low frequencies below 1 to 2 kHz but may be within normal limits, or even lower than normal. Figure 1–9 (Panels B, F) shows five examples from Shahnaz et al. (2009) that illustrate these three outcomes. In the typical otosclerotic ear, Shahnaz et al. (2009) found increased reflectance (decreased absorbance) below 1 kHz, compared to a normal group. This is due to increased stiffness of the middle-ear at the stapes, which increases the total stiffness of the middle ear measured at the eardrum. They found that WAI was a better predictor than tympanometry, with an 82% hit rate and 17% false alarm rate, and identified 500 Hz as a good band for detecting otosclerotic ears. They also identified a subgroup of ears where reflectance was lower than normal (absorbance was higher than normal) below 1 kHz. A combination of WAI and tympanometry was able to detect all otosclerotic ears for the price of a higher false-alarm rate.

Tympanic Membrane (TM) Perforations or Pressure Equalization (PE) Tubes

TM perforations or pressure-equalization (PE) tubes tend to produce power absorbance curves that are highly variable across frequency, with high absorbance at low frequencies and a large equivalent ear canal volume, typically much greater than 3 cc. It can be difficult to differentiate a noisy test with an acoustic leak from a TM perforation; however, repeated testing should reveal a stable pattern for PE tubes or a perforation. This variation across ears is apparent in Figure 1–9 (Panels C, G) where five ears with varying degrees of perforation show low reflectance at low frequencies but not much consistency.

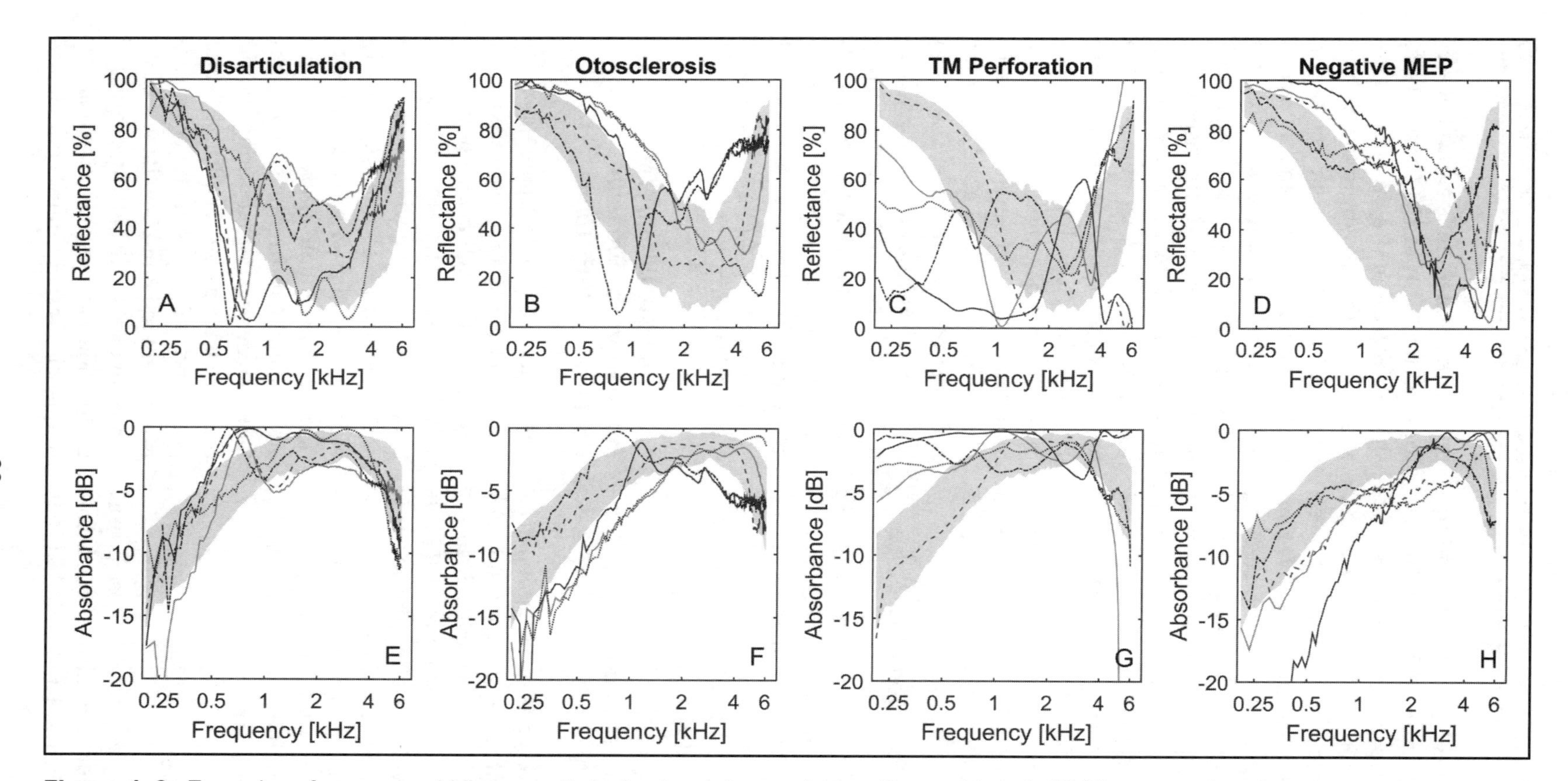

Figure 1–9. Examples of common middle-ear pathologies in adults, overlaid on Rosowski et al. (2012) norms, plotted as power reflectance (%) and absorbance level (dB re 100% absorbance). **A, E:** ossicular discontinuity showing the characteristic notch below 1 kHz (three randomly selected examples from Nakajima). **B, F:** otosclerosis (five randomly selected from Nakajima et al. (2012). **C, G:** TM perforation (five examples from unpublished data). **D, H:** NMEP ranging from approximately −384 to −65 daPa (five examples from Robinson et al. [in press]).

These data come from an unpublished study (Feeney, Hunter, Jeng, & Lapsley Miller).

As in the case of ossicular discontinuity, although the absorbance is elevated due to a TM perforation, this power is *not* transmitted to the inner ear but is dissipated in the middle-ear cavity (Voss, Rosowski, Merchant, & Peake, 2001). The resonant frequency at which we see dissipation is related to the middle-ear cavity size *and* the size of the hole (these two parameters can be modeled as a *Helmholtz resonator* (Voss et al., 2001), which is what you get when you blow over the top of an empty bottle).

Larger perforations are detectable with otoscopy, but smaller perforations may be hard to visualize. In a cadaveric ear, smaller perforations were more apparent in the WAI response than larger ones (Nakajima, Rosowski, Shahnaz, & Voss, 2013; Voss et al., 2012), presenting as very low reflectance around 1 kHz.

Negative Middle-Ear Pressure (NMEP) and Eustachian Tube Dysfunction (ETD) in Adults

Abnormal NMEP can occur from eustachian tube dysfunction (ETD) and it typically increases the stiffness of the middle ear. On average, NMEP causes a decrease in the power absorbance level for low- to mid-frequencies, below 2 to 4 kHz, with large intersubject variation in NMEP effects. Across ears and NMEP levels, the absorbance is most sensitive to NMEP near 1 kHz (Robinson et al., in press; Shaver & Sun, 2013). Middle-ear pressure cannot be directly estimated from WAI as is the case with tympanometry (i.e., from tympanic peak pressure readings). The degree of absorbance change is associated with degree of NMEP, but this is most noticeable when considering changes in a subject or in large group averages—it is not easy to detect the degree of NMEP from single measurements unless it is severe. It is important to evaluate the presence of NMEP as it can affect other measurements, especially OAEs (even with NMEP within a clinically normal range). Figure 1–9 (Panels D, H) illustrates five examples of ears with NMEP where normal-hearing experimental subjects induced NMEP using a Toynbee maneuver (closing mouth and pinching nostrils, then swallowing) (Robinson et al., in press). Tympanometry indicated NMEP ranging from approximately −384 to −65 daPa.

Differential Diagnosis of Conductive Hearing Loss (CHL) in Adults

It should now be apparent that although some middle-ear pathologies are easy to detect from normal, in some cases they can be difficult to distinguish because the effects on WAI are similar, despite having different causes. For instance, otosclerosis and NMEP can both cause an increased stiffness as seen at the eardrum even though that stiffness is generated in different ways. From what we know so far about WAI, it is not always possible to *differentially* diagnose every middle-ear disorder. However, in conjunction with other tests (such as air-bone gap [ABG]), different disorders can be teased out, including some conditions where it had been previously difficult, expensive, or not possible.

An excellent example is CHL with an intact TM and aerated middle ear, which can be associated with three conditions: ossicular fixation (usually from otosclerosis), ossicular discontinuity, and superior semicircular canal dehiscence (SCD) (Nakajima et al., 2012). These three conditions are challenging to differentially diagnose in the clinic and may require surgery or expensive tests to fully investigate. Nakajima et al. (2012) showed how to use WAI and an ABG audiogram to aid in differential diagnosis in an office setting.

Ossicular discontinuity and otosclerosis are described above. SCD is often referred to as a "third window" lesion of the inner ear (i.e., in addition to the oval and round windows of the cochlea), caused by a space where there is bone loss (Merchant & Rosowski, 2008). This affects the reflection of sound from the cochlea. SCD ears tend to show an abnormal peak in the absorbance around 1 kHz (Nakajima et al., 2012), which is smaller, wider, and higher in frequency than typically seen with ossicular discontinuity.

The absorbance level difference (ALD) used by Nakajima et al. (2012) for differential diagnosis was calculated by subtracting the band-averaged absorbance level (in decibels) over 0.6 to 1 kHz from the mean normative absorbance level (averaged over the same range) from the companion study of normal ears by Rosowski et al. (2012) The ABG they used for the differential diagnosis was defined as the average gap between 1 and 4 kHz (this separates out the SCD cases where the ABG is most apparent at frequencies less than 1 kHz). This test is applicable to patients presenting with CHL, defined as >10 dB ABG on pure-tone audiometry (averaged over 500, 1,000, 2,000 Hz or 250, 500, 1,000 Hz), and with an intact TM and aerated middle ear. Nakajima et al. (2012) found:

- Ears with ALD (0.6–1 kHz) <1 dB and ABG (1–4 kHz) >10 dB were associated with stapes fixation.
- Ears with ABG (1–4 kHz) ≤10 dB were associated with SCD.
- Ears with ALD (0.6–1 kHz) ≥1 dB and ABG (1–4 kHz) >20 dB were associated with ossicular discontinuity.

In this study, sensitivity and specificity were good (stapes fixation: 86%/100%, ossicular discontinuity: 83%/96%, and SCD: 100%/95% for sensitivity and specificity, respectively). These numbers are based on $N = 31$ ears. These results suggest further study is warranted to refine the differential diagnostic potential of WAI and audiometry.

The Past and the Future of Wideband Acoustic Immittance (WAI)

There is a long history of effort in developing measurement methods and instrumentation for middle-ear evaluation. It took 30 years for tympanometry to become part of the standard clinical battery for hearing evaluation. In light of the high newborn hearing screening false-positive rate and the advantages of digital technology, the development of WAI as a clinical tool for middle-ear evaluation was launched with the support of National Institute on Deafness & Other Communication Disorders (NIDCD) funding in the mid-1990s. The first commercial research WAI measurement system was made available in 2003, and the first clinical instrument with Food and Drug Administration (FDA) clearance was available in 2006. As each WAI measurement provides an abundance of information compared to tympanometry, WAI opened wide a window into middle-ear assessment.

There are several important clinical applications that WAI can support. Some clinical applications of WAI go beyond the scope of this chapter. For instance, WAI may be used to assess middle-ear acoustic reflex. It has been demonstrated that the middle-ear reflex threshold obtained using WAI is lower than that obtained using standard tympanometry (Schairer, Feeney, & Sanford, 2013). Another area of application is detection of traumatic brain injury; WAI measurements may be compared before and after a traumatic incident, or used to monitor recovery (Voss et al., 2010).

In this chapter, we discussed two important clinical applications of WAI: RETFPL stimulus calibration, and noninvasive middle-ear assessment and diagnosis. Stimulus level adjustment based on FPL provides individualized in-the-ear calibrations, such that acoustic stimuli may be delivered to the inner ear at the intended sound pressure level. This will enable more accurate and repeatable hearing assessments for all audiologic tests, and more accurate performance for hearing instruments (e.g., hearing aids). Additionally, if an RETFPL calibration is used for OAE measurement, WAI is acquired without any extra effort and may be used to inform the diagnosis.

WAI may currently be combined with other audiologic tests to provide quick differential diagnosis for hearing screening and for advanced hearing assessment. Many studies have systematic changes in WAI between normal and pathological middle ears, and made recommendations for detection of certain pathologies. This is especially valuable for pathologies that require middle-ear surgery, where the ability to more accurately identify the pathology and its degree before surgery will improve the surgery preparation and patient care (in some cases, unnecessary exploratory surgery can be avoided). WAI can also be used for monitoring middle-ear status pre- and post-surgery.

Though many studies to date considered only the power absorbance or reflectance, due to unknown phase contributions of the residual ear canal between the probe and TM, many strategies for estimating the complex TM reflectance have been proposed based on physical models of the ear canal. Analysis of the TM reflectance phase is expected to improve differential diagnosis in the future. In addition to advances in modeling middle-ear function, larger normative populations are needed to improve clinical viability of WAI for differential diagnosis of middle-ear pathology.

The clinical applications of WAI are many, and the technology is ready.

References

Abur, D., Horton, N. J., & Voss, S. E. (2014). Intrasubject variability in power reflectance. *Journal of the American Academy of Audiology, 25*(5), 441–448.

Aithal, S., Kei, J., & Driscoll, C. (2014a). Wideband absorbance in Australian Aboriginal and Caucasian neonates. *Journal of the American Academy of Audiology, 25*(5), 482–494.

Aithal, S., Kei, J., & Driscoll, C. (2014b). Wideband absorbance in young infants (0–6 months): A cross-sectional study. *Journal of the American Academy of Audiology, 25*(5), 471–481.

Aithal, S., Kei, J., Driscoll, C., & Khan, A. (2013). Normative wideband reflectance measures in healthy neonates. *International Journal of Pediatric Otorhinolaryngology, 77*(1), 29–35.

Allen, J. B. (1986). Measurement of eardrum acoustic impedance. In J. B. Allen, J. L. Hall, A. E. Hubbard, S. T. Neely, & A. Tubis (Eds.), *Peripheral auditory mechanisms* (pp. 44–51). New York, NY: Springer-Verlag.

Allen, J. B., & Fahey, P. F. (1992). Using acoustic distortion products to measure the cochlear amplifier gain on the basilar membrane. *Journal of the Acoustical Society of America, 92*(1), 178–188.

Allen, J. B., Jeng, P. S., & Levitt, H. (2005). Evaluation of human middle ear function via an acoustic power assessment. *Journal of Rehabilitation Research and Development, 42*(4 Suppl. 2), 63–78.

Aron, M., Floyd, D., & Bance, M. (2015). Voluntary eardrum movement: A marker for tensor tympani contraction? *Otology & Neurotology, 36*(2), 373–381.

Bance, M., Makki, F. M., Garland, P., Alian, W. A., van Wijhe, R. G., & Savage, J. (2013). Effects of tensor tympani muscle contraction on the middle ear and markers of a contracted muscle. *Laryngoscope, 123*(4), 1021–1027.

Beers, A. N., Shahnaz, N., Westerberg, B. D., & Kozak, F. K. (2010). Wideband reflectance in normal Caucasian and Chinese school-aged children and in children with otitis media with effusion. *Ear and Hearing, 31*(2), 221–233.

Beranek, L. L. (1949). *Acoustic measurements.* New York, NY: John Wiley.

Burdiek, L. M., & Sun, X. M. (2014). Effects of consecutive wideband tympanometry trials on energy absorbance measures of the middle ear. *Journal of Speech, Language, and Hearing Research, 57*(5), 1997–2004.

Ellison, J. C., Gorga, M., Cohn, E., Fitzpatrick, D., Sanford, C. A., & Keefe, D. H. (2012). Wideband acoustic transfer functions predict middle-ear effusion. *Laryngoscope, 122*(4), 887–894.

Feeney, M. P., Grant, I. L., & Marryott, L. P. (2003). Wideband energy reflectance measurements in adults with middle-ear disorders. *Journal of Speech, Language, and Hearing Research, 46*(4), 901–911.

Feeney, M. P., Hunter, L. L., Kei, J., Lilly, D. J., Margolis, R. H., Nakajima, H. H., . . . Voss, S. E. (2013). Consensus statement: Eriksholm workshop on wideband absorbance measures of the middle ear. *Ear and Hearing, 34*(Suppl. 1), 78S–79S.

Feeney, M. P., & Keefe, D. H. (1999). Acoustic reflex detection using wide-band acoustic reflectance, admittance, and power measurements. *Journal of Speech, Language, and Hearing Research, 42*(5), 1029–1041.

Feeney, M. P., & Keefe, D. H. (2001). Estimating the acoustic reflex threshold from wideband measures of reflectance, admittance, and power. *Ear and Hearing, 22*(4), 316–332.

Feldman, A. S. (1976). Tympanometry—Procedures, interpretations and variables. In A. S. Feldman & L. A. Wilber (Eds.), *Acoustic impedance and admittance: The measurement of middle ear function* (pp. 105–155). Baltimore, MD: Williams & Wilkins.

Gorga, M. P., Neely, S. T., Ohlrich, B., Hoover, B., Redner, J., & Peters, J. (1997). From laboratory to clinic: A large scale study of distortion product otoacoustic emissions in ears with normal hearing and ears with hearing loss. *Ear and Hearing, 18*(6), 440–455.

Groon, K. A., Rasetshwane, D. M., Kopun, J. G., Gorga, M. P., & Neely, S. T. (2015). Air-leak effects on ear-canal acoustic absorbance. *Ear and Hearing, 36*(1), 155–163.

Hunter, L. L., Bagger-Sjoback, D., & Lundberg, M. (2008). Wideband reflectance associated with otitis media in infants and children with cleft palate. *International Journal of Audiology, 47*(Suppl. 1), S57–S61.

Hunter, L. L., Feeney, M. P., Lapsley Miller, J. A., Jeng, P. S., & Bohning, S. (2010). Wideband reflectance in newborns: Normative regions and relationship to hearing-screening results. *Ear and Hearing, 31*(5), 599–610.

Hunter, L. L., Tubaugh, L., Jackson, A., & Propes, S. (2008). Wideband middle ear power measurement in infants and children. *Journal of the American Academy of Audiology, 19*(4), 309–324.

ISO. (1997). *389-2:1997, Acoustics—Reference zero for the calibration of audiometric equipment—Part 2: Reference equivalent threshold sound pressure levels for pure tones and insert earphones* (Vol. ISO 389-2:1997). Geneva, Switzerland: Author.

Jerger, J. (1970). Clinical experience with impedance audiometry. *Archives of Otolaryngology, 92*(4), 311–324.

Keefe, D. H. (1984). Acoustical wave propagation in cylindrical ducts: Transmission line parameter approximations for isothermal and nonisothermal boundary conditions. *The Journal of the Acoustical Society of America, 75*(1), 58–62.

Keefe, D. H. (2007). Influence of middle-ear function and pathology on otoacoustic emissions. In M. S. Robinette & T. J. Glattke (Eds.), *Otoacoustic emissions: Clinical applications* (3rd ed., pp. 163–196). New York, NY: Thieme.

Keefe, D. H., Bulen, J. C., Arehart, K. H., & Burns, E. M. (1993). Ear-canal impedance and reflection coefficient in human infants and adults. *Journal of the Acoustical Society of America, 94*(5), 2617–2638.

Keefe, D. H., Hunter, L. L., Patrick Feeney, M., & Fitzpatrick, D. F. (2015). Procedures for ambient-pressure and tympanometric tests of aural acoustic reflectance and admittance in human infants and adults. *Journal of the Acoustical Society of America, 138*(6), 3625.

Keefe, D. H., Ling, R., & Bulen, J. C. (1992). Method to measure acoustic impedance and reflection coefficient. *Journal of the Acoustical Society of America, 91*(1), 470–485.

Keefe, D. H., Sanford, C. A., Ellison, J. C., Fitzpatrick, D. F., & Gorga, M. P. (2012). Wideband aural acoustic absorbance predicts conductive hearing loss in children. *International Journal of Audiology, 51*(12), 880–891.

Keefe, D. H., & Schairer, K. S. (2011). Specification of absorbed-sound power in the ear canal: Application to suppression of stimulus frequency otoacoustic emissions. *Journal of the Acoustical Society of America, 129*(2), 779.

Kei, J., Sanford, C. A., Prieve, B. A., & Hunter, L. L. (2013). Wideband acoustic immittance measures: Developmental characteristics (0 to 12 months). *Ear and Hearing, 34*(Suppl. 1), 17S–26S.

Kemp, D. T. (1978). Stimulated acoustic emissions from within the human auditory system. *Journal of the Acoustical Society of America, 64*(5), 1386–1391.

Lewis, J. D., & Neely, S. T. (2015). Non-invasive estimation of middle-ear input impedance and efficiency. *Journal of the Acoustical Society of America, 138*(2), 977.

Lüscher, E., & Zwislocki, J. J. (1947). The delay of sensation and the remainder of adaptation after short pure-tone impulses on the ear. *Acta Oto-Laryngologica, 35*, 428–455.

Lynch, T. J., III, Nedzelnitsky, V., & Peake, W. T. (1982). Input impedance of the cochlea in cat. *Journal of the Acoustical Society of America, 72*(1), 108–130.

Lynch, T. J., III, Peake, W. T., & Rosowski, J. J. (1994). Measurements of the acoustic input impedance of cat ears: 10 Hz to 20 kHz. *Journal of the Acoustical Society of America, 96*(4), 2184–2209.

Margolis, R. H., Saly, G. L., & Keefe, D. H. (1999). Wideband reflectance tympanometry in normal adults. *Journal of the Acoustical Society of America, 106*(1), 265–280.

Merchant, S. N., & Rosowski, J. J. (2008). Conductive hearing loss caused by third-window lesions of the inner ear. *Otology & Neurotology, 29*(3), 282–289.

Metz, O. (1946). The acoustic impedance measured on normal and pathological ears. *Acta Oto-Laryngologica, 63*(Suppl.), 3–254.

Møller, A. R. (1960). Improved technique for detailed measurements of the middle ear impedance. *The Journal of the Acoustical Society of America, 32*(2), 250–257.

Møller, A. R. (1983). *Auditory physiology*. New York, NY: Academic Press.

Nakajima, H. H., Pisano, D. V., Roosli, C., Hamade, M. A., Merchant, G. R., Mahfoud, L., . . . Merchant, S. N. (2012). Comparison of ear-canal reflectance and umbo velocity in patients with conductive hearing loss: A preliminary study. *Ear and Hearing, 33*(1), 35–43.

Nakajima, H. H., Rosowski, J. J., Shahnaz, N., & Voss, S. E. (2013). Assessment of ear disorders using power reflectance. *Ear and Hearing, 34*(Suppl. 1), 48S–53S.

Neely, S. T., & Gorga, M. P. (1998). Comparison between intensity and pressure as measures of sound level in the ear canal. *Journal of the Acoustical Society of America, 104*(5), 2925–2934.

Neely, S. T., Stenfelt, S., & Schairer, K. S. (2013). Alternative ear-canal measures related to absorbance. *Ear and Hearing, 34*(Suppl. 1), 72S–77S.

Parent, P., & Allen, J. B. (2010). Time-domain "wave" model of the human tympanic membrane. *Hearing Research, 263*(1–2), 152–167.

Prieve, B. A., Vander Werff, K. R., Preston, J. L., & Georgantas, L. (2013). Identification of conductive hearing loss in young infants using tympanometry and wideband reflectance. *Ear and Hearing, 34*(2), 168–178.

Puria, S., & Allen, J. B. (1991). A parametric study of cochlear input impedance. *Journal of the Acoustical Society of America, 89*(1), 287–309.

Puria, S., & Allen, J. B. (1998). Measurements and model of the cat middle ear: Evidence of tympanic membrane acoustic delay. *Journal of the Acoustical Society of America, 104*(6), 3463–3481.

Rasetshwane, D. M., & Neely, S. T. (2011). Inverse solution of ear-canal area function from reflectance. *Journal of the Acoustical Society of America, 130*(6), 3873–3881.

Robinson, S. R., Nguyen, C. T., & Allen, J. B. (2013). Characterizing the ear canal acoustic impedance and reflectance by pole-zero fitting. *Hearing Research, 301*, 168–182.

Robinson, S. R., Thompson, S., & Allen, J. B. (in press). Effects of negative middle ear pressure on wideband acoustic immittance in normal-hearing adults. *Ear and Hearing*, doi:10.1097/AUD.0000000000000280.

Rosowski, J. J., Nakajima, H. H., Hamade, M. A., Mahfoud, L., Merchant, G. R., Halpin, C. F., & Merchant, S. N. (2012). Ear-canal reflectance, umbo velocity, and tympanometry in normal-hearing adults. *Ear and Hearing, 33*(1), 19–34.

Sanford, C. A., & Brockett, J. E. (2014). Characteristics of wideband acoustic immittance in patients with middle-ear dysfunction. *Journal of the American Academy of Audiology, 25*(5), 425–440.

Sanford, C. A., Keefe, D. H., Liu, Y. W., Fitzpatrick, D., McCreery, R. W., Lewis, D. E., & Gorga, M. P. (2009). Sound-conduction effects on distortion-product otoacoustic emission screening outcomes in newborn infants: test performance of wideband acoustic transfer functions and 1-kHz tympanometry. *Ear and Hearing, 30*(6), 635–652.

Schairer, K. S., Feeney, M. P., & Sanford, C. A. (2013). Acoustic reflex measurement. *Ear and Hearing, 34*(Suppl. 1), 43S–47S.

Scheperle, R. A., Neely, S. T., Kopun, J. G., & Gorga, M. P. (2008). Influence of in situ, sound-level calibration on distortion-product otoacoustic emission variability. *Journal of the Acoustical Society of America, 124*(1), 288–300.

Shahnaz, N. (2008). Wideband reflectance in neonatal intensive care units. *Journal of the American Academy of Audiology, 19*(5), 419–429.

Shahnaz, N., Bork, K., Polka, L., Longridge, N., Bell, D., & Westerberg, B. D. (2009). Energy reflectance and tympanometry in normal and otosclerotic ears. *Ear and Hearing, 30*(2), 219–233.

Shahnaz, N., Feeney, M. P., & Schairer, K. S. (2013). Wideband acoustic immittance normative data: Ethnicity, gender, aging, and instrumentation. *Ear and Hearing, 34*(Suppl. 1), 27S–35S.

Shanks, J. (1988). Tympanometry. *Journal of Speech and Hearing Disorders, 53*(4), 354–377.

Shaver, M. D., & Sun, X. M. (2013). Wideband energy reflectance measurements: Effects of negative middle ear pressure and application of a pressure compensation procedure. *Journal of the Acoustical Society of America, 134*(1), 332–341.

Shaw, E. A. (1980). The acoustics of the external ear. In G. A. Studebaker & I. Hochberg (Eds.), *Acoustical factors affecting hearing aid performance* (pp. 109–125). Baltimore, MD: University Park Press.

Siegel, J. H. (1994). Ear-canal standing waves and high-frequency sound calibration using otoacoustic emission probes. *Journal of the Acoustical Society of America, 95*(5, Pt. 1), 2589–2597.

Souza, N. N., Dhar, S., Neely, S. T., & Siegel, J. H. (2014). Comparison of nine methods to estimate ear-canal stimulus levels. *Journal of the Acoustical Society of America, 136*(4), 1768–1787.

Stepp, C. E., & Voss, S. E. (2005). Acoustics of the human middle-ear air space. *Journal of the Acoustical Society of America, 118*(2), 861–871.

Terkildsen, K., & Nielsen, S. S. (1960). An electroacoustic impedance measuring bridge for clinical use. *Archives of Otolaryngology, 72*(3), 339–346.

Thompson, S. (2013). *Impact of negative middle ear pressure on distortion product otoacoustic emissions* (Doctoral dissertation). City University of New York, New York, NY.

Voss, S. E., Adegoke, M. F., Horton, N. J., Sheth, K. N., Rosand, J., & Shera, C. A. (2010). Posture systematically alters ear-canal reflectance and DPOAE properties. *Hearing Research, 263*(1–2), 43–51.

Voss, S. E., & Allen, J. B. (1994). Measurement of acoustic impedance and reflectance in the human ear canal. *Journal of the Acoustical Society of America, 95*(1), 372–384.

Voss, S. E., Horton, N. J., Woodbury, R. R., & Sheffield, K. N. (2008). Sources of variability in reflectance measurements on normal cadaver ears. *Ear and Hearing, 29*(4), 651–665.

Voss, S. E., Merchant, G. R., & Horton, N. J. (2012). Effects of middle-ear disorders on power reflectance measured in cadaveric ear canals. *Ear and Hearing, 33*(2), 207–220.

Voss, S. E., Rosowski, J. J., Merchant, S. N., & Peake, W. T. (2001). Middle-ear function with tympanic-membrane perforations. II. A simple model. *The Journal of the Acoustical Society of America, 110*(3), 1445–1452.

Voss, S. E., Stenfelt, S., Neely, S. T., & Rosowski, J. J. (2013). Factors that introduce intrasubject variability into ear-canal absorbance measurements. *Ear and Hearing, 34*(Suppl. 1), 60S–64S.

West, W. (1928). Measurements of the acoustical impedances of human ears. *Post Office Electrical Engineers' Journal, 21*, 293.

Withnell, R. H., Jeng, P. S., Waldvogel, K., Morgenstein, K., & Allen, J. B. (2009). An in situ calibration for hearing thresholds. *Journal of the Acoustical Society of America, 125*(3), 1605–1611.

Zwislocki, J., & Feldman, A. S. (1970). Acoustic impedance of pathological ears. *ASHA Monographs, 15*, 1–42.

APPENDIX 1–A
Reflective Terminology

This table defines the terminology related to acoustic reflectance. The umbrella term *wideband acoustic immittance* (WAI) refers to all flavors of impedance, admittance, and reflectance. Most impedance concepts may be understood in terms of reflectance, which is conceptually equivalent to impedance but a more intuitive construction. When transforming an impedance (or admittance) to reflectance, one must first normalize it by the ear canal characteristic resistance $r_0 = pc/A(x)$, where r_0 is the density of the air, c is the speed of sound, and A is the cross-sectional area of the ear canal. Working with the normalized impedance (admittance) simplifies the interpretation of WAI data. Furthermore, it reduces the variability across subjects, since the area is best estimated when the data is taken, based on the size of the probe tip used for the measurement. These measures may be expressed either in the time domain (e.g., $p(t,x)$), which is real, or in the frequency domain (e.g., $P(f,x)$), which is complex. The term "mho" is "ohm" spelled backwards.

Physical Characteristic	Term	Function Representation	Unit
pressure	sound pressure	$p(t,x)$ or $P(f,x)$ [Pa]	pascal
	sound pressure level (SPL)	$20 \log_{10} \left(\frac{P(f,x)}{P_{ref}}\right)$ [dB SPL] $P_{ref} = 20 * 10^{-6}$ [Pa]	decibel
velocity	particle velocity (SVL)	$v(t,x)$ or $V(f,x)$ [m/s]	meter per second
	volume velocity	$U(f,x) = A(x)V(f,x)$ [m^3/s]	cubic meter per second
power	sound intensity	$\mathbb{i}(t,x) = p(t,x)\, v(t,x)$ [W/m^2]	watt per square meter
	sound intensity level (SIL)	$10 \log_{10} \left(\frac{\mathbb{i}(t,x)}{\mathbb{i}_{ref}}\right)$ [dB SIL] $\mathbb{i}_{ref} = 10^{-12}$ [Pa]	decibel
	sound power	$p(t,x) = \mathbb{i}(t,x)A(x)$ [W] or $P(f,x)$ [W]	watt
energy	sound energy	$\varepsilon(t,x) = \int_{-\infty}^{t} \mathbb{i}(\tau,x)A(x)d\tau$ [J]	joule
complex reflectance $\Gamma(f)$	reflectance magnitude	$\lvert\Gamma(f,x)\rvert$	ratio (dimensionless)
	reflectance phase	$\angle\Gamma(f,x)$ [rad]	radian
	reflectance group delay (phase slope)	$\tau_{\Gamma}(f,x) = \frac{-1}{2\pi} \frac{d}{df} \angle\Gamma(f,x)$ [s]	second
	power reflectance	$\lvert\Gamma(f,x)\rvert^2$	ratio (dimensionless)
	power absorbance	$1 - \lvert\Gamma(f,x)\rvert^2$	ratio (dimensionless)
	power absorbance level	$10\log_{10}(1 - \lvert\Gamma(f,x)\rvert^2)$ [dB]	decibel

continues

Appendix 1–A. *continued*

Physical Characteristic	Term	Function Representation	Unit
complex admittance $Y(f)$	admittance magnitude	$\lvert Y(f,x) \rvert$ [S or ℧]	siemens or mho
	admittance phase	$\angle Y(f,x)$ [rad]	radian
	conductance (real part)	$G(f,x) = Re\{Y(f,x)\}$ [S or ℧]	siemens or mho
	susceptance (imaginary part)	$B(f,x) = Im\{Y(f,x)\}$ [S or ℧]	siemens or mho
complex impedance $Z(f)$	impedance magnitude	$\lvert Z(f,x) \rvert$ [Ω]	ohm
	impedance phase	$\angle Z(f,x)$ [rad]	radian
	resistance (real part)	$R(f,x) = Re\{Z(f,x)\}$ [Ω]	ohm
	reactance (imaginary part)	$X(f,x) = Im\{Z(f,x)\}$ [Ω]	ohm

CHAPTER 2

Otoacoustic Emissions: Measurement, Modeling, and Applications

Glenis Long and Bastian Epp

Review of Cochlear Processing and Otoacoustic Emissions

We start with a brief review of otoacoustic emissions (OAEs) with a focus on the definition and classification of OAEs. More details about the history and fundamentals of OAEs can be found in Kemp (2008).

Hearing begins when sounds are transmitted into the cochlea. OAEs reverse this process. They are sounds generated in the cochlea, measured in the ear canal, and considered a by-product of a nonlinear and active process in the cochlea, sometimes referred to as the "cochlear amplifier." The active process is responsible for the remarkable sensitivity of the auditory system and is commonly attributed to outer hair cell (OHC) motility (Ashmore, 2008).

Low-level vibrations are amplified by the process, increasing the system's sensitivity (Dallos, 2008). Since the amount of energy that can be provided by the active process is limited, the amplification saturates gradually with increasing stimulus levels. This results in a compressive and nonlinear input/output characteristic, which, on the one hand, helps to increase the effective dynamic range of stimulus levels that can be processed in the inner ear, and, on the other hand, can produce nonlinear distortion. The amplification occurs near the best frequency region of each stimulus (i.e., the region that is most sensitive to the frequencies contained in the stimulus). This frequency-specific amplification improves the ability of the system to determine the frequency components of stimuli. Since amplification is an active process, it requires an energy source. The energy source for this amplification is present in the form of the endocochlear potential, which is maintained by the vascular supply to the inner ear stria vascularis (Guinan et al., 2012). When the OHCs are damaged, amplification, degree of nonlinearity, and frequency resolution are reduced, leading to higher thresholds, reduced dynamic range, and poorer frequency resolution (Olson et al., 2012).

The amount of basilar membrane (BM) vibration amplification is connected to the magnitude of OAE recorded in the ear canal, similar to the dependence of BM vibration amplitude on the intensity of an external stimulus (Olson et al., 2012). Anything that modifies either the OHC transduction or the endocochlear potential will reduce the amplification of the BM vibration and hence the magnitude of the OAE. This connection can be used to investigate cochlear processing, since the pattern of growth of the OAEs reflects many of the properties of the basilar membrane response. Most types of OAEs grow nearly linearly for very low-level stimuli and then saturate at higher levels, following the compressive characteristics of basilar membrane (BM) processing, which is most compressive for intermediate input signal levels (Janssen & Müller, 2008).

The amount of amplification at a specific region of the cochlea depends on the velocity of the basilar membrane (BM). This might be caused by stimulation at the best frequency, or by frequencies remote from that region. If several traveling waves each triggered by different frequency components overlap, the magnitude of the BM vibration around one region is influenced by vibrations that reach their maxima in other nearby regions. This offset in vibration will reduce the vibration-level dependent gain and consequently the overall response to the tones. This phenomenon is a property of the cochlear nonlinearity and is known as "suppression," or "two-tone suppression" in the case of two-tone stimuli. Together with the compressive characteristic of BM vibration, it plays a role in many experimental paradigms to measure OAEs.

Classification of OAEs

OAEs can be classified into different types. One type can be detected in the absence of any deliberate acoustic stimulation. These spontaneous otoacoustic emissions (SOAEs) are self-sustained and can be regarded as sine waves jittered both in amplitude and phase. They provide a strong indication that the ear is capable of generating, or at least sustaining, spontaneous vibratory energy in the cochlea, leading to the SOAE in the ear canal. The prevalence of SOAE in humans depends on the procedures used to measure them. The most sensitive studies have detected SOAEs in about 70% of normal ears, and more in females than males, both in adults and in children (reviewed in Bright, 2007). SOAEs also interact with vibrations triggered by external stimuli. When an external stimulus is close in frequency and intense enough, it will force the vibration underlying the SOAE to oscillate for a short period in phase with the vibration frequency of the external stimulus. This nonlinear interaction is referred to as entrainment. Some clinical devices (ILO) use this property and measure SOAEs by entraining them with clicks. These types of OAEs have been called synchronized SOAEs or SSOAEs.

Other types of OAEs are evoked by external stimulation (EOAE). These are present in nearly all normal-hearing ears but are absent in ears with significant cochlear damage. EOAEs are often further categorized according to the stimuli used to evoke them. When the stimuli are short in duration (transient), the OAEs are referred to as transient evoked otoacoustic emissions (TEOAE). TEOAEs are evoked by broadband stimuli and hence excite and provide information about a broad region in the cochlea. TEOAEs were the first to be discovered (Kemp, 1978). Since they represent a type of "echo" from the inner ear, they were initially referred to as "Kemp-echoes."

OAEs can also be evoked by very narrowband stimuli-like tones. The OAE evoked by a single tone is called SFOAE (stimulus frequency otoacoustic emissions). Due to the narrowband nature of tones, SFOAEs have been thought to provide information from a limited region along the cochlea, and dominantly, but not exclusively, from around the peak in the best frequency region. When evoked by two or more external tones (called primaries), the nonlinearity in the cochlea generates additional frequency components at predictable frequencies. OAEs at these additional frequencies are called distortion product otoacoustic emissions (DPOAE). Since DPOAEs are a product of the cochlear nonlinearity, evaluation of DPOAEs provides information about the state of this nonlinearity.

Generation Mechanisms of OAEs

An effort has been made to identify the generation mechanisms underlying OAEs. The two main contributors assumed to contribute are nonlinear distortion and linear reflection. Once acoustic energy in the form of distortion is generated in the cochlea, its effect is equivalent to as if it would be an external tone. The energy travels to other regions in the cochlea, where it can be reflected and be transmitted back through the middle ear to ear canal and be detected by the OAE probe. This is particularly true for DPOAE, but also other types of OAEs can be described by these mechanisms. For DPOAEs, the distortion product generated by the interaction of the two primaries travels in two directions. Some of the energy travels directly out into the ear canal, and some of it travels to the distortion product's best frequency place before also being reflected and traveling back into the ear canal. The former component is referred to as generator component, while the latter is referred to as reflection component. In the ear canal, these two components mix. Differences in travel time of the two components in the cochlea lead to differences in the phase. Depending on their relative phase, they add in or out of phase, resulting in fluctuations in the level of the OAE, referred to as fine structure.

Measurement Issues. There are many other sounds that can be detected in the ear canal like heartbeat, breathing, swallowing, and external noise. Since almost all OAEs are low-level signals, it can be challenging to separate the OAE from this background noise. The measurement paradigm and the analysis applied has a big effect on our ability to detect OAEs and the interpretation of the result. Since OAE recordings contain such a considerable noise floor, the absence of measureable OAEs should not lead to claims that OAEs are absent, but rather that no OAE could be detected above the noise floor—a difference when it comes to diagnosis.

A helpful tool to separate the acoustic energy in the ear canal into different frequencies is spectral analysis. Most investigators use the Fast Fourier Transform (FFT), providing both amplitude and phase of the spectral components. The resolution of the FFT, and thus the noise floor, depend on the duration of the digitized recording. When using an FFT, it is essential that the FFT analysis has sufficient frequency resolution to separate all components. The analysis via FFT makes no explicit assumptions about the frequencies present and calculates the energy at resolved frequencies. If the length of the signal to be analyzed is chosen in a favorable way, and the OAE frequencies coincide with a resolved frequency, then the analysis via FFT provides a very efficient bandpass filter for that particular frequency. Sometimes components are generated that are not expected and aligned with a resolved frequency. This can contaminate estimates of OAE level and phase. Phase is an important factor in OAE generation. However, most clinicians do not consider this information. Including phase information could potentially provide additional information of, for example, the region of generation.

A number of additional techniques can be applied to reduce the noise floor and thus improve the signal-to-noise ratio, including:

- Artifact rejection—a subset of the recording with high peaks in the time signal originating from, for example, swallowing noises, can be rejected from the analysis. One has to be careful, though, to reject noise and not exclude valid data.
- Spectral averaging—spectral analysis can be done on multiple subsets of the data and the results can be averaged. This does not reduce the noise floor, but does decrease the variance of the noise floor, making it easier to detect OAE near the noise floor (e.g., Popelka et al., 1995).
- Temporal averaging—when the OAE is fixed in phase and the noise is random, time-locked temporal averaging of subsets reduces the noise floor, leaving the OAE unchanged. Temporal averaging improves the signal-to-noise ratio but does not reduce the variance of the noise floor.

A combination of artifact reject, temporal averaging, and spectral averaging provides the optimal low noise floor with low variance (e.g., Popelka et al., 1995).

In addition to the poor signal-to-noise ratio, another challenge for recording OAEs is the calibration of the stimulus used to evoke them. Acoustically, the ear canal acts like a semi-open tube. Placement of an OAE probe in the ear canal will result in destructive and constructive interference and consequently attenuation and amplification of single frequencies in the ear canal. Since ear canals differ, and the position of the probe can vary, is it impossible to use a unified value for all listeners. This needs to be taken into consideration when interpreting the level of the OAEs using the absolute level (Siegel, 2007). There exist a number of different calibration routines that have been described in literature (reviewed in Souza et al., 2014).

Modeling of Otoacoustic Emissions

Models of OAEs are a frequently used tool in research. They allow modification of systems that would be hard to realize experimentally. In addition, models can be used to overcome experimental limitations like, for example, exposure time, stimulation level, or simply problems with availability of test subjects. This possibility to extensively test the model finally allows the design of new hypotheses and accurately designed experiments.

Models can be understood as entities, imitating certain aspects of the system under investigation. These descriptions are always simplified rather than complete. First, not all aspects of the system are known with complete certainty. Second, unnecessarily complex models are hard to evaluate and the complexity might obscure the interpretations. A good model, where "good" is to be understood as "suitable," needs to be as simple as possible, but at the same time as complex as required. For example, for certain applications, a head might be adequately modeled as a sphere, while others may require a description of additional details like, for example, the pinnae. When designing a model, it is essential that the simplifications and the connected limitations are understood and explicitly stated. Violation of these limitations might lead to overstretching of the model and lead to conclusions that are outside the accessible range of the model and hence are implausible.

There are different types of models of the hearing system, such as conceptual models, functional models, and physiologic models, as well as partial mixtures of all three. Conceptual models typically focus on the connection and interaction of processes on a purely conceptual level. Functional models implement a process on a very simple or very abstract level without necessarily imposing strict limitations on the plausibility of the method used. Physiologic models are restrained by inherent limitations of the system under investigation, for example, the anatomical structure or available resources. The choice of the model class depends entirely on the purpose of the model. It is just as unreasonable to try to model all known properties, as it is to use conceptual models to understand details of physiologic processes.

Existing models are capable of clarifying various aspects of OAEs (e.g., Epp et al., 2010; Liu & Neely 2010; Mauermann, Uppenkamp, van Hengel, & Kollmeier, 1999; Moleti et al., 2012; Verhulst et al., 2012). Some of these models aim at simulating OAEs as they are measured experimentally in the ear canal (i.e., as a sound waveform). Others describe the processing in the cochlea and leave out propagation of the energy into the ear canal. Strictly speaking, OAEs are defined as sound signals measurable in the ear canal. But because a vibration needs to be present in the cochlea before the OAE can be detected in the ear canal, one might argue that modeling the generation mechanisms of OAEs without modeling the propagation into the ear canal is acceptable. Models that describe cochlear processing based on biophysical principles, and not necessarily OAEs, have contributed significantly to our current understanding of OAE generation (De Boer, 1983; Neely & Kim, 1986;

Shera, 2003; Strube, 1985; Talmadge et al., 1998; Van Hengel et al., 1996).

An OAE Model Step by Step

The steps for designing a model capable of simulating OAEs follow. The numerical aspects of solving the formulated equations will not be considered; they can be found in the literature (Duifhuis, 2012; Elliott et al., 2007). The described models approach each cochlear segment as an entity. They do not consider or differentiate the details of structure and processing within the organ of Corti or explore the interaction of the organ of Corti with the BM. Such approaches are commonly referred to a "macromechanical." More detailed models of the organ of Corti are referred to as "micromechanical" models. Macromechanical models have successfully been used to model OAEs. Micromechanical models currently demand too much computational complexity to account for processing of the whole cochlea. First attempts are, however, made to include micromechanics into models aimed toward the simulation of OAEs (Meaud & Lemons, 2015).

When designing a model of OAEs, the initial step is to consider the generation mechanisms underlying the OAEs to be modeled. The second step is to decide which animal should be modeled. This choice determines the range of frequencies processed by the model. For a solution of the model with numerical models, the next step is to discretize the model into single segments (Figure 2–1). The number of segments determines the frequency resolution. Since the goal is to approximate a continuous system, the number of segments should be large enough to avoid unwanted artifacts, which are generated when the discretization is too coarse. In electrical engineering, a model with these properties is termed transmission line, since

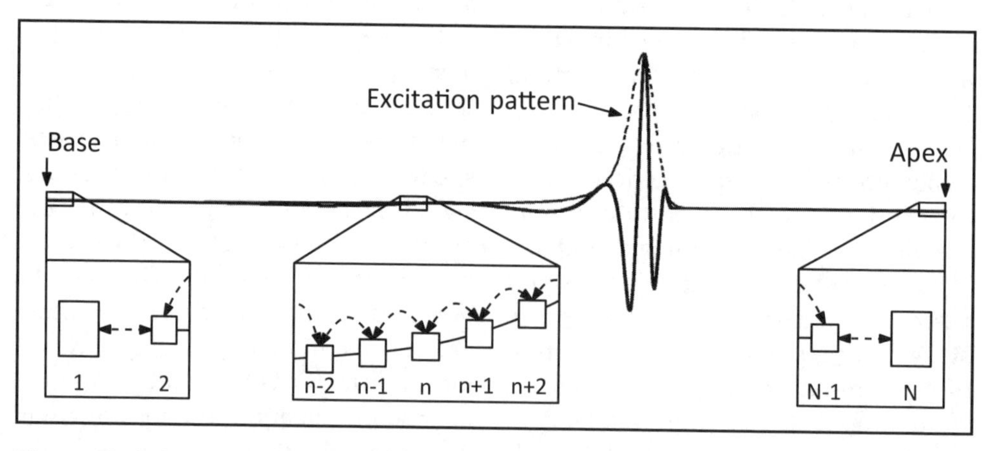

Figure 2–1. Traveling wave (*solid line*) and corresponding excitation pattern (*dashed line*) in a continuous and a discretized model of the basilar membrane. In the discretized model, the continuous membrane is divided into segments (2, …, $N-1$). Neighbored segments interact through an inertial connection to model the fluid in the cochlear duct. At the base and the apex, boundary conditions are implemented as impedances to model the behavior of the middle ear (1) and the helicotrema (N).

it is well suited to model wave propagation along cables. For the application of analytical methods, the model does not require the discretization step. For nonlinear systems, these analytical solutions, are, however, unknown in most instances, while a large number of methods exist to solve the discretized models on a computer.

Once these decisions have been made, a model can be designed. Since each segment is an oscillating system, the equation of motion for each segment can be formulated as a second-order differential equation. This equation describes the interaction between three elements: an inertial element (a mass), a compliant element (a spring), and a resistive element (a damper). Each of these elements reacts differently to an applied force. The spring will react to the applied displacement while the damping will react to the velocity, i.e., change of displacement. The mass in turn will react to the acceleration, i.e., changes of velocity. The fact that each of these elements reacts to different aspects of the driving force means that they react to the driving force with a different relative phase. The velocity is the first derivative of the displacement and shifted by 90 degrees relative to the displacement. The acceleration is the derivative of the velocity and hence shifted by 180 degrees. This interplay between the resulting forces leads to the desired oscillatory dynamics and to resonant behavior. The spring and the mass are required to obtain an oscillating and resonating system, and the ratio of their values defines the resonance frequency. The amount of damping determines the frequency selectivity and maximum amplitude of the oscillator. Low values of damping lead to high amplitudes for a very narrow frequency band, while large values reduce maximum amplitude and result in a less selective oscillator.

A formal description of the response of such an oscillating system is the impedance. The impedance can be seen as a frequency-dependent resistance, which also introduces a phase shift. The magnitude of the impedance reflects the effect of the impedance on the amplitude, and the resulting phase can be obtained by evaluation of the argument, or phase, of the complex number. In contrast to a resistance, the impedance is complex-valued that makes a change in phase possible. In a cochlear model, the impedance is the ratio of the driving pressure relative to the resulting segment velocity. Assume a constant driving pressure for all frequencies. A segment with some resonance frequency will react to driving pressures well below the resonance frequency with a low-amplitude response in phase with the stimulus, similar to moving one end of a spring with a mass attached to the other end very slowly up and down—the mass essentially follows the movement. This is due to the high impedance where the denominator of the ratio (the resulting velocity) is small compared to the nominator (the driving pressure). With increasing frequency, but same amplitude of driving pressure, the mass will start to oscillate with larger amplitude and to lag behind the movement of the driving point. The impedance drops since the velocity in the denominator is increasing and hence the ratio is getting smaller. At the resonance frequency, the mass will be moving up when the spring is moved down and vice versa—both the mass and the driving pressure compress the spring at the same time. The velocity will be at maximum and hence the impedance very small. At higher frequencies, the velocity decreases again and the impedance increases.

The description until now includes only a single segment (i.e., one point on the BM). In order to enable the system to respond to all desired frequencies, it is required to combine all the segments. To achieve a realistic distribution of frequencies along the length of the cochlea, the spatial domain (the length of the cochlea) needs to be linked to the frequency domain. A common assumption is that the frequencies are spaced exponentially from high to low frequencies in the direction of base to apex. This projection is called a place-frequency map. It can be achieved by varying the ratio between mass and stiffness of the segments according to an exponential function of the distance from the base. Under the assumption that the mass of each segment is constant, an exponentially decaying spring constant will lead to such an exponentially varying place-frequency map. This means that the spring is getting less and less stiff from the oval window toward the helicotrema. To achieve propagation of a traveling wave, one needs to introduce coupling between the segments. The cochlea segment is surrounded by the cochlea fluid, which can be assumed to be incompressible. In a discretized model, the coupling can be introduced by coupling neighboring segments by another mass element, representing the amount of fluid being moved back and forth when neighboring segments oscillate. Due to this coupling, each segment communicates with its neighbors. Once the frequencies are mapped to the length of the cochlea and the segments are coupled, the system will show the desired response. Excited by a stimulus, a traveling wave will propagate from the base to the apex, and a number of segments will be displaced with different amplitudes. The pattern of amplitudes of all the segments to a stimulus is referred to as excitation pattern. The width of the resulting excitation pattern can then be controlled by the damping parameter, which is connected to the tuning of each segment. Tuning of an oscillator is often quantified as the quality factor Q. The quality factor describes the relation between the spectral bandwidth and the center frequency.

The first and last segments (1 and N in Figure 2–1) have only one nearest neighbor to which they can be coupled. On the side of the missing neighbor, they will encounter the boundaries of the system. The first segment plays an important role since it is coupled to the outside world, and hence to the OAE data obtained in the ear canal. In the cochlea, this boundary condition is the oval window. If the boundary condition does not allow a transmission of energy out of the cochlea, no OAE can be measured in the ear canal and all energy will be reflected back into the cochlea. In a listener, this could happen in the case of otosclerosis or in newborn babies when the middle ear is still filled with fluid. If, however, all energy is transmitted to the ear canal, only a subset of OAEs could be simulated, because many of the OAEs are assumed to depend on multiple reflections within the cochlea. With the most apical segment, the boundary condition at the helicotrema can be modeled. In the cochlea, the helicotrema connects the scala media with the scala tympani by a small opening. For very low frequencies, the helicotrema can be seen as a short circuit where the fluid can pass through without resistance (i.e., without any pressure difference between the two scalae). Most models include the helicotrema as a short circuit that has little impact except when very low-frequency OAEs are considered. At this point, the model is almost complete. So far, the model is passive and composed of linear components, with very smooth transitions of mechanical properties from base to apex.

Think about a "panoramic view" of the cochlea when stimulated with a pure tone. The width of the excitation pattern depends on the damping of each segment. It turns out that excitation patterns can be sharpened by reducing the damping, but the shape will not correspond to physiologic measures of BM motion. In order to obtain the amplitude and "broad and tall" excitation patterns (Zweig, 1991), an active process is required that is somewhat frequency selective. This means that an additional element could be added to the mass, spring, and damping elements in the passive system providing the desired energy. The only element dissipating energy in the passive system is the damping term. One approach is to make the damping of the segments negative, which will revert the dissipation of energy into the injection of energy. The damping element will not drag on the mass but will push it each time it moves. If all segments of the cochlea have negative damping, this will result in spontaneous oscillation of all cochlear segments, giving potential instability and not lead to the desired excitation pattern. A successful way to include an active process is to inject the energy not at the place of maximum response but at a slightly basal region. This can be done by modification of the either the stiffness or the damping term in the impedance (Neely & Kim, 1986). The result is that injected energy is dissipated in the peak region oscillations, leading to a stable oscillation. Another formulation derived from animal data includes both negative damping as an active element and a time-delayed stiffness term that stabilizes the oscillations (Zweig, 1991). Including an active process will provide the amplification and the shape of the excitation patterns observed in healthy cochleae, which predicts the levels of the simulated OAE in the ear canal.

Our model is now active and linear and accounts for realistic excitation patterns for low input levels. Physiologic data show that the excitation patterns in the cochlea broaden and that there is a compressive growth with increasing input level. These aspects can be modeled by a level-dependent component, a nonlinearity. In the cochlea, this level-dependent behavior is correlated with the assumption that the amplification in the cochlea is less effective at high input levels than at low input levels, connected to the saturation of the active process.

One approach to implement nonlinearity is to make the model components nonlinear. A model with nonlinear components can only be approximated by a level-dependent impedance, because the nonlinearity will result in nonharmonic motion, and the concept of impedance as it is formulated in most textbooks assumes harmonic motion. In practice, this approximation turns out to be applicable (Verhulst et al., 2012), but certain tools like the solution of the equations using a Fourier transform can no longer be used. The nonlinearity can be designed to fit the input-output characteristics observed in cochlea processing. By using a level-dependent impedance formulation, the input-output characteristics of the cochlea can nicely be described as a trajectory in the so-called Laplace domain (Verhulst et al., 2012) by variation of the real and complex parts of the impedance. Alternatively, the nonlinear response can be achieved empirically by replacing the constants representing the damping and the stiffness in the equation of motion by functions that depend on either velocity or displacement of the corresponding segment. The former approach allows a clear selection of the input-output characteristics and control for stability. The latter approach might be easier to interpret in terms of the underlying damping and stiffness components, while at the same time being harder to keep stable. One consequence of this nonlinear processing is the generation of distortion components, an aspect found in data and that is evaluated in a number of OAE paradigms. Now the model will be able to account for nonlinear processing and generate distortion components.

Additional modifications are required to be able to account for all aspects of OAEs. Many OAEs show large variability, or fine structure, as a function of frequency. Hence, the data suggest that the amount of energy reflected back into the ear canal differs along the cochlea. With the very smooth transition of mechanical parameters as implemented so far, the traveling wave will not encounter any sudden changes in impedance, and hence the amount of reflected energy is independent of the frequency. In order to obtain more realistic reflections, the smooth transition of impedance from one segment to the other needs to be perturbed. In the model, this can be achieved by a stochastic variation of the place-frequency map, sometimes referred to as "irregularities," or "roughness" (Shera, 2015), by either a variation of the function describing the mass or the spring constants of each segment along the cochlea. The biophysical basis for this approach has not yet been identified, but since irregularity is common in biological systems, this assumption seems feasible.

Once a plausible parameter-set for the model is found, such a model can be very useful in systematic investigations of the generation of OAEs. It can also help to identify limitations of OAEs for audiologic purposes. As an example are OAEs only net manifestations of processing and interaction of the single segments in the cochlea. Hence, a reduced amplitude of an experimentally measured DPOAE can have multiple causes. It might be due to a reduced linearity in the cochlea, reducing the generator component stemming from nonlinear processing in the overlap region of the primaries. Or it might be due to an interaction of the reflection component from the best frequency region of the DPOAE frequency and the generator component. While there is experimentally no direct way of determining which of these explanations is correct, the reflection component can be removed in the model by removal of the roughness in the region of the reflection component (Mauermann et al., 1999). In summary, models of OAEs and cochlear processing can help to deepen our understanding of the mechanisms and role of cochlear processing in hearing and ultimately contribute to the development of novel and selective diagnosis procedures in audiology.

Applications of OAE

Potential and current applications of OAE come from their capacity to provide a noninvasive and efficient window into cochlear function in humans and other species. Knowledge about cochlear function is important, since the cochlea is the information bottleneck for the whole auditory system. Up to now, direct measures of cochlear mechanics (see Olson et al., 2012) are difficult, are expensive, and cannot be done in humans.

The major clinical application of OAE is neonatal screening with TEOAEs or DPOAEs, which aims to determine which neonates will have significant hearing loss (reviewed in Akinpelu et al., 2014). An ideal screening tool would accurately detect all hearing-impaired infants (sensitivity) and never diagnose hearing loss in normally hearing infants (false-alarm rate or 1 – specificity). Unfortunately, few diagnostic tests in any medical field can accomplish such a dichotomy without either type of error. All infants are born with fluid in the outer and middle ears, which not only impedes sound transmission into the cochlea but also impedes sound transmission from the cochlea to the ear canal, which can potentially lead to undetectable OAE. The longer the postnatal interval, the less likely it is that infants will fail an OAE screen. Ideally, all children who fail the screen should receive subsequent diagnostic investigations using other OAE or electrically evoked potentials to establish whether the loss was cochlear in origin (Akinpelu et al., 2014).

OAE screening is also used to screen school-age children in order to detect late-onset, acquired, and progressive loss. It is also used to screen potential participants in research to increase the probability that they have normal hearing. Some OAE screeners evaluate the probability of normal hearing at specific frequencies using established norms, while others give a general pass/fail recommendation for each ear.

OAE levels are also used to predict audiometric thresholds in nonresponding listeners. Most clinical devices use fixed stimulus levels and parameters to estimate OAE levels as a function of frequency. Experimentally established correlations between the OAE levels and audiometric thresholds permit then a partial estimation of audiometric thresholds. The correlation strength depends on frequency (reviewed in Rasetshwane et al., 2013, 2015) and is usually stronger when multiple OAE estimates are combined (multivariate estimates). Due to the nonlinear properties of the cochlea, the stimuli influence the vulnerability of the resulting OAE to hearing loss. Responses to low-level stimuli are most influenced by the active process and thus provide a sensitive indication of the state of the active process and hence a potential hearing loss, but can be difficult to measure due to signal-to-noise issues. OAEs are easier to measure when obtained from a range of stimulus levels. The resulting input/output functions provide information about supra-threshold nonlinear cochlear processing and hence the level dependence of the active process. It has been shown that the slope of the growth of OAEs with stimulus level depends on stimulus level and cochlear health (Neely et al., 2009), which is consistent with the cochlear nonlinearity described in the modeling section.

For many applications, contamination by noise is a major problem when evaluating low-level OAEs. OAE thresholds are often predicted using models of the expected OAE growth with stimulus level, for example, DPOAE (cf. Boege & Janssen, 2002) and TEOAE (cf. Mertes & Goodman, 2013). However, all OAEs show variations with frequency known as fine structure stemming from the interaction of at least two components coming from different cochlear regions or stemming from different cochlear functions. This can lead to input/output functions, which do not fit predicted nonlinear OAE properties. Consequently, threshold estimates can be improved by separating the generator and the reflection DPOAE components (Dalhoff, Turcanu, Vetešník, & Gummer, 2013; Mauermann & Kollmeier, 2004) and different latency TEOAE response (Mertes & Goodman, 2013; Sisto et al., 2015).

Because different OAEs are generated from different cochlear regions and are produced by different cochlear mechanisms, they may be differentially affected by cochlear pathology. For example, the DPOAE reflection component has been shown to be more vulnerable to aspirin consumption than the generator component (Rao & Long, 2011). Another study showed that DPOAE generator components from well-controlled diabetic teenagers were similar to those of control teenagers, but the reflection components from the diabetic participants were significantly smaller than those from the normal control group (Spankovich, 2010). Traditional clinical OAE measurements do not distinguish between the different OAE generation mechanisms. If, however, the two components have differential sensitivity, it is possible that combining information about different components or different OAE types may aid diagnosis (cf. Konrad-Martin et al., 2012) by providing extra information.

Although the audiometric thresholds have become the "gold standard" for evaluating cochlear health, they are not sensitive to mild cochlear damage and might hence permit detection of cochlear damage later than it might be possible using OAEs. Individuals with lower than normal TEOAE but normal audiometric thresholds were more likely to experience threshold shifts after noise exposure (Marshall et al., 2009). Some OAE properties are correlated with some supra-threshold processes and thus might potentially be used to better understand and predict supra-threshold perception and mild damage to the active process. It has been shown that OAE input/output properties can help model loudness perception (TEOAE [Epstein & Florentine, 2005] and DPOAE [Rasetshwane et al., 2013]).

Even though OAEs selectively probe cochlear mechanics, they are not completely independent of retro-cochlear processes. Cochlear mechanics, and hence OAEs, can be modulated by the olivocochlear fibers originating in the brainstem. These efferent fibers innervate the outer hair cells, reducing the active process. This system is not yet well understood, but it seems as if one major role of this system is to make it easier to detect sounds in noise, which is a common problem in individuals with auditory processing problems (see Guinan, 2014, for a review).

Hearing impairment is not always connected to reduced or absent OAEs. There are some individuals with increased audiometric thresholds but have normal indicating healthy cochlear processing, but a retro-cochlear impairment. This combination of an audiogram and OAE aids the diagnosis of some less common types of hearing loss. Prior to the use of OAEs, audiologists could not determine whether hearing loss stemmed from damage to the OHC or to problems such as damage to the inner hair cell synapse or the auditory nerve. Individuals with impaired hearing and associated reduced auditory brainstem responses are now diagnosed as having auditory neuropathy (Giraudet & Avan, 2012). The recent discovery of what has been sometimes been called "hidden hearing loss" (reviewed in Plack et al., 2014), where responses to near-threshold stimuli are normal, neural responses to higher level stimuli are reduced while OAE levels continue to grow. Anatomical and functional evidence reveals that this stems from degeneration of the synapse between the inner hair cell (IHC) and the auditory nerve fibers that respond best to high-level stimuli, and thus are essential for coding intensity fluctuations at higher levels as required in speech in noise detection.

Final Comments

This chapter introduced the concepts underlying OAEs and models to better understand their generation and underlying mechanisms. Models of OAE generation provide a useful approach to study noninvasively the dynamics inside the cochlea and to build novel hypotheses and experimental paradigms, which can lead to new experimental and clinical tools. While OAEs have contributed to a large extent to our understanding of cochlear processing, they have not yet been fully incorporated into audiologic procedures. Since OAEs are the only reliable tool available to exclusively study cochlear processing not affected by any neural components, they have considerable potential to transfer basic knowledge into clinical procedures and research paradigms. Such inclusion will lead not only to more sensitive and precise diagnosis tools but also expand our knowledge about the processing of sound in the cochlea and the information passed on to retro-cochlear stages of the auditory pathway. In addition to clinical practice, such knowledge has implications for the development and fitting of hearing assistive devices. Currently, the lack of detailed knowledge about cochlear processing limits the resulting benefit and quality of such prosthetic devices—a point that hopefully will be improved for the benefit of people with hearing deficits.

References

Akinpelu, O. V., Peleva, E., Funnell, W. R. J., & Daniel, S. J. (2014). Otoacoustic emissions in newborn hearing screening: A systematic review of the effects of different protocols on test outcomes. *International Journal of Pediatric Otorhinolaryngology, 78*(5), 711–717. http://doi.org/10.1016/j.ijporl.2014.01.021

Ashmore, J. (2008). Cochlear outer hair cell motility. *Physiological Reviews, 88*(1), 173–210.

Boege, P., & Janssen, T. (2002). Pure-tone threshold estimation from extrapolated distortion product otoacoustic emission I/O-functions in normal and cochlear hearing loss ears. *The Journal of the Acoustical Society of America, 111*(4), 1810–1818.

Bright, K. E. (2007). Spontaneous otoacoustic emissions in populations with normal hearing sensitivity. In M. S. Robinette & T. J. Glattke (Eds.), *Otoacoustic emissions* (pp. 69–86). New York, NY: Thieme Medical.

Dalhoff, E., Turcanu, D., Vetešník, A., & Gummer, A. W. (2013). Two-source interference as the major reason for auditory-threshold estimation error based on DPOAE input-output functions in normal-hearing subjects. *Hearing Research, 296*, 67–82.

Dallos, P. (2008). Cochlear amplification, outer hair cells and prestin. *Current Opinion in Neurobiology, 18*(4), 370–376.

De Boer, E. (1983). No sharpening? A challenge for cochlear mechanics. *The Journal of the Acoustical Society of America, 73*(2), 567–573. doi:10.1121/1.389002

De Boer, E. (1997). Connecting frequency selectivity and nonlinearity for models of the cochlea. *Auditory Neuroscience, 3*(4), 377–388.

Duifhuis, H. (2012). *Cochlear mechanics: Introduction to a time domain analysis of the nonlinear cochlea.* Berlin, Germany: Springer Science & Business Media.

Elliott, S. J., Ku, E. M., & Lineton, B. (2007). A state space model for cochlear mechanics. *The Journal of the Acoustical Society of America, 122*(5), 2759–2771. doi:10.1121/1.2783125

Epp, B., Verhey, J. L., & Mauermann, M. (2010). Modeling cochlear dynamics: Interrelation between cochlea mechanics and psychoacoustics. *The Journal of the Acoustical Society of America, 128*(4), 1870–1883. doi:10.1121/1.3479755

Epstein, M., & Florentine, M. (2005). Inferring basilar-membrane motion from tone-burst otoacoustic emissions and psychoacoustic measurements. *The Journal of the Acoustical Society of America, 117*(1), 263–274. http://doi.org/10.1121/1.1830670

Giraudet, F., & Avan, P. (2012). Auditory neuropathies: Understanding their pathogenesis to illuminate intervention strategies. *Current Opinion in Neurology, 25*(1), 50–56.

Guinan, J. J. (2014). Olivocochlear efferent function: issues regarding methods and the interpretation of results. *Frontiers in Systems Neuroscience, 8*, 142. http://doi.org/10.3389/fnsys.2014.00142

Guinan, J. J., Jr., Salt, A., & Cheatham, M. A. (2012). Progress in cochlear physiology after Békésy. *Hearing Research, 293*(1–2), 12–20. http://doi.org/10.1016/j.heares.2012.05.005

Janssen, T., & Müller, J. (2008). Otoacoustic emissions as a diagnostic tool in a clinical context. In G. Manley, R. Fay, & A. Popper (Eds.), *Active processes and otoacoustic emissions in hearing* (Vol. 30, pp. 421–460). New York, NY: Springer. http://doi.org/10.1007/978-0-387-71469-1

Kemp, D. T. (1978). Stimulated acoustic emissions from within the human auditory system. *The Journal of the Acoustical Society of America, 64*(5), 1386–1391. doi:10.1121/1.382104

Kemp, D. (2008). Otoacoustic emissions: concepts and origins. In G. Manley, R. Fay, & A. Popper (Eds.), *Active processes and otoacoustic emissions in hearing* (Vol. 30, p. 1). New York, NY: Springer. http://doi.org/10.1007/978-0-387-71469-1

Konrad-Martin, D., Reavis, K. M., Mcmillan, G. P., & Dille, M. F. (2012). Multivariate DPOAE metrics for identifying changes in hearing: Perspectives from ototoxicity monitoring. *International Journal of Audiology, 51*(Suppl. 1), S51–S62.

Liu, Y.-W., & Neely, S. T. (2010). Distortion product emissions from a cochlear model with nonlinear mechanoelectrical transduction in outer hair cells. *The Journal of the Acoustical Society of America, 127*(4), 2420–2432. doi:10.1121/1.3337233

Marshall, L., Miller, J. A. L., Heller, L. M., Wolgemuth, K. S., Hughes, L. M., Smith, S. D., & Kopke, R. D. (2009). Detecting incipient inner-ear damage from impulse noise with otoacoustic emissions. *The Journal of the Acoustical Society of America, 125*(2), 995–1013.

Mauermann, M., & Birger, K. (2004). Distortion product otoacoustic emission (DPOAE) input/output functions and the influence of the second DPOAE source. *The Journal of the Acoustical Society of America, 116*(4), 2199–2212.

Mauermann, M., Uppenkamp, S., van Hengel, P. W., & Kollmeier, B. (1999). Evidence for the distortion product frequency place as a source of distortion product otoacoustic emission (DPOAE) fine structure in humans: I. Fine structure and higher-order DPOAE as a function of the frequency ratio f2/f1. *The Journal of the Acoustical Society of America, 106*(6), 3473–3483. Retrieved from http://www.ncbi.nlm.nih.gov/pubmed/10615687

Meaud, J., & Lemons, C. (2015). Nonlinear response to a click in a time-domain model of the mammalian ear. *The Journal of the Acoustical Society of America, 138*(1), 193–207.

Mertes, I. B., & Goodman, S. S. (2013). Short-latency transient-evoked otoacoustic emissions as predictors of hearing status and thresholds). *The Journal of the Acoustical Society of America, 134*(3), 2127–2135.

Moleti, A., Botti, T., & Sisto, R. (2012). Transient-evoked otoacoustic emission generators in a nonlinear cochlea. *Journal of the Acoustical Society of America, 131*(4), 2891–2903. doi:10.1121/1.3688474

Neely, S. T., Johnson, T. A., Kopun, J., Dierking, D. M., & Gorga, M. P. (2009). Distortion-product otoacoustic emission input/output characteristics in normal-hearing and hearing-impaired human ears. *The Journal of the Acoustical Society of America, 126*(2), 728–738. http://doi.org/10.1121/1.3158859

Neely, S. T., & Kim, D. O. (1986). A model for active elements in cochlear biomechanics. *Journal of the Acoustical Society of America, 79*(5), 1472–1480. doi:10.1121/1.393674

Olson, E. S., Duifhuis, H., & Steele, C. R. (2012). Von Békésy and cochlear mechanics. *Hearing Research, 293*(1–2), 31–43. http://doi.org/10.1016/j.heares.2012.04.017

Plack, C. J., Barker, D., & Prendergast, G. (2014). Perceptual consequences of "hidden" hearing loss. *Trends in Hearing, 18,* 1–11. http://doi.org/10.1177/2331216514550621

Popelka, G., Karzon, R., & Arjmand, E. (1995). Growth of the 2f1-f2 distortion product otoacoustic emission for low-level stimuli in human neonates. *Ear and Hearing, 16*(2), 159–165.

Rao, A., & Long, G. R. (2011). Effects of aspirin on distortion product fine structure: Interpreted by the two-source model for distortion product otoacoustic emissions generation. *The Journal of the Acoustical Society of America, 129*(2), 792–800.

Rasetshwane, D. M., Fultz, S. E., Kopun, J. G., Gorga, M. P., & Neely, S. T. (2015). Reliability and clinical test performance of cochlear reflectance. *Ear and Hearing, 36*(1), 111–124. http://doi:org/10.1097/AUD.0000000000000089

Rasetshwane, D. M., Neely, S. T., Kopun, J. G., & Gorga, M. P. (2013). Relation of distortion-product otoacoustic emission input-output functions to loudness. *The Journal of the Acoustical Society of America, 134*(1), 369–383. http://doi.org/10.1121/1.4807560

Robles, L., & Ruggero, M. A. (2001). Mechanics of the mammalian cochlea. *Physiological Reviews, 81*(3), 1305–1352.

Shera, C. (2003). Mammalian spontaneous otoacoustic emissions are amplitude-stabilized cochlear standing waves. *Journal of the Acoustical Society of America, 114*(1), 244. doi:10.1121/1.1575750

Shera, C. A. (2015). The spiral staircase: Tonotopic microstructure and cochlear tuning. *Journal of Neuroscience, 35*(11), 4683–4690. doi:10.1523/JNEUROSCI.4788-14.2015

Siegel, J. H. (2007). Calibrating otoacoustic emission probes. In M. S. Robinette & T. J. Glattke (Eds.), *Otoacoustic emissions: Clinical application* (3rd ed., pp. 403–427). New York, NY: Thieme Medical.

Sisto, R., Moleti, A., & Shera, C. A. (2015). On the spatial distribution of the reflection sources of different latency components of otoacoustic emissions. *The Journal of the Acoustical Society of America, 137*(2), 768–776. http://doi.org/10.1121/1.4906583

Souza, N. N., Dhar, S., Neely, S. T., & Siegel, J. H. (2014). Comparison of nine methods to estimate ear-canal stimulus levels. *The Journal of the Acoustical Society of America, 136*(4), 1768–1787. http://doi.org/10.1121/1.4894787

Spankovich, C. (2010). *Early indices of auditory pathology in young adults with type-1 diabetes* (Doctoral dissertation). Vanderbilt University, Nashville, TN.

Strube, H. W. (1985). A computationally efficient basilar-membrane model. *Acta Acustica United With Acustica, 58*(4), 207–214.

Talmadge, C. L., Tubis, A., Long, G. R., & Piskorski, P. (1998). Modeling otoacoustic emission and hearing threshold fine structures. *The Journal of the Acoustical Society of America, 104*(3 Pt 1), 1517–1543. Retrieved from http://www.ncbi.nlm.nih.gov/pubmed/9745736

Van Hengel, P. W., Duifhuis, H., & van den Raadt, M. P. (1996). Spatial periodicity in the cochlea: the result of interaction of spontaneous emissions? *The Journal of the Acoustical Society of America, 99*(6), 3566–3571. doi:10.1121/1.414955

Verhulst, S., Dau, T., & Shera, C. A. (2012). Nonlinear time-domain cochlear model for transient stimulation and human otoacoustic emission. *The Journal of the Acoustical Society of America, 132*(6), 3842–3848. doi:10.1121/1.4763989

Zweig, G. (1991). Finding the impedance of the organ of Corti. *The Journal of the Acoustical Society of America, 89*(3), 1229–1254. doi:10.1121/1.400653

CHAPTER 3

The Audiogram: What It Measures, What It Predicts, and What It Misses

Anthony T. Cacace and Robert F. Burkard

In human testing, probably the most frequently used and most fundamental test in audiology is the pure-tone air and bone-conduction audiogram. As a metric of hearing sensitivity (i.e., a representation of hearing threshold as a function of frequency), the audiogram has received worldwide acceptance. In addition to assessing auditory thresholds for pure-tone stimuli and quantifying the degree of hearing loss, it also provides a baseline and comparison for other associated tests (i.e., word recognition testing; spondee thresholds), helps to differentiate different types of peripheral hearing loss (conductive, sensorineural, mixed), and aids in judging the effectiveness of different treatment modalities (surgical procedures, effects of pharmaceutical agents such as ototoxic medications, etc.). Terminology is important in this area so that confusion and misrepresentation do not result. Therefore, throughout this chapter, we use the term "sensitivity" to refer to threshold phenomena.

It's So Simple: Yet So Complex

In the construction of the audiogram, we make the assumption that by selectively sampling thresholds at different frequencies (places) along the basilar membrane within the inner ear, we are assessing the functional status of sensory cells and neurons at these locations, never losing sight of the fact that with psychoacoustic (behavioral) testing, we are measuring the output of the entire system. In a typical clinical context, the audiogram is measured at octave intervals (one point per octave) from 0.25 to 8.0 kHz. The notion here is that as auditory threshold increases (i.e., get worse), the amount of damage to the sensory epithelium and neural structures at those places that were sampled within the inner ear also increase. When plotted on a graph, the *y*-axis of the audiogram is represented in decibels hearing level (dB HL) and the *x*-axis is represented as frequency; both axes are logarithmic scales. For those with a clinical interest, plots of different audiograms, with different degrees and types of hearing loss, can be found in Chapter 7 of this book.

Knowledge of the traveling wave within the inner ear and the peak of this biomechanical event take on importance in our interpretation of these data; particularly relevant is when there is moderate-to-severe hearing loss in the lower frequencies and when higher frequency thresholds are in the normal range (more on this later) and/or when disease processes can affect the biomechanics (stiffness gradient) along the basilar membrane, from base to apex, like in endolymphatic hydrops. Furthermore, in behavioral studies designed to assess auditory thresholds, relatively long duration pure tones (≥200-ms sine waves) with rise/fall times typically ≥50 ms are utilized as stimulus parameters to maintain the frequency specificity of this metric. Use of long rise/fall times, for all intents and purposes, eliminates acoustic switching transients from contaminating/confounding threshold measures, particularly if hearing loss is present. Furthermore, due to temporal summation (integration) (Zwislocki, 1960), stimulus duration at or above 200 ms is considered an infinitely long signal. Thus, in theory, as stimulus duration exceeds 200 ms, auditory thresholds should not improve. There is also a trading/inverse relationship between threshold and stimulus duration (i.e., thresholds increase as stimulus duration decreases). Consequently, there is approximately a 10-dB threshold difference for every decade change in duration (i.e., between a 20-ms and 200-ms tone). The difference in threshold as a function of duration has been used clinically to delineate site-of-lesion; this particular test has come to be known as brief-tone audiometry (Wright, 1978).

While the value of the pure-tone audiogram is clear, what it predicts in terms of functional value and what it misses are concerns that we plan to review. Thus, the pluses and minuses of this metric are noteworthy and will be explored in more detail herein. Furthermore, there are both national and international technical standards that specify the types of earphones, ear cushions, bone conduction oscillators, forces to be applied to bone conduction band, earphone headsets, and artificial ears (ear simulators) to be used in these measurements so that repeatable measures can be obtained within and between clinical settings both in the United States and in different countries throughout the world.

Calibration and Standardization

In order for the audiogram obtained anywhere in the world and to be more or less comparable, standards have been developed by various agencies or groups. In the United States, we often use standards developed by the American National Standards Institute (ANSI). For the most part, these ANSI standards are technically equivalent with the international standards created by the International Organization for Standardization (ISO) and the International Electrotechnical Commission (IEC). The ANSI standards most relevant to audiometry begin with "S3." There are four ANSI standards technical committees that address issues related to acoustics, shock, and vibration: S1 Acoustics, S2 Shock and Vibration, S3 Bioacoustics, and S12 Noise. As we are reviewing technical features about the audiogram, S3.6 is highly relevant to this discussion. This standard describes the technical details for calibrating an audiometer. There are definitions of terms used, which is critically important for the purposes of calibration. For example, sound pressure level (SPL) is defined as $10\log_{10}(P/P_{ref})^2$. The definition needs to be technically accurate. We need to subscript the "log" in order to note that it is the base 10 (Briggsian) logarithm, rather than the natural (Naperian) logarithm. "P" refers to the sound pressure, which is now, for the most part defined in meter, kilogram, seconds (MKS units) and uses the Pascal (Pa) as the unit of pressure. In older textbooks in audiology, it was defined in centimeters, grams, seconds (cgs) units, or dynes/cm^2 (microbar). It is called a microbar because it is one millionth of barometric pressure. "P_{ref}" refers to the reference pressure, which in air and in MKS units is 0.00002 Pa (or 20 μPa), while in cgs units, this is 0.0002 dynes/cm^2 (microbar). Just to make this brief tutorial slightly more complicated, the reference pressure in water is 1 μPa, which would be 0.00001 microbar. At this point, you are probably thinking, "Who cares?" Well, you should care. Because if your audiometer were calibrated to the 1 μPa (i.e., the "water") reference, your acoustic output would differ by 26 dB ($10\log_{10}(1/20)^2 = -26$ dB) for a given hearing level "HL" dial setting than if calibrated correctly to the 20 μPa reference. Although somewhat arbitrary, the 20 μPa reference is critical in ensuring that the audiograms collected across the United States (indeed, the world, as the ANSI audiometer Standard S3.6 is in harmony with the comparable international standard) are comparable. A more detailed and comprehensive treatment of calibration and standards relevant to audiologists can be found in Burkard (2014, 2015).

As we review what the audiogram measures, what it predicts, and what it misses, we consider changes in threshold below 1.0 kHz and above 8.0 kHz. This delineation is arbitrary but useful.

Frequency-Dependent Effects Below 1.0 kHz

In a clinical context and for accuracy of measurement, audiometric thresholds below 1.0 kHz are a good place to start since it is important to know when individuals actually "feel vibrations" of the auditory stimulus rather than actually hear them via the auditory sensory modality. Early work in this area was championed by Boothroyd and Cawkwell (1970). At high stimulus levels, vibrations can be detected by sensory structures in the skin (Pacinian corpuscles) rather than hair cells within the inner ear. The mechanoreceptive Pacinian system has a narrow bandwidth that is maximally sensitive to vibration at ~40 Hz (0.04 kHz) and does not respond above 1.0 kHz (see Verrillo, 1975, for a review), assuming of course that stimulus safeguards are in place like center-surround elements of the mechano-stimulator when vibration testing is performed.

As the degree of hearing loss increases, particularly in the lower frequency range, and when higher frequency thresholds are in the normal or near-normal range (≤25 dB HL), some interpretive issues also become apparent. First, as we noted above, it is entirely possible that the identification of a moderate or moderate-to-severe hearing loss at the very low frequency range (i.e., 0.125 kHz, 0.250 kHz) can actually be based on vibrotactile sensation, rather than auditory sensation. Additionally, it is also possible that the extent of hearing loss in the lower frequency range can be underestimated (not accurately represented) due to an upward spread of excitation of the traveling wave as stimulus intensity is increased. This has been shown by Thornton and Abbas (1980) and Turner et al. (1983) in a series of clever experiments. To validate the concept of upward spread of excitation, psychophysical tuning curves were used with the probe frequency centered in low-frequency hearing loss area of interest. If the upward spread of excitation was the correct interpretation, then the tip of the tuning curve, rather than being at the place of maximum low-frequency hearing loss, would shift to a more basal (higher frequency) location. In many but not all instances, so-called "mistuned" tuning curves were observed, thus validating the upward spread of excitation hypothesis.

Upward Spread of Excitation

Georg von Békésy earned the Nobel Prize in Medicine or Physiology for his groundbreaking work on basilar-membrane mechanics (see Békésy, 1960). In addition to demonstrating that the cochlear partition performs a mechanical Fourier transform on the input stimulus, he found that high frequencies are encoded near the base, with low frequencies progressively encoded more apically. This important work also demonstrated that low-frequency sounds, at moderate and high stimulus levels, produced substantial displacement of more basal regions of the cochlear partition. This upward (basal) spread of excitation provides a mechanistic explanation why you cannot have a severe or profound low-frequency hearing loss with normal or near-normal thresholds in the middle-to-high frequency range. This is further explicated if we look at auditory nerve fiber tuning curves. In this context, shallow tuning is observed on the low-frequency tail of the tuning curves, which means that if a threshold at, for example, 4.0 kHz is 10 dB SPL, the threshold of that high-frequency auditory nerve fiber will only be perhaps 40 dB higher in threshold at 0.5 kHz than it was at 4.0 kHz. Thus, in this instance, with a profound hearing loss at 0.5 kHz (perhaps because of a "cochlear dead zone"—see below), one would get a 50-dB SPL threshold simply because more basal cochlear regions are responding to this low-frequency stimulation. In contrast, the high-frequency slope of tuning curves can be very steep, and thus one could have completely normal hearing at 4.0 kHz but have a severe-to-profound hearing loss at 8.0 kHz, which truly reflects the hearing thresholds at both the 4.0 kHz and 8.0 kHz regions. Let us propose a subject with a flat hearing loss in one ear (60-dB SPL thresholds from 0.5–8.0 kHz), and in the other, the 0.5-kHz threshold is at 60 dB SPL, improving to 10 dB SPL at 4.0 kHz and 8.0 kHz. In the latter ear, the threshold at 0.5 kHz is actually due to the basal spread of excitation from the 2.0-kHz region. In this subject, at 0.5 kHz, the subject could also have diplacusis, because in one ear, the pitch would be that of a 0.5-kHz stimulus, while in the latter ear, it would be to 2.0 kHz.

Dehiscence Syndromes: So-Called Inner Ear Conductive Hearing Loss (i.e., the Third Window Phenomenon)

First described by Minor et al. (1998, 2000, 2003) and subsequently confirmed by other representative case studies by this group (e.g., Carey et al., 2000; Hirvonen et al., 2001, 2003), superior canal dehiscence (SCD) refers to the condition where bone is missing, often over the top of the superior semicircular canal. Consequently, this bony dehiscence makes the superior semicircular canal more sensitive/vulnerable to acoustic stimuli where loud noises can produce vertigo and even blurred vision. With respect to the audiogram, this disorder has the unique distinction as showing air-bone gaps (bone conduction being better than air conduction) in the lower frequency range despite the fact that there is no apparent transmission loss in the outer or middle ear. This unusual "conductive" hearing loss is attributable to changes in inner ear biomechanics. In contrast to a conductive loss resulting from a middle ear condition like otosclerosis, the pattern of audiometric test results from SCD is counterintuitive, most notably (as noted above) by manifesting low-frequency air-bone gaps with "supra-normal" (better) thresholds for bone conduction than one might expect, particularly in the presence of (1) intact acoustic stapedius reflexes, (2) observable otoacoustic emissions (despite the presence of conductive hearing loss), and (3) intact cervical vestibular evoked myogenic potentials that are typically absent in true conductive hearing loss (for example, in otosclerosis). Merchant and Rosowski (2008) hypothesized that this is due to "impedance differences between the scala vestibuli and the scala tympani side of the cochlear partition, which in turn, is due to a difference between the impedance of the oval and round windows, respectively. This inequality leads to a pressure difference across the cochlear partition, resulting in motion of the basilar membrane that leads to the perception of bone-conducted sound. A pathologic third window on the vestibular side of the cochlear partition increases the pressure difference between the 2 sides of the cochlear partition by lowering the impedance on the vestibuli side, thereby improving the cochlear response to bone conduction" (Merchant & Rosowski, 2008, p. 4).

The Tonndorf Model and the Pathophysiology of Hearing Loss in Ménière's Disease

Ménière's disease is a clinical entity characterized by fluctuating sensorineural hearing loss, tinnitus, and vestibular-related symptoms (e.g., vertigo). While exact mechanisms underlying the disease process are not completely understood, increases in endolymphatic pressure/volume (endolymphatic hydrops) in the inner ear have been attributed to some of the symptoms described above. Such effects can also produce distinct patterns of the audiogram that may help to predict different stages in the disease process along with several key anatomical and physiologic effects underlying these changes (e.g., Tonndorf, 1968). Briefly, changes in endolymphatic pressure and volume alter the biomechanical properties of the basilar membrane, resulting in varying degrees of low-frequency hearing loss that can fluctuate in the acute stages of the disease and change permanently as the disease progresses to a more chronic stage. A key element of the Tonndorf model is the elastic properties of Reissner's membrane. Presumably, as long as Reissner's membrane remains intact and elastic, hearing will fluctuate over time, predominantly but not exclusively in the lower frequency range. But once this elasticity is lost, the hearing loss pattern changes and the audiometric profile assumes a flat configuration. While this description may appear somewhat simplistic in nature, it does capture the essence of the audiometric effects and may be pathognomonic of the disease process, helping professionals to follow the course of the disease and possibly to decide when medical or surgical treatment is most advisable. A diagram of the audiometric effects noted above is shown in Figure 3–1.

Frequency-Dependent Effects Above 8.0 kHz

Fausti and colleagues (1993, 1999) showed that audiometric testing above 8.0 kHz (herein called "ultrahigh frequency" hearing) can detect detrimental effects of ototoxic medication before it enters the standard audiometric range and impairs speech recognition/perception

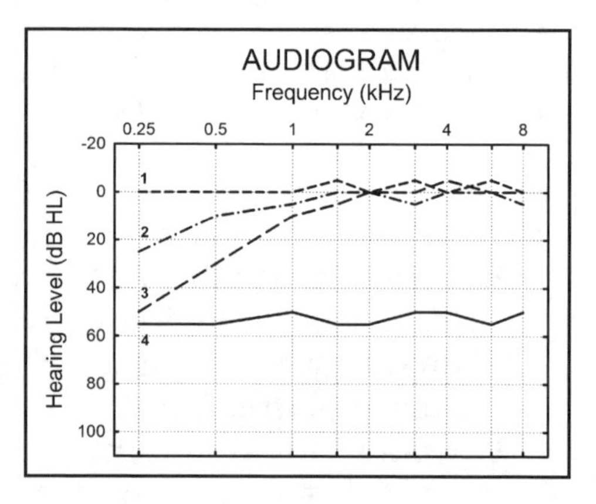

Figure 3–1. Graphic representation of a group of schematized audiograms in Ménière's disease where there is a reduction and fluctuations in low-frequency thresholds (Plots 2 and 3) corresponding to increases in endolymphatic pressure/volume. This typically occurs in the acute stages of the disease. Over time, thresholds in the higher frequencies also worsen until there appears to be a flat loss (Plot 4). This flat loss configuration corresponds to a chronic stage of the disease process where Reissner's membrane loses it elasticity. Other corresponding features/symptoms include diplacusis and roaring tinnitus.

abilities. If ultrahigh frequency assessment is performed repeatedly while a patient is being administered ototoxic medications (e.g., while being given aminoglycosides for bacterial infections or during chemotherapy using antineoplastic agents), then if ultrahigh frequency thresholds start to show adverse effects of treatment (i.e., where auditory thresholds are getting worse), the treatment regimen could possibly be modified in an attempt to minimize loss of hearing in the traditional audiometric frequency range. However, these are medical decisions where changes in treatment may or may not be possible. It is also noteworthy that ANSI S3.6-2010 includes reference equivalent threshold sound pressure levels (RETSPLs: i.e., 0 dB HL on the audiometer dial) for frequencies up to 20.0 kHz. It is further noted that audiometric testing at these frequencies requires the use of special circumaural earphones, several of which are no longer available commercially. Also, the clinician should be cognizant of calibration issues for the audiometric frequencies above 8.0 kHz; this topic, however, while very important, is beyond the scope of this discussion.

Prognostic Value of Ultrahigh Frequency Audiometric Testing

In addition to its diagnostic value, ultrahigh frequency audiometric testing has also played an important role in understanding other curious auditory phenomena found in the literature. For example, Berlin and colleagues (Berlin et al., 1978; Collins et al., 1981) observed a subgroup of individuals with "deaf-like speech" and severe-to-profound pure-tone hearing loss in the standard audiometric range but with excellent sibilant articulation abilities. How is this possible? It turns out that these individuals had normal or near-normal ultrahigh frequency hearing thresholds (above 8.0 kHz) and could hear the pops, bursts and temporal features of some of the relevant speech sounds. Collins et al. (1981) also investigated auditory signal processing in a single individual based on psychoacoustic measures, including difference limens for frequency and intensity, gap detection, rate matching, temporal integration, and psychoacoustical tuning curves in the 13.0-kHz range. In addition, measures of speech production and perception were made when the patient was using an experimental "transpositional" hearing aid designed to transpose speech range frequencies into the region of better ultrahigh frequency residual hearing. Results indicated that the individual could efficiently process signals that fall within the band of best hearing and that ultrahigh frequency audiometric hearing may be useful in speech communication. Thus, this type of hearing loss would have gone undetected without pure-tone testing above 8.0 kHz and the use of the appropriate high frequency headphones.

Cochlear Dead Zones

As explained above, due to basal spread of excitation, it is quite possible that limited regions of the cochlea could be denuded of hair cells or be missing auditory nerve fibers (popularly called cochlear dead zones) and that the apical tail of the tuning curves in cochlear regions immediately basal to these dead zones responds to acoustic stimulation, at moderate stimulation levels. Many relevant studies by Moore and colleagues and others (e.g., Cox et al., 2012; Moore, 2001; Moore & Alcantara, 2001; Zhang et al., 2014) have tackled this important area of investigation.

Hidden Hearing Loss

How many auditory nerve fibers are needed in order to perceive a puretone stimulus at threshold? This important question has been addressed over five decades ago by Professor Harold Schuknecht, in experiments in cats. As we will point our later, this information forms the basis and limitations of the audiogram and information dealing with so-called hidden hearing loss (see Schuknecht & Woellner, 1955), where it was found that ~75% of spiral ganglion cells can be missing and auditory thresholds can still be normal but above this percentage point, threshold changes become evident. Indeed, these observations have contributed to Schuknecht's "Neural Presbycusis" category of age-related hearing loss, which is characterized by poorer monosyllabic word recognition performance than would have been expected from the degree of threshold elevation observed (Schuknecht, 1974). It is also important to point out that "synaptopathy" and its effects on hearing has, until very recently, been a theoretical construct in humans. However, in recent studies using immunohistochemical staining protocols tagged with fluorescent markers (fluorophores with optimal excitation at 405, 488, 568, and 647 nm) and confocal microscopy outfitted with appropriate lasers, Viana et al. (2015) and Liberman and Liberman (2015) found that in post mortem human tissue (normal individuals with presbycusis), there were fewer synapses on the remaining hair cells than would be predicted from spiral ganglion cell counts. This work, although limited in nature, suggests that aging humans may have the same loss of synapses that has been described experimentally in animal models following noise exposure.

In mice, there is clear evidence that noise exposure at frequencies above the traditional human audiometric frequency range can lead to substantial loss of auditory nerve fibers while having comparatively little effect on ABR threshold but showing a substantial decrease in ABR amplitude at moderate-to-high stimulus levels. Minimal loss of auditory evoked potential (AEP) threshold but a substantial decrease in AEP amplitude above threshold has also been observed in chinchillas with selective inner hair cell (IHC) loss following administration of the antineoplastic agent carboplatin (Harrison, 1998; Qiu et al., 2000). As most auditory nerve fibers innervate IHCs, it should be no surprise that loss of IHCs or loss of auditory nerve fibers should each result in similar changes in threshold and in suprathreshold responses. Using near-field responses from the auditory nerve, inferior colliculus, and auditory cortex, Qiu et al. (2000) found a drop in amplitude of both the auditory nerve and inferior colliculus responses, but little change in threshold, when there was a partial loss of IHCs. These results are similar to the ABR changes attributed to synaptopathy by Kujawa and Liberman (2009, 2015). Surprisingly, Qiu et al. (2000) also found an enhancement (increase) in the response from the auditory cortex at moderate-to-high stimulus levels, which could be attributed to adaptive plasticity at the cortical level. Although the responses from the auditory nerve were reduced in amplitude, there is reportedly little change in the amplitude of the cochlear microphonic and otoacoustic emissions when the outer hair cells are spared (Harrison, 1998; Qiu et al., 2000).

In addition to synaptopathy and carboplatin-induced IHC loss in specific animal models, there is also a clinical entity in humans that shows aberrant/desynchronized and/or absent ABR (or CAP or acoustic reflex), with normal cochlear microphonics (CMs) or OAEs, and auditory thresholds often in or near the normal range which is known as auditory neuropathy (AN) or auditory neuropathy/auditory dyssynchrony (AN/AD) disorder (see Cacace & Burkard, 2009; Cacace & Pinheiro, 2011; Sininger & Starr, 2001). For reasons that elude the authors of this chapter, the term "auditory neuropathy spectrum disorder" has also been used. Our current ability to differentiate IHC loss, synaptopathy, and loss/damage to auditory nerve fibers is not perfect, work in this area is ongoing (e.g., Santarelli et al. 2008; 2015), but further research is needed. At the time this chapter is being written, the identification of the true site-of-lesion in a given individual with AN/AD or IHC dysfunction remains a challenging proposition (see Cacace & Burkard, 2009, for a more detailed discussion of AN/AD and site-of-lesion). Also pertinent to the area of hidden hearing loss is the concept of audiometric "fine structure." This highly relevant topic provides the entré into the next important question.

Does Everyone Have Hidden Hearing Loss? Is It a Question of Resolution?

As we noted in the beginning of this chapter, the classical audiogram typically samples frequencies at one point per octave from 0.25 kHz to 8.0 kHz and in some instances in half-octave steps as a way to better define precipitous hearing loss in the higher frequency range, namely at 3.0 and 6.0 kHz. Obviously, this leaves a lot of sensory epithelium (between octave spacing) unassessed and functionally undefined with respect to threshold. Clearly, the undefined nature of these in between-octave frequencies is directly related to the resolution by which thresholds are assessed.

Implications From Audiometric Fine Structure

The microstructure or fine structure of the audiogram refers to those audiometric thresholds in between the standard octave frequencies but at a much higher frequency resolution (spacing) than typically used in a clinical assessment strategy (Long, 1984). While this work can be tedious and very time-consuming, it brings to light the fact that the normal audiogram has an irregular shape with notable peaks and valleys (troughs) which can have a range of 15 dB or more (Lee & Long, 2012). As these data reveal, threshold/frequency profiles are not just a straight line as the audiogram seems to imply. It is also of interest that Schaette and McAlpine (2011) find that individuals with normal audiometric thresholds at standard test frequencies can also have tinnitus. This is an interesting phenomenon that should also be considered or at least discussed in the context of audiometric fine structure analysis.

What are some factors that can contribute to these irregular frequency-threshold profiles in between the standard octave frequencies? One notable consideration is the presence of spontaneous otoacoustic emissions (SOAEs), which can alter the fine structure (Lee & Long, 2012; Long & Tubis, 1988). SOAEs are self-generated events generated from the cochlea or can be synchronized by a stimulus. Also, Kemp initially showed some undulating features if a higher level of frequency resolution was used. In his early work (summarized in Kemp, 2003), Kemp looked at high-frequency resolution distortion product otoacoustic emissions (DPOAEs) and found the same peaks and troughs in DPOAE amplitude as reported behaviorally. This "fine structure" of the DPOAEs is thought to result from the interaction between the two generation sources of DPOAEs: one near the f2 frequency and the other at the DPOAE frequency place (e.g., Lonsbury-Martin & Martin, 2002). A detailed description of mechanisms underlying otoacoustic emission generation is beyond the scope of this chapter but is aptly covered by Long and Epp (Chapter 2, this book).

Bone Conduction and Air-Bone Gaps

Another issue often raised by student clinicians pertains to whether bone conduction thresholds can be worse than air conduction thresholds, assuming of course that the instrumentation is working properly and there are no issues related to calibration. It is often assumed (in theory) that bone conduction thresholds cannot be worse than air conduction thresholds. The logic underlying this assertion relates to the fact that if the air conduction threshold is a measure of the whole system (the conductive, sensory, and neural systems) and if bone conduction thresholds represent just part of the system (i.e., the conductive system), then bone conduction thresholds can never be worse than air conduction thresholds. However, as Studebaker (1967) points out, we must take into consideration sampling variability and statistical issues.

There are a number of challenges in pure-tone audiometry that are related to various sources of variability. Take, for example, if you test a subject's air-conduction thresholds in a clinic in Detroit. Then, that individual moves to Buffalo and has his or her hearing tested again. Would you expect the audiometric thresholds to be exactly the same, for all frequencies and for both ears? If so, in the vast majority of cases, you are going to be disappointed. There are a number of reasons for this. Generally speaking, even using the exact same equipment, one would expect thresholds at a given frequency to vary by ±5 dB, simply due to test-retest variability, attentional issues, the probabilistic nature of threshold estimation, transducer placement, etc. Second, the calibration guidelines in the audiometer standard allow for some variation in calibration in the level of the calibration, which is ±3 dB to 5.0 kHz, and ±5 dB at higher frequencies. So even using exactly the same equipment and transducers, each calibrated to meet the S3.6-2010 standard, there is a 6-dB (below 5 kHz) or 10-dB (above 5.0 kHz) range in stimulus level on transducer output that stills meet the calibration guidelines specified in ANSI S3-6. Third, the normative data for RETSPLs (i.e., 0 dB HL) of the various transducers/couplers were obtained on different subjects, and hence differences in median threshold could be due to different central tendencies in the threshold distributions in each group of "normal-hearing subjects". Each of the couplers used in S3.6-2010 has a different transfer function that will have a unique effect on each individual ear it is coupled with. If we assume that this across-coupler/earphone variability is again ±5 dB, then for low frequencies, we can assume that the various sources of error are independent and (using a formula reminiscent of adding uncorrelated sounds) estimate the variability as $(\pm5\text{ dB} + \pm3\text{ dB} + \pm5\text{ dB})^{.5} = \sim\pm8$ dB, which is more than a 15-dB range. Now, if we want to consider air-bone gaps, we also need to first consider that each of those threshold estimations (air and bone, for a given ear and a given frequency) is subject to the same variability, which in the above example would give us up to a 30-dB range. This would not happen often, but it could happen. Next, one has to consider that since we emerged from the sea many millions of years ago and replaced our gills with a middle ear, most of the time (except perhaps when we are swimming or have a fixed stapes due to otosclerosis), we do not hear via bone conduction. Although it is recognized that couplers and artificial ears are imperfect representations of the point impedance and transfer function of a given individual's ear, the artificial mastoid is even a poorer representation in this regard. The exception with respect to air-conduction measurements made on earphones is the Zwislocki ear simulator (aka, the Zwislocki coupler; Zwislocki, 1980). This particular ear simulator evolved as part of a working group of the Committee on Hearing, Bioacoustics, and Biomechanics of the National Research Council to evaluate standard couplers in use at the time (see Zwislocki, 1967 for an overview). The working group made important recommendations and showed that the coupler designed by Zwislocki was a clear advance to the standard 20-cc or 6-cc couplers.

With respect to bone conduction, the truth of the matter is that there is great variation in the mechanics of individual skulls (think about how different the mechanical properties of an adult human are compared to a newborn infant). With the variability described above, it is truly a wonder how often an air-bone gap does illuminate the presence of a conductive hearing loss. The variability issues in air and bone conduction thresholds are discussed in a far more detailed and scholarly manner in Studebaker (1967). Furthermore, Margolis and Moore (2011) and Margolis et al. (2013) have shown that the commonly reported false air-bone gap at 4.0 kHz (i.e., an air-bone gap in the absence of a conductive hearing loss) can be avoided if the reference equivalent threshold force level (RETFL) for bone conduction threshold at 4.0 kHz (i.e., the dB HL value for bone conduction) is decreased by ~14 dB. Changing RETSPLs in the relevant national and international standards will no doubt be a long, drawn-out and contentious process.

Issues Related to Speech Perception

You cannot always accurately estimate a given subject's (or patient's) speech perception abilities from their audiogram. Suprathreshold speech perception tests are strongly dependent on numerous factors that are largely independent of audiometric threshold, including the level of the speech signals, the type of speech signal (e.g., syllables, words, sentences), whether they are presented in quiet or noise (and the signal-to-noise ratio if presented with a noise background), audiometric threshold, the type of hearing loss (conductive, sensory, AN/AD), and certain cognitive abilities of the listener. Even with very similar audiograms, different patients can have quite disparate word recognition scores and discrimination abilities. We live in a noisy world, and for the clinical audiologist, it is very common to hear patients or spouses opine that they can hear just fine but cannot "understand" what people are saying, especially in a noisy setting (e.g., at a bar or notably at a faculty meeting). Killion and Niquette (2000) and Killion (2002) discuss issues such as underlying IHC and OHC loss, as well as the relationship between the audiogram and speech-in-noise performance and conclude that the hearing of speech-in-noise is not strongly dependent on the audiogram. We must conclude from this work that if you are interested in determining how well or how poorly an individual processes speech-in-noise, then you need to measure his or her performance directly.

Summary

The audiogram is the gold-standard test of hearing ability. American and international standards dictate strict guidelines for the acoustic calibration of the audiometric equipment, which makes it possible to compare audiometric testing that occurs not only in the United States but anywhere across the globe. Nonetheless, response interpretation is made much more complicated by the variability that arises from several factors, including the output of the transducers, between-subject variability in the threshold measures themselves, and the effects of auditory pathology on the estimates of threshold across frequency. Accurate interpretation of the audiogram requires scientific knowledge of acoustics, the anatomy/physiology of the auditory system, and a detailed understanding of the pathophysiology of a myriad of disorders that affect the auditory system and brain. It requires substantial clinical experience to interpret the audiogram and to make an accurate site-of-lesion determination and/or develop effective intervention plans. The pure-tone audiogram cannot, in isolation, inform the clinician how the patient will perform on suprathreshold tests, including complex signals such as speech in quiet and/or noise; thus, testing must be performed directly in order to obtain a more complete picture of the patient's hearing abilities.

References

Békésy, G. V. (1960). *Experiments in hearing*. New York, NY: McGraw-Hill.

Berlin, C. I., Wexler, K. F., Jerger, J. F., Halperin, H. R., & Smith, S. (1978). Superior ultra-audiometric hearing: A new type of hearing loss which correlates highly with unusually good speech in the "profoundly deaf." *Otolaryngology, 86*, 111–116.

Boothroyd, A., & Cawkwell, S. (1970). Vibrotactile thresholds in pure tone audiometry. *Acta Otolaryngologica, 69*, 381–387.

Burkard, R. F. (2014). Standards and Calibration. Part 1: Standards process, physical principles, pure tone and speech audiometry. *Seminars in Hearing, 35*, 267–360.

Burkard, R. F. (2015). Standards and Calibration. Part 2: Brief stimuli, immittance, amplification, and vestibular assessment. *Seminars in Hearing, 36*, 1–74.

Cacace, A. T., & Burkard, R. F. (2009). Auditory neuropathy: Bridging the gap between basic science and current clinical concerns. In A. T. Cacace & D. J. McFarland (Eds.), *Current controversies in central auditory processing disorders (CAPD)* (pp. 305–343). San Diego, CA: Plural.

Cacace, A. T., & Pinheiro, J. M. B. (2011). The mitochondrial connection in auditory neuropathy. *Audiology and Neurotology, 16*, 398–413.

Carey, J. P., Minor, L. B., & Nager, G. T. (2000). Dehiscence of thinning of bone overlying the superior semicircular canal in a temporal bone survey. *Archives of Otolaryngology-Head and Neck Surgery, 126*, 137–147.

Collins, M. J., Cullen, J. K., Jr., & Berlin, C. I. (1981). Auditory signal processing in a hearing-impaired subject with residual ultra-audiometric hearing. *Audiology, 20*, 347–361.

Cox, R. M., Johnson, J. A., & Alexander, G. C. (2012). Implications of high-frequency cochlear dead regions for fitting hearing aids to adults with mild to moderately-severe hearing loss. *Ear and Hearing, 33*, 573–587.

Fausti, S. A., Henry, J. A., Helt, W. J., Phillips, D. S., Frey, R. H., Noffsinger, D., . . . Fowler, C. G. (1999). An individualized, sensitive frequency range for early detection of ototoxicity. *Ear and Hearing, 20*, 497–505.

Fausti, S. A., Henry, J. A., Schaffer, H. I., Olson, D. J., Frey, R. H., & Bagby, G. C. (1993). High-frequency monitoring for early detection of cisplatin ototoxicity. *Archives of Otolaryngology-Head and Neck Surgery, 119*, 661–666.

Harrison, R. (1998). An animal model of auditory neuropathy. *Ear and Hearing, 19*, 355–361.

Hirvonen, T. P., Carey, J. P., Liang, C. J., & Minor, L. B. (2001). Superior canal dehiscence: Mechanisms of pressure sensitivity in a chinchilla model. *Archives of Otolaryngology-Head and Neck Surgery, 127*, 1331–1336.

Hirvonen, T. P., Weg, N., Zinreich, S. J., & Minor, L. B. (2003). High-resolution CT findings suggest a developmental abnormality underlying superior canal dehiscence syndrome. *Acta Otolaryngologica, 23*, 477–481.

Kemp, D. T. (2003). *The OAE story*. Hatsfield, Hertz, UK: Otodynamics.

Kemp, D. T. (2010). Otoacoustic emissions and evoked potentials. In P. Fuchs (Ed.), D. Moore (Chief, Ed.), *The Oxford handbook of auditory science: The ear* (pp. 93–137). Oxford, UK: The Oxford Library of Psychology.

Killion, M. C. (2002). New thinking on hearing in noise: A generalized articulation index. *Seminars in Hearing, 23*, 57–75.

Killion, M. C., & Niquette, P. A. (2000). What can the pure-tone audiogram tell us about a patient's SNR loss? *The Hearing Journal, 53*, 46–53.

Kujawa, S. G., & Liberman, M. C. (2009). Adding insult to injury: Cochlear nerve degeneration after "temporary" noise-induced hearing loss. *Journal of Neuroscience, 29*, 14077–14085.

Kujawa, S. G., & Liberman, M. C. (2015). Synaptopathy in the noise-exposed and aging cochlea: Primary neural degeneration in acquired sensorineural hearing loss. *Hearing Research, 330*, 191–199.

Lee, J., & Long, G. R. (2012). Stimulus characteristics which lessen the impact of threshold fine structure on estimates of hearing status. *Hearing Research, 283*, 24–32.

Liberman, L., & Liberman, M. C. (2015, Summer). Immunostaining and confocal microscopy applied to the analysis of afferent and efferent synapses in the human organ of Corti. *Newsletter of the NIDCD National Temporal Bone, Hearing and Balance Pathology Resource Registry*, pp. 1–4.

Long, G. R. (1984). The microstructure of quiet and masked thresholds. *Hearing Research, 15*, 73–87.

Long, G. R., & Tubis, A. (1988). Investigations into the nature of the association between threshold microstructure and otoacoustic emissions. *Hearing Research, 36*, 125–138.

Lonsbury-Martin, B., & Martin, G. (2002). Distortion product otoacoustic emissions. In M. Robinette & T. Glattke (Eds.), *Otoacoustic emissions* (2nd ed., pp. 116–142). New York, NY: Thieme.

Margols, R. H., Eikelboom, R. H., Johnson, C., Ginter, S. M., Swanepoel, D. W., & Moore, B. C. J. (2013). False air-bone gaps at 4 kHz in listeners with normal hearing and sensorineural hearing loss. *International Journal of Audiology, 52*, 526–532.

Margolis, R. H., & Moore, B. C. (2011). AMTAS(®): Automated method for testing auditory sensitivity: III. sensorineural hearing loss and air-bone gaps. *International Journal of Audiology, 50*, 440–447.

Merchant, S. N., & Rosowski, J. J. (2008). Conductive hearing loss caused by third-window lesions of the inner ear. *Otology & Neurotology, 29*, 282–289.

Minor, L. B. (2000). Superior canal dehiscence syndrome. *American Journal of Otology, 21*, 9–19.

Minor, L. B., Carey, J. P., Cremer, P. D., Lustig, L. R., & Streubel, S-O. (2003). Dehiscence of bone overlying the superior canal as a cause of apparent conductive hearing loss. *Otology & Neurotology, 24*, 270–278.

Minor, L. B., Solomon, D., Zinreich, J. S., & Zee, D. S. (1998). Sound and/or pressure-induced vertigo due to bone dehiscence of the superior semicircular canal. *Archives of Otolaryngology-Head and Neck Surgery, 124*, 249–258.

Moore, B. C. J. (2001). Dead regions in the cochlea: Diagnosis, perceptual consequences, and implications for the fitting of hearing aids. *Trends in Amplification, 5*, 1–34.

Moore, B. C. J., & Alcantara, J. I. (2001). The use of psychophysical tuning curves to explore dead regions in the cochlea. *Ear and Hearing, 22*, 268–278.

Qiu, C. X., Salvi, R., Ding, D., & Burkard, R. F. (2000). Inner hair cell loss can lead to enhancement of responses from auditory cortex in unanesthetized chinchillas. *Hearing Research, 139*, 153–171.

Santarelli, R., del Castillo, I., Cama, E., Scimemi, P., & Starr, A. (2015). Audibility, speech perception and processing of temporal cues in ribbon synaptic disorders due to OTOF mutations. *Hearing Research*, 330, 200–212.

Santarelli, R., Starr, A., Michalewski, H., & Arslan, E. (2008). Neural and receptor cochlear potentials obtained by transtympanic electrocochleography in auditory neuropathy. *Clinical Neurophysiology, 119*, 1028–1041.

Schaette, R., & McAlpine, D. (2011). Tinnitus with a normal audiogram: Physiological evidence of hidden hearing loss and computational model. *Journal of Neuroscience, 31*, 13452–13457.

Schuknecht, H. F. (1974). *Pathology of the ear*. Cambridge, MA: Harvard University Press.

Schuknecht, H. F., & Woellner, R. C. (1955). An experimental and clinical study of deafness from lesions of the cochlear nerve. *Journal of Laryngology, 69*, 75–97.

Sininger, Y., & Starr, A. (2001). Auditory neuropathy. In *A new perspective on hearing disorders*. San Diego, CA: Singular.

Studebaker, G. A. (1967). Intertest variability and the air-bone gap. *Journal of Speech and Hearing Disorders, 32*, 82–86.

Thornton, A. R., & Abbas, P. J. (1980). Low-frequency hearing loss: Perception of filtered speech, psychophysical tuning curves, and masking. *Journal of the Acoustical Society of America, 67*, 638–643.

Tonndorf, J. (1968). Pathophysiology of the hearing loss in Meniere's disease. *Otolaryngology Clinics of North America, 1*, 375–388.

Turner, C., Burns, E. M., & Nelson, D. A. (1983). Pure tone pitch perception and low-frequency hearing loss. *Journal of the Acoustical Society of America, 73*, 966–975.

Verrillo, R. T. (1975). Cutaneous sensation. In B. Scharf (Ed.), *Experimental sensory psychology* (pp. 150–184). Glenview, IL: Scott, Foresman and Company.

Viana, L., O'Malley, J., Burgess, B., Jones, D., Oliviera, C., Santos, F., . . . Liberman, M. C. (2015). Cochlear neuropathy in human presbycusis: Confocal analysis of hidden hearing loss in post-mortem tissue. *Hearing Research, 327*, 78–88.

Wright, H. (1978). Brief-tone audiometry. In J. Katz (Ed.), *Handbook of clinical audiometry* (2nd ed., pp. 218–232). Baltimore, MD: Williams & Wilkins.

Zhang, T., Dorman, M. F., Gifford, R., & Moore, B. C. J. (2014). Cochlear dead regions constrain the benefit of combining acoustic stimulation with electric stimulation. *Ear and Hearing, 35*, 410–417.

Zwislocki, J. J. (1960). Theory of temporal auditory summation. *Journal of the Acoustical Society of America, 32*, 1046–1060.

Zwislocki, J. J. (1967). *Couplers for calibration of earphones*. Report W. G. 48, National Academy of Sciences—National Research Council, Committee on Hearing, Bioacoustics, and Biomechanics, Washington, D. C.

Zwislocki, J. J. (1980). An ear simulator for acoustic measurements: Rationale, principles, and limitations. In G. A. Studebaker & I. Hochberg (Eds.), *Acoustical factors affecting hearing aid performance* (pp. 127–147). Baltimore, MD: University Park Press.

CHAPTER 4

Contemporary Issues in Vestibular Assessment

Faith W. Akin, Owen D. Murnane, and Kristal Mills Riska

The primary function of the vestibular system is to maintain postural and gaze stability. Housed within the bony and membranous labyrinth of the inner ear, the vestibular system is composed of five sensory end organs that detect and code head motion: three semicircular canals (SCCs) and two otolithic organs. The five end organs provide sensory input for two important vestibular-mediated reflex pathways: the vestibulo-ocular reflex (VOR) and the vestibulo-spinal reflex (VSR). Loss of vestibular function can result in vertigo (the illusion of movement), oscillopsia (blurred vision during head movement), postural instability, and/or motion intolerance. Historically, clinical assessment of vestibular function has focused on measurement of the VOR during activation of the horizontal semicircular canal (one of the five vestibular sensory organs) via caloric irrigation or rotational stimuli. In fact, the bithermal caloric test, considered the mainstay of the vestibular battery, has remained largely unchanged since it was first described by Fitzgerald and Hallpike in 1942. Peripheral vestibular dysfunction, however, can occur in one or both labyrinths, in one or both branches of the vestibular nerve, and in one or more vestibular sensory organs (Agrawal et al., 2013; Ernst et al., 2005; McCaslin et al., 2011; Murray et al., 2007; Pelosi et al., 2013; Schönfeld et al., 2010).

In the past two decades, vestibular research has focused on the development and utility of clinical tests to assess the other four vestibular sensory organs and their pathways. New tests of vestibular function include the video head impulse test (vHIT), cervical vestibular evoked myogenic potentials (cVEMPs), ocular vestibular evoked myogenic potentials (oVEMPs), and subjective visual vertical (SVV). The vHIT assesses VOR function for all three pairs of semicircular canals, and the cVEMP, oVEMP, and SVV provide a clinical measure of otolith organ function. Figure 4–1 shows the number of peer-reviewed articles published on each of these tests over the past 20 years. The purpose of this chapter is to explore contemporary tests of vestibular function and to highlight gaps in understanding where applicable. An overview of the relevant anatomy and physiology for the tests described in this chapter is provided; however, the interested reader is directed to Baloh and Honrubia (2001) and Goldberg et al. (2012) for a more detailed description.

The vestibular sensory organs are oriented for three-dimensional coding of head motion. Specifically, the three SCCs (anterior, posterior, horizontal) sit nearly orthogonal to each other, whereas the utricle is oriented horizontally and the saccule is oriented vertically. The vestibular end organs are innervated by the vestibular branch of the vestibulocochlear nerve (CN VIII). The horizontal and anterior SCCs, utricle, and portions of the saccule are innervated by the superior division of the vestibular nerve, and the posterior SCC and the majority of the saccule are innervated by the inferior division of the vestibular nerve.

The VOR produces compensatory eye movements to maintain gaze stability. The VOR pathway includes the sensory receptors in the peripheral vestibular system (SCCs or otolith organs), a central integrator (i.e., the vestibular nuclei in the brainstem), and a motor output (i.e., the extraocular muscles). The angular VOR is driven by inputs from the SCCs and capable of encod-

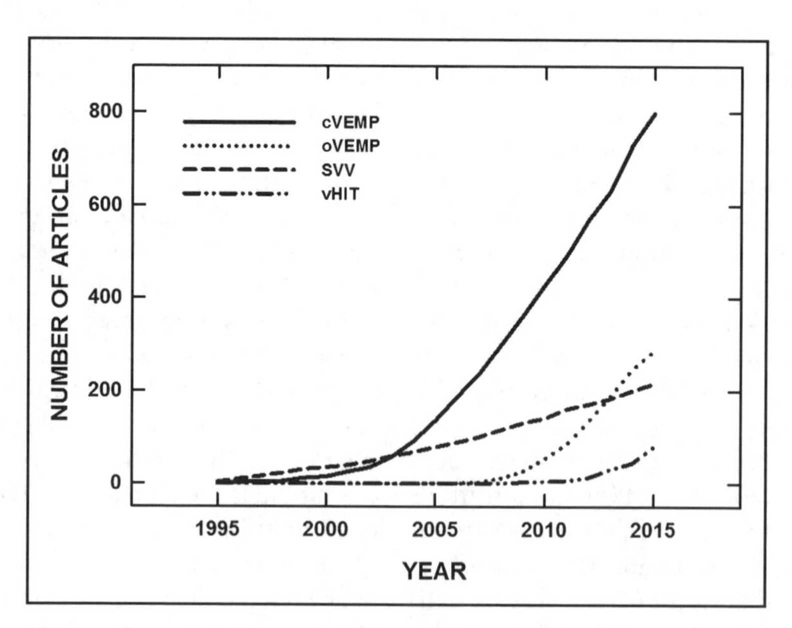

Figure 4–1. Cumulative frequency of articles published on cervical vestibular evoked myogenic potential (cVEMP), ocular vestibular evoked myogenic potential (oVEMP), subjective visual vertical (SVV), and video head impulse test (vHIT) as indexed in PubMed between 1995 and 2015.

ing motion across a range of frequencies. Acting similar to a high pass filter, frequencies between 1 to 6 Hz are optimally encoded (Minor et al., 1999). The SCCs are oriented orthogonally to each other and each SCC is paired with a complementary SCC on the opposite side of the head (i.e., the right and left horizontal SCC, the right anterior SCC and left posterior SCC, and the left anterior and right posterior SCCs). These co-planar pairs operate in a push-pull manner such that when one SCC of the pair is excited (i.e., increased neural firing), the other is inhibited (i.e., decreased neural firing). The sensory information from the peripheral end organ is relayed to the vestibular nuclei and ultimately the extraocular muscles via the reflex pathway. The activation of the extraocular muscles produces a compensatory eye movement equal and opposite to the head movement. The activation of the VOR is based on two important physiological principles (Ewald's laws): (1) Eye movements occur in the plane of the stimulated canal, and (2) excitatory responses have a larger dynamic range than inhibitory responses (e.g., Baloh et al., 1977; Cohen et al., 1964; Estes et al., 1975; Ewald, 1892; Goldberg & Fernandez, 1971). The vHIT, discussed below, is based on these two principles.

The otolith organs contribute to the translational/linear VOR as they are sensitive to linear acceleration including gravitational forces that act when the head is tilted (Goldberg & Fernandez, 1975). The activation pattern for the translational VOR is complex given the polarization vectors associated with the otolith organs and the modulation of the response based on viewing distance (e.g., Angelaki, 2004; Angelaki & McHenry, 1999; Lindeman, 1969; Tomko & Paige, 1992).

The otolith organs also provide significant sensory contributions to two descending pathways that produce changes in posture by contraction of antigravity muscles, which help to keep the head and body centered and stabilized: the medial vestibulospinal tract and the lateral vestibulospinal tract. Both tracts originate in the vestibular nuclei with the medial tract descending from the medial vestibular nucleus and the lateral tract from the lateral vestibular nucleus. The lateral vestibulospinal tract carries information throughout the entire length of the spinal cord in a highly organized rostral to caudal projection pattern. For example, the cervical regions receive input from the rostral region of the lateral vestibular nucleus, whereas lumbosacral regions receive input from the caudal region of the lateral vestibular nucleus (e.g., Baloh & Honrubia, 2001; Goldberg et al., 2012). Conversely, the medial vestibulospinal tract primarily travels along the descending medial longitudinal fasciculus and projects to interneurons of the cervical region of the spinal cord. Of particular interest for this chapter is the sacculocollic pathway as it relates to the recording of cVEMPs. In addition to sensing linear acceleration and gravity, the otolith organs have retained a phylogenetically conserved response to sound at high intensity levels. To record a cVEMP, the saccule is activated by sound and afferent information is relayed from the sensory end organ (saccule) via the inferior branch of the vestibulocochlear nerve to the lateral and medial vestibular nucleus. The efferent portion of the reflex travels along the VSR pathway and synapses at the accessory motor nucleus (CN XI), which, in turn, projects to the sternocleidomastoid muscle (Colebatch & Halmagyi, 1992; Fitzgerald, Comerfield, & Tuffery, 1982; McCue & Guinan, 1995; Murofushi & Curthoys, 1997).

Video Head Impulse Test

The video head impulse test (vHIT) assesses the dynamic function of parallel semicircular canal (SCC) pairs using abrupt, unpredictable, passive horizontal or vertical head rotations. The test is based on the bedside Head Impulse test (HIT) first described by Halmagyi and Curthoys (1988). Because the three parallel SCC pairs (left anterior and right posterior SCCs or LARP, the right anterior and left posterior SCCs or RALP, and the right and left horizontal SCCs) are nearly orthogonal to each other, a head impulse in the plane of one pair will stimulate mainly that pair and not the other two SCC pairs. Moreover, the asymmetric response of primary vestibular afferents dictates that the VOR during a canal-plane impulse toward a particular SCC is driven mainly by that SCC and not by its co-planar counterpart. For both anatomic and physiologic reasons, therefore, the vHIT assesses the function of each SCC separately. In healthy individuals, the angular VOR ensures gaze stability during head rotations by generating eye movements that are equal and opposite the head rotation. The gain of the VOR (eye velocity/head velocity) in healthy subjects, therefore, approaches unity. In contrast, the patient with a vestibular loss is unable to maintain gaze on the target during ipsilesional head rotation. Instead, the eyes move with the head (due to the deficient VOR) and are taken off target so that at the end of the head rotation, the patient makes a voluntary corrective saccade to refixate the visual target. The corrective saccade is visible to the examiner and, therefore, is called an overt saccade. The observation of an overt saccade during the bedside HIT is an indirect sign of horizontal SCC hypofunction on the side to which the head was rotated. The bedside HIT, therefore, is a subjective test and there is no objective measure of the corrective saccade or VOR gain. The outcome of the test is based on the examiner's subjective visual observation of the presence or absence of an overt saccade; if the corrective saccades occur *during* the head rotation (covert saccades), then it is unlikely that they will be observed by the examiner and thereby increase the likelihood of a false-negative result (Weber, MacDougall, Halmagyi & Curthoys, 2009). The false-negative rate of the bedside HIT in patients with peripheral vestibular disorders has been estimated at 14% based on the rate of occurrence of isolated covert saccades detected with the vHIT (Blödow et al., 2013). Similarly, a disassociation between the outcomes of the horizontal vHIT and the bedside HIT was observed in 32% of patients with peripheral vestibular disorders (Pérez-Fernández et al., 2012). The clinical application of the bedside HIT has largely been limited to the evaluation of horizontal SCC function and has not been used routinely for assessment of the vertical SCCs.

The eye movement response can be recorded during the HIT using either magnetic field scleral search coils or high-speed video-oculography (e.g., Aw et al., 1996; Cremer et al., 1998; MacDougall et al., 2009). Both methods provide a measure of VOR gain and record both covert and overt corrective saccades. Although the scleral search coil technique (Robinson, 1963) is considered the gold standard for recording eye movement, it has been used primarily as a research tool and is considered too invasive, expensive, and time intensive for routine clinical use. In contrast, the vHIT is noninvasive, safe, comfortable, fast, relatively inexpensive, and portable.

The vHIT instrumentation consists of a high-speed (~250 frames/s) monocular (or binocular) digital infrared video camera, a laptop computer, and software. Depending on the vHIT device, the camera is either embedded in head-worn goggles or mounted on a tripod facing the patient (e.g., Bartl et al., 2009; Blödow et al., 2013; MacDougall et al., 2009; MacDougall et al., 2013a, 2013b; Murnane et al., 2014; Pérez-Fernández et al., 2012; Ulmer & Chays, 2005; Ulmer et al., 2011; Weber et al., 2009). Head movement is recorded by an inertial measurement unit (triaxial linear accelerometer and gyroscopes) mounted on the head-worn goggles, or the change in the angle of head position during the head impulse is recorded by an external camera. The prototype of at least one vHIT device (Otometrics Impulse) has been validated with comparable results obtained for simultaneous video and magnetic field scleral search coil eye movement recordings in normal controls and in selected patients with well-defined vestibular losses for head impulses in the horizontal and vertical planes (MacDougall et al., 2009; MacDougall et al., 2013a, 2013b).

The stimulus for the vHIT consists of a manual, passive (clinician moves patient's head), unpredictable, brisk head rotation with a peak angular velocity of ~150 to 400°/s and a peak angular acceleration of ~2,000 to 6,000°/s^2. To perform the vHIT, patients are seated and instructed to maintain their gaze on an earth-fixed visual target located at a distance of ~1 m straight ahead at eye level. For vHIT devices employing head-worn goggles, eye position is calibrated using projected targets from a goggle-mounted laser immediately prior to testing. The clinician stands behind the patient, places both hands on top of the head (or on each side of the face at the jaw line), and manually rotates the head abruptly and unpredictably to the left or right through a small angle (10 to 20°) in the horizontal plane to stimulate the left and right horizontal SCCs. To test either of the co-planar vertical canal pairs (RALP or LARP), the patient's head is first turned ~30 to 40° relative to the trunk, which aligns the vertical canal pair with the trunk's sagittal plane (Migliaccio & Cremer, 2011). The clinician places one hand on top of the head and the other hand under the chin and rotates the head either forward (stimulates the anterior canal) or back (stimulates the posterior canal).

The current commercially available vHIT devices provide the examiner with immediate visual and/or audio feedback regarding the adequacy of the head impulse. If the head impulse meets the criterion of the manufacturer's data collection/processing algorithm, then the eye movement response is accepted and the gain of the VOR is calculated. In general, vHIT data collection/processing algorithms include a minimum velocity/acceleration for the head impulse and maintenance of eye tracking during the head impulse. It is important, however, for clinicians to manually inspect the "accepted" head velocity and eye velocity waveforms for the presence of artifacts (Mantokoudis et al., 2015). VOR gain is calculated as either position gain or velocity gain. Position gain has been calculated as the ratio of the area under the desaccaded eye velocity waveform (from the start of the head impulse to the second zero-crossing) to the area under the head velocity waveform during the same time interval (MacDougall et al., 2009; MacDougall et al., 2013a, 2013b). Velocity gain is calculated as the ratio of eye velocity to head velocity measured at single post-head impulse onset times (e.g., Mossman et al., 2015). Figure 4–2 shows the head and eye velocity waveforms obtained from a normal subject.

Several recent studies have reported normative VOR gain data for horizontal (Li et al., 2015; Matiño-Soler et al., 2014; McGarvie et al., 2015; Mossman et al., 2015) and vertical (McGarvie et al., 2015) head impulses in relatively large samples of healthy community-dwelling individuals across a wide age span (~10 to 90+ years of age) and over a range of peak head velocities (~70 to ~250°/s). Overall, the results of these studies indicated a small decrease in VOR gain at high peak head velocities but no significant decrease in VOR gain

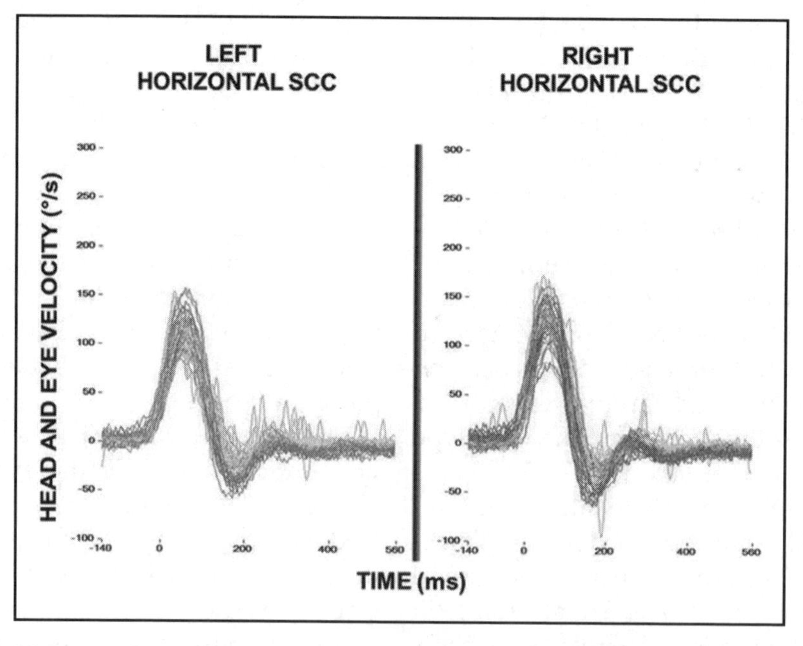

Figure 4–2. Head velocity (black) and corresponding eye velocity (gray) waveforms obtained for multiple head impulses in the horizontal plane from a normal subject. Head and eye velocity waveforms obtained for leftward horizontal head impulses are on the left and head (red) and eye velocity waveforms obtained for rightward horizontal head impulses are on the right. The mean (±SD) VOR gains were 0.92 (0.08) and 1.03 (0.06) for the left and right horizontal head impulses, respectively.

as a function of age. The average VOR gains collapsed across age ranged from ~0.94 to 1.06 for the horizontal SCCs with slightly lower average values for the vertical SCCs. The lower limit (mean VOR gain minus 2 SD) for normal VOR gain was ~0.80 for the horizontal SCCs and ~0.60 to ~0.70 for the vertical canals.

The vHIT has increasingly been used in the assessment of patients with vestibular disorders, including vestibular neuritis, Ménière's disease, benign paroxysmal positional vertigo, vestibular schwannoma, vestibular migraine, and bilateral vestibular loss (e.g., Albernaz & Maia, 2014; Bartolomeo et al., 2013; Blödow et al., 2013; Curthoys, 2012; Pérez- Fernández et al., 2012; Taylor et al., 2015; Ulmer et al., 2011; Walther & Blödow, 2013; Zulueta-Santos et al., 2014). Figure 4–3 shows the head velocity and corresponding eye velocity waveforms for multiple head impulses in the horizontal plane obtained from a patient with an uncompensated right

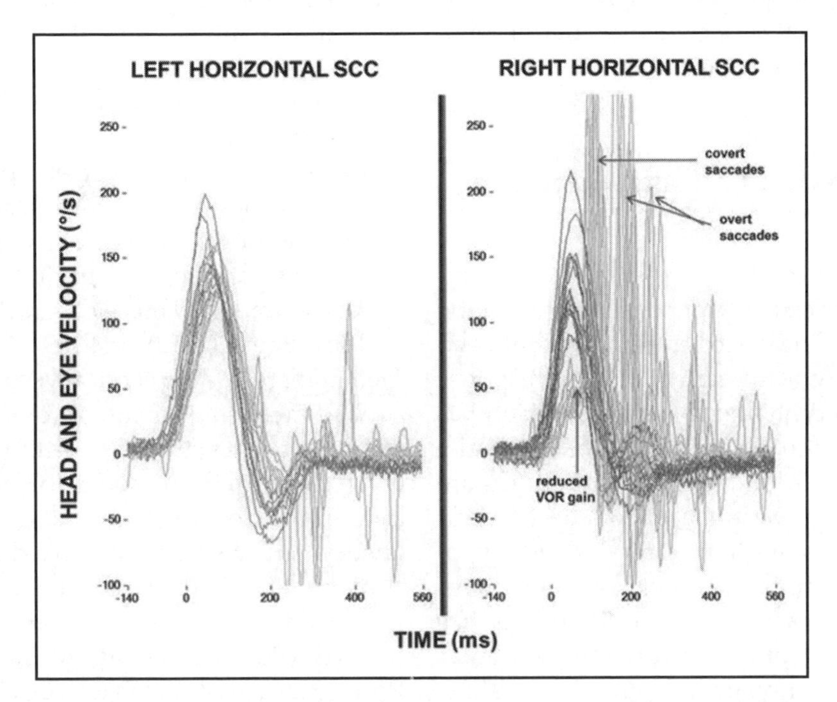

Figure 4–3. Head velocity (black) and corresponding eye velocity (gray) waveforms for multiple head impulses in the horizontal plane obtained from a patient with an uncompensated right unilateral vestibular loss. Patient had left-beating spontaneous nystagmus, 100% unilateral caloric weakness AD, and an asymmetry AD on rotary chair testing. Head and eye velocity waveforms obtained for left horizontal head impulses are on the left and head and eye velocity waveforms obtained for right horizontal head impulses are on the right. VOR gain was WNL for left horizontal head impulses, whereas gain was reduced for right (ipsilesional) horizontal head impulses. In addition, both overt and covert saccades (*arrows*) were recorded for right horizontal head impulses.

unilateral vestibular loss. The vHIT has been used successfully in the vestibular assessment of children as young as 3 years of age (Hamilton et al., 2015; Hülse et al., 2015) and has proved to be a sensitive test in discriminating central (stroke) from peripheral (vestibular neuritis) causes of the acute vestibular syndrome in the emergency department (Newman-Toker et al., 2008; Newman-Toker, Kerber, et al., 2013; Newman-Toker, Tehrani, et al., 2013). The application of the vHIT in combination with cervical and ocular vestibular evoked myogenic potentials makes it possible to assess the function of all five vestibular sensory organs.

Tests of Otolith Function

The two otolith organs (the saccule and the utricle) provide sensory information about linear acceleration, head tilt, and gravity. Although the VOR response to head tilt and linear acceleration has been used experimentally as a measure of otolith function, these methods are often cumbersome, uncomfortable, and not sensitive to unilateral otolith hypofunction. The clinical assessment of otolith organ function, therefore, has focused primarily on the vestibular evoked myogenic potential, a short latency electromyogram produced by the otolithic response to sound or vibration. Behavioral and neurophysiologic studies demonstrate that the vestibular system, the saccule in particular, is sensitive to sound. In fact, the saccule serves as the organ of hearing in several nonmammalian species (Moffat & Capranica, 1976; Popper & Fay, 1973). Studies in mammals have demonstrated that otolith irregular afferents are activated by high-level air-conducted sound and bone conduction vibration (e.g., Curthoys et al., 2006; Curthoys et al., 2012; McCue & Guinan, 1994, 1995, Murofushi et al., 1996a; Murofushi & Curthoys, 1997; Young et al., 1977). In contrast, very few (if any) canal regular afferents respond to sound or vibration in animals with intact labyrinths (Carey et al., 2004; Curthoys et al., 2006; Zhu et al., 2011). In mammals, saccular afferents have a strong projection to cervical muscles via the sacculocollic reflex and a weaker projection to the ocular motor system. In contrast, utricular neurons have strong projections to extraocular motor neurons (for review, see Uchino & Kushiro, 2011).

Vestibular evoked myogenic potentials are vestibular responses to sound or vibration originating from the otolith organs and recorded from the cervical or extraocular muscles. Colebatch and Halmagyi (1992) first described a method of recording cervical vestibular evoked myogenic potentials (cVEMPs) from surface electrodes over the tonically contracted sternocleidomastoid (SCM) muscles. More recently, short-latency ocular vestibular evoked myogenic potentials (oVEMPs) have been recorded from surface electrodes below the eyes (Rosengren et al., 2005; Todd, Rosengren, Aw, & Colebatch, 2007).

Cervical Vestibular Evoked Myogenic Potentials (cVEMPs)

cVEMPs are evoked by either air conduction or bone conduction acoustic stimuli, skull taps, or galvanic stimulation and characterized by a short-latency biphasic waveform. Sustained contraction of the SCM muscle is required during cVEMP acquisition because the response is elicited during a brief period of inhibition or interruption of the underlying motor unit firing of the SCM muscle (Colebatch & Rothwell, 2004). The cVEMP is dependent on the integrity of vestibular afferents as the response is abolished after inferior vestibular nerve section but preserved in subjects with severe-to-profound sensorineural hearing loss (e.g., Colebatch & Halmagyi, 1992; Colebatch, Halmaggyi; & Skuse, 1994). Evidence in humans indicates that the saccule is the sensory receptor for the cVEMP as it remains intact in individuals who have discrete genetic mutations of the cochlea and semicircular canals but normal otolith organs (e.g., Sheykholeslami & Kaga, 2002). Using an intraoperative procedure in humans, Basta et al. (2005) demonstrated that electrical stimulation of the inferior vestibular nerve (but not the superior vestibular nerve) evoked a cVEMP ipsilateral to the stimulation. The cVEMP is mediated by a predominantly ipsilateral reflex pathway that includes transduction at the saccule, the inferior branch of the vestibular nerve, the vestibular nuclei, the descending medial vestibulospinal tract, the accessory nucleus, the accessory nerve, and the motor neurons of the SCM muscle.

The electrode montage used to record cVEMPs includes a noninverting electrode on the upper one third to midpoint of the SCM muscle, an inverting electrode at the sternoclavicular junction, and a ground electrode on the forehead. This electrode configuration results in a biphasic waveform characterized by a positive peak (P1 or p13) at ~13 ms and a negative peak (N1 or n23) at ~23 ms poststimulus onset (Figure 4–4). The SCM muscle

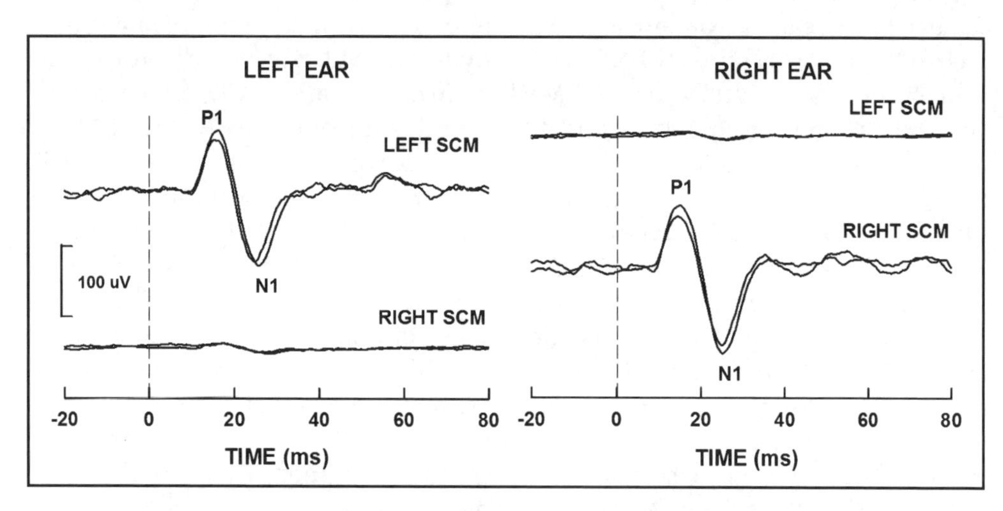

Figure 4–4. Two-channel cervical vestibular evoked myogenic potentials recorded in a healthy individual using 500-Hz tone bursts presented at 89 dB nHL via air conduction. The left panel shows cVEMPs recorded during unilateral activation of the left SCM muscle and the air conduction stimulus presented to the left ear. The right panel shows cVEMPs recorded during unilateral activation of the right SCM muscle and the air conduction stimulus presented to the right ear. The dashed lines indicate the stimulus onset. The response was bandpass filtered using a 10-Hz high-pass cutoff and a 1-Hz low-pass cutoff.

is activated unilaterally with a head turn or bilaterally with the head lifted while the subject is in the supine position. In healthy individuals, the largest cVEMP amplitudes for air conduction (AC) stimuli are obtained for tone burst frequencies from 400 to 1,000 Hz (e.g., Akin, Murtname, & Proffitt, 2003; Park et al., 2010; Todd et al., 2000; Welgampola & Colebatch, 2001), and the frequency characteristics are consistent with the neurophysiologic finding that acoustically responsive afferent fibers in the mammalian inferior vestibular nerve are most sensitive to frequencies between 500 and 1,000 Hz (McCue & Guinan, 1995; Murofushi, Matsuzaki, & Wu, 1999). The cVEMP amplitude is proportional to both stimulus level and the EMG level produced during the SCM muscle activation; however, cVEMP latency is independent of both factors (e.g., Akin et al., 2003; Akin et al., 2004; Colebatch et al., 1994; Lee Lim, Clouston, Sheean, & Yiannikas, 1995; Ochi, Ohashi, & Nishino, 2001).

In healthy individuals, AC cVEMP thresholds range from 100 to 120 dB pSPL for 500-Hz tone bursts (Akin et al., 2003; Isaradisaikul et al., 2012; Janky & Shepard, 2009; Maes et al., 2009). A normal middle ear system is required to conduct the AC stimulus to the inner ear and stimulate the saccular macula. AC cVEMPs, therefore, are typically reduced or absent in patients with conductive hearing loss, whereas BC cVEMPs are typically present in these patients (Sheykholeslami, Murofushi, Kermany, & Kaga, 2000; Welgampola, Rosengren, Halmagyi, & Colebatch, 2003; Zhou et al., 2012).

Due to the large intersubject variability of the cVEMP amplitude, clinical interpretation of the cVEMP has focused primarily on amplitude or threshold asymmetries between the right and left ears as an indication of the likely side of pathology. For interaural comparisons, the cVEMP amplitude is corrected by dividing the absolute peak-to-peak amplitude by the mean rectified EMG or by ensuring symmetrical SCM activation using a biofeedback method during the recording (e.g., Akin & Murnane, 2001; Colebatch et al., 1994).

The diagnostic utility of the cVEMP has been described for audiovestibular disorders, including vestibular neuritis, Ménière's disease, superior semicircular canal dehiscence (SSCD), and vestibular Schwannoma (e.g., Brantberg, Bergenius, & Tribukait, 1999; de Waele, Tran Ba Huy, Diard, Freyss, & Vidal, 1999; Matsuzaki, Murofushi, & Mizuno, 1999; Murofushi, Halmagyi, Yavor, & Colebatch, 1996b). An absent cVEMP or a significant interaural amplitude difference is the most common abnormality in vestibular-related disorders. The main exception is SSCD, which is associated with abnormally low cVEMP thresholds (e.g., Brantberg et al., 1999).

Although the clinical use of cVEMPs has focused primarily on peripheral vestibular disorders, cVEMP findings have also been described in individuals with neurologic pathology. Because the cVEMP pathway includes the vestibular nuclei and medial vestibulospinal tract, cVEMP abnormalities are most common in individuals with lower brainstem lesions but can also occur with some cerebellar lesions (Heide et al., 2010; Itoh et al., 2001; Kirchner et al., 2011; Pollak, Kushnir, & Stryjer, 2006). cVEMP abnormalities in individuals with central vestibular involvement are most often characterized as absent or delayed responses (Oh et al., 2015). In individuals with multiple sclerosis, prolonged latencies are the most common finding and decreased amplitude is less common (Alpini, Pugnetti, Caputo, & Cesarani, 2005; Gabelic et al., 2013; Shimizu, Murofushi, Sakurai, & Halmagyi, 2000; Versino et al., 2002).

cVEMP has been used to examine the impact of noise, blast exposure, and traumatic brain injury on the vestibular system. Studies in humans suggest that the saccule may be particularly susceptible to noise or blast-related damage (Akdogan et al., 2009; Akin et al., 2012; Hsu et al., 2008; Kerr & Byrne, 1975; Wang et al., 2006). Emerging evidence suggests that cVEMPs are more likely to be absent in individuals with more severe noise-induced hearing loss than in individuals with mild noise-induced hearing loss (Akin et al., 2012). Similarly, recent studies have demonstrated that absent cVEMPs are prevalent in individuals who experience dizziness or balance problems following a head injury and occur more often than horizontal SCC dyfunction (Akin & Murnane, 2011; Lee et al., 2011).

Age-related anatomic and physiologic degenerative changes in the vestibular system are well documented, and numerous studies using cVEMP have demonstrated a functional age-related change to the sacculocollic pathway (e.g., Akin et al., 2011; Basta et al., 2005; Johnsson, 1971; Lopez, Honrubia, & Baloh, 1997; Richter, 1980; Su, Huang, Young, & Cheng, 2004; Welgampola & Colebatch, 2001b). In individuals over the age of 60, cVEMP thresholds are higher and the response rate (presence) and amplitude declines (e.g., Akin et al., 2011; Maes et al., 2010; Singh et al., 2014; Su et al., 2004; Welgampola & Colebatch, 2001b). The effect of age on cVEMP latency is less clear as several studies have shown no effect of age (Akin et al., 2011; Basta et al., 2005; Su et al., 2004; Welgampola & Colebatch, 2001b), whereas others have shown an age-related increase in latency (Brantberg et al., 2007; Lee et al., 2008; Maes et al., 2010; Singh et al., 2014).

To date, there have been ~800 peer-reviewed published articles on the cVEMP over the past 20 years; however, there are several obstacles to widespread clinical use in the United States (Nelson et al., 2016). First, although the cVEMP is easily recorded using commercially available evoked potential instrumentation, measurement of SCM muscle electromyography is necessary to provide valid interaural comparisons of cVEMP amplitude. Several evoked potential manufacturers have developed software to monitor and/or correct for asymmetrical SCM muscle activation; and, the marketing of one of these instruments has only recently been approved by the U.S. Food and Drug Administration (FDA). Furthermore, reimbursement for cVEMP as a clinical procedure has been impacted by the delay in FDA approval. Finally, it is unclear if abnormal otolith organ function impacts postural stability and participation in activities of everyday living (Basta et al., 2005; Fujimoto et al., 2010; McCaslin et al., 2011; Murray et al., 2007).

Ocular Vestibular Evoked Myogenic Potentials (oVEMPs)

oVEMPs are recorded from surface electrodes below the eyes and characterized by an initial negative (excitatory) peak at 10 to 12 ms and a subsequent positive peak at 15 to 20 ms (Rosengren et al., 2005; Rosengren, Jombik, Halmagyi, & Colebatch, 2009; Todd, Rosengren, Aw, & Colebatch, 2007). Similar to the cVEMP, the oVEMP is a myogenic response that is recorded by activating the otoliths with either an air conduction, bone conduction, or galvanic stimulus and is present in individuals with profound sensorineural hearing loss but intact vestibular function and absent in patients with vestibulopathy (Chihara, Iwasaki, Ushio, & Murofushi, 2007; Iwasaki et al., 2007; Iwasaki et al., 2009; Rosengren et al., 2005). The largest oVEMPs in response to AC stimuli have been obtained from electrodes located just beneath the eye contralateral to the stimulus ear with the patient's gaze fixed upward on a stationary visual target (Govender, Rosengren, & Colebatch, 2009; Murnane et al., 2011). The oVEMP is absent in patients with exenteration (removal) of the eye and extraocular muscles but present in patients with exenteration of the eye but intact extraocular muscles (Chihara et al., 2009). During simultaneous recording of the oVEMP in humans using needle (single-unit) and surface electrodes, Weber et al. (2012) determined that the myogenic origin of the initial negative peak of the oVEMP is excitation of the inferior oblique extraocular muscles. In addition, they confirmed that the oVEMP was mediated by a crossed pathway. These results indicate that the oVEMP is produced by synchronous excitatory activity in the extraocular muscles and mediated by a predominantly contralateral otolith-ocular pathway.

The use of bone conduction (BC) vibration to elicit oVEMPs (BC oVEMPs) has gained attention because the BC oVEMPs have a higher prevalence rate and larger amplitude than AC oVEMPs (Rosengren et al., 2011). An audiometric BC oscillator (Radioear B-71) applied to the mastoid was initially used to elicit BC oVEMPs; however, larger responses have subsequently been obtained using a Bruel and Kjaer 4810 Mini-Shaker applied to the midline forehead at Fz or a tendon hammer used to tap the midline forehead (e.g., Iwasaki et al., 2007; Rosengren et al., 2005; Rosengren et al., 2011; Todd, Rosengren, & Colebatch, 2008). Bone conduction vibration at Fz produces a linear acceleration (jerk stimulus) at both mastoids (Iwasaki et al., 2008). In contrast to the oVEMPs obtained using monaural AC stimuli, bone conduction stimuli applied to Fz produce bilaterally symmetric oVEMPS in normal controls (Figure 4–5; e.g., Iwasaki et al., 2008; Rosengren et al., 2011). Similar to the oVEMPs obtained using monaural AC stimuli and consistent with a crossed reflex pathway, BC oVEMPs (stimulus applied to midline forehead) are absent from the eye contralateral to the lesion in patients with unilateral vestibular neurectomy (e.g., Iwasaki et al., 2007).

Human and animal studies suggest that BC oVEMPs are predominantly mediated by the utricle and the superior division of the vestibular nerve. The superior vestibular nerve contains all utricular afferents and very few saccular afferents (De Burlet, 1929). Clinical studies have demonstrated that BC oVEMPs are absent in patients with vestibular neuritis affecting the superior vestibular nerve (e.g., Iwasaki et al.,

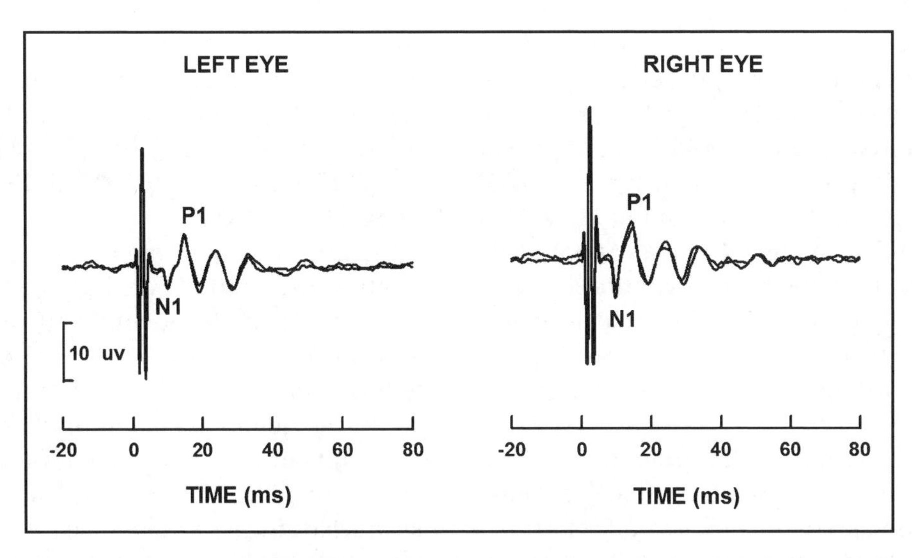

Figure 4–5. Bone conduction (BC) ocular vestibular evoked myogenic potentials (oVEMPs) recorded in a healthy individual during upward gaze. The BC stimulus was a 500-Hz tone burst (157 dB peak FL) produced with a Bruel and Kjaer Mini-Shaker (model 4810) placed at Fz. The left panel shows oVEMPs recorded from the left eye, and the right panel shows the response recorded from the right eye using a maximum vertical gaze elevation. The large amplitude artifact in the recording at 0 s represents the stimulus onset. The response was bandpass filtered using a 10-Hz high-pass cutoff and a 1,000-Hz low-pass cutoff.

2009; Manzari et al., 2010). In contrast, AC cVEMPs are often intact in patients with superior vestibular neuritis (Govender et al., 2011). In addition, Curthoys and colleagues (2006) demonstrated that bone conduction vibration activate otolithic vestibular neurons in the utricular macula. These findings provide support for a utricular/superior vestibular nerve origin for BC oVEMPs and suggest that the response provides complementary diagnostic information relative to the AC cVEMP.

Similar to the cVEMP, AC and BC oVEMP amplitude and threshold are dependent on the stimulus level and EMG level (i.e., magnitude of the vertical gaze angle) (Murnane et al., 2011; Rosengren et al., 2013). In contrast to cVEMPs, however, oVEMP does not require correction or monitoring of background EMG level. Recorded during upward gaze in the midline, the extraocular EMG level and the relationship between the recording electrodes and the relevant muscles is likely comparable for the two sides.

Because the oVEMP is mediated by a predominantly crossed pathway, patients with vestibular disorders may have absent or decreased amplitude oVEMP recorded beneath the eye contralateral to the stimulated ear (e.g., Chihara et al., 2007). In patients with SSCD, oVEMPs have abnormally large suprathreshold amplitudes and abnormally low thresholds (Rosengren et al., 2005; Rosengren, Aw, Halmagyi, Todd,

& Colebatch, 2008) showed that oVEMP amplitude provided better separation between normals and SSCD patients than oVEMP threshold. In a comparison study examining oVEMP and cVEMP responses to various stimuli, Janky et al. (2013) demonstrated that AC oVEMPs provided the best separation between patients with SSCD and healthy controls. oVEMP does not require correction or monitoring of background EMG level.

In summary, the preponderance of evidence suggests that cVEMPs obtained with air conduction stimuli reflect primarily saccular and inferior vestibular nerve function, whereas oVEMPs obtained with bone conduction and air conduction stimuli reflect primarily utricular and superior vestibular nerve function. The recording of BC oVEMPs for clinical testing presents several instrumentation-related obstacles, including the limited output of the audiometric B-71 oscillator and the calibration of the Bruel and Kjaer Mini-Shaker and tendon hammers.

Subjective Visual Vertical and Centrifugation

The subjective visual vertical (SVV) is a psychophysical measure that has been used to assess the perception of spatial orientation or tilt in individuals with peripheral and central vestibular disorders. SVV can be assessed by asking an individual to adjust a luminous line to the vertical or horizontal (subjective visual vertical or horizontal) in a lightproof environment. The SVV is calculated as the angle between the line (perceptual vertical) and true vertical (gravitational vertical). In an upright static position, healthy individuals align the linear marker within ~2° of true vertical (0°) (e.g., Akin et al., 2011; Böhmer & Rickenmann, 1995; Brodsky et al., 2015; Friedman, 1970; Neal, 1926; Zwergal et al., 2009).

Because the otolith organs serve as gravity sensors and contribute to the perception of spatial orientation (earth verticality), a unilateral vestibular deafferentation causes a tonic imbalance in the neuronal resting discharge that results in an ocular torsion (a torsional deviation of the eyes) toward the affected side (Curthoys et al., 1991). Thus, the otolith organs of the contralateral (healthy) ear "push" the SVV up to 20° to the opposite, diseased side (e.g., Böhmer & Rickenmann, 1995).

There are several limitations to the clinical use of the static SVV (i.e., performed in an upright stationary position). First, the SVV is a spatial orientation task that is dependent on multimodal sensory input; therefore, static SVV can be affected by factors other than peripheral vestibular disease. Furthermore, patients with brainstem or cerebellar lesions can show ocular torsion and offset of the static SVV (Dietrich & Brandt, 1993; Mossman & Halmagyi, 1997). In general, lower brainstem lesions involving the vestibular nucleus can result in an ipsilesional tilt of the SVV, whereas upper brainstem lesions involving the interstitial nucleus of Cajal or cerebellar lesions can result in contralesional tilts (e.g., Baier et al., 2012; Dieterich & Brandt, 1993). In addition, abnormal static SVV following unilateral vestibular hypofunction is analogous to spontaneous nystagmus of the VOR/ semicircular canals (Böhmer & Rickenmann, 1995; Curthoys, Dai, & Halmagyi, 1991). That is, static SVV returns to normal (≤2°) following vestibular compensation; therefore, the test is less sensitive to chronic vestibular disorders than acute vestibular disorders (e.g., Böhmer & Rickenmann, 1999).

To improve the sensitivity of the SVV test, off-axis rotation has been used to unilaterally stimulate the otolith organs using centrifugation during measurement of the SVV. To perform this test, an individual is positioned upright and rotated at a constant velocity (240–400°/s) with the test ear positioned off-axis (typically 7–8 cm) and the nontest ear positioned on-axis (0 gravito-inertial force or GIF) (Figure 4–6). In an upright position, the utricle is positioned horizontally (3.5–4 cm from midline) and senses linear acceleration in the interaural plane. During constant velocity rotation, the VOR response from the horizontal semicircular canals extinguishes and the off-axis rotation creates a centrifugal force (or linear acceleration) that stimulates the utricle positioned off-axis. That is, only the test (off-axis) ear is stimulated during centrifugation.

During unilateral centrifugation, the SVV tilts away from the side that is positioned off-axis. That is, the SVV tilts leftward when an individual is offset to the right of the axis of rotation, and the SVV tilts rightward when the head is offset left. In healthy individuals, the magnitude of the SVV tilt is opposite and symmetrical for offset right and offset left centrifugation. Wetzig and colleagues (1990) demonstrated that the magnitude of the SVV was proportional to the eccentricity of the rotary chair and that the direction of the chair's rotation (i.e., clockwise vs. counterclockwise) did not change the direction or magnitude of the SVV.

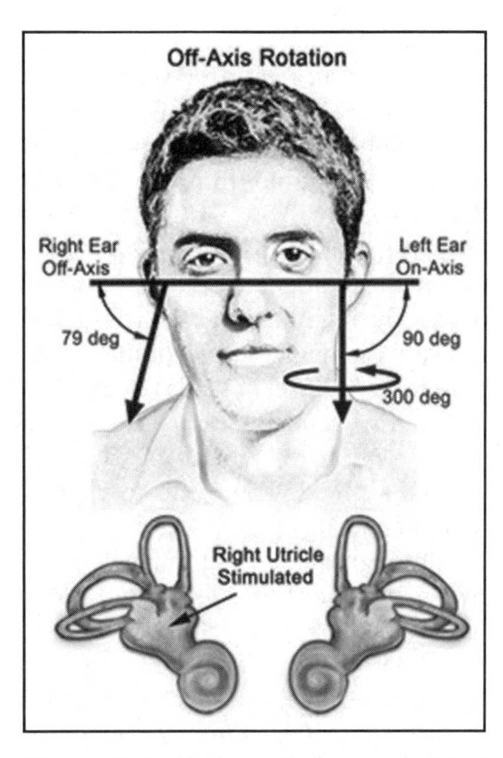

Figure 4–6. A schematic representation of utricular stimulation during off-axis rotation (unilateral centrifugation) for measurement of the subjective visual vertical. Constant off-axis rotation (unilateral centrifugation) with the right ear offset 7 cm from the center and the nontest ear positioned on center; only the right ear is stimulated, resulting in increased GIF to the test ear compared to the nontest (left) ear (right utricle: GIF = 999.3 cm/s^2 or 1.019 g; left utricle: GIF = 980.7 cm/s^2 or 1 g) resulting in a perceived gravity angle of 11° relative to vertical (90° – 11° = 79°) on the right (test) side and 0° (90° – 0° = 90°) on the left (nontest) side. Adapted from Clark et al. (2003).

Individuals with unilateral vestibular hypofunction exhibit an SVV asymmetry during stimulation of the involved side with off-axis yaw rotation (Böhmer & Mast, 1999; Clarke, Shönfeld, Hamann, & Scherer, 2001). Specifically, when the lesioned ear is centrifuged, the SVV does not shift because the utricle does not respond to the GIF (Wetzig et al., 1990). The SVV during unilateral centrifugation, therefore, may be a more sensitive test of otolith organ function than the static SVV test (Clarke et al., 2001). Schönfeld and Clark (2011) proposed that patient response patterns of SVV during unilateral centrifugation indicate the degree of vestibular compensation. Valko et al. (2011) found that SVV during centrifugation was slightly less sensitive than bone conduction oVEMPs in determining utricular dysfunction.

Although the normal ranges for the SVV in the static position are well established, there are few published normative data for the SVV test during unilateral centrifugation (Akin et al., 2011; Clark et al., 2001; Janky & Shepard, 2011; Schönfeld & Clark, 2011). Furthermore, the high cost of the off-axis rotational chair is prohibitive for some clinical settings. Other methods used to stimulate the vestibular system during SVV have included static head tilt and galvanic vestibular stimulation (e.g., Valko et al., 2011; Zink et al., 1998).

Although new otolith organ tests are available, they are used less widely than horizontal canal tests, and one reason may be that it is unclear if abnormal otolith organ function impacts a person negatively in terms of maintaining balance, participating in activities of everyday living, and the incidence of falls. The standard of care for many patients with vestibular dysfunction is vestibular rehabilitation, which includes gaze stability exercises based primarily upon principles of vestibular adaptation of the VOR (or hSCC). There is some evidence that combined otolith and canal dysfunction does not negatively impact rehabilitation outcomes (Murray et al., 2010); however, individuals with isolated otolith dysfunction were not included in the study. Further work is needed to determine the effect of otolith organ dysfunction on postural stability, the implications for clinical recommendations, and rehabilitation strategies.

Conclusion

Until recently, vestibular clinical assessment was limited to electronystagmography (or videonystagmography) and rotary chair testing to detect unilateral or bilateral vestibular loss of the horizontal semicircular canal/superior vestibular nerve. New tests of vestibular function supplement the vestibular test battery by providing diagnostic information across vestibular sensory organs and pathways. Specifically, the vHIT provides a clinical measure of VOR function for all six semicircular canals. cVEMPs using air conduction stimuli reflect saccular and inferior vestibular nerve function, and oVEMPs using both air and bone conduction stimuli likely reflect utricular and superior vestibular nerve function. Similarly, the SVV test during off-axis rotation may be a useful method to assess utricular function in patients complaining of dizziness and/or imbalance.

Peripheral vestibular loss can occur in one or both labyrinths, in one or both branches of the vestibular nerve, and in one or more vestibular sensory organs. It has been demonstrated that otolith abnormalities do not always correlate with semicircular canal involvement (e.g., Ernst et al., 2005; Iwasaki et al., 2015). That is, vestibular pathology can affect the otoliths (abnormal SVV and/or VEMP) and spare the horizontal semicircular canal (normal calorics), or vice versa. In addition, a vestibular pathology may affect both otolith and horizontal semicircular canal function. These findings are intriguing because both the horizontal semicircular canal and the utricle are innervated by the superior branch of the vestibular nerve. Thus, the addition of tests of otolith function and vHIT to the traditional vestibular test battery may provide the ability to differentiate vestibular end organ from vestibular nerve dysfunction.

References

Agrawal, Y., Bremova, T., Kremmyda, O., & Strupp, M. (2013). Semicircular canal, saccular and utricular function in patients with bilateral vestibulopathy: Analysis based on etiology. *Journal of Neurology, 260*(3), 876–883.

Akdogan, O., Selcuk, A., Take, G., Erdoğan, D., & Dere, H. (2009). Continuous or intermittent noise exposure, does it cause vestibular damage? An experimental study. *Auris Nasus Larynx, 36*(1), 2–6.

Akin, F. W., & Murnane, O. D. (2001). Vestibular evoked myogenic potentials: Preliminary report. *Journal of the American Academy of Audiology, 12*(9), 445–452.

Akin, F. W., & Murnane, O. D. (2011). Head injury and blast exposure: Vestibular consequences. *Otolaryngologic Clinics of North America, 44*(2), 323–334.

Akin, F. W., Murnane, O. D., Panus, P. C., Caruthers, S. K., Wilkinson, A. E., & Proffitt, T. M. (2004). The influence of voluntary tonic EMG level on the vestibular-evoked myogenic potential. *Journal of Rehabilitation Research and Development, 41*(3B), 473–480.

Akin, F. W., Murnane, O. D., Pearson, A., Byrd, S., & Kelly, J. K. (2011). Normative data for the subjective visual vertical test during centrifugation. *Journal of the American Academy of Audiology, 22*(7), 460–468.

Akin, F. W., Murnane, O. D., & Proffitt, T. M. (2003). The effects of click and tone-burst stimulus parameters on the vestibular evoked myogenic potential (VEMP). *Journal of the American Academy of Audiology, 14*(9), 500–509.

Akin, F. W., Murnane, O. D., Tampas, J. W., & Clinard, C. G. (2011). The effect of age on the vestibular evoked myogenic potential and sternocleidomastoid muscle tonic electromyogram level. *Ear and Hearing, 32*(5), 617–622.

Akin, F. W., Murnane, O. D., Tampas, J. W., Clinard, C., Byrd, S., & Kelly, J. K. (2012). The effect of noise exposure on the cervical vestibular evoked myogenic potential. *Ear and Hearing, 33*(4), 458–465.

Albernaz, P. L. M., & Maia, F. C. Z. (2014). The video head impulse test. *Acta Oto-Laryngologica, 134*(12), 1245–1250.

Alpini, D., Pugnetti, L., Caputo, D., & Cesarani, A. (2005). Vestibular evoked myogenic potentials in multiple sclerosis: A comparison between onset and definite cases. *International Tinnitus Journal, 11*(1), 48.

Angelaki, D. E. (2004). Eyes on target: What neurons must do for the vestibuloocular reflex during linear motion. *Journal of Neurophysiology, 92*(1), 20–35.

Angelaki, D. E., & Dickman, J. D. (2000). Spatiotemporal processing of linear acceleration: Primary afferent and central vestibular neuron responses. *Journal of Neurophysiology, 84*(4), 2113–2132.

Angelaki, D. E., & McHenry, M. Q. (1999). Short-latency primate vestibuloocular responses during translation. *Journal of Neurophysiology, 82*(3), 1651–1654.

Aw, S. T., Haslwanter, T., Halmagyi, G. M., Curthoys, I. S., Yavor, R. A., & Todd, M. J. (1996). Three-dimensional vector analysis of the human vestibuloocular reflex in response to high-acceleration head rotations: I. Responses in normal subjects. *Journal of Neurophysiology, 76*(6), 4009–4020.

Baier, B., Thömke, F., Wilting, J., Heinze, C., Geber, C., & Dieterich, M. (2012). A pathway in the brainstem for roll-tilt of the subjective visual vertical: Evidence from a lesion–behavior mapping study. *The Journal of Neuroscience, 32*(43), 14854–14858.

Baloh, R. W., & Honrubia, V. (2001). *Clinical neurophysiology of the vestibular system* (3rd ed.). New York, NY: Oxford University Press.

Baloh, R. W., Honrubia, V., & Konrad, H.R . (1977). Ewald's second law re-evaluated. *Acta Oto-Laryngologica, 83*(5–6), 475–479.

Bartl, K., Lehnen, N., Kohlbecher, S., & Schneider, E. (2009). Head impulse testing using video-nystagmography, *Annals of the New York Academy of Sciences, 1164*, 331–333.

Bartolomeo, M., Biboulet, R., Pierre, G., Mondain, M., Uziel, A., & Venail, F. (2013). Value of the video head impulse test in assessing vestibular deficits following vestibular neuritis. *European Archives of Oto-Rhino-Laryngology, 271*(4), 681–688.

Basta, D., Todt, I., & Ernst, A. (2005). Normative data for P1/N1-latencies of vestibular evoked myogenic potentials induced by air-or bone-conducted tone bursts. *Clinical Neurophysiology, 116*(9), 2216–2219.

Basta, D., Todt, I., Scherer, H., Clarke, A., & Ernst, A. (2005). Postural control in otolith disorders. *Human Movement Science, 24*(2), 268–279.

Blödow, A., Pannasch, S., & Walther, L. E. (2013). Detection of isolated covert saccades with the video head impulse test in peripheral vestibular disorders. *Auris Nasus Larynx, 40*(4), 348–351.

Böhmer, A., & Mast, F. (1998). Chronic unilateral loss of otolith function revealed by the subjective visual vertical during off center yaw rotation. *Journal of Vestibular Research: Equilibrium and Orientation, 9*(6), 413–422.

Böhmer, A., & Rickenmann, J. (1995). The subjective visual vertical as a clinical parameter of vestibular function in peripheral vestibular diseases. *Journal of Vestibular Research: Equilibrium and Orientation, 5*(1), 35–45.

Brantberg, K., Bergenius, J., & Tribukait, A. (1999). Vestibular-evoked myogenic potentials in patients with dehiscence of the superior semicircular canal. *Acta Oto-Laryngologica, 119*(6), 633–640.

Brantberg, K., Granath, K., & Schart, N. (2007). Age-related changes in vestibular evoked myogenic potentials. *Audiology and Neurotology, 12*, 247–253.

Brodsky, J. R., Cusick, B. A., Kenna, M. A., & Zhou, G. (2015). Subjective visual vertical testing in children and adolescents. *The Laryngoscope*. Advance online publication. doi:10.1002/lary.25389

Carey, J. P., Hirvonen, T. P., Hullar, T. E., & Minor, L. B. (2004). Acoustic responses of vestibular afferents in a model of superior canal dehiscence. *Otology and Neurotology, 25*(3), 345–352.

Chihara, Y., Iwasaki, S., Ushio, M., Fujimoto, C., Kashio, A., Kondo, K., . . . Murofushi, T. (2009). Ocular vestibular-evoked myogenic potentials (oVEMPs) require extraocular muscles but not facial or cochlear nerve activity. *Clinical Neurophysiology, 120*(3), 581–587.

Chihara, Y., Iwasaki, S., Ushio, M., & Murofushi, T. (2007). Vestibular-evoked extraocular potentials by air-conducted sound: Another clinical test for vestibular function. *Clinical Neurophysiology, 118*(12), 2745–2751.

Clarke, H., Schönfeld, U., Hamann, C., & Scherer, A. (2001). Measuring unilateral otolith function via the otolith-ocular response and the subjective visual vertical. *Acta Oto-Laryngologica, 121*(545), 84–87.

Cohen, B., Suzuki, J. I., Shanzer, S., & Bender, M. B. (1964). Semicircular canal control of eye movements. In M. B. Bender (Ed.), *The oculomotor system*, (pp. 163–177). New York, NY: Harper & Row.

Colebatch, J. G., & Halmagyi, G. M. (1992). Vestibular evoked potentials in human neck muscles before and after unilateral vestibular deafferentation. *Neurology, 42*(8), 263–274.

Colebatch, J. G., Halmagyi, G. M., & Skuse, N. F. (1994). Myogenic potentials generated by a click-evoked vestibulocollic reflex. *Journal of Neurology, Neurosurgery and Psychiatry, 57*(2), 190–197.

Colebatch, J. G., & Rothwell, J. C. (2004). Motor unit excitability changes mediating vestibulocollic reflexes in the sternocleidomastoid muscle. *Clinical Neurophysiology, 115*(11), 2567–2573.

Cremer, P. D., Halmagyi, G. M., Aw, S. T., Curthoys, I. S., McGarvie, L. A., Todd, M. J., . . . Black, R. A. (1998). Semicircular canal plan head impulses detect absent function of individual semicircular canals. *Brain, 121,* 699–716.

Curthoys, I. S. (1987). Eye movements produced by utricular and saccular stimulation. *Aviation, Space, and Environmental Medicine, 58*(9 Pt. 2), A192–A197.

Curthoys, I. S. (2012). The interpretation of clinical tests of peripheral vestibular function. *The Laryngoscope, 122*(6), 1342–1352.

Curthoys, I. S., Blanks, R. H. I., & Markham, C. H. (1977). Semicircular canal functional anatomy in cat, guinea pig and man. *Acta Oto-Laryngologica, 83*(1–6), 258–265.

Curthoys, I. S., Dai, M. J., & Halmagyi, G. M. (1991). Human ocular torsional position before and after unilateral vestibular neurectomy. *Experimental Brain Research, 85*(1), 218–225.

Curthoys, I. S., Kim, J., McPhedran, S. K., & Camp, A. J. (2006). Bone conducted vibration selectively activates irregular primary otolithic vestibular neurons in the guinea pig. *Experimental Brain Research, 175*(2), 256–267.

Curthoys, I. S., & Vulovic, V. (2011). Vestibular primary afferent responses to sound and vibration in the guinea pig. *Experimental Brain Research, 210*(3–4), 347–352.

Curthoys, I. S., Vulovic, V., Sokolic, L., Pogson, J., & Burgess, A. M. (2012). Irregular primary otolith afferents from the guinea pig utricular and saccular maculae respond to both bone conducted vibration and to air conducted sound. *Brain Research Bulletin, 89*(1), 16–21.

De Burlet, H. M. (1929). Zur vergleichenden Anatomie der Labyrinth innervation. *Journal of Comparative Neurology, 47,* 155–169.

de Waele, C., Huy, P. T. B., Diard, J. P., Freyss, G., & Vidal, P. P. (1999). Saccular dysfunction in Meniere's disease. *Otology and Neurotology, 20*(2), 223–232.

Dieterich, M., & Brandt, T. (1993). Ocular torsion and tilt of subjective visual vertical are sensitive brainstem signs. *Annals of Neurology, 33*(3), 292–299.

Ernst, A., Basta, D., Seidl, R. O., Todt, I., Scherer, H., & Clarke, A. (2005). Management of posttraumatic vertigo. *Otolaryngology-Head and Neck Surgery, 132*(4), 554–558.

Ernst, A., Todt, I., Seidl, R. O., Eisenschenk, A., Blödow, A., & Basta, D. (2006). The application of vestibular-evoked myogenic potentials in otoneurosurgery. *Otolaryngology-Head and Neck Surgery, 135*(2), 286–290.

Estes, M. S., Blanks, R. H., & Markham, C. H. (1975). Physiologic characteristics of vestibular first-order canal neurons in the cat: I. Response plane determination and discharge characteristics. *Journal of Neurophysiology, 38*(5), 1232–1249.

Ewald, R. (1892). *Physiologische Untersuchungen ueber das Endorgan des Nervus Octavus.* Wiesbaden, Germany: Bergmann.

Fernandez, C., & Goldberg, J. M. (1976). Physiology of peripheral neurons innervating otolith organs of the squirrel monkey: III. Response dynamics. *Journal of Neurophysiology, 39*(5), 996–1008.

Fitzgerald, G., & Hallpike, C. S. (1942). Studies in human vestibular function: I. Observations on the directional preponderance ("Nystagmusbereitschaft") of caloric nystagmus resulting from cerebral lesions. *Brain, 65,* 115–137.

Fitzgerald, M. J., Comerford, P. T., & Tuffery, A. R. (1982). Sources of innervation of the neuromuscular spindles in sternomastoid and trapezius. *Journal of Anatomy, 134*(Pt 3), 471.

Friedmann, G. (1970). The judgement of the visual vertical and horizontal with peripheral and central vestibular lesions. *Brain, 93*(2), 313–328.

Fujimoto, C., Murofushi, T., Chihara, Y., Ushio, M., Yamaguchi, T., Yamasoba, T., & Iwasaki, S. (2010). Effects of unilateral dysfunction of the inferior vestibular nerve system on postural stability. *Clinical Neurophysiology, 121*(8), 1279–1284.

Gabelic, T., Krbot, M., Šefer, A. B., Išgum, V., Adamec, I., & Habek, M. (2013). Ocular and cervical vestibular evoked myogenic potentials in patients with multiple sclerosis. *Journal of Clinical Neurophysiology, 30*(1), 86–91.

Goldberg, J. M., & Fernandez, C. (1971). Physiology of peripheral neurons innervating semicircular canals of squirrel monkey: I. Resting discharge and response to constant angular accelerations. *Journal of Neurophysiology, 34*(4), 635–660.

Goldberg, J. M., & Fernandez, C. (1975). Responses of peripheral vestibular neurons to angular and linear accelerations in the squirrel monkey. *Acta Oto-Laryngologica, 80*(1–6), 101–110.

Goldberg, J. M., Wilson, V. J., Cullen, K. E., Angelaki, D. E., Broussard, D. M., Büttner-Ennever, J. A., . . . Minor, L. B. (2012). *The vestibular system: A sixth sense.* New York, NY: Oxford University Press.

Govender, S., Rosengren, S. M., & Colebatch, J. G. (2009). The effect of gaze direction on the ocular vestibular evoked myogenic potential produced by air-conducted sound. *Clinical Neurophysiology, 120*(7), 1386–1391.

Govender, S., Rosengren, S. M., & Colebatch, J. G. (2011). Vestibular neuritis has selective effects on air- and bone-conducted cervical and ocular vestibular evoked myogenic potentials. *Clinical Neurophysiology, 122*(6), 1246–1255.

Halmagyi, G. M., & Curthoys, I. S. (1988). A clinical sign of canal paresis. *Archives of Neurology, 45*(7), 737–739.

Hamilton, S. S., Zhou, G., & Brodsky, J. R. (2015). Video head impulse testing (VHIT) in the pediatric population. *International Journal of Pediatric Otorhinolaryngology, 79*(8), 1283–1287.

Heide, G., Luft, B., Franke, J., Schmidt, P., Witte, O. W., & Axer, H. (2010). Brainstem representation of vestibular evoked myogenic potentials. *Clinical Neurophysiology, 121*(7), 1102–1108.

Herdman, S. J., Blatt, P., Schubert, M. C., & Tusa, R. J. (2000). Falls in patients with vestibular deficits. *Otology and Neurotology, 21*(6), 847–851.

Hsu, W. C., Wang, J. D., Lue, J. H., Day, A. S., & Young, Y. H. (2008). Physiological and morphological assessment of the saccule in guinea pigs after noise exposure. *Archives of Otolaryngology–Head & Neck Surgery, 134*(10), 1099–1106.

Hülse, R., Hörmann, K., Servais, J. J., Hülse, M., & Wenzel, A. (2015). Clinical experience with video head impulse test in children. *International Journal of Pediatric Otorhinolaryngology, 79*(8), 1288–1293.

Isaradisaikul, S., Navacharoen, N., Hanprasertpong, C., & Kangsanarak, J. (2012). Cervical vestibular-evoked myogenic potentials: Norms and protocols. *International Journal of Otolaryngology, 2012,* 913515.

Isu, N., Graf, W., Sato, H., Kushiro, K., Zakir, M., Imagawa, M., & Uchino, Y. (2000). Sacculo-ocular reflex connectivity in cats. *Experimental Brain Research, 131*(3), 262–268.

Itoh, A., Kim, Y.S., Yoshioka, K., Kanaya, M., Enomoto, H., Hiraiwa, F., & Mizuno, M. (2001). Clinical study of vestibular-evoked myogenic potentials and auditory brainstem responses in patients with brainstem lesions. *Acta Otolaryngologica, 545,* 116–119.

Iwasaki, S., Chihara, Y., Smulders, Y. E., Burgess, A. M., Halmagyi, G. M., Curthoys, I. S., & Murofushi, T. (2009). The role of the superior vestibular nerve in generating ocular vestibular-evoked myogenic potentials to bone conducted vibration at Fz. *Clinical Neurophysiology, 120*(3), 588–593.

Iwasaki, S., Fujimoto, C., Kinoshita, M., Kamogashira, T., Egami, N., & Yakisoba, T. (2015). Clinical characteristics of patients with abnormal ocular/cervical vestibular evoked myogenic potentials in the presence of normal caloric responses. *Annals of Otology, Rhinology, and Laryngology, 124*(6), 458–465.

Iwasaki, S., McGarvie, L. A., Halmagyi, G. M., Burgess, A. M., Kim, J., Colebatch, J. G., & Curthoys, I. S. (2007). Head taps evoke a crossed vestibulo-ocular reflex. *Neurology, 68*(15), 1227–1229.

Iwasaki, S., Smulders, Y. E., Burgess, A. M., McGarvie, L. A., MacDougall, H. G., Halmagyi, G. M., & Curthoys, I. S. (2008). Ocular vestibular evoked myogenic potentials to bone conducted vibration of the midline forehead at Fz in healthy subjects. *Clinical Neurophysiology, 119*(9), 2135–2147.

Janky, K. L., Nguyen, K. D., Welgampola, M., Zuniga, M. G., & Carey, J. P. (2013). Air-conducted oVEMPs provide the best separation between intact and superior canal dehiscent labyrinths. *Otology & Neurotology, 34*(1), 127–134.

Janky, K. L., & Shepard, N. (2009). Vestibular evoked myogenic potential (VEMP) testing: Normative threshold response curves and effects of age. *Journal of the American Academy of Audiology, 20*(8), 514.

Janky, K. L., & Shepard, N. (2011). Unilateral centrifugation: utricular assessment and protocol comparison. *Otology & Neurotology, 32*(1), 116-121.

Johnsson, L. G. (1971). Degenerative changes and anomalies of the vestibular system in man. *The Laryngoscope, 81*(10), 1682–1694.

Jombik, P., & Bahýl, V. (2005). Short latency disconjugate vestibulo-ocular responses to transient stimuli in the audio frequency range. *Journal of Neurology, Neurosurgery and Psychiatry, 76*(10), 1398–1402.

Kerr, A. G., & Byrne, J. E. T. (1975). Concussive effects of bomb blast on the ear. *The Journal of Laryngology and Otology, 89*(2), 131–144.

Kirchner, H., Kremmyda, O., Hüfner, K., Stephan, T., Zingler, V., Brandt, T., . . . Strupp, M. (2011). Clinical, electrophysiological, and MRI findings in patients with cerebellar ataxia and a bilaterally pathological head-impulse test. *Annals of the New York Academy of Sciences, 1233*(1), 127–138.

Lee, J. D., Park, M. K., Lee, B. D., Park, J. Y., Lee, T. K., & Sung, K. B. (2011). Otolith function in patients with head trauma. *European Archives of Oto-Rhino-Laryngology, 268*(10), 1427–1430.

Lee, S. K., Cha, C. I., Jung, T. S., Park, D. C., & Yeo, S. G. (2008). Age-related differences in parameters of vestibular evoked myogenic potentials. *Acta Oto-Laryngologica, 128*(1), 66–72.

Lee Lim, C., Clouston, P., Sheean, G., & Yiannnikas, C. (1995). The influence of voluntary EMG activity and click intensity on the vestibular click evoked myogenic potential. *Muscle and Nerve, 18*(10), 1210–1213.

Li, C., Layman, A. J., Geary, R., Anson, E., Carey, J. P., Ferrucci, L., & Agrawal, Y. (2015). Epidemiology of vestibulo-ocular reflex function: Data from the Baltimore Longitudinal Study of Aging. *Otology and Neurotology, 36*(2), 267.

Lindeman, H. H. (1969). *Form and structure of the statoconial membranes and the cupulae.* Berlin, Germany: Springer.

Lopez, I., Honrubia, V., & Baloh, R. W. (1997). Aging and the human vestibular nucleus. *Journal of Vestibular Research, 7*, 77–85.

MacDougall, H. G., McGarvie, L. A., Halmagyi, G. M., Curthoys, I. S., & Weber, K. P. (2013a). Application of the video head impulse test to detect vertical semicircular canal dysfunction. *Otology and Neurotology, 34*(6), 974–979.

MacDougall, H. G., McGarvie, L. A., Halmagyi, G. M., Curthoys, I. S., & Weber, K. P. (2013b). The video head impulse test (vHIT) detects vertical semicircular canal dysfunction. *PLoS One, 8*(4), e61488.

MacDougall, H. G., Weber, K. P., McGarvie, L. A., Halmagyi, G. M., & Curthoys, I. S. (2009). The video head impulse test: Diagnostic accuracy in peripheral vestibulopathy. *Neurology, 73,* 1134–1141.

Maes, L., Dhooge, I., D'haenens, W., Bockstael, A., Keppler, H., Philips, B., . . . Vinck, B. M. (2010). The effect of age on the sinusoidal harmonic acceleration test, pseudorandom rotation test, velocity step test, caloric test, and vestibular-evoked myogenic potential test. *Ear and Hearing, 31*(1), 84–94.

Maes, L., Vinck, B. M., De Vel, E., D'haenens, W., Bockstael, A., Keppler, H., . . . Dhooge, I. (2009). The vestibular evoked myogenic potential: A test-retest reliability study. *Clinical Neurophysiology, 120*(3), 594–600.

Mangabeira Albernaz, P. L., & Zuma e Maia, F. C. (2014). The video head impulse test. *Acta Oto-Laryngologica, 134*(12), 1245–1250.

Mantokoudis, G., Saber Tehrani, A. S., Kattah, J. C., Eibenberger, K., Guede, C. I., Zee, D. S., & Newman-Toker, D. E. (2015). Quantifying the vestibulo-ocular reflex with video-oculography: Nature and frequency of artifacts. *Audiology and Neurotology, 20*(1), 39–50.

Manzari, L., Tedesco A., Burgess, A. M., & Curthoys, I. S. (2010). Ocular vestibular-evoked myogenic potentials to bone-conducted vibration in superior vestibular neuritis show utricular function. *Otolaryngology-Head and Neck Surgery, 143,* 274–280.

Matiño-Soler, E., Esteller-More, E., Martin-Sanchez, J.C., Martin-Sanchez, J.M., & Perez-Fernandes, N. (2014). Normative data on angular vestibulo-ocular responses in the yaw axis measured using the video head impulse test. *Otology and Neurotology, 36,* 466–471.

Matsuzaki, M., Murofushi, T., & Mizuno, M. (1999). Vestibular evoked myogenic potentials in acoustic tumor patients with normal auditory brainstem responses. *European Archives of Oto-Rhino-Laryngology, 256*(1), 1–4.

McCaslin, D. L., Jacobson, G. P., Grantham, S. L., Piker, E. G., & Verghese, S. (2011). The influence of unilateral saccular impairment on functional balance performance and self-report dizziness. *Journal of the American Academy of Audiology, 22*(8), 542–549.

McCue, M. P., & Guinan, J. J. (1994). Acoustically responsive fibers in the vestibular nerve of the cat. *The Journal of Neuroscience, 14*(10), 6058–6070.

McCue, M. P., & Guinan, J. J. (1995). Spontaneous activity and frequency selectivity of acoustically responsive vestibular afferents in the cat. *Journal of Neurophysiology, 74*(4), 1563–1572.

McGarvie, L. A., MacDougall, H. G., Halmagyi, G. M., Burgess, A. M., Weber, K. P., & Curthoys, I. S. (2015). The video head impulse test (vHIT) of semicircular canal function–age-dependent normative values of VOR gain in healthy subjects. *Frontiers in Neurology, 6,* 154. doi:10.3389/fneur.2015.00154

Migliaccio, A. A., & Cremer, P. D. (2011). The 2D modified head impulse test: A 2D technique for measuring function in all six semi-circular canals. *Journal of Vestibular Research, 21*(4), 227–234.

Minor, L. B., Lasker, D. M., Backous, D. D., & Hullar, T. E. (1999). Horizontal vestibulo-ocular reflex evoked by high-acceleration rotations in the squirrel monkey: I. Normal responses. *Journal of Neurophysiology, 82*(3), 1254–1270.

Moffat, A. J., & Capranica, R. R. (1976). Auditory sensitivity of the saccule in the American toad (*Bufo americanus*). *Journal of Comparative Physiology, 105*(1), 1–8.

Mossman, B., Mossman, S., Purdie, G., & Schneider, E. (2015). Age dependent normal horizontal VOR gain of head impulse test as measured with video-oculography. *Journal of Otolaryngology-Head and Neck Surgery, 44*(1), 1–8.

Mossman, S., & Halmagyi, G. M. (1997). Partial ocular tilt reaction due to unilateral cerebellar lesion. *Neurology, 49*(2), 491–493.

Murnane, O. D., Akin, F. W., Kelly, K. J., & Byrd, S. (2011). Effects of stimulus and recording parameters on the air conduction ocular vestibular evoked myogenic potentials. *Journal of the American Academy of Audiology, 22*(7), 469–480.

Murnane, O., Mabrey, H., Pearson, A., Byrd, S., & Akin, F. (2014). Normative data and test-retest reliability of the SYNAPSYS video head impulse test. *Journal of the American Academy of Audiology, 25*(3), 244–252.

Murofushi, T., & Curthoys, I. S. (1997). Physiological and anatomical study of click-sensitive primary vestibular afferents in the guinea pig. *Acta Oto-Laryngologica, 117*(1), 66–72.

Murofushi, T., Curthoys, I. S., & Gilchrist, D. P. (1996a). Response of guinea pig vestibular nucleus neurons to clicks. *Experimental Brain Research, 111*(1), 149–152.

Murofushi, T., Halmagyi, G. M., Yavor, R. A., & Colebatch, J. G. (1996b). Absent vestibular evoked myogenic potentials in vestibular neurolabyrinthitis: An indicator of inferior vestibular nerve involvement? *Archives of Otolaryngology-Head and Neck Surgery, 122*(8), 845–848.

Murofushi, T., Matsuzaki, M., & Wu, C. H. (1999). Short tone burst–evoked myogenic potentials on the sternocleidomastoid muscle: Are these potentials also of vestibular origin? *Archives of Otolaryngology-Head and Neck Surgery, 125*(6), 660–664.

Murray, K. J., Hill, K. D., Phillips, B., & Waterston, J. (2007). The influence of otolith dysfunction on the clinical presentation of people with a peripheral vestibular disorder. *Physical Therapy, 87*(2), 143–152.

Murray, K. J., Hill, K. D., Phillips, B., & Waterston, J. (2010). Does otolith organ dysfunction influence outcomes after a customized program of vestibular rehabilitation? *Journal of Neurologic Physical Therapy, 34*(2), 70–75.

Neal, E. (1926). Visual localization of the vertical. *American Journal of Psychology, 37*(2), 287–291.

Nelson, M. D., Akin, F. W., Riska, K. M., Andreson, K., & Stamps, S. (2016). Vestibular assessment and rehabilitation: Ten-year survey trends of audiologist's opinions and practice. *Journal of the American Academy of Audiology, 27*(2), 126–140.

Newman-Toker, D. E., Kattah, J. C., Alvernia, J. E., & Wang, D. Z. (2008). Normal head impulse test differentiates acute cerebellar strokes from vestibular neuritis. *Neurology, 70*(24, Pt. 2), 2378–2385.

Newman-Toker, D. E., Kerber, K. A., Hsieh, Y. H., Pula, J. H., Omron, R., Saber Tehrani, A. S., . . . Kattah, J. C. (2013). HINTS outperforms ABCD2 to screen for stroke in acute continuous vertigo and dizziness. *Academic Emergency Medicine, 20*(10), 986–996.

Newman-Toker, D. E., Tehrani, A. S. S., Mantokoudis, G., Pula, J. H., Guede, C. I., Kerber, K. A., . . . Kattah, J. C. (2013). Quantitative video-oculography to help diagnose stroke in acute vertigo and dizziness toward an ECG for the eyes. *Stroke, 44*(4), 1158–1161.

Ochi, K., Ohashi, T., & Nishino, H. (2001). Variance of vestibular-evoked myogenic potentials. *The Laryngoscope, 111*(3), 522–527.

Oh, S. Y., Kim, H. J., & Kim, J. S. (2015). Vestibular-evoked myogenic potentials in central vestibular disorders. *Journal of Neurology*, 1–11. http://dx.doi.org/10.1016/j.clinph.2014.12.021

Park, H. J., Lee, I. S., Shin, J. E., Lee, Y. J., & Park, M. S. (2010). Frequency-tuning characteristics of cervical and ocular vestibular evoked myogenic potentials induced by air-conducted tone bursts. *Clinical Neurophysiology, 121*(1), 85–89.

Pelosi, S., Schuster, D., Jacobson, G. P., Carlson, M. L., Haynes, D. S., Bennett, M. L., . . . Wanna, G. B. (2013). Clinical characteristics associated with isolated unilateral utricular dysfunction. *American Journal of Otolaryngology, 34*(5), 490–495.

Pérez-Fernández, N., Gallegos-Constantino, V., Barona-Lleo, L., & Manrique-Huarte,

R. (2012). Clinical and video-assisted examination of the vestibulo-ocular reflex. *Acta Otorrinolaringológica Española, 63*(6), 429–435.

Pollak, L., Kushnir, M., & Stryjer, R. (2006). Diagnostic value of vestibular evoked myogenic potentials in cerebellar and lower-brainstem strokes. *Neurophysiologie Clinique/Clinical Neurophysiology, 36*(4), 227–233.

Popper, A. N., & Fay, R. R. (1973). Sound detection and processing by teleost fishes: A critical review. *The Journal of the Acoustical Society of America, 53*(6), 1515–1529.

Richter, E. (1980). Quantitative study of human Scarpa's ganglion and vestibular sensory epithelia. *Acta Oto-Laryngologica, 90*(1–6), 199–208.

Robinson, D. A. (1963). A method of measuring eye movement using a scleral search coil in a magnetic field. *IEEE Transactions on Biomedical Engineering, 10*(4), 137–145.

Rosengren, S. M., Aw, S. T., Halmagyi, G. M., Todd, N. M., & Colebatch, J. G. (2008). Ocular vestibular evoked myogenic potentials in superior canal dehiscence. *Journal of Neurology, Neurosurgery and Psychiatry, 79*(5), 559–568.

Rosengren, S. M., Colebatch, J. G., Straumann, D., & Weber, K. P. (2013). Why do oVEMPs become larger when you look up? Explaining the effect of gaze elevation on the ocular vestibular evoked myogenic potential. *Clinical Neurophysiology, 124*(4), 785–791.

Rosengren, S. M., Govender, S., & Colebatch, J. G. (2011). Ocular and cervical vestibular evoked myogenic potentials produced by air- and bone-conducted stimuli: Comparative properties and effects of age. *Clinical Neurophysiology, 122*(11), 2282–2289.

Rosengren, S. M., Jombik, P., Halmagyi, G. M., & Colebatch, J. G. (2009). Galvanic ocular vestibular evoked myogenic potentials provide new insight into vestibulo-ocular reflexes and unilateral vestibular loss. *Clinical Neurophysiology, 120*(3), 569–580.

Rosengren, S. M., Todd, N. M., & Colebatch, J. G. (2005). Vestibular-evoked extraocular potentials produced by stimulation with bone-conducted sound. *Clinical Neurophysiology, 116*(8), 1938–1948.

Schönfeld, U., & Clarke, A. H. (2011). A clinical study of the subjective visual vertical during unilateral centrifugation and static tilt. *Acta Oto-Laryngologica, 131*(10), 1040–1050.

Schönfeld, U., Helling, K., & Clarke, A. H. (2010). Evidence of unilateral isolated utricular hypofunction. *Acta Otolaryngologica, 130*(6), 702–707.

Sheykholeslami, K., & Kaga, K. (2002). The otolithic organ as a receptor of vestibular hearing revealed by vestibular-evoked myogenic potentials in patients with inner ear anomalies. *Hearing Research, 165*(1), 62–67.

Sheykholeslami, K., Kermany, M. H., & Kaga, K. (2001). Frequency sensitivity range of the saccule to bone-conducted stimuli measured by vestibular evoked myogenic potentials. *Hearing Research, 160*(1), 58–62.

Sheykholeslami, K., Murofushi, T., Kermany, M. H., & Kaga, K. (2000). Bone-conducted evoked myogenic potentials from the sternocleidomastoid muscle. *Acta Oto-Laryngologica, 120*(6), 731–734.

Shimizu, K., Murofushi, T., Sakurai, M., & Halmagyi, M. (2000). Vestibular evoked myogenic potentials in multiple sclerosis. *Journal of Neurology Neurosurgery Psychiatry, 69*, 276–277.

Singh, N. K., Kashyap, R. S., Supreetha, L., & Sahana, V. (2014). Characterization of age-related changes in sacculocolic response parameters assessed by cervical vestibular evoked myogenic potentials. *European Archives of Oto-Rhino-Laryngology, 271*(7), 1869–1877.

Su, H. C., Huang, T. W., Young, Y. H., & Cheng, P. W. (2004). Aging effect on vestibular evoked myogenic potential. *Otology and Neurotology, 25*(6), 977–980.

Suzuki, J. I., Tokumasu, K., & Goto, K. (1969). Eye movements from single utricular nerve stimulation in the cat. *Acta Oto-Laryngologica, 68*(1–6), 350–362.

Taylor, R. L., Kong, J., Flanagan, S., Pogson, J., Croxson, G., Pohl, D., & Welgampola, M. S. (2015). Prevalence of vestibular dysfunction in patients with vestibular schwannoma using video head-impulses and vestibular-evoked potentials. *Journal of Neurology, 262*(5), 1228–1237.

Todd, N. P. M., Cody, F. W., & Banks, J. R. (2000). A saccular origin of frequency tuning in myogenic vestibular evoked potentials? Implications for human responses to loud sounds. *Hearing Research, 141*(1), 180–188.

Todd, N. P. M., Rosengren, S. M., Aw, S. T., & Colebatch, J. G. (2007). Ocular vestibular evoked myogenic potentials (OVEMPs) produced by air- and bone-conducted sound. *Clinical Neurophysiology, 118*(2), 381–390.

Todd, N. P., Rosengren, S. M., & Colebatch, J. G. (2008). Ocular vestibular evoked myogenic potentials (OVEMPs) produced by impulsive transmastoid accelerations. *Clinical Neurophysiology, 119*(7), 1638–1651.

Todd, N. P., Rosengren, S. M., & Colebatch, J. G. (2009). A utricular origin of frequency tuning to low-frequency vibration in the human vestibular system? *Neuroscience Letters, 451*(3), 175–180.

Tomko, D. L., & Paige, G. D. (1992). Linear vestibuloocular reflex during motion along axes between nasooccipital and interaurala. *Annals of the New York Academy of Sciences, 656*(1), 233–241.

Uchino, Y., & Kushiro, K. (2011). Differences between otolith- and semicircular canal-activated neural circuitry in the vestibular system. *Neuroscience Research, 71*(4), 315–327.

Uchino, Y., Sasaki, M., Sato, H., Bai, R., & Kawamoto, E. (2005). Otolith and canal integration on single vestibular neurons in cats. *Experimental Brain Research, 164*(3), 271–285.

Uchino, Y., Sasaki, M., Sato, H., Imagawa, M., Suwa, H., & Isu, N. (1996). Utriculoocular reflex arc of the cat. *Journal of Neurophysiology, 76*(3), 1896–1903.

Ulmer, E., Bernard-Demanze, L., & Lacour, M. (2011). Statistical study of normal canal deficit variation range. Measurement using the head impulse test video system. *European Annals of Otorhinolaryngology, Head and Neck Diseases, 128*(5), 278–282.

Ulmer, E., & Chays, A. (2005). Curthoys and Halmagyi head impulse test: An analytical device [article in French]. *Annales d'Otolaryngologie et de Chirurgie Cervicofaciale, 122*(2), 84–90.

Valko, Y., Hegemann, S. C., Weber, K. P., Straumann, D., & Bockisch, C. J. (2011). Relative diagnostic value of ocular vestibular evoked potentials and the subjective visual vertical during tilt and eccentric rotation. *Clinical Neurophysiology, 122*(2), 398–404.

Versino, M., Colnaghi, S., Callieco, R., Bergamaschi, R., Romani, A., & Cosi, V. (2002). Vestibular evoked myogenic potentials in multiple sclerosis patients. *Clinical Neurophysiology, 113*(9), 1464–1469.

Walther, L. E., & Blödow, A. (2013). Ocular vestibular evoked myogenic potential to air conducted sound stimulation and video head impulse test in acute vestibular neuritis. *Otology and Neurotology, 34*(6), 1084–1089.

Wang, Y. P., Hsu, W. C., & Young, Y. H. (2006). Vestibular evoked myogenic potentials in acute acoustic trauma. *Otology and Neurotology, 27*(7), 956–961.

Weber, K. P., MacDougall, H. G., Halmagyi, G. M., & Curthoys, I. S. (2009). impulsive testing of semicircular-canal function using video-oculography. *Annals of the New York Academy of Sciences, 1164*(1), 486–491.

Weber, K. P., Rosengren, S. M., Michels, R., Sturm, V., Straumann, D., & Landau, K. (2012). Single motor unit activity in human extraocular muscles during the vestibulo-ocular reflex. *The Journal of Physiology, 590*(13), 3091–3101.

Welgampola, M. S., & Colebatch, J. G. (2001a). Characteristics of tone burst-evoked myogenic potentials in the sternocleidomastoid muscles. *Otology and Neurotology, 22*(6), 796–802.

Welgampola, M. S., & Colebatch, J. G. (2001b). Vestibulocollic reflexes: Normal values and the effect of age. *Clinical Neurophysiology, 112*(11), 1971–1979.

Welgampola, M. S., Rosengren, S. M., Halmagyi, G. M., & Colebatch, J. G. (2003). Vestibular activation by bone conducted sound. *Journal of Neurology, Neurosurgery, and Psychiatry, 74*, 771–778.

Wetzig, J., Reiser, M., Martin, E., Bregenzer, N., & von Baumgarten, R. J. (1990). Unilateral centrifugation of the otoliths as a new method to determine bilateral asymmetries of the otolith apparatus in man. *Acta Astronautica, 21*(6), 519–525.

Young, E. D., Fernandez, C., & Goldberg, J. M. (1977). Responses of squirrel monkey vestibular neurons to audio-frequency sound and head vibration. *Acta Oto-Laryngologica, 84*(1–6), 352–360.

Zhou, G., Poe, D., & Gopen, Q. (2012). Clinical use of vestibular evoked myogenic potentials in the evaluation of patients with air-bone gaps. *Otology and Neurotology, 33*(8), 1368–1374.

Zhu, H., Tang, X., Wei, W., Mustain, W., Xu, Y., & Zhou, W. (2011). Click-evoked responses in vestibular afferents in rats. *Journal of Neurophysiology, 106*(2), 754–763.

Zink, R., Bucher, S. F., Weiss, A., Brandt, T., & Dieterich, M. (1998). Effects of galvanic vestibular stimulation on otolithic and semicircular canal eye movements and perceived vertical. *Electroencephalography and Clinical Neurophysiology, 107*(3), 200–205.

Zulueta-Santos, C., Lujan, B., Manrique-Huarte, R., & Perez-Fernandez, N. (2014). The vestibulo-ocular reflex assessment in patients with Ménière's disease: Examining all semicircular canals. *Acta Oto-Laryngologica, 134*(11), 1128–1133.

Zwergal, A., Rettinger, N., Frenzel, C., Dieterich, M., Brandt, T., & Strupp, M. (2009). A bucket of static vestibular function. *Neurology, 72*(19), 1689–1692.

CHAPTER 5

Genetics of Deafness: In Mice and Men

Mirna Mustapha and Avril Genene Holt

In this chapter, we discuss heritable hearing loss with an emphasis on age-related hearing impairment (ARHI). We begin by providing a brief overview of congenital hearing loss. We then touch upon challenges and recent developments in deciphering the genetic basis of deafness. While we do not go into detail about the discovery of any particular gene, we do provide some insight into how the gene discovery process is implemented. Finally, we present support for the concept of using similar as well as new technologies, to identify genes responsible for ARHI. This is becoming more important as our population continues to live longer, and with no major human studies in this area to date, the need for early diagnosis and interventions allowing the restoration of function is critically high. Together, novel genomic technologies, mouse genetic tools, and patients with positive family histories will yield a breakthrough in the discovery of ARHI genes. These genes could be a major player in the future of personalized medicine with the ability to have a major impact on the quality of life for the elderly in our society.

Congenital Deafness

Approximately one of every 1,000 infants is born deaf. For half of those with congenital deafness, the cause is genetic. Genetic congenital deafness can be divided into syndromic and nonsyndromic hearing loss. When other impairments are present in addition to hearing loss, the deafness is classified as syndromic. However, if the only evident impairment is hearing loss, then the deafness is termed nonsyndromic. Roughly, 70% of congenital deafness is nonsyndromic. The genes responsible for nonsyndromic deafness can be classified based on their patterns of inheritance. The patterns are either dominant or recessive. In the case of dominant inheritance, one copy of an abnormal gene is sufficient to develop a phenotype, whereas with recessive inheritance, two abnormal copies are necessary. Another, more rare pattern of inheritance, mitochondrial inheritance, involves transmission of deafness related genes from mother to child. Understanding the pattern of inheritance facilitates the way in which causative genes are identified.

Challenges

While cases of familial deafness were first described in the 1930s, not until 1994 was the first deafness-related gene identified. Some of the difficulties in progress stemmed from the fact that deafness is characterized by high genetic heterogeneity. That is, mutations in several different genes produce the same phenotype. In addition, intracommunity marriages between those affected with deafness can lead to the segregation of more than one mutated gene. This leads to a more complex pattern of inheritance, making the causative gene harder to localize. Linkage analyses consist of localizing a specific gene to a chromosomal region (locus) by measuring the recombination rates between phenotypic and genetic markers. Two conventional methods were employed to overcome those early challenges and identify deafness-related genes: consanguineous deaf families with numerous affected and nonaffected members and mouse models of deafness.

Using these tools, approximately 70 genes involved in nonsyndromic deafness have been reported along with more than 130 corresponding loci (http://hereditaryhearingloss.org). Of these identified genes, 30% to 50% of nonsyndromic recessive deafness is due to mutations in the first discovered gene (DFNB1). The DFNB1 gene encodes for a gap junction protein (connexin 26). To date we have been directly involved in the identification of a number of genes, including those involved in syndromic deafness, such as SANS, affecting stereocilia differentiation and the retina in Usher syndrome. Of the nonsyndromic genes, Whirlin and Otoferlin (Mburu et al., 2003; Roux et al., 2006; Yasunaga et al., 1999) provide excellent examples of how mouse and human genetics studies can work synergistically to gain insight into the pathophysiology of deafness. The majority of the heritable forms of hearing impairment are caused by mutations in specific genes producing defects in mechanical stimulation or hair cell transduction (i.e., Whirlin) (Mburu et al., 2003). One of the more rare forms of genetic deafness affects the transmission of electrical signals from the inner hair cells (IHCs) to the brain via spiral ganglion neurons (auditory nerve), a process that is still poorly understood. One of the first mutated genes identified as critical for neurotransmission is Otoferlin.

Auditory Neuropathy

"Auditory neuropathy" is the term used to codify a primary degeneration of the auditory nerve. In humans, the auditory neuropathy construct is based on a body of evidence derived from behavioral, electroacoustic, electrophysiological and histopathologic data. Auditory neuropathy (AN) is characterized by intact frequency selectivity (normal outer hair cell function) and abnormal neural activity emanating from the auditory nerve. Diseases associated with AN include Friedreich's ataxia, Mohr-Tranebjaerg syndrome (aka, deafness-dystonia-optic neuropathy

syndrome), hereditary motor and sensory neuropathies (aka, Charcot-Marie-Tooth disease), autosomal dominant optic atrophy, and hyperbilirubinemia. In fact, all of the conditions noted above (except for hyperbilirubinemia), are syndromic types of AN that are linked directly or indirectly to mitochondrial dysfunction (Campuzano et al., 1996; Dürr et al., 1996; Kaplan, 1999; Koehler et al., 1999; Shy, 2004). In addition, there are mutations in genes such as pejvakin (PJVK), that result in non-syndromic AN in humans (Delmaghani et al., 2006).

However, in recent years, there have been other conditions that clinicians and even some basic scientists confuse with AN that now is referred to as "*Auditory synptopathy*". Auditory synptopathy (AS) is the term used to define dysfunction of IHC ribbon synapses. In humans, this would translate into normal OAEs, the absence of acoustically evoked ABRs and profound sensorineural hearing loss. Examples of these constituent disorders, include the Otoferlin (OTOF) and vesicular glutamate transporter 3 (VGLUT3) mutations. The pattern of test results observed between OTOF, VGLUT3, in comparison to AN, are functionally and fundamentally different from this condition. First of all, OTOF and VGLUT3 are presynaptic anomalies localized to the IHC and have clear differences in terms of their phenotypic expression. In animal models with the OTOF mutation, evidence indicates that it is a disorder of "*exocitosis*" and that OTOF is the calcium sensor within the IHC (Roux et al., 2006; Johnson & Chapman, 2010; Ramakrishnan et al., 2014). Specifically, if there is no output from the cell in terms of neurotransmitter release (exocitosis), the result is a "*severe-to-profound*" sensorineural hearing loss. However, because all other aspects of the cell are functionally normal, including OHC function, OAEs will be present and normal but ABRs will be absent. In contrast to AN, the absence of ABRs is the result of lack of output from the hair cells and not from desynchronized firing patterns from the auditory nerve. Electrically evoked ABRs (eABRs) are typically present and robust since spiral ganglion cell counts are largely unaffected. With the VGLUT3 mutation, glutamate does not get loaded into synaptic vesicles of the IHC, but otherwise, all other aspects of the cellular machinery is normal. Upon acoustic stimulation and when the cell is depolarized, synaptic vesicles are released from the cell into the synaptic cleft. But since there is no glutamate loaded in the vessicles, there will be no postsynaptic neural excitation, and again, profound neuro-sensory hearing loss results. While the mechanisms are different, OTOF and VGLUT3, they both have a similar outcome; profound neuro-sensory hearing loss (Seal et al., 2008). Another distinction between OTOF and VGLUT3 mutations, relates to the fact that that VGLUT3 mutant animals develop seizures.

Based on these findings, it should be of no surprise that these types of phenomena have their greatest impact on newborn hearing screening programs. But the pattern of results that can be obtained can cause confusion and possible misinterpretation depending on the tests being performed. For example, in newborn infant hearing screening and follow up exams, two tests are typically performed based on acoustic stimulation; OAEs (transient, TEOAEs; or distortion product, DPOAEs) and ABRs. Because the OHCs are otherwise normal in these conditions, OAEs will be intact and normal but ABRs will appear absent. If we stopped here, this pattern of results would give the appearance of an AN. However, as noted above, ABRs are absent not because auditory nerve fibers produce a desynchronized response, but because there is no output from the cell. The relevant test of interest here that needs to be performed next is the electrically evoked ABRs (eABRs). In the case of OTOF and VGLUT3, eABRs are robust and normal in appearance.

Treatment options for the inner hair cell dysfunctions, like OTOF and VGLUT3 mutations are noteworthy. Cochlear implants (CIs) would be the treatment of choice since presurgical neural responses from eABRs are robust and indicate a modest number of spiral ganglion cells that can be activated by the implant. Consequently, the future is bright for children like this, assuming of course that no cognitive impairment(s) exists. In contrast, hearing aids or CIs in AN are not an entirely viable option; AN typically shows degeneration over time, and whereas some positive results might be manifest initially by these treatment options, poor performance is typically the end result (Giraudet & Avan, 2012) and alternative treatments need to be investigated.

The Value of Mouse Models

Mice are the premier mammalian organism for studying genetics, gene-environment interactions, disease pathophysiology, and initial preclinical testing of therapeutic approaches. Studies in mice have been the foundation of our understanding of gene function in mammals. An excellent example stems from otoferlin, one of the human deafness genes. The relevance of this gene was first revealed in a family affected with profound sensorineural deafness. However, the inaccessibility of the inner ear, lack of access to human tissue, and lack of appropriate cell culture models led us to develop a mouse model to more fully understand the role of this gene in hearing. Using the mouse model, we discovered that while OTOF is required for the presynaptic neurotransmitter release in hair cells, the auditory nerve function is intact.

Deciphering the site-of-lesion in this form of auditory synaptopathy has improved clinical diagnosis and provided people carrying this mutation with the option of a cochlear implant as a therapeutic approach to restore their hearing.

More recently, the use of mouse genetics led to the discovery of an important role for cell adhesion molecules. Study of the cell adhesion molecules, thrombospodin 1 and 2 (TSP1 and TSP2), provided the first evidence of these types of synaptogenic molecules in the development and function of the inner ear. Lack of TSP1 and TSP2 results in progressive and age-related loss of auditory and vestibular function caused by fewer functional synapses responding to stimuli. Based on the knowledge gleaned from the mouse studies, we propose that TSP1 and TSP2 are candidates for screening in patients with inner ear dysfunction of unknown etiology.

Age and Noise-Related Auditory Neuropathy

Presbycusis, or age-related hearing impairment (ARHI), is the most common sensory deficit in the elderly. The key feature of this progressive condition is a problem with hearing and speech discrimination in noisy environments, which is a characteristic of impaired hearing ability in aging ears. Loss of speech discrimination can cause isolation and in more severe cases can result in depression and cognitive decline, especially in the aging population. The causative factors for presbycusis are both genetic and environmental. While environmental risk factors for ARHI have been well documented, the identification of genetic factors has been more problematic.

Recent findings suggest that substantial degeneration of auditory nerve fibers and their connections to the IHCs occurs with exposure to loud noise as well as with ARHI. This type of auditory neuropathy is also called hidden hearing impairment because it is *invisible* or not detectable with the usual clinical hearing tests.

In the mammalian cochlea, two major subtypes of auditory neurons exist: type I SGNs are myelinated and carry all the auditory information from the organ of Corti IHCs to the auditory brainstem. More than 90% of all auditory neurons are type I SGNs. In sharp contrast, unmyelinated type II SGNs innervate OHCs, which are responsible for cochlear amplification and sharp frequency tuning. The role of type II SGNs is poorly understood, and it has been speculated that they might be involved in auditory pain and perhaps in hyperacusis.

The mature IHC is innervated by multiple type I SGN fibers that are anatomically and physiologically diverse. During development, each type I SGN adopts morphologic and electrophysiologic properties such as low or high thresholds to enable the neuron to interpret and transmit complex sound stimuli. Recent studies have shown that at least two subtypes of type I SGNs exist that display distinct electrophysiologic properties. Low-threshold type I SGNs display high spontaneous discharge rates, whereas high-threshold type I SGNs have low spontaneous rates (Goetz et al., 2014). Single fiber recording studies demonstrated that in both noise-exposed and aging rodents, there is a selective reduction in the percentage of high threshold SGNs resulting first in synapse degeneration —AS, followed by delayed death of SGNs causing AN (Furman, Kujawa, & Liberman, 2013; Schmiedt, Mills, & Boettcher, 1996). There is no direct evidence, however, as to why high threshold SGNs are preferentially affected by noise exposure as compared to the low threshold SGN population. In addition, mechanisms that are involved in the differentiation of type I SGNs into low- and high-threshold neurons are also unknown. Understanding the source of type I SGN heterogeneity and differential vulnerability to damage and/or aging is essential for understanding the transition from AS to AN disease processes.

Noise Insult

Recent studies provide evidence that noise exposure resulting in a temporary threshold shift (TTS) can produce AS (Hickox & Liberman, 2014; Kujawa & Liberman, 2015; Lin, Furman, Kujawa, & Liberman, 2011; Mendus et al., 2014). The functional and morphologic results of a noise exposure producing a TTS mimics the effects seen in ARHI (e.g., reduced amplitude of ABR wave 1 and reduced afferent synapse numbers). This is the case when referring to people or "wild-type" animals with no mutations in genes with susceptibility to noise. In contrast, mutations in a single gene can increase susceptibility to noise, turning what would normally be a noise exposure producing a TTS into one that produces a permanent threshold shift (PTS). Taken together, noise exposures can have differential effects on individuals, depending on their genetic makeup. In support of this concept, recent findings from our group suggest that even when genes from the same family are mutated, they can be differentially affected by noise. Noise exposure in Thrombospondin mutants (TSP1 and TSP2) revealed a TTS in the TSP1; however, TSP2 mutants exhibited increased susceptibility to noise exposure resulting in a PTS-related synaptopathy. The PTS was observed at the 16-kHz frequency correlated to the synapse loss in the corresponding cochlear region. Together the results from these studies indicate that TSPs are important not only for synaptic formation during development but also for maintenance of cochlear function. Moreover, individual forms of TSP may protect the cochlea in a frequency-specific manner.

Although prevalence and etiology of adult AS/AN is not well understood, recent studies in rodents suggest that the largest proportion of acquired or age-related AS/AN is due to noise-induced synapse and type I SGN degeneration (Kujawa & Liberman, 2015).

Genetic Tools

Over the last decade, linkage analysis and positional cloning have been successful for identification of monogenic forms of deafness genes (hereditaryhearingloss.org). There are several limitations to applying this Mendelian-type strategy to clone genes for common, complex diseases such as ARHI. Mutations in Mendelian disorders are characterized by strong genotype-phenotype correlations, whereas the genetic determinants of complex diseases are susceptibility mutations rather than disease mutations. Thus, individuals who carrythe susceptibility mutation may not be affected unless exposed to environmental insults such as noise.

Unlike the study of congenital hearing impairment, for ARHI, performing linkage analysis followed by sequencing-based screening has been hampered due to the unavailability of several elderly affected and nonaffected family members.

Few genome-wide association studies (GWASs) have been performed in humans. One of the main finding from these studies is that, like many complex traits, ARHI is polygenic, involving mutations in several genes, often with variants of limited effect size (Fransen et al., 2015; Ohmen et al., 2014).

Onset, progression, and severity of complex diseases such as ARHI are determined by interactions between genetic and environmental factors. Therefore, adjustment for environmental risk factors, considering potential gene-gene interaction pathways and sorting families with similar ARHI phenotypes, will help identify causative genetic mutations. Different from the approaches used in the past, the high-throughout genomic technologies offered to us today will also facilitate the discovery of ARHI genes using small outbred families with distinct phenotypes.

The development of novel genomic technologies will enhance our understanding of the molecular physiology of hearing and deafness. These new technologies will have an impact on ARHI and noise-induced hearing impairment that will be far reaching. Eventually, the direct benefit of these genomic technologies will be the availability of comprehensive genetic diagnosis.

Conclusion

ARHI is the most common sensory deficit in the elderly and accounts for nearly a third of all instances of hearing impairment. As the population in developed countries is generally ageing, we can expect a proportionate increase in age-related disorders such as ARHI. Therefore, there is an urgent need to develop novel therapeutic strategies that can be used to prevent, delay, or ameliorate ARHI, given that few options are available to patients who find hearing aids unsuitable. Having an early clinical and/or genetic diagnostic for ARHI would greatly facilitate development of effective therapies. One such early marker might be the reduced ABR wave-1 amplitude that is interpreted as a possible decline in auditory nerve activity.

References

Campuzano, V., Montermini, L., Molto, M. D., Pianese, L., Cossee, M., Cavalcanti, F. . . . Pandolfo, M. (1996). Friedreich's ataxia: autosomal recessive disease caused by an intronic GAA triplet repeat expansion. Science 271, 1423–1427.

Delmaghani, S., del Castillo, F. J., Michel, V., Leibovici, M., Aghaie, A., Ron, U. . . . Petit, C. (2006). Mutations in the gene encoding pejvakin, a newly identified protein of the afferent auditory pathway, cause DFNB59 auditory neuropathy. *Nature Genetics, 38*, 770–778.

Dürr, A., Cossee, A., Agrid, Y., Campuzano, V., Mignard, C., Penet, C. . . . Koenig, M. (1996). Clinical and genetic abnormalities in patients with Friedreich's ataxia. *The New England Journal of Medicine, 335*, 1169–1175.

Fransen, E., Bonneux, S., Corneveaux, J. J., Schrauwen, I., Di Berardino, F., White, C. H., . . . Friedman, R. A. (2015). Genome-wide association analysis demonstrates the highly polygenic character of age-related hearing impairment. *European Journal of Human Genetics, 23*, 110–115.

Furman, A. C., Kujawa, S. G., & Liberman, M. C. (2013). Noise-induced cochlear neuropathy is selective for fibers with low spontaneous rates. *Journal of Neurophysiology, 110*, 577–586.

Giraudet, F., & Avan, P. (2012). Auditory neuropathies: understanding their pathogenesis to illuminate intervention strategies. *Current Opinion in Neurology, 24*, 50–56.

Goetz, J. J., Farris, C., Chowdhury, R., & Trimarchi, J. M. (2014). Making of a retinal cell: Insights into retinal cell-fate determination. *International Review of Cell and Molecular Biology, 308*, 273–321.

Hickox, A. E., & Liberman, M. C. (2014). Is noise-induced cochlear neuropathy key to the generation of hyperacusis or tinnitus? *Journal of Neurophysiology, 111*, 552–564.

Johnson, C. P., & Chapman, E. R. (2010). Otoferlin is a calcium sensor that directly regulates SNARE-mediated membrane fusion. *Journal of Cell Biology, 191*, 187–197.

Kaplan, J. (1999). Friedreich's ataxia is a mitochondrial disorder. *Proceedings of the National Academy of Science USA, 28*, 10948–10949.

Koehler, C. M., Leuenberger, D., Merchant, S., Renold, A., Junne, T., & Schatz, G. (1999). Human deafness dystonia syndrome is a mitochondrial disease. Proceedings of the National Academy of Science USA, 96, 2141–2146.

Kujawa, S. G., & Liberman, M. C. (2015). Synaptopathy in the noise-exposed and aging cochlea: Primary neural degeneration in acquired sensorineural hearing loss. *Hearing Research, 330,* 191–199.

Lin, H. W., Furman, A. C., Kujawa, S. G., & Liberman, M. C. (2011). Primary neural degeneration in the guinea pig cochlea after reversible noise-induced threshold shift. *Journal of the Association for Research in Otolaryngology, 12,* 605–616.

Mburu, P., Mustapha, M., Varela, A., Weil, D., El-Amraoui, A., Holme, R. H., . . . Brown, S. D. (2003). Defects in whirlin, a PDZ domain molecule involved in stereocilia elongation, cause deafness in the whirler mouse and families with DFNB31. *Nature Genetics, 34,* 421–428.

Mendus, D., Sundaresan, S., Grillet, N., Wangsawihardja, F., Leu, R., Muller, U., . . . Mustapha, M. (2014). Thrombospondins 1 and 2 are important for afferent synapse formation and function in the inner ear. *European Journal of Neuroscience, 39,* 1256–1267.

Ohmen, J., Kang, E. Y., Li, X., Joo, J. W., Hormozdiari, F., Zheng, Q. Y., . . . Friedman, R. A. (2014). Genome-wide association study for age-related hearing loss (AHL) in the mouse: A meta-analysis. *Journal of the Association for Research in Otolaryngology, 15,* 335–352.

Ramakrishnan, N. A., Drescher, M. J., Morley, B. J., Kelley, P. M., & Drescher, D. G. (2014). Calcium regulates molecular interactions of Otoferlin with soluble NSF attachment protein receptor (ARE) proteins required for hair cell exocytosis. *Journal of Biological Chemistry, 289,* 8750–8766.

Roux, I., Safieddine, S., Nouvian, R., Grati, M. H., Simmler, M-C., Bahloul, A., . . . Petit, C. (2006). Otoferlin, defective in a human deafness form, is essential for exocytosis at the auditory ribbon synapse. *Cell, 127,* 277–289.

Ruel, J., Emery, S., Nouvian., R., Bersot, T., Amilhon, B., Van Rybroek, J. M. . . . Puell, J. L. (2008). Impairment of SLC17A8 encoding vesicular glutamate transporter-3, VGLUT3, underlies nonsyndromic deafness DFNA25 and inner hair cell dysfunction in null mice. *The American Journal of Human Genetics, 83,* 278–292.

Schmiedt, R. A., Mills, J. H., & Boettcher, F. A. (1996). Age-related loss of activity of auditory-nerve fibers. *Journal of Neurophysiology, 76,* 2799–2803.

Seal, R. P., Akil, O., Yi, E., Weber, C. M., Grant, L., Yoo, J. . . . Edwards, R. H. (2008). Sensorineural deafness and seizures in mice lacking vesicular glutamate transporter 3. *Neuron, 57,* 263–275.

Shy, M. E. (2004). Charcot-Marie-Tooth disease: an update. *Current Opinion in Neurology, 17,* 579–585.

Van Camp, G., & Smith, R. (2015). *Hereditary hearing loss.* Retrieved from http://www.hereditaryhearingloss.org

Varga, R., Avenarius, M. R., Kelley, P. M., Keats, B. J., Berlin, C. I., Hood, L. J., Kimberling, W. J. (2006). OTOF mutations revealed by genetic analysis of hearing loss families including a potential temperature sensitive auditory neuropathy allele. *Journal of Medical Genetics, 43,* 576–581.

Yasunaga, S., Grati, M., Cohen-Salmon, M., El-Amroui, A., Mustapha, M., Salem, N., . . . Petit, C. (1999). A mutation in OTOF, encoding otolerlin, FER-1-like protein, causes DFNB9, a nonsyndromic form of deafness. *Nature Genetics, 21,* 363–369.

CHAPTER 6

Molecular-Based Measures for the Development of Treatment for Auditory System Disorders

Important Transformative Steps Toward the Treatment of Tinnitus

Avril Genene Holt, Catherine A. Martin, Antonela Muca, Angela R. Dixon, and Magnus Bergkvist

In this chapter, we use gene expression methodology to characterize and target auditory system function/dysfunction at the molecular level. We provide a brief overview of tinnitus and discuss developments in hearing loss and tinnitus research, with regard to gene expression-based studies. Given the sheer number of steps involved in the gene expression process, particularly with respect to auditory system dysfunction, we restrict this chapter to a brief overview of relevant topics. We then emphasize the significance of gene expression-based tinnitus studies, by providing results that demonstrate how differential gene expression, specifically for N-methyl-D-aspartate (NMDA) receptors, is important in the study of hearing loss and tinnitus. Finally, we propose the combination of gene expression results with transformative technologies, including manganese-enhanced magnetic resonance imaging (MEMRI), functionalized nanoparticles (NPs), repetitive transcranial magnetic stimulation (rTMS), and optogenetics, in order to drive the discovery of effective treatment options. The goal is to provide a unique vantage point when applying the gene expression process to complex auditory system disorders, like tinnitus, which can be challenging to study, in an effort to inspire thinking that is "outside of the box." This approach necessitates constructive collaborations with individuals within several notable scientific disciplines, bridging molecular biology, nanobioscience, physics, engineering, and audiology/hearing science. Their expertise and the clever use of new technology will be the fundamental factors that transform the field.

Overview of Tinnitus

Tinnitus is the perception of a buzzing or ringing sound in the absence of an external acoustic stimulus. Tinnitus affects more than 15% of the people in the United States and is the number one compensated military service–related disability, costing over $2.75 billion annually (AMVETS, 2015). Despite the pressing need, there are currently no broadly effective treatments for tinnitus. Several studies have suggested that tinnitus is a condition of maladaptive homeostatic plasticity. Homeostatic plasticity is a negative feedback process by which neuronal activity is altered by intrinsic and/or synaptic means in an effort to maintain or regain a target level of activity. Both hearing loss and tinnitus have been reported to be associated with changes in spontaneous neuronal activity (Brozoski et al., 2007; Cacace, Lovely, McFarland, et al., 1994; Cacace, Lovely, Winter, et al., 1994; Cacace et al., 2014; Holt et al., 2010; Lockwood et al., 1998). These hearing loss and tinnitus-related changes in spontaneous neuronal activity have been reported in several brain regions (Brozoski et al., 2002; Brozoski et al., 2007; Francis & Manis, 2000; Hirsch & Oertel, 1988; Holt et al., 2010; Kalappa et al., 2014; Kaltenbach & Afman, 2000; Koerber et al., 1966; Longenecker & Galazyuk, 2011; Salvi et al., 1978), including the dorsal and ventral cochlear nucleus (DCN and VCN) and the inferior colliculus (IC). Noise overstimulation can result in both spatial and temporal differences in spontaneous neuronal activity. For example, 5 days after noise overstimulation, there are sustained increases in spontaneous neuronal activity in the DCN (Kaltenbach & Afman, 2000) that are believed to play an important role in tinnitus (Finlayson & Kaltenbach, 2009). More recent evidence suggests that hyperactivity within IC neurons may be an equally important factor for the emergence of noise-induced tinnitus since increased neuronal activity in the IC occurs 4 to 12 hours following noise exposure, a timescale that is consistent with the emergence of noise-induced tinnitus (Mulders & Robertson, 2013). Since IC neurons are upstream of CN neurons, a change in neuronal activity in the CN may result in robust changes in the IC.

Gene Expression to Appraise Auditory Function

Gene expression is the process by which information encoded in a gene is converted into functional products, including various ribonucleic acids (RNAs) or proteins, ultimately giving rise to structures found in organisms. Taken in the context of noise-induced tinnitus, for example, reactive changes within cells (changes in neuronal activity) at different sites in the periphery (hair cells, supporting cells, primary neurons), brainstem, and cortex can be assessed by quantifying reactive gene levels and protein products reflective of these changes. Currently, the gold standard in gene expression quantification is polymerase chain reaction (PCR). This widely accepted technique allows for the amplification of specific segments of deoxyribonucleic acid (DNA). Through amplification, this method allows detection of the presence of specific genes in specific tissues even if they are present in only small quantities. This is especially important when examining neurotransmitter-related gene products in the central nervous system (CNS) as the levels of expression are often relatively low for these critical functional elements. However, use of amplification for comparisons between two different conditions (e.g., with and without tinnitus) dictates the use of caution, since many rounds of amplification may result in nonlinear amplification and counteract the ability to make comparisons across groups. Before detailing the usefulness of gene expression-based studies in discerning tinnitus-related mechanisms, we first review general principles of PCR and discuss useful variants of the technique.

Polymerase Chain Reaction (PCR)

As a molecular biology technique, PCR can be used to amplify specific regions of DNA, with scientists further manipulating the amplification products (copies of DNA) to gain information about the function of a specific tissue or mechanism for a condition. For example, to determine whether a gene is actually expressed in a particular tissue, PCR can be applied to RNA such as messenger RNA (mRNA). However, unlike double-stranded DNA, single-stranded RNA is prone to degradation. Therefore, before PCR amplification, the mRNA is first converted into complementary DNA (cDNA), in a process called reverse transcription, commonly referred to as RT-PCR. The PCR reaction requires a combination of four key components: thermostable DNA polymers, nucleotide triphosphates (dNTPs —building units of DNA), gene-specific primers, and a DNA or cDNA sample. The amplification of the cDNA is achieved by varying the temperature during three successive steps (one cycle), which are denaturation, annealing, and elongation. First, during denaturation, high heat is used to separate the two strands of DNA by breaking the hydrogen bonds between them. Second, during the annealing process, the DNA is cooled, allowing gene-specific primers to bind to each template strand of DNA. Third, during elongation, the temperature is elevated to activate Taq polymerase to drive formation of a complete DNA strand (polymerization). A PCR reaction can be divided into exponential, linear, and plateau phases. During the exponential phase, each cycle results in a doubling of the amount of DNA. Next, the reaction proceeds to the linear phase, a time when the reaction is slowing down due to depletion of reagents. After approximately 30 to 40 cycles, usually the plateau phase has been reached and the reaction has stopped, indicative of the fact that all PCR reagents are exhausted.

At this point in order to quantify gene expression levels, samples are subjected to gel-electrophoresis, allowing separation of nucleic acids based upon their size and charge.

As a variant of PCR, quantitative real-time PCR (qRT-PCR) allows collection of data throughout the amplification process during each cycle of the PCR reaction. In addition, fluorophores are added to the PCR mixture to detect levels of replicated materials. To collect data, a qRT-PCR instrument is necessary.

Common fluorescence-based qRT-PCR methods employ either SYBR Green or TaqMan assays. SYBR Green is a commonly used affordable dye that binds to the minor groove of DNA and only fluoresces upon binding to double-stranded DNA (dsDNA). A major shortcoming of the SYBR Green assay is a lack of sensitivity, since SYBR Green binds all dsDNA, not only newly formed dsDNA. Alternatively, TaqMan probe consists of an oligonucleotide probe covalently conjugated to a fluorophore reporter at the 5′ end and a non-fluorescent quencher (Q) at the 3′ end. The TaqMan probe hybridizes to DNA templates at regions flanked by forward and reverse primer pairs. Upon cleavage by Taq DNA polymerase, the fluorophore is dissociated from the quencher and can fluoresce (due to decreased proximity to the quencher) upon laser, activation. With the TaqMan approach, the fluorescent signal intensity is proportional only to the amount of newly synthesized dsDNA, an advantage that allows increased sensitivity (detection of small quantities of specific gene products in a given tissue).

While conventional PCR is useful for detection of the expression of specific genes in given tissues, the method can take days to obtain results and has low sensitivity and resolution. Alternatively, qRT-PCR is useful for quantification and thus comparison of gene expression levels. Limitations of qRT-PCR with TaqMan assays include the inability to make comparisons across probes (genes).

Western Blotting to Evaluate Differential Protein Production

Once differential expression of a particular gene is demonstrated with PCR, a complementary technique, Western blotting, can be performed to determine the presence of the corresponding protein in specific cells or tissues, their relative quantities, and state (e.g., phosphorylated). For Western blotting, first proteins that are isolated from tissues of interest are denatured and coated in negative charge. Next, the proteins are separated by weight using gel electrophoresis. Afterward, the proteins are transferred from the gel to a membrane, for labeling with protein-specific antibodies and then visualized and quantified using chemiluminescence.

Western blotting is an effective method for assessing the presence and production level of individual proteins in specific tissues. Since the antibody concentration for detection of the protein of interest can be scaled, proteins can often be detected even when the protein sample size is small. However, Western blotting is dependent upon the availability of specific antibodies for reliable detection. In addition, transfer of larger proteins to the membrane can prove problematic.

Application of Gene Expression to the Central Auditory System

As a general principle, homeostatic mechanisms regulate various facets of neuronal activity. Auditory dysfunction often results in changes in neuronal activity. For example, in the auditory brainstem, depressed auditory input (i.e., from peripheral hearing loss) can strengthen excitatory synaptic transmission (Kotak et al., 2005; Kotak & Sanes, 2002; Muly et al., 2004; Svirskis et al., 2002; Vale & Sanes, 2002) and weaken inhibitory transmission (Kotak et al., 2005; Suneja, Benson, et al., 1998; Suneja, Potashner, et al., 1998; Vale & Sanes, 2002), resulting in increased neuronal activity. Consequently, changes in neuronal activity can result in or result from differential expression of neurotransmitter-related genes. In fact, perturbing of normal hearing status can involve a spatiotemporal-dependent continuum of gene expression levels, particularly as hearing thresholds transition to new set points (in the case of permanent threshold shifts) or return to previous levels (in the case of temporary threshold shifts). Many previous studies have reported noise-induced changes in gene expression in the CN, IC, and auditory cortex (AC) (Campeau et al., 2002; Cui et al., 2007; Dong et al., 2010; Fyk-Kolodziej et al., 2015; Holt et al., 2005; Illing et al., 1999; Illing et al., 2002; Jakob et al., 2015; Luo et al., 1999; Nakagawa et al., 2000; Sato et al., 2000; Shimano et al., 2013; Smith et al., 2014; Sun et al., 2008; Suneja & Potashner, 2002; Tan et al., 2007; Wallhausser-Franke et al., 2003; Zhang et al., 2003).

In subsequent sections, we describe the function of one such family of neurotransmitter-related receptors, NMDA receptors. We discuss and provide examples of the relationship between noise-induced tinnitus and modulation of gene expression and protein levels for NMDA receptors.

Gene Expression of NMDA Receptors in the Auditory System

Glutamate is the primary neurotransmitter used at excitatory synapses in the central nervous system. This neurotransmitter can bind a variety of receptors, including both α-amino-3-hydroxy-5-methyl-4 isoxazolepropionic acid (AMPA), kainate, and NMDA receptors. The AMPA receptors are constitutively expressed, and depending upon the receptor subunit composition, can be delivered and inserted into the synapse through mechanisms that are not activity dependent. However, NMDA receptors are activated and delivered to the synapse in an activity-dependent manner.

Thus, NMDA receptors have been implicated in activity-dependent plasticity. These receptors play a role in changing the balance between excitation and inhibition with different receptor subunit combinations responsible for trafficking receptors with different deactivation times to the synapse (Cull-Candy et al., 2001; McIlhinney et al., 2003; Waxman & Lynch, 2005). Currently, there are three families of NMDA receptors that have been identified, NMDR1, NMDR2 (A-D), and NMDR3(A-B). Generally, a functional receptor consists of two NR1 subunits and two NMDR2 or NMDR3 subunits. Depending on the receptor subunit composition, the receptor can have various deactivation times, from fast, such as the case with NMDR2A, to slow, as with NMDR2D (5-second decay constant).

The expression of four NMDA receptor subunits NMDR2A-D has been modified in animal models. In addition to the generation of knockout mice, in which expression of the subunits has been lost, animal models have been developed in which both NMDR2B and NMDR2D have been overexpressed. In behavioral tests, overexpression of NMDR2B appears to enhance the ability to retain information leading to better long-term memory. High levels of expression of NMDR2D confer sensitivity to glycine and L-glutamate (see Mishina et al., 1993, for review). Overexpression of NMDR2D also retards the development of induced epileptic seizures. Since NMDR2D is implicated in the development of epilepsy (Bengzon et al., 1999; Okabe et al., 1998), a condition that is similar to tinnitus in that affected neurons are more easily excited and demonstrate increases in neuronal activity, NMDR2D may also play a role in tinnitus. In the case of noise-induced tinnitus, the elevated NMDR2D levels we observe may be a compensatory response as the auditory pathway tries to regain homeostasis.

In the auditory system, NMDR1 and NMDR2 (A-D) are expressed in the cochlea, CN, and IC (Niedzielski et al., 1997; Usami et al., 1995). These receptor subunits do not appear to be directly involved in synaptic transmission in the cochlea; however, they are implicated in synaptic repair after excitotoxicity (d'Aldin et al., 1997). Antagonists for NMDA receptors can protect hair cells from aminoglycoside (Basile et al., 1996), noise, and ischemia-induced toxicity (Duan et al., 2000; Pujol et al., 1992; Puel et al., 1994; Puel et al., 1995). When applied into the cochleae, NMDA receptors antagonists MK-801 (channel blocker), 7-chlorokynurenate (glycine-site antagonist), and gacyclidine (phencyclidine-site antagonist) strongly reduced the occurrence of false-positive responses induced by salicylate-generated tinnitus, i.e., they attenuated the tinnitus (Guitton et al., 2003).

A Study Examining Gene Expression of NMDA Receptors in Tinnitus Models

Models of Noise-Induced Tinnitus

In general, two basic models of noise-induced tinnitus have been developed in the rat: those that produce a permanent shift in hearing thresholds (PTS) and those that produce a temporary shift in hearing thresholds (TTS). In one animal model of tinnitus, Jastreboff et al. (1988) demonstrated that rats previously exposed to a narrowband noise still perceived sound even in the absence of an acoustic stimulus. Others (Zhang & Kaltenbach, 1998) have used this model to show that in the DCN of rats exposed to a similar intense narrowband noise, neurons demonstrated increased spontaneous activity when compared to non-noise-exposed animals. In this PTS model, the narrowband noise to which the animals were subjected generated an irreversible shift in auditory thresholds. In another tinnitus model developed by Bauer and Brozoski (2001), the duration of the noise exposure and the degree of cochlear damage were correlated with the behavioral characteristics of the tinnitus. In their TTS model, animals were exposed to a tone sufficient for a transient shift in auditory thresholds that reverted to baseline levels 5 days later. In both animal models, psychophysical data indicated the presence of tinnitus, suggesting that the perception of tinnitus is separate from hearing loss and that tinnitus led to increased neuronal activity in the DCN, but only days after the onset of the tinnitus.

While we have learned a great deal from the two aforementioned models of tinnitus, there are still many unanswered questions. Here, we determine the level of hair cell damage that is sustained in both animal models of tinnitus as well as the hearing status of the animals following noise exposure. Additionally, considering the mounting evidence that tinnitus results in a shift in the balance between excitation and inhibition in the CN and that tinnitus can occur even when there is no permanent threshold shift, we (1) assessed whether changes in gene expression levels of NMDA receptors occur in both animal models of tinnitus and (2) whether those changes are similar in direction and magnitude in both models. Our rationale being that those genes that met our criterion are more likely to be associated with tinnitus and not just hearing loss or acoustic trauma alone.

The degree of damage to the cochlea would be one predictor of the scale of expected changes in gene expression and ultimately neuronal activity, as the system attempts to restore normal output, possibly by modulating gain over time. We tested auditory brainstem responses (ABRs) to determine the effects of noise exposure on hearing loss in male Sprague Dawley rats belonging to either the TTS group, exposed to a 105-dB SPL 16-kHz octave band tone for 1 hour, or the PTS group, exposed to a 118-dB SPL 10-kHz 1/3-octave narrowband noise for 4 hours. Each noise exposure resulted in a different cochlear injury, and results were in agreement with earlier reports in the Bauer and Brozoski (2001) models.

To characterize anatomical changes resulting from noise overstimulation, we examined hair cell loss along the entire length of the cochlear tissue, by staining histologic preparation with 1% phalloidin (intrascalar). Results from ABRs and surface preparations of the cochleae (Figure 6–1) revealed minimal inner and outer hair cell loss (IHC, OHC) in control groups (0% IHC and 1.4% OHC), negligible hair cell loss in the TTS group (0.13% IHC and 0.45% OHC), and more substantial hair cell loss for the PTS group (8.2% IHC and 94.5% OHC). Although the results presented here are from animals exposed to a tone of similar frequency and intensity as that reported by the Bauer and Brozoski TTS tinnitus model, animals in the current study sustained less hair cell loss. The discrepancy in the results could be due to the fact that in the Bauer and Brozoski (2001) paradigm, the noise is introduced via a speculum into one ear while in the present study, the noise was introduced in free field conditions in nonanesthetized animals with a speaker placed above animal individually housed in wire mesh cages. Recently, TTS has been shown to result in long-term peripheral effects

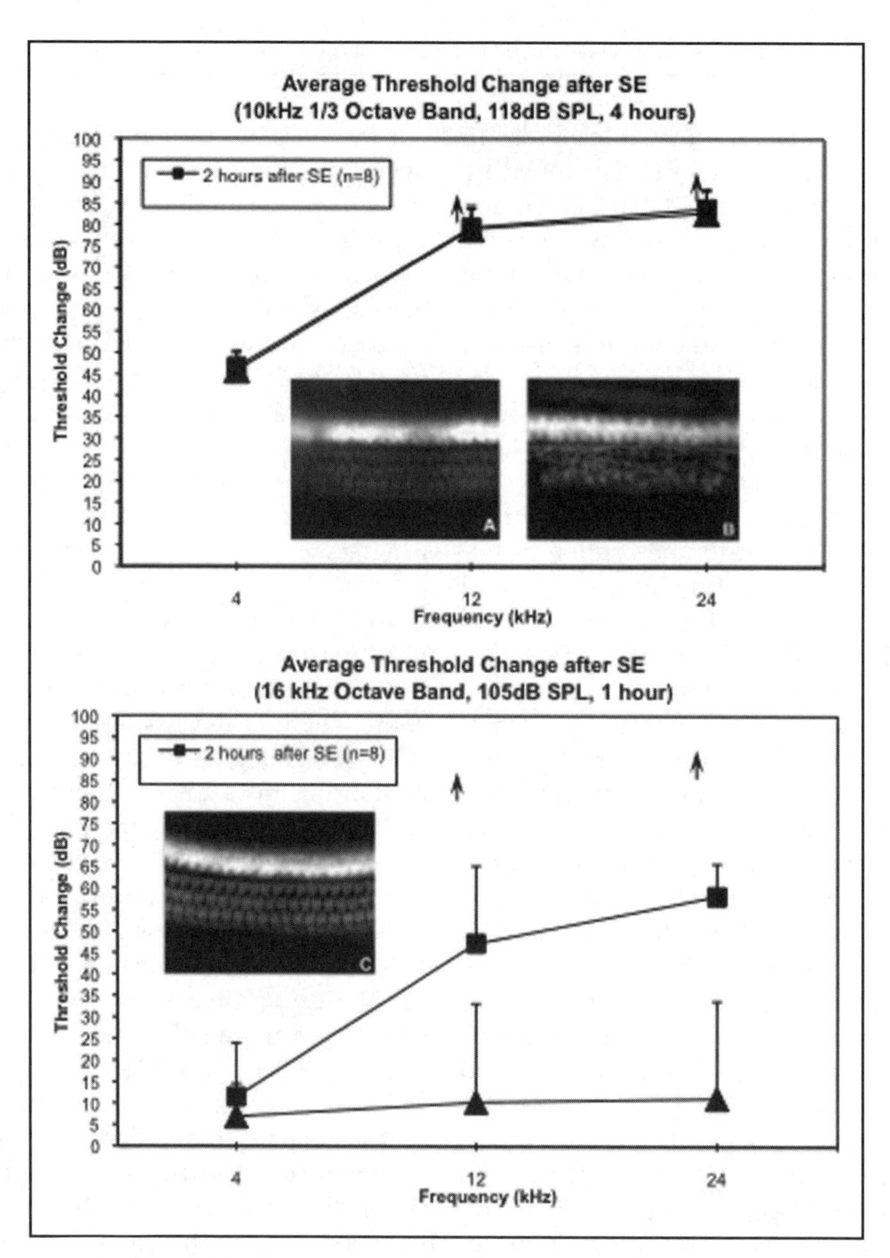

Figure 6–1. Auditory brainstem responses following (ABRs) and cochlear histology following PTS or TTS noise exposures. Auditory brainstem responses were examined both 2 hours and 4 days after the noise exposure. Hearing thresholds were elevated in both groups 2 hours following noise exposure. By 4 days, thresholds of the TTS noise group had returned to normal levels, whereas thresholds for the PTS group were higher at 4 days than at 2 hours following noise. The thresholds are expressed as threshold shifts from normal hearing thresholds. **A–C.** Photomicrographs of surface preparations of the organ of Corti 5 days after exposure to noise that caused either a PTS or TTS. **A.** The micrograph of cochlear hair cells in control animals. In unexposed animals, the cuticular plates of the three rows of outer hair cells are clearly seen. **B.** The PTS noise caused a loss of the majority of outer hair cells. **C.** The TTS noise exposure showed relatively intact outer hair cells.

even after the return of hearing (Knipper et al., 2013; Hickox & Liberman, 2014; Altschuler et al., 2015; Fernandez et al., 2015 ; Hickox & Liberman, 2014; Knipper et al., 2013). How these effects of diminished Wave 1 ABR amplitude and IHC synaptic loss compare over time with tinnitus susceptibility are important measures to consider.

A PTS or TTS Producing Noise Exposure Results in Changes in NMDA Receptor Gene Expression

One related feature that is common to both tinnitus models is an initial increase in neuronal activity, which may occur through a variety of mechanisms, including modulation of neurotransmitter receptors at the synapse in order to maintain homeostasis. Changes in NMDA receptor gene expression were examined at 5 days after noise-induced TTS or PTS. Five days is the time when Kaltenbach et al. (2000) demonstrated increased spontaneous neuronal activity over a large portion of the DCN following exposure to a PTS-producing noise.

We demonstrated differential gene expression of NMDA receptors from the CN of both TTS and PTS groups (Figure 6–2). For four of the genes, expression levels increased by more than 150% in each animal model (NMDAR1, NMDAR2B, NMDAR2C, NMDAR2D) with each showing a comparable magnitude of increase across both models. The genes showed increased expression in the cochlear nuclei from both models at a time when there was a return of normal hearing thresholds in the TTS model and while there was considerable hearing loss in the PTS model.

Specifically, gene expression levels for four of the NMDA receptor genes, NMDAR 1, NMDAR 2B, NMDAR 2C, and NMDAR 2D, increased following noise exposure. The NMDAR 2D subunit showed the largest increase in gene expression for both groups, with a 12.0-fold increase for the PTS group and a 14.1-fold increase in the TTS group. In addition, two subunits had

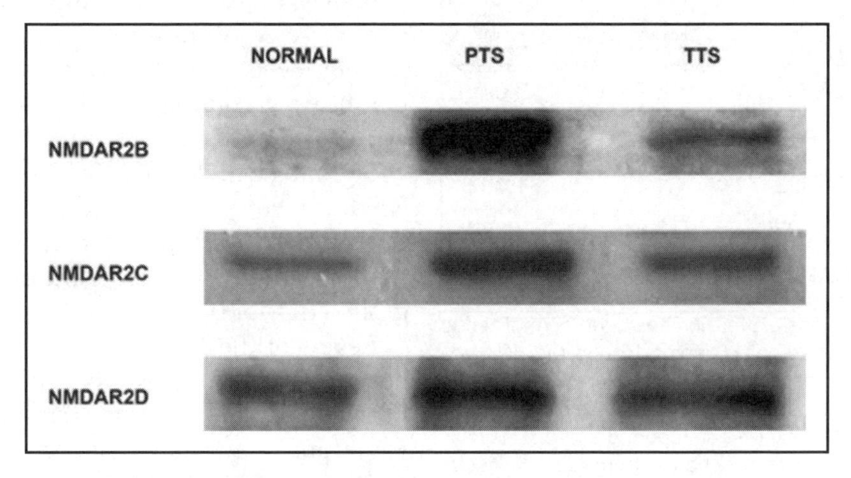

Figure 6–2. Western blot showing changes in NMDA receptor protein levels in the CN of normal and noise-exposed animals. PTS, permanent threshold shift; TTS, temporary threshold shift.

large increases that were similar when comparing noise exposure groups, with NMDAR 1 (3.3-fold PTS and 2.01-fold TTS) and NMDAR 2C (3.8-fold PTS and 3.6-fold TTS) showing at least 2-fold changes in expression. The NMDAR 2B subunit by comparison had very modest increases in each group (1.12-fold PTS and 1.49-fold TTS). The NMDAR 2A subunit had lower mRNA levels in the PTS condition (0.765-fold change) but had increased expression in the TTS condition (1.51-fold change). Thus, NMDAR 2A gene expression levels were more variable across the two noise exposure paradigms (Table 6–1). Taken together, these observations suggest that changes in NMDA receptor gene expression levels occurring in both models are not directly related to deafness and/or the amount of neuronal input at the time of assessment. These changes might therefore be associated with the tinnitus reported in both of these models and could provide clues to the underlying mechanisms.

Table 6–1. Quantitative RT-PCR Analysis of mRNA Levels for Five NMDA Receptor Subunits

	Gene Expression (fold change)	
	CN PTS	CN TTS
NMDAR1	3.26	2.06
NMDAR2A	0.77	1.51
NMDAR2B	1.12	1.49
LNMDAR2C	3.81	3.57
NMDAR2D	12.30	14.16

Note. Gene expression has been normalized to the housekeeping gene S16 and the data are reported as the ratio of gene expression from the noise-exposed groups to that of the normal group. Therefore, expression reported at the "1" level is interpreted as no change from normal. Above 1 indicates increased gene expression while expression below 1 is interpreted as a decrease.

In a previous report, decreases in NMDAR1 expression occurred in the CN following cochlear deafferentation (Sato et al., 2000). Since NMDAR1 subunits are required for the formation of a functional NMDA receptor, changes in the expression level of this subunit suggest changes in the total number of receptors expressed instead of a change in the function of the receptor. The NMDA receptors have been implicated in learning related plasticity (Sun et al., 2005) with decreases in the levels of NMDAR2A and NMDAR2B subunits in the auditory cortex after improved discrimination in an auditory learning paradigm.

As a complementary assessment, we also quantified protein levels for three NMDA receptors, NMDAR2B, NMDAR2C, and NMDAR2D (see Figure 6–2). We found that the CN from the TTS and PTS groups exhibited changes in protein levels for each of the NMDA receptor subunits tested. For NMDAR2B, both the PTS (3.4-fold) and the TTS (2.8-fold) noise exposure resulted in elevated protein levels when compared to normal-hearing controls (see Figure 6–2). The NMDAR2C was increased in the PTS group (1.8-fold) and the TTS group (1.9-fold). For the NMDAR2D subunit, protein levels were increased in the both the PTS group (1.25-fold) and the TTS group (1.21-fold).

Levels of the NMDA receptor have been reported to change in response to hearing loss. In the AC, auditory deprivation and noise trauma has been shown to increase NMDAR2A protein (Wang et al., 2005). Interestingly, our current results and previous studies have demonstrated increases in NMDAR2B expression with deafness, in the CN and superior olivary complex (Nakagawa et al., 2000).

Central NMDA Receptors May Represent Additional Sites for Tinnitus Attenuation and May Be Beneficial Across Tinnitus Models

Changes in all of the aforementioned genes that increased in both noise exposure models suggest compensatory mechanisms that increase the expression of specific genes when there is increased excitation such as that observed in the CN following the onset of noise induced tinnitus. Changes in the expression of these neurotransmission-related genes might be one method by which the CN tries to restore homeostasis following acoustic trauma.

NMDA Receptor Interactions

In addition to modulating the number and type of NMDA receptor subunits, tinnitus may also influence NMDA receptor interactions. Specifically, NMDA receptors form direct interactions with dopamine receptors (DRs) and DRs have a host of protein partners that influence synaptic output (neuronal activity). Therefore, NMDA receptors provide a considerable pool of targets from which to select for modulation toward a therapeutic end. Recent studies have demonstrated changes in dopamine receptor gene expression levels following cochlear destruction (Fyk-Kolodziej et al., 2015). Studying the relationship between NMDA receptors and DRs in the context of tinnitus is an area that is ripe for exploration (Fan et al., 2014).

Gene Expression as a Guide to Develop New Therapeutic Strategies for Tinnitus

Mechanisms Underlying the Perception of Tinnitus Are Poorly Understood

Several studies now contend that targeting of NMDA receptors represent a promising therapeutic strategy for treating noise-induced tinnitus (Bing et al., 2015; Guitton et al., 2004; Guitton & Dudai, 2007). While there are many compounds that may influence the percepts of tinnitus in various regions along the auditory pathways, developing a metric by which to screen these compounds and determine their site(s) of action is crucial. Learning more about the sites of action, the length of time effects persist, and changes in biomarkers associated with tinnitus will allow us to identify mechanisms underlying tinnitus and to develop an effective treatment. Changes in NMDA receptors support the idea of tinnitus-related changes in neuronal activity. This begs the question of whether tinnitus-related changes in neuronal activity only occur in the CN or whether other brain regions exhibit similar responses. One method that allows us to address this question in several brain regions at once is manganese enhanced resonance imaging (MEMRI).

Manganese-Enhanced MRI (MEMRI) Is a Powerful Tool for Measuring Neuronal Activity In Vivo

Manganese-enhanced MRI is a noninvasive imaging method used to assess neuronal activity. Manganese (Mn^{2+}) ions accumulate in active neurons via passage through voltage-gated calcium channels. Therefore, more active neurons have more Mn^{2+} accumulation. The efflux of Mn^{2+} from neurons is fairly slow (Itoh et al., 2008). For MEMRI, the paramagnetic Mn^{2+} ion is used as both a contrast agent and calcium channel probe that allows for the visualization of neuronal activity repeatedly in live animals. Activation of NMDA receptors increases Mn^{2+} uptake (Itoh et al., 2008). Feasibility of the use of MEMRI in auditory pathways for the study of noise-induced tinnitus has been demonstrated (Brozoski et al., 2007; Cacace et al., 2014; Holt et al., 2010). In fact, central administration of NMDA receptor antagonists has been shown to diminish tinnitus and MEMRI detectable increases in neuronal activity. Given that MEMRI is fairly noninvasive and translatable to the clinic, this innovative approach provides a metric for following changes in tinnitus-related neuronal activity over time to test the efficacy of treatments.

Disrupted homeostasis of ions, such as calcium, is hypothesized to play a substantial role in tinnitus. Increases in spontaneous and ensemble activity of the auditory nerve have been demonstrated in a model of tinnitus (Cazals et al., 1998; Evans & Borerwe, 1982; Martin et al., 1993; Muller et al., 2003; Schreiner et al., 1986; Stypulkowski, 1990). In addition, several studies show that calcium channel blockade can attenuate tinnitus percepts and decrease spontaneous neuronal excitability caused by salicylate (a drug that induces tinnitus). Despite these insights, methods that can probe tinnitus-related Ca^{2+} dysregulation in brain regions in vivo have not been available, leaving untested the hypothesis that Ca^{2+} dysregulation occurs with tinnitus in vivo. Demonstration of a link between tinnitus and neuronal ion dysregulation is expected to greatly aid in the development and application of interventions for tinnitus. Current methods for alleviating tinnitus are lacking. New and innovative approaches to treat tinnitus are needed. As biomarkers such as hyperactive brain regions, receptors that are differentially expressed, and mechanisms underlying tinnitus are identified, methods for treatment can be explored.

Repetitive Transcranial Magnetic Stimulation (rTMS) Modulates Neuronal Activity and Transiently Decreases Tinnitus Percepts

The current studies are relevant to human studies of tinnitus. Loudness of tinnitus has been associated with AC increases in neuronal activity (Lockwood et al., 1998; Reyes et al., 2002). Neurons in the AC were more excitable upon stimulation by noise in patients with both tinnitus and deafness (Lockwood et al., 1998). In animal studies, noise exposure can also result in increased spontaneous neuronal activity in the AC (Brozoski et al., 2007; Komiya & Eggermont, 2000). Recent studies in humans with tinnitus used rTMS to reduce activity-related neuronal asymmetry following tinnitus. Low-frequency rTMS resulted in reduced activity levels in the AC lasting several days with reduced tinnitus percepts. The rTMS method affects all of the neurons in a region below the stimulator probe and is therefore not very specific. In addition to identifying brain regions with differential changes in gene expression and hyperactivity, complementary techniques such as immunocytochemistry can be used to localize related proteins to specific neurons. Then those specific neurons can be manipulated to modulate activity. One such method for in vivo modulation of specific neurons is optogenetics.

Optogenetics for Modulating Neuronal Activity in Specific Neurons

As implied by its name, optogenetics is a method that couples genetic and optical techniques to acquire precise control over neuronal activity. By co-opting genetic modification techniques, neurons can be induced to express an array of light-sensitive membrane-bound ion pumps and channels, from the opsin family of microbes. Each type of opsin is sensitive to a specific wavelength of light. Therefore, transfection of opsin genes into neurons permits precise regulation of action potentials (Boyden et al., 2005). Optogenetics has been successfully applied to auditory neurons. The activity of spiral ganglion, cochlear nucleus, and auditory cortex neurons has been modulated via transfection of specific neurons with opsins and stimulation with light of particular wavelengths for either excitation or inhibition (e.g., Darrow et al., 2015; Hight et al., 2015; Kozin et al., 2015; Lima et al., 2009; Shimano et al., 2013).

In addition to rTMS and optogenetic treatment to modulate neuronal activity, pharmaceutical treatments may also prove beneficial. For example, when animals with psychophysical evidence of tinnitus were treated with the gamma-Aminobutric acid (GABA) agonist gabapentin, the drug successfully reduced tinnitus percepts (Bauer & Brozoski, 2001). However, the drug was not successful in human clinical trials. One reason may be due to the method of delivery. A method that allows delivery to sites that specifically demonstrate tinnitus-related functional changes might greatly improve the probability of success. Nanoparticle technology provides an opportunity to target specific cells based upon their gene expression profile.

Nanoparticles for Drug Delivery

The use of a nanoparticle (NP) platform for tinnitus treatment would represent a new and exciting avenue of research for tinnitus, a field in which the NP approach is still in the early stages of development. The goal would be to enable targeting and concentration of promising pharmacological agents in specific regions of interest. Development of an NP platform for use in vivo with animal models of tinnitus would be a good first step. Such a platform would ideally have the ability to pass over the blood-brain barrier (BBB), display targeting moieties (derived from gene expression studies) for localization to hyperactive regions, and provide means to monitor their localization in the brain. The internal core of the NP can be filled with a pharmaceutical agent to facilitate drug delivery. Ultimately, the goal is to develop targeted NPs carrying pharmacologic therapeutic cargo to improve treatment outcomes for individuals with tinnitus.

One possible NP platform that would allow targeting, localization, and monitoring may be the combination of bacteriophages and plant virus capsids. Capsids present a plethora of opportunities for surface ligand modification and cargo loading with numerous examples in the literature of their use in nanomaterial development (Bergkvist & Cohen, 2013). The MS2 coliphage for instance, is a ~∅30 nm NP that has been established as a targeted drug delivery platform and contrast agent (Brown et al., 2002; Cohen & Bergkvist, 2013; Hooker et al., 2008; Meldrum et al., 2010; Tong et al., 2009). A schematic of the capsid NP concept with dual targeting ligands for BBB transport and neuronal receptor recognition, with internal loading of small molecules, is shown in Figure 6–3.

By comparing a PTS model with a TTS model of noise overstimulation, we were able to learn more about NMDA receptor genes that are differentially expressed as a consequence of noise overstimulation and tinnitus. Increased gene expression levels in both models may provide clues to a common mechanism, such as tinnitus-induced calcium dysregulation producing increased spontaneous activity of neurons. In the presented study, gene expression levels for NMDA receptors, often associated with activity-dependent plasticity, were increased in both the PTS and TTS tinnitus models. Perhaps to understand more about the mechanisms of the central phenomena associated with these models, we will need to correlate changes in the periphery with central changes, exploit technologies that allow longitudinal comparisons, and devise delivery approaches to target specific

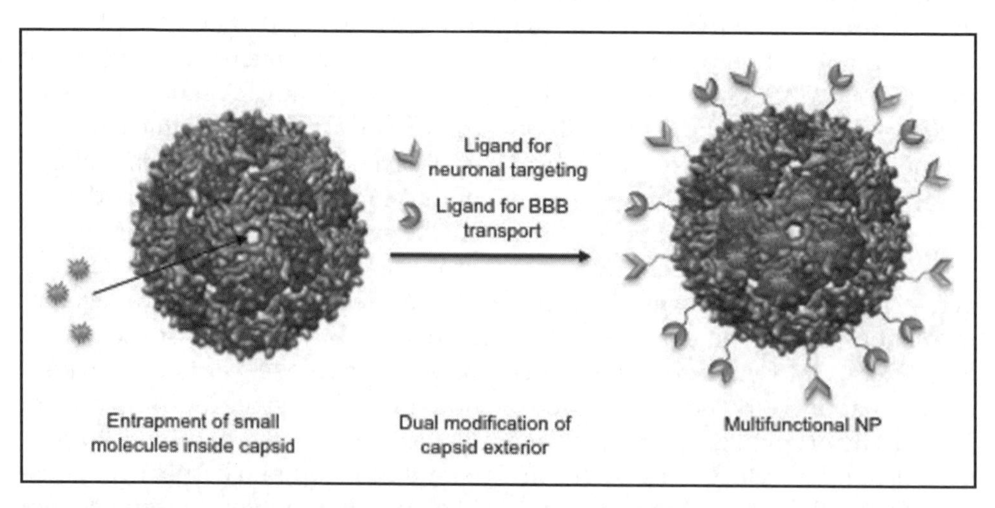

Figure 6–3. Concept illustration of capsid-based multifunctional NP platform where small molecules (imaging agents for localization and monitoring) can be loaded in the interior cavity and multiple ligands be conjugated to the exterior surface.

neurons in critical brain regions to successfully combat maladaptive homeostatic processes as tinnitus progresses.

In summary, this chapter stresses the importance of gene expression methods to detect auditory system dysfunction at the molecular level. One of the benefits of such an analysis is in developing novel detection tools and treatments, such as the use of MEMRI, functionalized nanoparticles, and optogenetics, using gene expression biomarkers as a guide. The ability to decorate the external surface of the nanoparticle with targeting moieties or endow specific neurons with the ability to respond to specific wavelengths of light provides just some of the examples of how gene expression results (i.e., NMDAR2D) can be used to identify, target, and alleviate tinnitus-related changes in neuronal activity. The novelty of our approach lies in the combination of these methods to advance research for combating auditory disorders, such as tinnitus, through the generation of sustainable treatments.

References

Altschuler, R. A., Dolan, D. F., Halsey, K., Kanicki, A., Deng, N., Martin, C., & Schacht, J. (2015). Age-related changes in auditory nerve-inner hair cell connections, hair cell numbers, auditory brain stem response and gap detection in UM-HET4 mice. *Neuroscience, 292,* 22–33.

AMVETS (American Veterans) DDAV, Paralyzed Veterans of American (Paralyzed Veterans), and Veterans of Foreign Wars of the United States (VFW). (2015). *The independent budget for the Department of Veterans Affairs: A comprehensive budget & policy document created by veterans for veterans.* Washington, DC: Department of Veterans Affairs.

Basile, A. S., Huang, J. M., Xie, C., Webster, D., Berlin, C., & Skolnick, P. (1996). N-methyl-D-aspartate antagonists limit aminoglycoside antibiotic-induced hearing loss. *Nature Medicine, 2,* 1338–1343.

Bauer, C. A., & Brozoski, T. J. (2001). Assessing tinnitus and prospective tinnitus therapeutics using a psychophysical animal model. *Journal of the Association for Research in Otolaryngology, 2,* 54–64.

Bengzon, J., Okabe, S., Lindvall, O., & McKay, R. D. (1999). Suppression of epileptogenesis by modification of N-methyl-D-aspartate receptor subunit composition. *European Journal of Neuroscience, 11,* 916–922.

Bergkvist, M., & Cohen, B. (2013). Virus-based nanotechnology. In Y. Xie (Ed.), *The nanobiotechnology handbook.* Boca Raton, FL: CRC Press.

Bing, D., Lee, S. C., Campanelli, D., Xiong, H., Matsumoto, M., Panford-Walsh, R., . . . Singer, W. (2015). Cochlear NMDA receptors as a therapeutic target of noise-induced tinnitus. *Cell Physiology and Biochemistry, 35,* 1905–1923.

Boyden, E. S., Zhang, F., Bamberg, E,, Nagel, G., & Deisseroth, K. (2005). Millisecond-timescale, genetically targeted optical control of neural activity. *Nature Neuroscience, 8,* 1263–1268.

Brown, W. L., Mastico, R. A., Wu, M., Heal, K. G., Adams, C. J., Murray, J. B., . . . Stockley, P. G., (2002). RNA bacteriophage capsid-mediated drug delivery and epitope presentation. *Intervirology, 45,* 371–380.

Brozoski, T. J., Bauer, C. A., & Caspary, D. M. (2002). Elevated fusiform cell activity in the dorsal cochlear nucleus of chinchillas with psychophysical evidence of tinnitus. *Journal of Neuroscience, 22,* 2383–2390.

Brozoski, T. J., Ciobanu, L., & Bauer, C. A. (2007). Central neural activity in rats with tinnitus evaluated with manganese-enhanced magnetic resonance imaging (MEMRI). *Hearing Research, 228,* 168–179.

Cacace, A. T., Brozoski, T., Berkowitz, B., Bauer, C., Odintsov, B., Bergkvist, M., . . . Holt, A. G., (2014). Manganese enhanced magnetic resonance imaging (MEMRI): A powerful new imaging method to study tinnitus. *Hearing Research, 311,* 49–62.

Cacace, A. T., Lovely, T. J., McFarland, D. J., Parnes, S. M., & Winter, D. F. (1994). Anomalous cross-modal plasticity following posterior fossa surgery: Some speculations on gaze-evoked tinnitus. *Hearing Research, 81,* 22–32.

Cacace, A. T., Lovely, T. J., Winter, D. F., Parnes, S. M., & McFarland, D. J. (1994). Auditory perceptual and visual-spatial characteristics of gaze-evoked tinnitus. *Audiology, 33,* 291–303.

Campeau, S., Dolan, D., Akil, H., & Watson, S. J. (2002). c-fos mRNA induction in acute and chronic audiogenic stress: Possible role of the orbitofrontal cortex in habituation. *Stress, 5,* 121–130.

Cazals, Y., Horner, K. C., & Huang, Z. W. (1998). Alterations in average spectrum of cochleoneural activity by long-term salicylate treatment in the guinea pig: a plausible index of tinnitus. *Journal of Neurophysiology, 80,* 2113–2120.

Cohen, B. A., & Bergkvist, M. (2013). Targeted in vitro photodynamic therapy via aptamer-labeled, porphyrin-loaded virus capsids. *Journal of Photochemistry and Photobiology B: Biology, 121,* 67–74.

Cui, Y. L., Holt, A. G., Lomax, C. A., & Altschuler, R. A. (2007). Deafness associated changes in two-pore domain potassium channels in the rat inferior colliculus. *Neuroscience, 149,* 421–433.

Cull-Candy, S., Brickley, S., & Farrant, M. (2001). NMDA receptor subunits: Diversity, development and disease. *Current Opinion in Neurobiology, 11,* 327–335.

d'Aldin, C. G., Ruel, J., Assie, R., Pujol, R., & Puel, J. L. (1997). Implication of NMDA type glutamate receptors in neural regeneration and neoformation of synapses after excitotoxic injury in the guinea pig cochlea. *International Journal of Developmental Neuroscience, 15,* 619–629.

Darrow, K. N., Slama, M. C., Kozin, E. D., Owoc, M., Hancock, K., Kempfle, J., . . . Lee, D. J. (2015). Optogenetic stimulation of the cochlear nucleus using channelrhodopsin-2 evokes activity in the central auditory pathways. *Brain Research, 1599,* 44–56.

Dong, S., Mulders, W. H., Rodger, J., Woo, S., & Robertson, D. (2010). Acoustic trauma evokes hyperactivity and changes in gene expression in guinea-pig auditory brainstem. *European Journal of Neuroscience, 31,* 1616–1628.

Duan, M., Agerman, K., Ernfors, P., & Canlon, B. (2000). Complementary roles of neurotrophin 3 and a N-methyl-D-aspartate antagonist in the protection of noise and aminoglycoside-induced ototoxicity. *Proceedings of the National Academy of Sciences USA, 97,* 7597–7602.

Evans, E. F., & Borerwe, T. A. (1982). Ototoxic effects of salicylates on the responses of single cochlear nerve fibres and on cochlear potentials. *British Journal of Audiology, 16,* 101–108.

Fan, X., Jin, W. Y., & Wang, Y. T. (2014). The NMDA receptor complex: A multifunctional machine at the glutamatergic synapse. *Frontiers in Cellular Neuroscience, 8,* 160.

Fernandez, K. A., Jeffers, P. W., Lall, K., Liberman, M. C., & Kujawa, S. G. (2015). Aging after noise exposure: Acceleration of cochlear synaptopathy in "recovered" ears. *Journal of Neuroscience, 35,* 7509–7520.

Finlayson, P. G., & Kaltenbach, J. A. (2009). Alterations in the spontaneous discharge patterns of single units in the dorsal cochlear nucleus following intense sound exposure. *Hearing Research, 256,* 104–117.

Francis, H. W., & Manis, P. B. (2000). Effects of deafferentation on the electrophysiology of ventral cochlear nucleus neurons. *Hearing Research, 149,* 91–105.

Fyk-Kolodziej, B. E., Shimano, T., Gafoor, D., Mirza, N., Griffith, R. D., Gong, T. W., & Holt, A. G. (2015). Dopamine in the auditory brainstem and midbrain: Co-localization with amino acid neurotransmitters and gene expression following cochlear trauma. *Frontiers in Neuroanatomy, 9,* 88.

Guitton, M. J., Caston, J., Ruel, J., Johnson, R. M., Pujol, R., & Puel, J. L. (2003). Salicylate induces tinnitus through activation of cochlear NMDA receptors. *Journal of Neuroscience, 23,* 3944–3952.

Guitton, M. J., & Dudai, Y. (2007). Blockade of cochlear NMDA receptors prevents long-term tinnitus during a brief consolidation window after acoustic trauma. *Neural Plasticity, 2007,* 80904.

Guitton, M. J., Wang, J., & Puel, J. L. (2004). New pharmacological strategies to restore hearing and treat tinnitus. *Acta Otolaryngologica, 124,* 411–415.

Hickox, A. E., & Liberman, M. C. (2014). Is noise-induced cochlear neuropathy key to the generation of hyperacusis or tinnitus? *Journal of Neurophysiology, 111,* 552–564.

Hight, A. E., Kozin, E. D., Darrow, K., Lehmann, A., Boyden, E., Brown, M. C., . . . Lee, A., (2015). Superior temporal resolution of Chronos versus channelrhodopsin-2 in an optogenetic model of the auditory brainstem implant. *Hearing Research, 322,* 235–241.

Hirsch, J. A., & Oertel, D. (1988). Intrinsic properties of neurones in the dorsal cochlear nucleus of mice, in vitro. *The Journal of Physiology, 396,* 535–548.

Holt, A. G., Asako, M., Lomax, C. A., MacDonald, J. W., Tong L., Lomax, M. I., . . . Altschuler, R. A., (2005). Deafness-related plasticity in the inferior colliculus: Gene expression profiling following removal of peripheral activity. *Journal of Neurochemistry, 93,* 1069–1086.

Holt, A. G., Bissig, D., Mirza, N., Rajah, G., & Berkowitz, B. (2010). Evidence of key tinnitus-related brain regions documented by a unique combination of manganese-enhanced MRI and acoustic startle reflex testing. *PLoS One, 5,* e14260.

Hooker, J. M., O'Neil, J. P., Romanini, D. W., Taylor, S. E., & Francis, M. B. (2008). Genome-free viral capsids as carriers for positron emission tomography radiolabels. *Molecular Imaging and Biology, 10,* 182–191.

Illing, R. B., Cao, Q. L., Forster, C. R., & Laszig R. (1999). Auditory brainstem: Development and plasticity of GAP-43 mRNA expression in the rat. *The Journal of Comparative Neurology, 412,* 353–372.

Illing, R. B., Michler, S. A., Kraus, K. S., & Laszig R. (2002). Transcription factor modulation and expression in the rat auditory brainstem following electrical intracochlear stimulation. *Experimental Neurology, 175,* 226–244.

Itoh, K., Sakata, M., Watanabe, M., Aikawa, Y., & Fujii, H. (2008). The entry of manganese ions into the brain is accelerated by the activation of N-methyl-D-aspartate receptors. *Neuroscience, 154,* 732–740.

Jakob, T. F., Doring, U., & Illing, R. B. (2015). The pattern of Fos expression in the rat auditory brainstem changes with the temporal structure of binaural electrical intracochlear stimulation. *Experimental Neurology, 266,* 55–67.

Jastreboff, P. J., Brennan, J. F., & Sasaki, C. T. (1988). An animal model for tinnitus. *Laryngoscope, 98,* 280–286.

Kalappa, B. I., Brozoski, T. J., Turner, J. G., & Caspary, D. M. (2014). Single unit hyperactivity and bursting in the auditory thalamus of awake rats directly correlates with behavioural evidence of tinnitus. *The Journal of Physiology, 592,* 5065–5078.

Kaltenbach, J. A., & Afman, C. E. (2000). Hyperactivity in the dorsal cochlear nucleus after intense sound exposure and its resemblance to tone-evoked activity: A physiological model for tinnitus. *Hearing Research, 140,* 165–172.

Kaltenbach, J. A., Zhang, J., & Afman, C. E. (2000). Plasticity of spontaneous neural activity in the dorsal cochlear nucleus after intense sound exposure. *Hearing Research, 147,* 282–292.

Knipper, M., Van Dijk, P., Nunes, I., Ruttiger, L., & Zimmermann, U. (2013). Advances in the neurobiology of hearing disorders: Recent developments regarding the basis of tinnitus and hyperacusis. *Progress in Neurobiology, 111,* 17–33.

Koerber, K. C., Pfeiffer, R. R., Warr, W. B., & Kiang, N. Y. (1966). Spontaneous spike discharges from single units in the cochlear nucleus after destruction of the cochlea. *Experimental Neurology, 16,* 119–130.

Komiya, H., & Eggermont, J. J. (2000). Spontaneous firing activity of cortical neurons in adult cats with reorganized tonotopic map following pure-tone trauma. *Acta Otolaryngologica, 120,* 750–756.

Kotak, V. C., Fujisawa, S., Lee, F. A., Karthikeyan, O., Aoki, C., & Sanes, D. H. (2005). Hearing loss raises excitability in the auditory cortex. *Journal of Neuroscience, 25,* 3908–3918.

Kotak, V. C., & Sanes, D. H. (2002). Postsynaptic kinase signaling underlies inhibitory synaptic plasticity in the lateral superior olive. *Journal of Neurobiology, 53,* 36–43.

Kozin, E. D., Darrow, K. N., Hight, A. E., Lehmann, A. E., Kaplan, A. B., Brown, M. C., . . . Lee, D. J., (2015). Direct visualization of the murine dorsal cochlear nucleus for optogenetic stimulation of the auditory pathway. *Journal of Visualized Experiments, 95,* 52426.

Lima, S. Q., Hromadka, T., Znamenskiy, P. & Zador, A. M. (2009). PINP: A new method of tagging neuronal populations for identification during in vivo electrophysiological recording. *PLoS One, 4,* e6099.

Lockwood, A. H., Salvi, R. J., Coad, M. L., Towsley, M. L., Wack, D. S., & Murphy, B. W. (1998). The functional neuroanatomy of tinnitus: Evidence for limbic system links and neural plasticity. *Neurology, 50,* 114–120.

Longenecker, R. J., & Galazyuk, A. V. (2011). Development of tinnitus in CBA/CaJ mice following sound exposure. *Journal of the Association for Research in Otolaryngology, 12,* 647–658.

Luo, L., Ryan, A. F., & Saint Marie, R. L. (1999). Cochlear ablation alters acoustically induced c-fos mRNA expression in the adult rat auditory brainstem. *The Journal of Comparative Neurology, 404,* 271–283.

Martin, W. H., Schwegler, J. W., Scheibelhoffer, J., & Ronis, M. L. (1993). Salicylate-induced changes in cat auditory nerve activity. *Laryngoscope, 103,* 600–604.

McIlhinney, R. A., Philipps, E., Le Bourdelles, B., Grimwood, S., Wafford, K., Sandhu, S., & Whiting, P. (2003). Assembly of N-methyl-D-aspartate (NMDA) receptors. *Biochemical Society Transactions, 31*(Pt. 4), 865–868.

Meldrum, T., Seim, K. L., Bajaj, V. S., Palaniappan, K. K., Wu, W., Francis, M. B., . . . Pines, A. (2010). A xenon-based molecular sensor assembled on an MS2 viral capsid scaffold. *Journal of the American Chemical Society, 132,* 5936–5937.

Milbrandt, J. C., Holder, T. M., Wilson, M. C., Salvi, R. J., & Caspary, D. M. (2000). GAD levels and muscimol binding in rat inferior colliculus following acoustic trauma. *Hearing Research, 147,* 251–260.

Mishina, M., Mori, H., Araki, K., Kushiya, E., Meguro, H., Kutsuwada, T., . . . Sakimura, K., (1993). Molecular and functional diversity of the NMDA receptor channel. *Annals of the New York Academy of Sciences, 707,* 136–152.

Mulders, W. H., & Robertson, D. (2013). Development of hyperactivity after acoustic trauma in the guinea pig inferior colliculus. *Hearing Research, 298,* 104–108.

Muller, M., Klinke, R., Arnold, W., & Oestreicher, E. (2003). Auditory nerve fibre responses to salicylate revisited. *Hearing Research, 183,* 37–43.

Muly, S. M., Gross, J. S., & Potashner, S. J. (2004). Noise trauma alters D-[3H]aspartate release and AMPA binding in chinchilla cochlear nucleus. *Journal of Neuroscience Research, 75,* 585–596.

Nakagawa, H., Sato, K., Shiraishi, Y., Kuriyama, H., & Altschuler, R. A. (2000). NMDAR1 isoforms in the rat superior olivary complex and changes after unilateral cochlear ablation. *Brain Research. Molecular Brain Research, 77,* 246–257.

Niedzielski, A. S., Safieddine, S., & Wenthold, R. J. (1997). Molecular analysis of excitatory amino acid receptor expression in the cochlea. *Audiology and Neurotology, 2,* 79–91.

Okabe, S., Collin, C., Auerbach, J. M., Meiri, N., Bengzon, J., Kennedy, M. B., . . . Ronald, D. G., (1998). Hippocampal synaptic plasticity in mice overexpressing an embryonic subunit of the NMDA receptor. *Journal of Neuroscience, 18,* 4177–4188.

Puel, J. L., Pujol, R., Tribillac, F., Ladrech, S., & Eybalin, M. (1994). Excitatory amino acid antagonists protect cochlear auditory neurons from excitotoxicity. *Journal of Comparative Neurology, 341,* 241–256.

Puel, J. L., Saffiedine, S., Gervais d'Aldin, C., Eybalin, M., & Pujol, R. (1995). Synaptic regeneration and functional recovery after excitotoxic injury in the guinea pig cochlea. *Comptes rendus de l'Académie des sciences III, 318,* 67–75.

Pujol, R., Puel, J. L., & Eybalin, M. (1992). Implication of non-NMDA and NMDA receptors in cochlear ischemia. *NeuroReport, 3,* 299–302.

Reyes, S. A., Salvi, R. J., Burkard, R. F., Coad, M. L., Wack, D. S., Galantowicz, P. J., & Lockwook, A. H., (2002). Brain imaging of the effects of lidocaine on tinnitus. *Hearing Research, 171,* 43–50.

Salvi, R. J., Hamernik, R. P., & Henderson, D. (1978). Discharge patterns in the cochlear nucleus of the chinchilla following noise induced asymptotic threshold shift. *Experimental Brain Research, 32,* 301–320.

Sato, K., Shiraishi, S., Nakagawa, H., Kuriyama, H., & Altschuler, R. A. (2000). Diversity and plasticity in amino acid receptor subunits in the rat auditory brain stem. *Hearing Research, 147,* 137–144.

Schreiner, C. E., Snyder, R. L., & Johnstone, B. M. (1986). Effects of extracochlear direct current stimulation on the ensemble auditory nerve activity of cats. *Hearing Research, 21,* 213–226.

Shimano, T., Fyk-Kolodziej, B., Mirza, N., Asako, M., Tomoda, K., Bledsoe, S., et al. (2013). Assessment of the AAV-mediated expression of channelrhodopsin-2 and halorhodopsin in brainstem neurons mediating auditory signaling. *Brain Research, 1511,* 138–152.

Smith, A. R., Kwon, J. H., Navarro, M., & Hurley, L. M. (2014). Acoustic trauma triggers upregulation of serotonin receptor genes. *Hearing Research, 315,* 40–48.

Stypulkowski, P. H. (1990). Mechanisms of salicylate ototoxicity. *Hearing Research, 46,* 113–145.

Sun, W., Mercado, E., III., Wang, P., Shan, X., Lee, T. C., & Salvi, R. J. (2005). Changes in NMDA receptor expression in auditory cortex after learning. *Neuroscience Letters, 374,* 63–68.

Sun, W., Zhang, L., Lu, J., Yang, G., Laundrie, E., & Salvi, R. (2008). Noise exposure-induced enhancement of auditory cortex response and changes in gene expression. *Neuroscience, 156,* 374–380.

Suneja, S. K., Benson, C. G., & Potashner, S. J. (1998). Glycine receptors in adult guinea pig brain stem auditory nuclei: Regulation after unilateral cochlear ablation. *Experimental Neurology, 154,* 473–488.

Suneja, S. K., & Potashner, S. J. (2002). TrkB levels in the cochlear nucleus after unilateral cochlear ablation: Correlations with post-lesion plasticity. *Brain Research, 957,* 366–368.

Suneja, S. K., Potashner, S. J., & Benson, C. G. (1998). Plastic changes in glycine and GABA release and uptake in adult brain stem auditory nuclei after unilateral middle ear ossicle removal and cochlear ablation. *Experimental Neurology, 151,* 273–288.

Svirskis, G., Kotak, V., Sanes, D. H., & Rinzel, J. (2002). Enhancement of signal-to-noise ratio and phase locking for small inputs by a low-threshold outward current in auditory neurons. *Journal of Neuroscience, 22,* 11019–11025.

Tan, J., Ruttiger, L., Panford-Walsh, R., Singer, W., Schulze, H., Kilian, S. B., Hadjab S., & Rohbock, K. (2007). Tinnitus behavior and hearing function correlate with the reciprocal expression patterns of BDNF and Arg3.1/arc in auditory neurons following acoustic trauma. *Neuroscience, 145,* 715–726.

Tong, G. J., Hsiao, S. C., Carrico, Z. M., & Francis, M. B. (2009). Viral capsid DNA aptamer conjugates as multivalent cell-targeting vehicles. *Journal of the American Chemical Society, 131,* 11174–11178.

Usami, S., Matsubara, A., Fujita, S., Shinkawa, H., & Hayashi, M. (1995). NMDA (NMDAR1) and AMPA-type (GluR2/3) receptor subunits are expressed in the inner ear. *NeuroReport, 6,* 1161–1164.

Vale, C., & Sanes, D. H. (2002). The effect of bilateral deafness on excitatory and inhibitory synaptic strength in the inferior colliculus. *European Journal of Neuroscience, 16,* 2394–2404.

Wallhausser-Franke, E., Mahlke, C., Oliva, R., Braun, S., Wenz, G., & Langner, G. (2003). Expression of c-fos in auditory and non-auditory brain regions of the gerbil after manipulations that induce tinnitus. *Experimental Brain Research, 153,* 649–654.

Wang, Z., Ruan, Q., & Wang, D. (2005). Different effects of intracochlear sensory and neuronal injury stimulation on expression of synaptic N-methyl-D-aspartate receptors in the auditory cortex of rats in vivo. *Acta Otolaryngologica, 125,* 1145–1151.

Waxman, E. A., & Lynch, D. R. (2005). N-methyl-D-aspartate receptor subtypes: Multiple roles in excitotoxicity and neurological disease. *Neuroscientist, 11,* 37–49.

Zhang, J. S., & Kaltenbach, J. A. (1998). Increases in spontaneous activity in the dorsal cochlear nucleus of the rat following exposure to high-intensity sound. *Neuroscience Letters, 250,* 197–200.

Zhang, J. S., Kaltenbach, J. A., Wang, J., & Kim, S. A. (2003). Fos-like immunoreactivity in auditory and nonauditory brain structures of hamsters previously exposed to intense sound. *Experimental Brain Research, 153,* 655–660.

APPENDIX 6–A
Materials and Methods

Animals and Experimental Design

Male Sprague Dawley rats (185–275 g; Charles River Laboratories, Wilmington, MA) were maintained on a 12-hour reversed light dark cycle with free access to food and water. Animals were randomly assigned to one of three experimental groups. Groups 1 and 2 (n = 80) were exposed to sounds that have previously been shown to produce behavioral correlates of tinnitus (details provided below). Group 3 comprised untreated, normal-hearing animals (n = 32). Animal studies were approved by the University of Michigan Committee on the Use and Care of Animals.

Noise Exposures

Rats in Groups 1 and 2 were individually housed and exposed to free field noise in a sound attenuating booth (Acoustic Systems-Chamber). Group 1 animals were exposed to a 118-dB SPL 10-kHz 1/3-octave band sound for 4 hours. Group 2 animals were exposed to a 105-dB SPL 16-kHz-octave band noise for 1 hour. The sound levels in the cage were measured using a calibrated Bruel and Kjaer Precision sound level meter, model 2203. Animals were sacrificed 5 days after noise exposure. Group 3 (normal-hearing) control animals remained in the animal housing facility under ambient noise conditions until the time of tissue collection.

Auditory Brainstem Response (ABR)

Hearing thresholds were determined in a sound attenuating booth using the Tucker Davis Technology digital signal processing software package (System II; BioSig, Tucker Davis, Florida). Pure tones were used as the stimulus. The tone was delivered through a Beyer speaker (Beyer, Germany) placed directly into the external auditory canal. The reference electrode was inserted below the pinna, the active electrode was placed on the top of the head, and the ground electrode was place below the contralateral pinna. Tones were presented to anesthetized animals (ketamine 50 mg/kg and xylazine 2 mg/kg i.m.). Before a given noise exposure, baseline ABR thresholds were measured at frequencies of 4, 12, and 24 kHz in at least 20% of the animals within each experimental group (n = 6–8 per group) to confirm that animals had normal hearing. Following the noise exposures, ABR thresholds were determined both at 2 hours and 4 days to assess the degree of hearing loss. ABR thresholds were determined by identifying the lowest intensity of a tone between 0 and 100 dB that elicited a response.

Histologic Assessment—Cochlea

Cochlear hair cell counts. Eight rats from each group were heavily anesthetized with a single i.p. dose of 350 mg/kg of Fatal Plus (Vortech Pharmaceuticals, Dearborn, MI) and then were transcardially perfused with 4% paraformaldehyde. Bullae were removed and cochleae received intrascalar perfusion with the same fixative. The left cochleae were processed for cytocochleograms. Cochleae were stained with 1% phalloidin (intrascalar) for 1 hour at room temperature and rinsed in phosphate-buffered saline (PBS). The otic capsule was removed and the cochlear spiral dissected and surface preparations were then viewed under epifluorescent illumination.

RNA Isolation

Rats were anesthetized with a lethal dose (0.8–1 mL) of Fatal Plus (Vortech Pharmaceuticals, Dearborn, MI) and decapitated (n = 24 per noise exposure group). Brains were rapidly removed and the CN dissected away from lateral surface of the medulla with fine-tipped forceps and placed in RNA*later* (Ambion, Austin, TX). The CN from rats in Groups 1 and 2 were divided into seven pools containing three to four animals each for qRT-PCR. The CN from rats of Group 3 (normal hearing) were divided into four pools of four animals each for qRT-PCR.

Total RNA was isolated from the tissues homogenized in TRIZOL (Invitrogen, Carlsbad, CA), using a modification of the standard TRIZOL RNA extraction protocol. Chloroform was added to the CN homogenate, and the mixture was vortexed and centrifuged. The aqueous phase was transferred to a Phase Lock Gel (Heavy) tube (Eppendorf/Brinkmann, Westbury, NY) and extracted with acid phenol:chloroform (1:1). Total RNA was precipitated from the aqueous phase with isopropanol, resuspended in RNA Storage Solution (Ambion, Austin, TX), and stored at −80°C. RNA concentration was determined by UV spectrophotometry. RNA quality was assessed on an RNA 6000 Nano LabChip (Agilent, Palo Alto, CA) using an Agilent 2100 Bioanalyzer to assess the integrity of the 18S and 28S ribosomal RNA subunit bands.

Real-Time RT-PCR (qRT-PCR)

Five NMDA receptor subunits (NMDAR1, NMDAR2A, NMDAR2B, NMDAR2C, NMDAR2D) were assessed. Total RNA (1 µg) was used to generate cDNA using Superscript III (Invitrogen) and oligo primers (Milbrandt et al., 2000). The cDNA was diluted to 200 µL and 2.5 µL was used for each qRT-PCR reaction. TaqMan Assays-on-Demand (Applied Biosystems, Foster City, CA) were used for qRT- PCR reactions. Microtiter plates (384-well) were read on an ABI PRISM 7900. The data were analyzed using ABI PRISM 7900 Sequence Detection System software (Applied Biosystems, Foster City, CA). Each gene was assayed in triplicate in every pool of RNA.

For each gene under each condition, we used the $2^{-\Delta_1\Delta_2 Ct}$ method to calculate the fold change in gene expression (Livak & Schmittgen, 2001). We determined the cycle threshold (Ct), a cycle number at which PCR products are increasing exponentially. To normalize the data, the threshold value for ribosomal subunit S16 was subtracted from threshold values for each gene at each noise exposure (Δ_1). Then, the control threshold value for each gene was subtracted from the corresponding experimental threshold values in the group (Δ_2). The final value was then expressed as logarithm to base 2. Using this formula ($2^{-(\Delta_1\Delta_2Ct)}$), we calculated the relative differences between experimental and control groups as fold change. The average Ct values were calculated for each pool of every experimental group. Each pool was normalized to the average Ct value for the S16 ribosomal protein for each sample and then compared with the expression value of the control group.

Western Blot

Rats were deeply anesthetized with a lethal dose (0.8–1 mL) of Fatal Plus (Vortech Pharmaceuticals, Dearborn, MI) and decapitated (*n* = 24 per noise exposure group). Brains were rapidly removed and the CN dissected away from lateral surface of the medulla with fine-tipped forceps. Cochlear nuclei were isolated from either untreated normal control rats (*n* = 8) or rats exposed to a 118-dB SPL 10-kHz-1/3 octave band noise for 4 hours (*n* = 8) or animals exposed to a 105-dB SPL, 16-kHz-octave band noise for 1 hour (*n* = 8). Each CN from every animal was placed in a separate microcentrifuge tube containing 30 μL of lysis buffer prepared as previously described (Ptok et al., 1993): PBS containing 1% nonidet P-40 nonionic detergent (IGEPAL CA-630; Sigma) and 1× Complete Protease Inhibitor (Roche Applied Science, Indianapolis, IN), containing 1 mM EDTA. Cochlear nuclei were homogenized with a disposable plastic pestle pellet homogenizer (Kontes Glass, Vineland, NJ) and lysed for 45 min on ice. Cellular debris was removed by centrifugation and the supernatant removed and stored at –70°C until ready for use. Protein concentrations were determined using a DC Protein Assay Kit (Bio-Rad, Hercules, CA).

For Western blot, 7 to 21 μg of total CN protein was loaded onto a 12% sodium dodecyl sulfate (SDS)–polyacrylamide gel. Prior to loading, samples were diluted in Laemmli sample buffer (Bio-Rad) with 5% β-mercaptoethanol and denatured by boiling for 5 minutes. Molecular weight was determined using a MagicMark XP Protein Standards (Invitrogen). Gels were run for 1 hour at 130 V followed by an overnight transfer at 35 mAmp onto a nitrocellulose membrane (Bio-Rad). Immunoreactions were performed using the following primary antibodies: NMDAR 2B (1/500; Chemicon, Temecula, CA), NMDAR 2C (1/1000; Advanced Immunochemical Inc., Long Beach, CA), NMDAR 2D (1/20; Santa Cruz Biotechnology, Santa Cruz, CA), Tubulin (1/1,000; Abcam, Cambridge, MA), and a 1/10,000 dilution of horseradish peroxidase–labeled goat anti-rabbit immunoglobulin G, heavy and light chains (IgG (H+L)) secondary antibody (Bio-Rad). Immunolabeled protein bands were visualized by enhanced chemiluminescence (ECL) (Amersham Biosciences, Buckinghamshire, UK), followed by autoradiography. Protein levels were quantified by densitometry of scanned autoradiographs and integrated optical density was obtained using Metamorph software (Molecular Devices Corp., Sunnyvale, CA). Data were normalized to internal standards (untreated normal controls) on each gel and presented as the fold change relative to control protein levels.

Analysis

The resulting data were examined using StatView statistical analysis software (SAS Institute, Cary, NC). Data are presented as means ± SEM (standard error of the mean). One-way analysis of variance testing was performed to determine the significance of observed differences. Scheffe's F procedure was used for post hoc comparisons. Significance at the 95% confidence interval was represented as $p \leq 0.05$.

CHAPTER 7

Medical and Surgical Treatment of Inner Ear Disease

Lawrence R. Lustig

The cochlea is a highly complex, tightly regulated sensory end organ responsible for the conversion of environmental sound into a meaningful auditory neuronal response. Like all such highly ordered systems, with this complexity comes a vulnerability to many intrinsic and extrinsic phenomena leading to temporary or permanent hearing loss. This chapter provides a basic foundation of how the clinician approaches the cochlea and various pathologies that can affect it, leading to hearing loss.

Clinical Hearing Assessment

An accurate, objective and repeatable measure of one's ability to hear is a critical initial step in the evaluation of someone with hearing loss. Hearing loss can occur at any stage of the auditory pathway, from the ear canal to the auditory cortex. While simple otoscopy can identify obvious pathology in the external and middle ear, the ability to determine the cause and nature of hearing loss is often limited by physical exam alone. Thus, subjective and objective measures of hearing have become paramount in the clinical evaluation of the cochlea. With contemporary techniques, it is now possible to evaluate the functional integrity of the entire auditory pathway, from the tympanic membrane to cerebral cortex. Such testing, including, pure-tone threshold and speech testing, immittance testing, and auditory brainstem response testing, allows clinicians to characterize and localize the nature of hearing loss in most cases. An additional important electrical measure of auditory function, otoacoustic emissions, also has important and useful clinical applications, though this is discussed separately in another chapter.

Pure-Tone Audiometry

Pure-tone threshold audiometry, the most fundamental of all diagnostic hearing tests, is a behaviorally based measure of hearing that is used to distinguish between sensorineural (cochlear and central auditory pathways) and conductive (external and middle ear) types of hearing loss, representing a critical clinical distinction. The primary goal of pure-tone testing is to obtain a representation of the softest sound intensities one can hear across the frequency spectrum. These data can then be compared to well-established normative population-based standards to determine the nature and degree of one's hearing loss.

To obtain pure-tone thresholds, auditory stimuli are presented in a soundproof chamber to eliminate confounding background noise. In general, air conduction thresholds are obtained to determine actual levels of hearing while bone conduction thresholds are also obtained to determine true cochlear function; differences between these values represent conductive forms of hearing loss, caused by pathologic processes impeding sound transduction to the cochlea. With air conduction, stimuli consist of pure tones at octave levels presented most commonly from 25 to 8000 Hz while patients wear either an earphone or headphone. In some special circumstances (e.g., ototoxicity screens), higher frequencies are tested (Gordon, Phillips, Helt, Konrad-Martin, & Fausti, 2005). The patient indicates whenever a stimulus is heard, allowing the audiologist to record the softest intensity the patient is able to hear for each frequency. Bone conduction testing is identical, except that instead of headphones, a bone oscillator is placed on the mastoid of the test ear (immediately behind the auricle) and results are recorded at only 500 Hz, 1000 Hz, 2000 Hz, and 4000Hz (American National Standards Institute, 1978). Various methods have been developed for systematic tone presentations, such as the modified Hughson-Westlake technique, allowing an accurate and efficient recording of pure-tone thresholds (Beiter & Talley, 1976; Carhart & Jerger, 1959).

For widespread clinical consistency, the results of an audiologic evaluation are commonly plotted on an audiogram, which depicts *frequencies* (represented in hertz, Hz) on the x-axis and *intensities* (represented in decibels, dB) on the y-axis (Figure 7–1). This visual representation of an individual's hearing allows a rapid assessment of a pattern of hearing loss and ability to readily compare one audiogram to another, or monitor changes over time. To aid in standardization, graphic representation and symbols were recommended in 1974 by American Speech and Hearing Association and later revised by the American National Standards Institute (ANSI, 1978). Hearing loss is also commonly described by degree of severity and in relation to the intensity or decibel level of the threshold response (Table 7–1), allowing terms such as "severe" or "moderate" hearing loss to have consistent meanings across clinical settings (Clark, 1981; Goodman, 1965).

Another useful recorded value is the pure-tone average (PTA), which is simply the average of threshold responses obtained at 500 Hz, 1000 Hz, and 2000 Hz, key frequencies within the speech range (Carhart, 1971; Fletcher, 1950). This measure provides a rapid way of imparting, in a single number, a basic score of hearing.

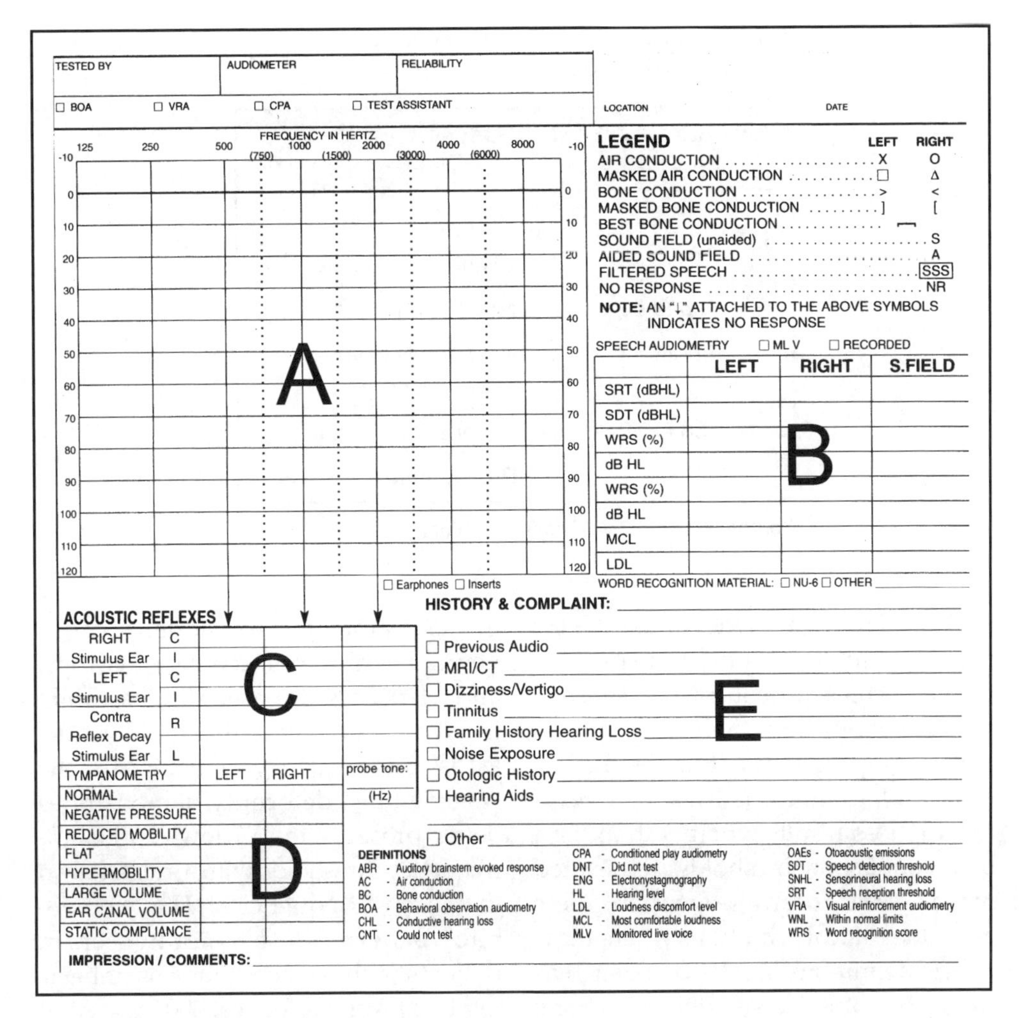

TESTED BY | AUDIOMETER | RELIABILITY

☐ BOA ☐ VRA ☐ CPA ☐ TEST ASSISTANT

LOCATION DATE

FREQUENCY IN HERTZ

125 250 500 1000 2000 4000 8000

(750) (1500) (3000) (6000)

-10 0 10 20 30 40 50 60 70 80 90 100 110 120

A

LEGEND LEFT RIGHT

AIR CONDUCTION . . . . . X O

MASKED AIR CONDUCTION . . . . . □ Δ

BONE CONDUCTION . . . . . > <

MASKED BONE CONDUCTION . . . . .] [

BEST BONE CONDUCTION . . . . .

SOUND FIELD (unaided) . . . . . S

AIDED SOUND FIELD . . . . . A

FILTERED SPEECH . . . . . SSS

NO RESPONSE . . . . . NR

NOTE: AN "↓" ATTACHED TO THE ABOVE SYMBOLS INDICATES NO RESPONSE

SPEECH AUDIOMETRY ☐ ML V ☐ RECORDED

	LEFT	RIGHT	S.FIELD
SRT (dBHL)			
SDT (dBHL)			
WRS (%)			
dB HL		B	
WRS (%)			
dB HL			
MCL			
LDL			

☐ Earphones ☐ Inserts

WORD RECOGNITION MATERIAL: ☐ NU-6 ☐ OTHER

ACOUSTIC REFLEXES

RIGHT Stimulus Ear	C			
	I			
LEFT Stimulus Ear	C		C	
	I			
Contra Reflex Decay Stimulus Ear	R			
	L			

TYMPANOMETRY	LEFT	RIGHT	probe tone: (Hz)
NORMAL			
NEGATIVE PRESSURE			
REDUCED MOBILITY			
FLAT		D	
HYPERMOBILITY			
LARGE VOLUME			
EAR CANAL VOLUME			
STATIC COMPLIANCE			

HISTORY & COMPLAINT:

☐ Previous Audio

☐ MRI/CT

☐ Dizziness/Vertigo

☐ Tinnitus

☐ Family History Hearing Loss

☐ Noise Exposure

☐ Otologic History

☐ Hearing Aids

☐ Other

E

DEFINITIONS

ABR - Auditory brainstem evoked response
AC - Air conduction
BC - Bone conduction
BOA - Behavioral observation audiometry
CHL - Conductive hearing loss
CNT - Could not test
CPA - Conditioned play audiometry
DNT - Did not test
ENG - Electronystagmography
HL - Hearing level
LDL - Loudness discomfort level
MCL - Most comfortable loudness
MLV - Monitored live voice
OAEs - Otoacoustic emissions
SDT - Speech detection threshold
SNHL - Sensorineural hearing loss
SRT - Speech reception threshold
VRA - Visual reinforcement audiometry
WNL - Within normal limits
WRS - Word recognition score

IMPRESSION / COMMENTS:

Figure 7–1. A typical audiometric examination sheet. The single sheet includes the pure-tone audiogram (region "A"), a standardized measure of hearing based upon pure-tone thresholds. Hearing is charted for each frequency, between 250 and 8000 Hz at the lowest intensity (in dB) heard by the listener. Universally accepted symbols to document the pure-tone scores are noted in the upper right corner (marked "Legend"). Speech and word understanding scores are charted alongside the audiogram (region "B"). Acoustic reflexes and tympanometry results are documented in regions C and D, respectively. Many examination sheets also contain an area for clinical information ("E"). Commonly used definitions and abbreviations are also commonly included, and here noted in the lower right corner ("Definitions"). By plotting all data on a single sheet, the clinician can rapidly assess a patient's degree of hearing loss and the nature of the hearing loss.

Patterns of Hearing Loss

Hearing loss is grouped under two broad categories: *Conductive* and *Sensorineural,* determined by comparison of the air and bone conduction threshold results. The difference in thresholds between air and bone conduction is termed the *air-bone gap* and determines the degree of conductive hearing loss

Table 7–1. Severity of Hearing Loss by Pure-Tone Threshold

Average Threshold Level (dB)	Suggested Description
−10 to 15	Normal hearing
16 to 25	Slight hearing loss
26 to 40	Mild hearing loss
41 to 55	Moderately severe hearing loss
56 to 70	Severe hearing loss
>71	Profound hearing loss

Source: Clark, 1981; Goodman, 1965.

(CHL). CHL occurs when a pathologic process impedes sound transmission to the cochlea. In contrast, a sensorineural hearing loss (SNHL) is due to pathologic processes within the cochlea or central auditory pathways. A *mixed* hearing loss results when both air and bone conduction thresholds are elevated in the presence of an air-bone gap. There are some clinical conditions where these classic definitions break down, such as in cochlear otosclerosis, whereby a bony pathology results in a mixed or SNHL (Ramsden et al., 2007).

In addition to the type of hearing loss, there are several classic audiometric configurations that are commonly seen, essentially describing the shape or slope of the pure-tone thresholds (Figure 7–2). A *down-sloping* pattern implies worse high-frequency than low-frequency thresholds, and is commonly seen with such conditions as age-related hearing loss (presbycusis), noise-induced hearing loss, and ototoxicity. A *rising*, or up-sloping pattern denotes poorer thresholds in the low-frequency range and is commonly seen in such pathologies as sudden sensorineural hearing loss, cochlear hydrops, and Ménière's disease. Flat losses are indicative of equal threshold drops across the frequency range. Lastly, the "cookie-bite" deformity demonstrates better preservation of low- and high-frequency thresholds with a greater drop in the mid-frequencies. The "cookie-bite" pattern is a classically seen in patients with congenital forms of hearing loss (Wong & Bance, 2004).

Speech Audiometry

While pure-tone thresholds give an indication of the cochlea's ability to detect sounds and at what intensity levels, they do not provide important qualitative information regarding sound perception, and thus by itself, the pure-tone audiogram is an incomplete measure of hearing. A common refrain of patients with hearing loss is that "I can hear, but I cannot understand." Speech audiometry provides this additional qualitative hearing assessment, and

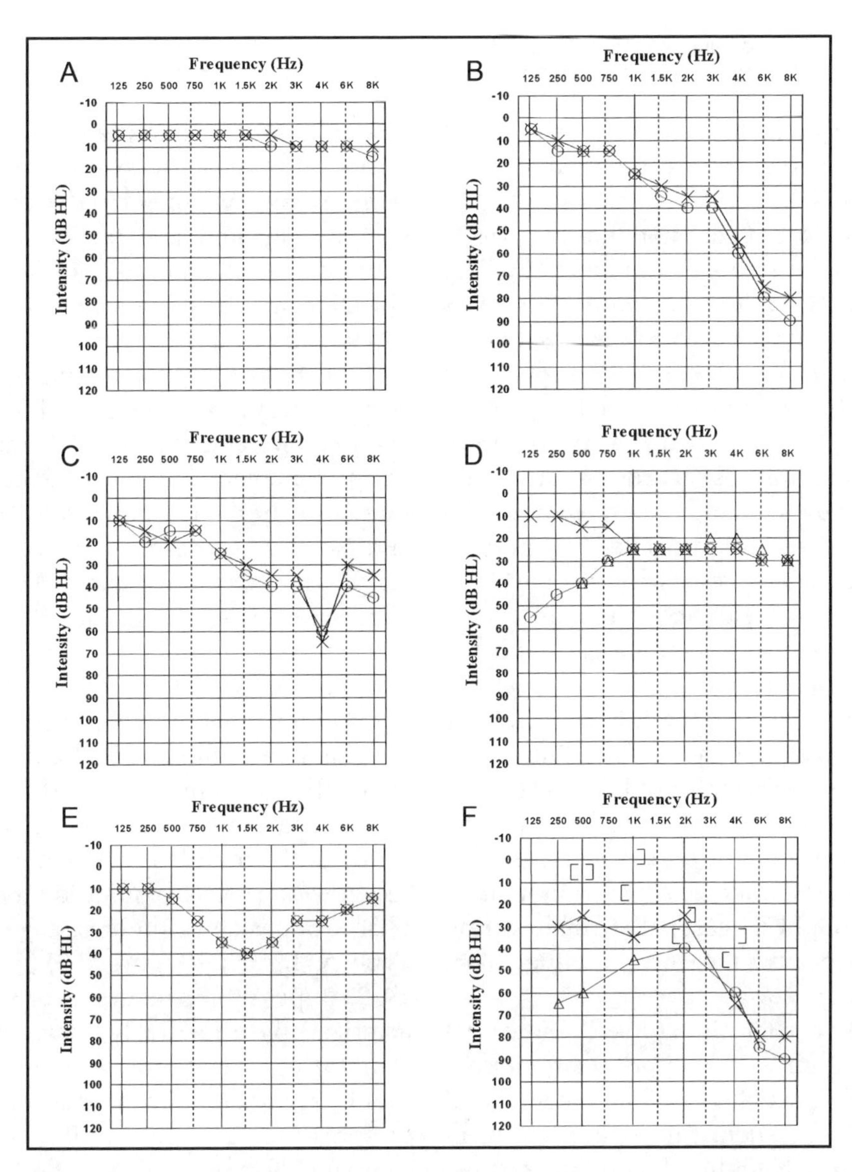

Figure 7–2. Commonly seen patterns of hearing loss audiometry on audiometry. **A.** Normal hearing. With pure-tone air conduction thresholds under 25 dB for all frequencies tested, this would be considered hearing within the normal range. By convention, the left ear is represented by an "X" at each frequency while the right ear is represented by an "O." In this case, both the left and right ear hear nearly identically, and thus overlapping patterns are seen. **B.** Down-sloping sensorineural hearing loss is the most common pattern of hearing loss seen, and is typically seen in age-related hearing loss. **C.** A "notch" at 4 kHz is commonly seen in cases of noise-induced hearing loss, due to the susceptibility of this range of hearing to noise damage. **D.** An up-sloping pattern of hearing loss, here seen in the right ear, is commonly seen in cases of endolymphatic hydrops or sudden sensorineural hearing loss. **E.** A "cookie-bite" deformity is commonly seen in cases of congenital hearing loss, whereby the mid-frequencies demonstrate a greater threshold shift than both the low and high frequencies. **F.** An example of a "mixed" hearing loss of both sensorineural and conductive hearing losses in the same ear. The bone conduction thresholds are represented by the open boxes while the air conduction thresholds by the standard "X" and "O" markings. Differences between the air and bone conduction thresholds represents the air-bone gap and the degree of conductive hearing loss. This patient exhibits a moderate low-frequency conductive hearing loss with a moderate to severe high-frequency sensorineural loss.

offering a better view of how a patient with hearing loss is functioning (Penrod, 1994).

The speech reception threshold (SRT) is defined as the softest level at which a patient can repeat a word. Since this measure typically correlates closely with pure-tone thresholds, it is one objective way to determine the accuracy of the pure-tone audiometry results (Penrod, 1994). In this test, words are presented either by live voice or a recording. The presented words are called "spondees," two-syllable words with an equal stress on each syllable (e.g., "baseball"). After the patients are familiarized with the words at an audible intensity level, the test proceeds as the intensity level is decreased until the words are no longer audible and a threshold is determined. Differences between the SRT and the PTA (average of pure-tone thresholds at 500, 1000, and 2000 Hz) of more than 6 dB might suggest patient malingering or inaccurate measures of either score (Carhart, 1971).

After determining the SRT, the next test is the speech discrimination score (SDS), calculated as the percentage of words a patient can repeat at a comfortable listening level. In contrast to spondees used in SRT, phonemes, phonetically balanced single syllables words, are used for SDS. The more challenging word list provides additional insight into patient functioning and is perhaps the best prognostic indicator of successful use of hearing aids (Dillon, 1982; Festen & Plomp, 1983).

Immittance Testing

Acoustic immittance testing, in the most broad sense, objectively measures the ease with which sound is transmitted through the tympanic membrane and middle ear. Two principal tests are employed in immittance measurements: *tympanometry* and the *acoustic reflex*. The objective nature of immittance measures makes it a particularly valuable tool when testing those unable to give accurate responses, such as in infants and toddlers, and the medically infirm.

Tympanometry measures the changes that occur to the tympanic membrane and ossicles as a result of a change in air pressure in the ear canal. The test allows the measure of canal volume as well as peak compliance and middle ear pressure. Canal volume measures are useful for determining obstruction or tympanic membrane perforation. In contrast, peak compliance and middle ear air pressure are used to assess the status of the middle ear. Three widely recognized patterns of middle ear compliance or peak pressure are recognized, Type A, B, or C (Figure 7–3) (Jerger, 1970). Type A tympanograms are characterized by a relatively sharp maximum at or near 0 mm, and are commonly seen ears with normal middle ear pressure. A subtype of the Type A tympanogram, termed As (s indicates "stiff"), can be seen in patients with otosclerosis. The type B tympanogram shows little or no maximum, with a flat peak. Type B tympanograms are typically found in ears with serous otitis media. In the Type C tympanogram, the maximum is shifted to the left of zero by negative pressure in the middle ear, and is commonly seen in cases of Eustachian tube dysfunction (Onusko, 2004).

The stapedius muscle reflex, or acoustic reflex, is another important component of immittance testing, and represents stapedius muscle contrac-

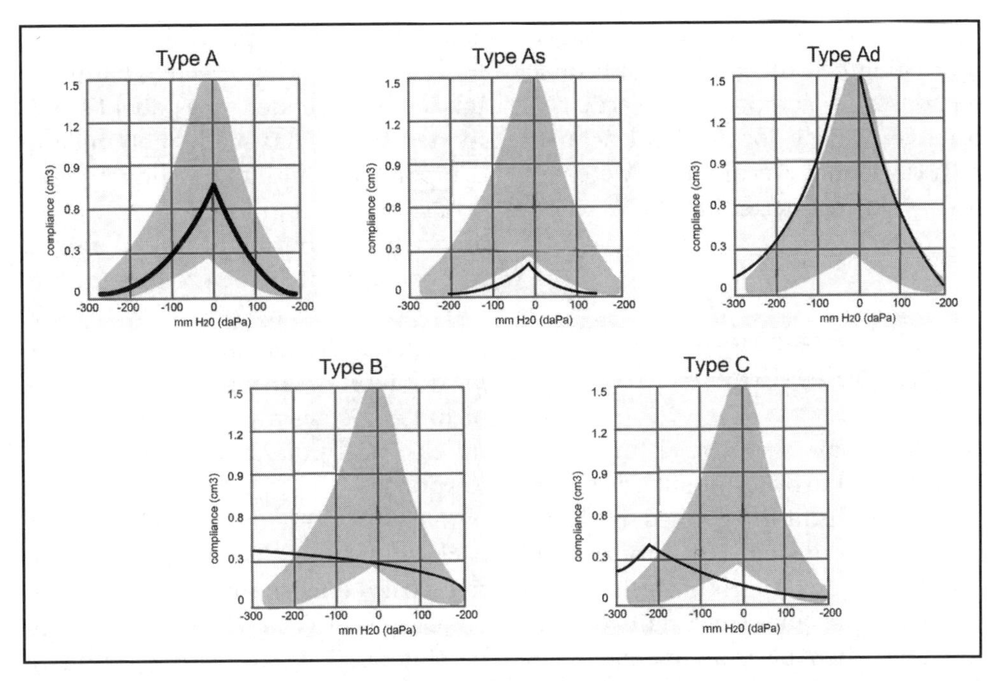

Figure 7–3. The tympanogram measures peak compliance of the middle ear. Three widely recognized patterns of middle ear compliance or peak pressure are recognized, types A, B, and C (J. F. Jerger, 1970) The Type A tympanogram is characterized by a relatively sharp maximum at or near 0 mm, and are commonly seen ears with normal middle ear pressure. A subtype of the Type A tympanogram, termed As (s indicates "stiff"), can be seen in patients with otosclerosis. Another subtype of the Type A tympanogram, type Ad, occurs with an overly compliant tympanic membrane as seen with atelectasis (an overly "floppy" eardrum). The Type B tympanogram shows little or no maximum, with a flat peak. Type B tympanograms are typically found in ears with serous otitis media. In the type C tympanogram, the maximum is shifted to the left of zero by negative pressure in the middle ear, and is commonly seen in cases of Eustachian tube dysfunction (Onusko, 2004).

tion in response to a loud auditory signal. This reflex is tested using ipsilateral or contralateral stimulation, and the reflex measured when the stapedial muscle contracts (measured by sensing the altered compliance of the tympanic membrane). The acoustic reflex threshold is the lowest intensity level presented where a change in compliance can be observed. Testing is typically performed at 500, 1000, 2000, and 4000 Hz, with thresholds in the 70- to 100-dB range. Pathologies that reduce or abolish the acoustic reflex include any form of conductive hearing loss, severe SNHL, or retrocochlear lesions such as tumors of the auditory nerve or pathway (Jerger, 1970). One example in which the acoustic reflex has gained contemporary relevance is the recently described entity of superior semicircular canal dehiscence (Minor et al., 2003). A common presenting finding in this condition is a conductive hearing loss; however, it is not a true conductive hearing loss but likely the presence of a third mobile window from the dehiscence that leads to an *apparent*

CHL. The acoustic reflex can thus readily distinguish this entity (with intact responses) as compared to a true conductive hearing loss (absent acoustic reflex) (Halmagyi et al., 2003; Merchant, Rosowski, & McKenna, 2007; Minor et al., 2003).

Acoustic Brainstem Response Testing

Auditory brain response testing (ABR; also referred to by terms such as BAER, brainstem auditory evoked response or BAEP, brainstem auditory evoked potential) is an important electrical test to measure cochlear and retrocochlear function. The ABR is an averaged surface-recording of the sound within the activated auditory pathway beginning in the cochlea. The objective nature of this test, independent of a patient's ability to respond, has made it an integral component of auditory testing in newborns, infants, and toddlers, as well as any patient who might not be able to respond accurately to pure-tone testing.

The ABR is measured with surface electrodes on the skull (usually mastoid process and vertex). Clicks or tone-pips are presented while neural electrical potentials are measured. Responses are averaged over large numbers of presentations to eliminate background noise. Normally, the first five positive measured peaks that can be recorded (termed Waves I through V) (Figure 7–4), occurring within a 10-ms timeframe, are used for clini-

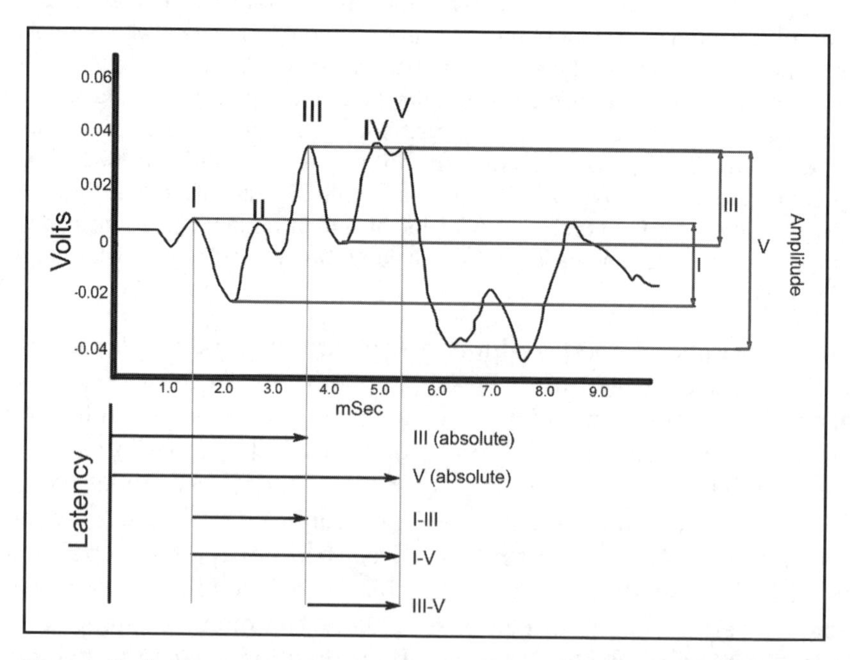

Figure 7–4. The acoustic brainstem response (ABR) test. Five repeatable waves are present in the normal ABR, Waves I to V. Standard measures of ABR include amplitude of Waves I, III, and V (demonstrated graphically on the right of the ABR waveform), absolute latency of Waves III and V, and latencies between Waves I and III, Waves I and V, and Waves III and V (shown graphically below the waveform).

cal analysis. Waves I and II are felt to represent synchronized activity with the cochlea and cochlear nerve (Moller, Jho, Yokota, & Jannetta, 1995). Wave III is believed to be due to activity within the cochlear nucleus, while Wave IV represents activation of the superior olivary complex. Wave V is thought to represent activity within the lateral lemniscus of the brainstem (Moller et al., 1995). It is Wave V that appears to be the most susceptible to hearing loss; Wave V latency has been shown to increase with increasing hearing loss, becoming absent in severe cases (Bauch & Olsen, 1986; Coats, 1978). In contrast, Wave I tends to be the most stable in sensorineural hearing loss (Galambos & Hecox, 1978; Jerger & Johnson, 1988).

Because of its consistency across patients and its ability to measure the retrocochlear auditory pathways, the ABR has become an important tool in clinical diagnosis. The two principal electrical abnormalities that are analyzed on the ABR are wave amplitude and latency changes (Jewett & Williston, 1971). Changes in peak amplitudes (as measured from the baseline to the wave peak), as well as changes in time interval between the similar contralateral wave (*relative* latency) or beyond accepted normative data (*absolute* latency) may indicate a form of retrocochlear (between the cochlea and auditory cortex) pathology, triggering the need for further evaluation with magnetic resonance imaging (MRI). However, ABR abnormalities are not absolutely indicative of central pathology, since a variety of other factors can influence ABR results (Hall, 1992; Hood, 1988).

Screening newborns and infants for hearing loss is the most common application of ABR testing, which is now universally applied in varying algorithms along with otoacoustic emissions (Kaye et al., 2006). The next most common application of ABR is its use in the diagnosis of vestibular schwannomas, the most common tumor affecting the auditory nerve. ABR has been cited as a highly sensitive (>90%) tool in the identification of tumors of the eighth nerve pathway (Eggermont, Don, & Brackmann, 1980; Josey, Glasscock, & Musiek, 1988; Montaguti, Bergonzoni, Zanetti, & Rinaldi Ceroni, 2007; Musiek, 1982). The presence of a Wave I and absence of Waves II to V is the most specific finding for a vestibular schwannoma, though one must be wary for both false-positive (>80%) and false-negative (12%–18%) results (Weiss, Kisiel, & Bhatia, 1990; D. Wilson, Hodgson, Gustafson, Hogue, & Mills, 1992). With the ubiquity of MRI and its clearly superior diagnostic capabilities, the utility of ABR has thus been questioned by many clinicians. Since an abnormal ABR warrants further evaluation with an MRI, many have suggested to simply avoid the ABR altogether and obtain an MRI in all cases of concern for retrocochlear lesions (such as with asymmetrical sensorineural hearing loss). The primary rationale for ABR has been its favorable cost-utility as part of the clinical workup for vestibular schwannoma (Chandrasekhar, Brackmann, & Devgan, 1995).

Radiographic Evaluation of the Ear

Computed tomography (CT) and MRI give the clinician an extraordinary ability to localize pathology within the inner ear and auditory pathway. The decision to pursue one or both of these imaging modalities in patients with hearing loss depends upon a number of factors, including type of hearing loss (conductive or sensorineural), age, and degree of hearing asymmetry, to name but a few.

CT imaging of the temporal bones remains an important clinical tool for evaluating several pathologic processes (Figure 7–5). The strengths of the CT scan include its superior bone detail, ability to detect bony erosive processes (e.g., cholesteatoma, tumor), spongiotic changes in the otic capsule (e.g., otosclerosis), and congenital abnormalities (e.g., aural atresia, congenital ossicular fixation). High-resolution CT detail is sufficient to evaluate the status of the ossicles and general morphology of the cochlea and vestibular organs. With the addition of contrast, CT can also identify some inflammatory processes within the inner ear and mastoid, as well as tumors of the facial or auditory nerves as small as 5 mm (Jackler & Dillon, 1988). However, as compared to MRI, CT is inferior at identifying soft tissue lesions such as tumors of the ear or skull base. However, in cases where an implanted device would contraindicate the use of MRI, or when someone was unable to physically sit still for the 30-minute exam due to such issues as anxiety or claustrophobia, a CT can often be sufficient.

MRI is one of the most important modalities clinicians have to evaluate otologic and retrocochlear pathology

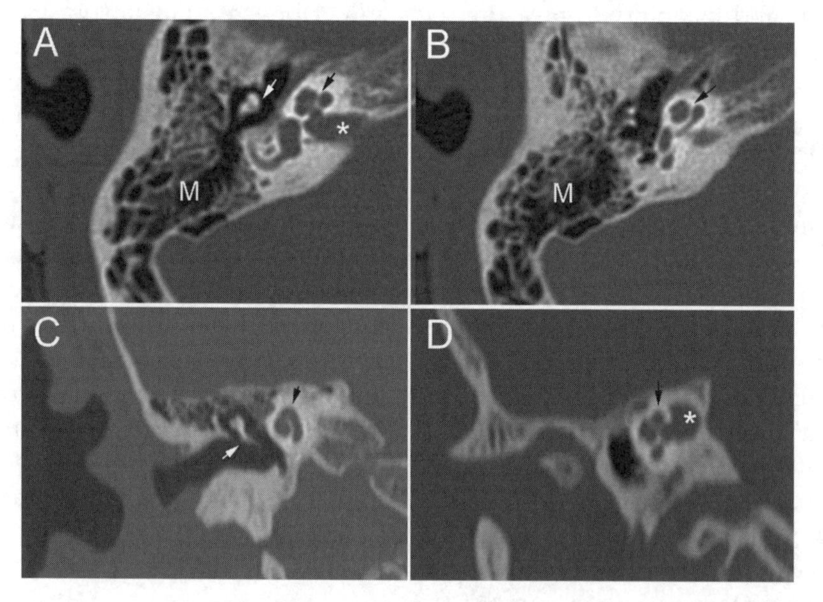

Figure 7–5. Normal computed tomography (CT) of the cochlea. The CT is superior to MRI in delineating the anatomy of the temporal bone. There are two axial (*A, B*), a coronal (*C*), and oblique (*D*) views. In these images, the cochlea (*black arrows*) can be seen, including the basal, mid, and apical turns. The ossicles can also be visualized (*white arrows*), as can the internal auditory canal (*asterisk*). The mastoid air cell system (M) is also clearly visualized by CT imaging.

(Figure 7–6). The image itself, in contrast to an x-ray–based modality like CT, is based upon mobile proton density and the time required for proton reorientation within a magnetic field. Varying degrees of water content and vascularity thus account for differences in signal (Jackler & Dillon, 1988). As a result, bony structures do not produce a signal on MRI, limiting its usefulness within the temporal bone where most structures are osseous. However, due to its superior detail of all soft tissue structures, MRI with contrast dye such as Gadolinium–DPTA is clearly superior than CT in identifying inflammatory processes (e.g., labyrinthitis) or neoplasms (e.g., vestibular schwannoma) within the temporal bone or along the central auditory pathways (Figure 7–7) (Jackler & Dillon, 1988). Newer imaging modalities and stronger magnets have also led to improving resolution of the cochlea itself. Most contemporary scanners can adequately visualize the turns of the cochlea, the presence of fluid within the cochlea, and the presence or absence of the auditory nerve.

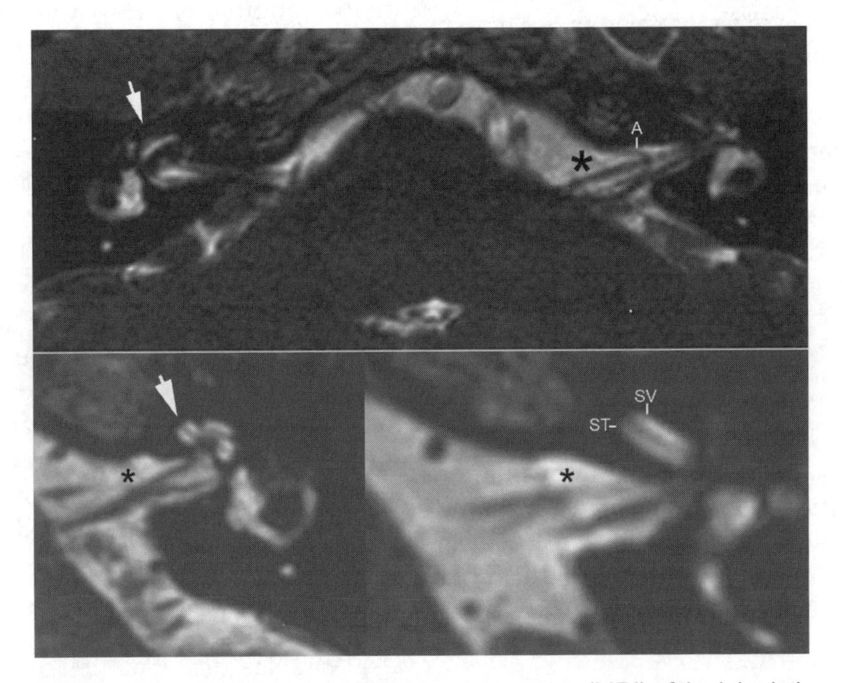

Figure 7–6. Normal magnetic resonance imaging (MRI) of the labyrinth and internal auditory canal. Images such as these, using a 3D Fiesta protocol, highlight fluid-filled structures such as the cochlea, while bone generates no signal (*black*). In this image, there is a view of the basal (*upper panel, arrow*) and mid and apical (*lower left, white arrow*) regions of the fluid-filled cochlear scalae. The left cochleovestibular and facial nerves traveling within the internal auditory canal (IAC) can be seen in relief against the cerebrospinal fluid (*asterisk*), along with a loop of the anterior inferior cerebellar artery ("A" upper panel) coursing through the IAC. In the lower left panel, one can differentiate between the scala tympani (ST) and scala vestibule (SV) within the basal turn of the cochlea.

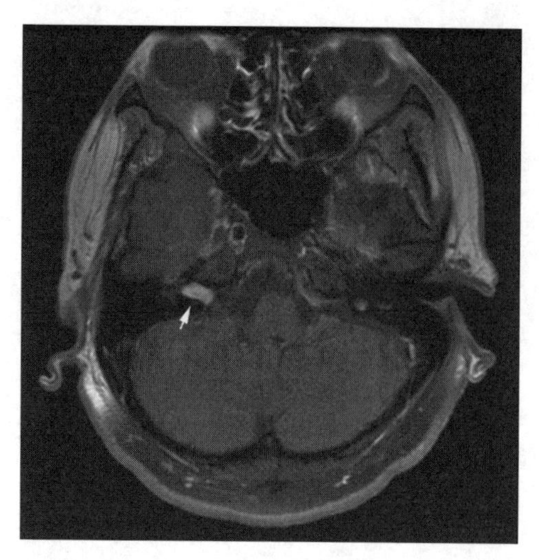

Figure 7–7. Vestibular Schwannoma. Various retrocochlear pathologies can present with asymmetrical sensorineural hearing loss. In most such cases, an MRI can definitively diagnose such a disorder. In this case, a patient with a right-sided sensorineural hearing loss underwent an MRI with contrast dye enhancement (gadolinium), leading to a diagnosis of a vestibular schwannoma located within the internal auditory canal (*arrow*).

These improvements have made MRI an invaluable tool in the evaluation of patients with severe to profound hearing for the suitability of cochlear implantation (Papsin, 2005; Trimble, Blaser, James, & Papsin, 2007).

Clinical Disorders of the Cochlea

Sensorineural Hearing Loss

Hearing loss is widely recognized as one of the most common disorders experienced by humans. Estimates of the prevalence of hearing loss are fraught with inaccuracies for a number of reasons however, including patient denial of a hearing loss and the difficulty of getting population-based audiometric data on a sufficient scale (Gates, Cooper, Kannel, & Miller, 1990). There is also a wide variability in hearing loss across different populations, varying by socioeconomic status (for example noise exposure in factory workers), making generalizations difficult (Davis, Stephens, Rayment, & Thomas, 1992).

Universal newborn infant hearing screening has facilitated data collection in this population, providing better estimates than in adults. Data indicate that congenital or prelingual severe to profound losses occur in approximately 0.5 to 3 per 1,000 live births (Augustson, Nilson, & Engstrand, 1990; Fortnum, Davis, Butler, & Stevens; Morgan & Canalis, 1991; Prager, Stone, & Rose, 1987; Riko, Hyde, & Alberti, 1985; Smith, Zimmerman, Connolly, Jerger, & Yelich, 1992). In an older pediatric population of 3- to 10-year-olds, estimates of serious hearing impairment were 1.1 cases per 1,000 children, with ~90% of cases being sensorineural (Van Naarden, Decoufle, & Caldwell, 1999; Wake et al., 2006). Data suggest that prevalence rates are higher in boys than girls (1.23 vs. 0.95 per 1,000 individuals) (Van Naarden et al., 1999). Factors that increase the risk for congenital or delayed-onset sensorineural hearing loss include a family history of hearing loss, congenital or central nervous system infections, ototoxic drug exposure, congenital deformities involving the head and neck, birth trauma, minority ethnicity, lower socioeconomic status, and other conditions often related to prematurity that prompt admission to an intensive care nursery (Davis et al., 1992).

Hearing loss in adults is widespread, being the third-most self-reported health problem in individuals over the age of 65 (arthritis and high blood pressure being first and second, respectively) (Havlik, 1986). New-onset hearing loss in patients between adolescence and their 50s is predominantly due to such pathologic disorders as loud noise exposure, trauma, otosclerosis, Ménière's disease, ototoxicity, idiopathic sudden sensorineural loss, and tumors (predominantly vestibular schwannomas). Noise exposure appears to be one of the most common cause of sensorineural hearing loss in this age group, predominantly due to occupational exposure to hazardous noise levels in such fields as farming, trucking, and heavy industry (Department of Labor, 1981; Green, 2002; Mrena, Ylikoski, Makitie, Pirvola, & Ylikoski, 2007). Because most of the workers in these professions are men, this is likely why men of working age constitute the fastest growing population of hearing-impaired individuals.

After 50, however, the prevalence of hearing loss accelerates dramatically, with age-related hearing loss and presbycusis contributing the most to this increase. Approximately 25% of patients between the ages 51 and 65 years have hearing thresholds greater than 30 dB (normal range being 0–20 dB) in at least one ear (Davis et al., 1992). Patients in this age group cannot be relied upon for accurate self-reported diagnosis of hearing loss; while about one third of persons over 65 and half of those over 85 will admit to a hearing loss, rigorous audiometric screening would identify an additional 15% of individuals with significant hearing loss in these age groups (Gates et al., 1990; Havlik, 1986; Mulrow & Lichtenstein, 1991).

A 1996 population study by the National Institutes of Health identified over 28 million Americans who are deaf or hearing impaired (National Institute on Deafness and Other Communication Disorders, 1996). Additional studies have demonstrated a greater than 50% increase in the number of patients with hearing loss between 1971 and 1991 (Dobie, 1993). As our population continues to age, the incidence and prevalence of hearing loss will also continue to grow, with estimates as high as 14% (Department of Labor, 1981; Dobie, 1993; Dobie, Katon, Sulliva, & Sakai, 1989).

Combining these data, one can estimate that approximately 1 in 10 Americans has a significant enough impairment to have a deficit in hearing conversational speech, while 1 in 100 has a loss severe enough to prevent the use of conventional hearing aids.

Noise-Induced Hearing Loss

Noise trauma is the second-most common cause of sensory hearing loss. Sounds exceeding 85 dB are potentially injurious to the cochlea, especially with prolonged exposure. Noise-induced hearing loss typically manifests in the high frequencies, particularly from 3 to 6 kHz, where a typical "notch" may be seen with audiometric testing. In contrast to presbycusis, recovery is seen at 8 kHz. With repeated noise exposure, however, hearing loss extends to adjacent frequencies, which makes differentiation from presbycusis difficult. Noise exposure alone does not produce a hearing loss greater than 75 dB in the higher frequencies and 40 dB in the lower frequencies. While it has traditionally been believed that hearing loss due to noise exposure does not progress once the exposure is discontinued, recent research suggests that early noise exposure predisposes to more progressive age-related hearing loss and that even isolated noise exposure can lead to significant neural damage (Fernandez, Jeffers, Lall, Liberman, & Kujawa, 2015).

Noise-induced hearing loss presents in two forms. Acoustic trauma is the sudden permanent hearing loss in response to a single, intense noise exposure, such as a blast or an explosion. Impulses typically range around 140 dB and cause direct mechanical injury to the hair cells of the cochlea. The degree of hearing loss is variable ranging from mild to profound and is often associated with new-onset tinnitus. Some improvement may be expected in the following days. Tympanic membrane perforation or ossicular disruption may cause a conductive hearing loss that accompanies the sensorineural loss.

Chronic noise-induced hearing loss, in contrast to acoustic trauma, represents the accumulation of cochlear neural damage from long-term exposure to lower-level noise. The principal mechanisms include occupational or industrial exposures, weapons, and excessively loud music.

The development of chronic noise-induced hearing loss is a two-stage process. First, the exposure to loud noise produces a temporary threshold shift, which is a brief hearing loss that completely resolves after a period of auditory rest, typically 16 to 48 hours. A permanent threshold shift occurs after repeated exposure to sufficiently intense noise. This second stage leads to an irreversible hearing loss through permanent hair cell damage. However, it should be noted that animal studies have documented that noise exposures that produce temporary threshold shifts can by themselves cause significant and permanent neural damage (Fernandez et al., 2015; Jensen, Lysaght, Liberman, Qvortrup, & Stankovic, 2015).

The Occupational Health and Safety Administration (OSHA) has set a permissible exposure limit for workers of 90 dB for 8 hours per day. The OSHA standard uses a 5-dB exchange rate, meaning for each increase in noise level by 5 dB, time exposure must be cut in half (e.g., a worker can be exposed to 100 dB of noise for a maximum of 2 hours). To limit occupational noise exposure, OSHA recommends assessment of noise levels, administrating controls, engineering controls, and the use of personal hearing protection devices. Among the latter, proper fitting earmuffs and earplugs can dampen noise intensity by 45 and 30 dB, respectively.

Recent research, however, has thrown these concepts around noise-induced hearing loss into question. A number of animal studies have emerged that suggest that even noise exposures that produce "temporary" threshold shifts are associated with long-term neuronal damage that could further exacerbate age-related hearing loss (Fernandez et al., 2015). It is very possible that many of the standards developed may be need to be revised as this research progresses.

Ototoxicity and Drug-Induced Hearing Loss

Modern medicines have greatly added to our overall quality of life, longevity, and ability to positively affect the course of numerous diseases. The "flip-side" to this benefit, however, is the potential side effects and complications inherent in nearly all drugs. Ototoxicity refers to the ability of a drug or substance to damage the inner ear, causing a functional impairment such as tinnitus, or loss of hearing or balance. There are currently several classes of drugs that are known to cause ototoxicity: salicylates, aminoglycoside antibiotics, macrolide antibiotics, vancomycin, loop diuretics, and certain classes of chemotherapeutic agents (Table 7–2).

Salicylates

Salicylates such as acetylsalicylic acid (ASA), more commonly referred to as

Table 7–2. Drugs With Reported Ototoxic and Vestibulotoxic Potential

<table>
<tr><td rowspan="3">Antibiotics</td><td>Aminoglycosides</td><td>Amikacin
Dibekacin
Gentamicin
Kanamycin
Neomycin
Netlimycin
Sisomicin
Streptomycin
Tobramycin</td></tr>
<tr><td>Macrolides</td><td>Erythromycin
Clarithromycin
Azithromycin</td></tr>
<tr><td colspan="2">Vancomycin</td></tr>
<tr><td>Loop Diuretics</td><td colspan="2">Azosemide
Bumetenide
Ethacryinic acid
Furosemide
Indapamide
Piretamide
Triflocin</td></tr>
<tr><td>Salicylates</td><td colspan="2">Acetylsalicylic acid (aspirin)</td></tr>
<tr><td colspan="3">Quinine</td></tr>
<tr><td rowspan="2">Antineoplastic drugs</td><td>Platinum compounds</td><td>Cisplatin
Carboplatin</td></tr>
<tr><td>DFMO</td><td></td></tr>
</table>

aspirin, were some of the earliest known ototoxic agents. In high doses, ASA can cause sensorineural hearing loss, tinnitus, and vertigo. In fact, it was precisely these side effects that were instrumental in determining appropriate dosages of ASA for rheumatic fever and gout in the beginning of the 1900s (Marchese-Ragona, Marioni, Marson, Martini, & Staffieri, 2007; Myers & Bernstein, 1965). While early studies showed that serum ASA concentrations of 20 to 50 mg/dL result in hearing losses of up to 30 dB, concentrations as low as 11 mg/dL have also been shown to be ototoxic in some individuals, with reversible tinnitus occurring at dosages much lower than this (Day et al., 1989).

Clinically, patients with ototoxicity due to salicylates will experience "supra-threshold" changes in hearing such as decreased temporal integration and resolution and impaired frequency selectivity (Boettcher & Salvi, 1991). These changes are generally reversible within 72 hours of drug cessation, however.

The exact mechanism of salicylate ototoxicity remains in question. Some theories of ASA-induced ototoxicity include altered levels of prostaglandins (Jung et al., 1988), inhibition of acetylcholinesterase activity of the efferent nerve endings in the organ of Corti (Ishii, Bernstein, & Bulogh, 1967), alternations in adenosine triphosphate levels in Reissner's membrane (Krzanowski & Matschinsky, 1971), and alterations in some transaminase and dehydrogenase systems (Silverstein et al., 1967). The exact anatomic portion of the inner that is affected by salicylates also remains in question. To date, histologic studies have shown that the stria vascularis, organ of Corti (including hair cells), and cochlear blood vessels all appear normal in animal models subjected to high doses of salicylates (Deer & Hunter-Duvar, 1982).

Physiologic studies of salicylate-induced effects on hearing have shown that the drug mimics the effects of efferent neuronal stimulation. By studying endocochlear potentials after high-dose ASA administration in adult cats, it has been shown that the compound action potential and summating potential were reduced, the endocochlear potential was unaffected, and the cochlear microphonic was increased (Stypulowski, 1990). It is inferred from these studies that salicylates may cause tinnitus by its direct effect on outer hair cells (controlled by the efferent neuronal pathways) and by selectively sparing the inner hair cells (afferent neuronal pathway). Other physiologic studies have implicated alterations in cochlear blood flow as a potential mechanism of salicylate-induced ototoxicity (Cazals, Li, Aurousseau, & Didier, 1988). Interestingly, salicylates, perhaps through their antioxidant properties, provide some degree of protection against Cispatinum ototoxicity (Hyppolito, de Oliveira, & Rossato, 2006).

Aminoglycosides

Because of the frequency of use, aminoglycosides are probably one of the most common causes of drug-induced ototoxicity. Because this class of antibiotics continues to remain highly effective against infection by gram-negative bacteria, knowledge about the presenting features of aminoglycoside ototoxicity is extremely important.

Aminoglycosides act by inhibition of bacterial ribosomal protein synthesis. Drugs within this class include streptomycin, neomycin, kanamycin, gentamicin, amikacin, tobramycin, netlimicin, dibekacin, and sisomicin. Of these, the most commonly used today in a form (intravenous) and dosage that may cause ototoxicity are gentamicin and tobramycin. Neomycin, used commonly in topical antibiotic ointments and oral bowel preparations, rarely achieves sufficient serum blood levels via these routes that would induce inner ear injury. The prevalence of ototoxicity differs for each antibiotic; ranging from 2% to 15% (Kahlmeter & Kahlager, 1984; Matz, 1993). An interesting feature of this class of drugs is that some tend to preferentially injure the balance organs (vestibulotoxic—streptomycin, gentamicin, and tobramycin), while others tend to have their most damaging effects on the organ of Corti (ototoxic—neomycin, kanamycin, amikacin, sisomycin). Both subclasses of medications, however, are able to injure either organ with a high enough dosage.

The mechanism of aminoglycoside-induced ototoxicity has been thoroughly studied, though many important details remain to be worked out (Rizzi & Hirose, 2007; Williams, Zenner, & Schacht, 1987). Initially, the antibiotic enters the apical region of the hair cell, likely via receptor-mediated endocytosis (Hashino & Shero, 1995). Because of their highly polar biochemical nature, the drug must be actively transported into the cell in an energy-dependent manner, partially dependent upon the structural protein myosin VIIA (Richardson et al., 1999). Within the hair cell, after being initially incorporated into lysosomes, the antibiotic will then lead to apoptosis via a mechanism that still remains to be worked out (Rizzi & Hirose, 2007).

Studies of patients with genetic mitochondrial disorders and aminoglycoside ototoxic sensitivity have also shed light on its potential mechanism of inner ear injury (Fischel-Ghodsian, 1999). Studies have shown that aminoglycosides may exert their detrimental effect through an alteration of mitochondrial protein synthesis. This exacerbates the inherent defect caused by the mutation, reducing the overall mitochondrial translation rate down to and below the minimal level required for normal cellular function (Guan, Fischel-Ghodsian, & Attardi, 2000).

Because only approximately 3% of an orally administered dose of an aminoglycoside is absorbed through the gastrointestinal tract, the preferred route of delivery for infections is intravenous. The half-life of the drug ranges from 3 to 15 hours under normal circumstances, but factors such as kidney disease can significantly alter the half-life (Lerner & Matz, 1980). As a result, whenever administering these drugs, serum peak and trough levels need to be monitored. However, the impact of an altered half-life or increased serum concentrations on the ototoxic potential is unclear. Though aminoglycosides are clearly ototoxic, a variety of studies have not shown a direct correlation between its ototoxicity and concentrations of the drug in serum or cochlear fluids (Henley & Schacht, 1988; Huang & Schacht, 1988; Kahlmeter & Kahlager, 1984; Lerner & Matz, 1980; Matz, 1993; Tran Ba Huy, Bernard, & Schacht, 1986). A consistent feature of aminoglycoside ototoxicity is the delayed onset between drug administration and onset of the hearing loss (Beaubian et al., 1990). This may be in part explained by the finding that a cytotoxic metabolite of the drug has been shown to cause death in hair cells (Huang & Schacht, 1988).

The histopathology of aminoglycoside-induced ototoxicity has been well documented. Gentamicin, tobramycin, and streptomycin preferentially act upon the balance organs, though in higher dosages will also injure the organ of Corti. These drugs have a predilection for Type I hair cells of the crista ampullaris, though later stages may also see destruction of Type II hair cells. Similarly, those drugs preferentially acting upon the organ of Corti (neomycin, kanamycin, amikacin) will selectively injure outer hair cells, while sparing inner hair cells. Furthermore, outer hair cells at the base appear to more susceptible than those at the apex. Inner hair cells are not completly spared and have been shown to be injured in several animal studies, along with other supporting structures (Hawkins, 1976; Hawkins & Engstrom, 1964; Rybak & Matz, 1986). Long-term effects of aminoglycoside ototoxicity include delayed secondary degeneration of spiral ganglion neurons (Kiang, Liberman, & Levine, 1976). While most of these findings have been shown in animal models, similar changes have been seen in human temporal bone specimens.

The clinical effects of aminoglycoside ototoxicity are variable. As previously stated, though drug levels of antibiotic do not clearly correlate with ototoxicity, patients with renal insufficiency are more susceptible. Initial auditory effects of ototoxicity are tinnitus and high-frequency sensorineural hearing loss. The lower frequencies become involved as the ototoxicity progresses. Because most audiograms will only test for hearing up through 8 kHz, if ototoxicity is suspected, special equipment or testing procedures at ultra-high frequencies may be required to detect early damage induced by the suspected antibiotic. Early identification of aminoglycoside ototoxicity is important, since stopping the antibiotic might halt further sensorineural hearing loss, though hearing loss can progress even after withdrawal of the medication (Lerner & Matz, 1980). Though the hearing loss sustained is sometimes partially reversible, once a loss has been present for 3 weeks, it is likely permanent (Rybak & Matz, 1986).

Vestibulotoxicity is common with the antibiotics gentamicin and streptomycin. For reasons that are not entirely known, these drugs preferentially act upon vestibular hair cells, and only in more advanced cases will the organ of Corti become involved. As in the cochlea, the obvious histologic deficit is loss of vestibular hair cells in the utricle, saccule, and semicircular canal ampullae (Koegel, 1985). Clinically, patients with vestibulotoxicity will experience vertigo, dysequilibrium, and oscillopsia. While in some cases the injury resulting from vestibulotoxicity can be compensated for, more often bilateral vestibular hypofunction as a result of aminoglycoside ototoxicity can be partially or completely disabling. As an aside, the vestibulotoxic potential of gentamicin is used *advantageously* to treat Ménière's disease by inducing a "chemical labyrinthectomy" through transtympanic application of the antibiotic (see the section including Ménière's disease for a further discussion of Ménière's treatment) (Nedzelski, Schessel, Bryce, & Pfleiderer, 1992).

Frequent assessment of vestibular function in patients undergoing intravenous gentamicin therapy is important to identify early vestibulotoxicity. The head thrust maneuver, a useful bedside clinical test that can accurately pick up early vestibulotoxicity, employs brief, high-acceleration horizontal head thrusts applied while instructing the patient to look carefully at the examiner's nose (Brock, Bellman, Yeomans, Pinkerton, & Pritchard, 1991; Goebel, 2001; Halmagyi, Curthoys, Colebatch, & Aw, 2005; Halmagyi, Yavor, & McGarvie, 1997). Damage to the vestibular-ocular reflex from aminoglycoside vestibulotoxicity will manifest as catch-up saccades of the patients' eyes as they try to maintain a fixed eye position during the rapid head rotation. Another useful test of vestibulotoxicity involves dynamic visual acuity testing. In this test, the results of a Snellen eye chart exam (with both eyes open) is compared to the exam while the patient's head is undergoing back and forth head turns (Herdman, Hall, Schubert, Das, & Tusa, 2007). If there is significant damage to the vestibulo-ocular reflex as a result of vestibulotoxicity, there will be a degradation of at least two lines on the eye exam. These highly sensitive tests can be an extremely useful part of the clinical armamentarium of physicians who administer aminoglycoside antibiotics.

Because of these potentially serious complications resulting from oto- and vestibulotoxicity, predisposing factors to their development have been sought. Duration of therapy >10 days, decreased pure-tone sensitivity, the presence of a severe underlying illness, a high total dose of drug, and high drug peak and trough levels have all been associated with an increased risk of ototoxicity (Brummett & Fox, 1989; Jackson & Arcieri, 1971). Thus, despite no clear correlation with serum levels of antibiotic, clinicians and pharmacists are strongly encouraged to closely monitor aminoglycoside peak and trough levels. Additionally, the patient's underlying renal condition needs to be taken into account by factoring in the creatinine clearance and adjusting the dosage appropriately.

As with most disorders, the best treatment of ototoxicity is prevention. A variety of compounds are currently being evaluated for the ability to ameliorate the effects of gentamicin-induced ototoxicity. To date, antioxidants have shown the greatest promise. These include salicylates (aspirin), vitamin E, n-acetylcysteine, deferoxamine, oleandrin, and minocycline, to name a few of the compounds that have been tried (Chen et al., 2007; Emanuele, Olivieri, Aldeghi, & Minoretti, 2007; Kharkheli, Kevanishvili, Maglakelidze, Davitashvili, & Schacht, 2007; Mostafa, Tawfik, Hefnawi, Hassan, & Ismail, 2007; Saxena, 2007; Wei et al., 2005).

Macrolide Antibiotics

Macrolide antibiotics, most notably erythromycin, have been reported to be ototoxic. Though not nearly as common as aminoglycoside-induced ototoxicity, high-dose erythromycin (oral and intravenous) has been reported to be associated with high-frequency sensorineural hearing loss in at least 35 cases (Brummett & Fox, 1989; Schweitzer & Olson, 1984). The newer macrolides, including azithromycin and clarithromycin, also have been reported to cause ototoxicity (Bizjak, Haug, Schilz, Sarodia, & Dresing, 1999; Wallace, Miller, Nguyen, & Shields, 1994). Those most susceptible include elderly individuals with liver or kidney disease. There also appears to be an association with Legionnaire's disease, most likely because the treatment for this disorder is high-dose erythromycin. There may also be central nervous system findings in patients with erythromycin ototoxicity, including diplopia, dysarthria, and psychiatric alterations. Fortunately, most cases of the ototoxicity and associated central findings are reversible once the drug is stopped. Guidelines for the prevention of erythromycin-induced ototoxicity have been put forth by Schweitzer and Olson (1984). These include a maximum daily dosage of 1.5 gm if there is associated elevation in the serum creatinine above 180 mol/L, pre- and posttreatment audiograms in all elderly patients with kidney and liver disease, and the exercise of great caution when administered along with other drugs with known ototoxic potential.

Vancomycin

The ototoxic potential of vancomycin, a powerful antibiotic used to treat aggressive staphylococcus infections such as methicillin-resistant *Staphylococcus aureus* (MRSA), is controversial (Brummett & Fox, 1989; R. E. Brummett, 1993; Klibanov, Filicko, DeSimone, & Tice, 2003). One problem in evaluating its ototoxic potential is that nearly all of the patients who have been reported to have hearing problems following vancomycin therapy have also been given an aminoglycoside during the course of treatment. Furthermore, animal studies have failed to demonstrate the ototoxic potential of vancomycin when given alone, while coadministration with gentamicin does cause increased inner ear damage as compared to aminoglycosides alone (R. E. Brummett et al., 1990). Lastly, a newborn hearing loss study in 2003 showed no association of vancomycin with sensorineural hearing loss (de Hoog et al., 2003). Thus, the data for vancomycin as a sole causative agent of ototoxicity are not strong. However, vancomycin may augment the ototoxic potential of aminoglycosides, though the precise mechanism for this potentiation is not clear.

Loop Diuretics

The loop diuretics are a family of drugs used for the treatment of kidney disease. These include the drugs azosemide, bumetanide, ethacrynic acid, furosemide, indacrinone, piretanide, and torasemide. Of these, by far the most commonly used is furosemide, also known by its trade name Lasix, while ethacrynic acid is rarely used. They are named loop diuretics because they act upon the loop of Henle in the kidney to inhibit the reabsorption of sodium and chloride. This allows water to osmotically follow the salts as they are excreted, leading to a diuresis. Ethacryinic acid is a more potent inhibitor of the ion transport than furosemide. With a half-life of approximately 30 minutes, furosemide is entirely eliminated by the kidney. Thus, individuals with renal insufficiency may have a markedly prolonged drug half-life in serum, increasing its ototoxic potential.

The precise mechanism of ototoxicity for loop diuretics has not been completely elucidated, though a variety of studies have implicated the stria vascularis of the inner ear as the principal site of injury (Bates, Beaumont, & Baylis, 2002; de Hoog et al., 2003; Shine & Coates, 2005). The loop diuretics are direct inhibitors of the $Na^{(+)}$-$K^{(+)}$- $2Cl^{-}$ co-transport system that also exists in the marginal and dark cells of the stria vascularis and is responsible for endolymph secretion (Humes, 1999; Marks & Schacht, 1983). This similarity may also be the link between agents with known ototoxicity and nephrotoxicity. In rodents, furosemide has been shown to inhibit several enzymatic pathways important for ion transport, one of the primary presumed functions of the stria vascularis (Rybak, 1982). This site of action has been bolstered by a variety of histopathologic studies as well, demonstrating a variety of degenerative changes in the stria vascularis following high-dose loop diuretic administration (Arnold, Nadol, & Weidauer, 1981; Rybak, 1982). Some less severe alterations have also been observed in the outer hair cells in the basal turn of the cochlea, but no injury has been identified in the cochlear nerve or spiral ganglion cells (Quick & Duvall, 1970). Within the balance organs, hair cells have shown mitochondrial abnormalities and increased granularity (Arnold et al., 1981).

Clinically, ototoxicity induced by loop diuretics may be transient or irreversible. Furthermore, hearing loss has been seen with both oral and intravenous dosages of the drugs. Fortunately, transient hearing loss appears to be more common, with hearing recovery taking place within 2 days of stopping the medication (Arnold et al., 1981). There are, however, clear cases of irreversible hearing loss with loop diuretics (Rybak, 1982). The infusion of loop diuretics has also been shown to suppress otoacoustic emissions and affect frequency selectivity (Martin, Jassir, Stagner, & Lonsbury-Martin, 1998). Rapid intravenous infusion of furosemide has been shown to increase the risk of the development of ototoxicity (Bosher, 1979). Other risk factors include administering the drug in those with renal insufficiency or premature infants (Bates et al., 2002; Gallagher & Jones, 1979; Shine & Coates, 2005).

Platinum Chemotherapeutic Compounds

Cisplatin (*Cis-diamminedichloroplatinum II*) and carboplatin (*Cis-diammin-1, 1-cyclobutane decarboxylate platinum II*) are potent antineoplastic drugs that have been used to successfully treat a variety of malignancies. An unfortunate side effect of this potentially life-saving medication, however, is ototoxicity in approximately 7% of recipients. The hearing loss is generally high frequency (>4000 Hz), bilateral, permanent, and accompanied by tinnitus (V. G. Schweitzer, 1993). The inner ear injury from platinum compounds is generally more severe in children, and has been reported as high as 82% in those undergoing autologous bone marrow transplantation (Brock et al., 1991; Parsons et al., 1998). Other risk factors for cisplatin ototoxicity include high cumulative dose of cisplatin, a history of noise exposure, a concomitant high doses of vincristine, another chemotherapeutic agent, and decreased serum albumin level, hemoglobin level, red blood cell count, and hematocrit (Blakley, Gupta, Myers, & Schwan, 1994; Bokemeyer et al., 1998; Chen et al., 2006). There also appears to be a genetic predisposition to cisplatinum-induced ototoxicity (Huang et al., 2007; Rybak, 2007). Patients undergoing cisplatin or carboplatin therapy therefore require audiometry prior to the onset of therapy, and at any suggestion of hearing loss or tinnitus.

It has been postulated that a possible mechanism of cisplatin ototoxicity is through the production of reactive oxygen species, which in turn leads to hair cell apoptosis. Inhibition of adenylate cyclase is one of the cellular effects that has been implicated (Bagger-Sjoback, Filipek, & Schacht, 1980). Cisplatin has also been shown to cause a rise in perilymph prostaglandins (L. Rybak & Matz, 1986). Other studies have concluded that blockage of outer hair cell ion transduction channels is responsible for the ototoxicity (McAlpine & Johnstone, 1990). Studies have implicated the inflammatory cytokines tumor necrosis factor-alpha (TNF-α) and interleukin-1-beta (IL-1β) and IL-6β (So et al., 2007).

The histopathologic changes in the inner ear as a result of cisplatin ototoxicity have been well described (Hinojosa, Riggs, Strauss, & Matz, 1995). Changes included loss of inner and outer hair cells in the basal turn of the cochlea, degeneration of the stria vascularis, and a significant decrease in spiral ganglion cells, predominantly in the upper turns.

There has been some success in limiting the ototoxic potential of platinum agents by coadministration of a variety of compounds, including antioxidants, caspase inhibitors, and inhibitors of apoptosis (Rybak, 2007). There has also been some success using sulfur- or sulfhydryl-containing compounds (thio compounds), which are thought to provide protection from cisplatin toxicity either by direct interaction between the cisplatin and the thio moiety, by displacing platinum from its site of toxic action, by preventing platinum from interfering with superoxide dismutase, or by scavenging of cisplatin-induced free radicals (Smoorenburg, De Groot, Hamers, & Klis, 1999).

Autoimmune Causes of Hearing Loss

Autoimmune inner ear disorders are a poorly understood group of disorders that can lead to loss of both hearing and balance. The rapid onset and resulting potentially severe disabilities of deafness and vertigo can be devastating to patients. However, prompt recognition and the institution of appropriate treatment can often reverse, or at least stabilize the hearing and balance deficit.

Like the barrier that exists between the brain and the blood, a similar "blood-labyrinthine" barrier also exists to maintain the ionic gradients of the inner ear (Harris & Ryan, 1995; Juhn, Hunter, & Odland, 2001). Both organs also lack an established lymphatic drainage pathway. The ear appears to be less immunoresponsive than the brain, however (Harris, Moscicki, & Ryan, 1997). When the ear does respond immunologically, the endolymphatic sac, which contains a majority of the immune cells of the inner ear, has been shown to play a major role in this response (Rask-Andersen & Stahle, 1980; Tomiyama & Harris, 1987). It is the inflammation associated with immune response that can produce a significant amount of cochlear injury, damage that can be partially mitigated by administering immunosuppressive agents (Darmstadt, Keithley, & Harris, 1990).

While there are clearly systemic autoimmune diseases that can also involve the ear, there are also autoimmune processes that solely affect the inner ear. The principal challenge for the clinician in either case is prompt diagnosis (Table 7–3). The first report on an immune-mediated hearing loss was by McCabe (1979). Since this seminal report, the diagnosis of autoimmune inner ear disorders has been aided by the availability of a Western blot assay toward a 68-Kd antigen that was associated with a positive response to steroid treatment for the disorder (Harris & Sharp, 1990). Though the 68-Kd protein has not been conclusively identified, evidence points to heat-shock protein (Billings, Keithley, & Harris, 1995). In a study of 279 patients with suspected autoimmune hearing loss, 90 (32%) were positive by Western blot, and there was a distinct female predominance in the positive group (63%), correlating with the clinical distribution of autoimmune hearing loss (Harris et al., 1997). The clinical utility of the test is questionable, however, since test results will often not dictate treatment, and patients with suspected hearing loss will usually receive the same treatment regardless of the test result (Bonaguri et al., 2007).

Clinically, patients with autoimmune inner ear disease will develop rapidly progressive, bilateral hearing loss over several weeks' time. There is also a roughly 2:1 female to male preponderance. Fluctuations in hearing are common during the course, as are the symptoms of aural fullness. Vertigo and disequilibrium are present in approximately one third of cases (Hughes, Freedman, Haberkamp, & Guay, 1996; Yehudai, Shoenfeld, & Toubi, 2006).

Approximately 30% of patients will also have manifestations of systemic autoimmune dysfunction (Hughes, Barna, Kinney, Calabrese, & Nalepa, 1988). Autoimmune diseases commonly associated with sensorineural hearing loss include rheumatoid arthritis, polyarteritis nodosa, Wegener's granulomatosis, Cogan's syndrome, Behçet's syndrome, relapsing polychondritis,

Table 7–3. Diagnostic Evaluation of a Patient With Suspected Autoimmune Hearing Loss

Blood Work	Additional Testing
Complete blood count with differential (CBC)	Urinalysis
Fluorescent treponemal antibody (FTA-ABS)	Soft tissue biopsy (e.g., temporal arteritis or minor salivary gland)
Rheumatoid factor (RF)	
Antinuclear Antibody (ANA)	
Anti–double-stranded DNA antibody	
Complement assay	
Immunoglobulin analysis by subclass	
Thyroid function tests	

and systemic lupus erythematosus ("lupus"). If one of these disorders is suspected, consultation with a rheumatologist is warranted.

Because of the potentially devastating effects of loss of hearing and balance, treatment of autoimmune inner ear disease consists of high-dose steroids. Though there is no consensus dosage of steroids, a common treatment regimen is 1 to 2 mg/kg/day (approximately 60 mg per day) of prednisone for 4 weeks (Harris et al., 1997). Patients whose hearing loss returns following withdrawal of steroids can be considered for a longer course of prednisone with a slow taper. For patients in whom autoimmune inner ear disease is strongly suspected and who either receive no benefit from steroids, develop side effects from steroid usage, or require long-term steroids to maintain hearing (>3 months) then consideration should be given to the use of cytotoxic medications such as methotrexate or cyclophosphamide (Berrettini et al., 1998; Matteson et al., 2000). In one of the few well-designed, controlled trials to date of autoimmune hearing loss, methotrexate was tested for its efficacy in this condition. The results, however, suggested that following an initial good response to steroids, methotrexate provided no benefit over placebo in maintaining hearing (Harris et al., 2003). However, the use of such medications requires close clinical monitoring of blood counts, liver enzymes, and renal function to detect the potential complications of liver and pulmonary interstitial fibrosis. For severe cases not responding to the above treatment regimens, the use of plasmapheresis has also been described, though cost and reimbursement factors are major obstacles in the use of this therapy (Luetje & Berliner, 1997).

More recently, a number of studies have investigated the use of inhibitors of tumor necrosis factor α (TNF-α) in the treatment of autoimmune inner ear disease. TNF-α is involved in the amplification of the adaptive immune response, and its inhibitors are commonly used in the treatment of a variety of nonmalignant autoimmune disorders. The utility and appropriate indications and mechanism of delivery of these compounds are still under investigation. Van Wijk documented nine patients with autoimmune inner ear disease (AIED), all of whom were successfully treated initially with oral steroids (Van Wijk, Staecker, Keithley, & Lefebvre, 2006). Five patients were unable to be weaned from steroids, while the remaining four had hearing loss following glucocorticoid cessation. All patients were treated with transtympanic infusions of infliximab over the round window niche weekly for 4 weeks. Four of five steroid-dependent patients were able to be tapered off oral steroid therapy without hearing loss, and three of the four remaining patients with hearing loss achieved hearing improvements comparable to treatment with systemic methylprednisolone. In contrast, Liu retrospectively evaluated the response to intravenous infliximab among eight patients with AIED refractory to a combination of steroid, methotrexate and cyclophosphamide therapies (Liu, Rubin, & Sataloff, 2011). These patients were given between 320 and 600 mg infliximab intravenously at 4- to 12-week intervals and found that none experienced an objective improvement in audiometric results. Similarly, a single-blinded placebo-controlled trial of 20 patients with AIED found that etanercept 25 mg subcutaneously twice weekly was no better than placebo over 8 weeks of treatment in achieving improvements in pure-tone averages or word recognition scores (Cohen, Shoup, Weisman, & Harris, 2005). There is also a report of sensorineural hearing loss among patients with autoimmune disease receiving systemic adalimumab, raising concern regarding the safety profile of TNF-α inhibitors in patients with hearing loss (Conway, Khan, & Foley-Nolan, 2011).

The use of rituximab has also been described in the treatment of AIED. Rituximab is a monoclonal antibody directed against the CD20 antigen, which results in a reduction in circulating and tissue B cells, and has been Food and Drug Administration (FDA) approved for the treatment of certain lymphomas. Cohen investigated in a pilot study seven patients with AIED who received an initial benefit after a 4-week course of prednisone therapy (Cohen et al., 2011). Patients received two injections of 1,000 mg separated by 2 weeks. Five of the seven patients met the study criteria of a 10-dB pure-tone average improvement in at least one ear or an improvement in word recognition scores by 12% or more.

Recent data have also emerged using the IL-1 receptor antagonist Anakinra for patients with AIED who become unresponsive to steroids (Vambutas et al., 2014). In a phase I/II open-label study, 70% of patients had improved hearing, which correlated with reductions in IL-1β plasma levels. Upon discontinuation of the therapy, three patients relapsed.

Idiopathic Sudden Sensorineural Hearing Loss

Perhaps one of the most frightening conditions for patients in otology is idiopathic sudden sensorineural hearing (ISSHL). In contrast to autoimmune inner ear disorders where hearing may be lost over the course of several months, patients with ISSHL may develop a partial or complete sensorineural hearing loss over the course of several hours, or may simply wake up one morning without warning with a "dead ear." The name (idiopathic) as well as the numerous theories of its etiology underscores how little is known about this condition (Finger & Gostian, 2006; Jaffe, 1973).

The incidence of ISSHL has been estimated to range from 5 to 20 per 100,000 persons per year, most commonly occurring in the sixth decade of life (Byl, 1984). This accounts for approximately 15,000 cases worldwide annually, or roughly 1% of all cases of sensorineural hearing loss (Byl, 1977). The disorder affects men and women equally. Unlike autoimmune inner ear disorders, bilateral involvement is rare, and simultaneous bilateral involvement is even rarer (Mattox & Simmons, 1977; Shaia & Sheehy, 1976). Patients most commonly present with a rapidly progressive unilateral sudden sensorineural hearing loss, often upon awakening (Mattox & Simmons, 1977). This is often accompanied by a sensation of aural fullness and tinnitus. Vertigo is seen in less than half of cases (Mattox & Simmons, 1977).

As the name implies, the etiology of ISSHL remains a mystery. The most commonly believed etiology for ISSHL is viral, despite no clearly established link between the two (Nosrati-Zarenoe, Arlinger, & Hultcrantz, 2007). This association comes from several lines of evidence: Nearly one third of patients presenting with sudden sensorineural hearing loss report a viral-like upper respiratory infection prior to the onset of their hearing loss (Mattox & Lyles, 1989). Furthermore, temporal bone histopathologic studies have implicated a viral etiology for ISSHL based upon postmortem findings (Beal, Hemenway, & Lindsay, 1967; Vasama & Linthicum, 2000). Despite these associations, however, no study has ever shown a clear relationship between viral titers and degree of hearing loss, and antiviral medication has not been shown to impact upon the hearing outcome in patients with ISSHL (Haberkamp & Tanyeri, 1999; Norris, 1988; Stokroos, Albers, & Tenvergert, 1998; Wilson, Byl, & Laird, 1980).

A vascular etiology has also been proposed for ISSHL. Vascular disorders such as leukemia and sickle cell disease have been shown to cause hearing loss, while cardiac bypass surgery also has a low risk of causing sensorineural hearing loss presumably due to embolic events intraoperatively (De la Cruz & Bance, 1998; Millen, Toohill, & Lehman, 1982; Van Prooyen-Keyzer, Sadik, Ulanovski, Parmantier, & Ayache, 2005; Walsted, Andreassen, Berthelsen, & Olesen, 2000). Further, vasospasm of the internal auditory artery during cerebellopontine angle surgery has also been shown to cause hearing loss (Mom, Telischi, Martin, Stagner, & Lonsbury-Martin, 2000). However, vasodilatory medications have not been shown to be clearly efficacious with regard to ISSHL, though are still frequently used for lack of a better alternative in many cases (Fetterman, Saunders, & Luxford, 1996; Fisch, 1983; Haberkamp & Tanyeri, 1999; Mattox, 1980; Norris, 1988). Additional etiologies proposed for ISSHL include diabetes mellitus, autoimmune, perilymphatic fistula, and cochlear hydrops (Adkisson & Meredith, 1990; Wilson et al., 1982).

A number of other potentially treatable disorders can mimic ISSHL, and these should be sought in the initial diagnostic evaluation of patients who present with a rapid onset hearing loss. In approximately one in eight patients with vestibular schwannoma, for example, patients will present with a sudden hearing loss mimicking ISSHL, and will often initially respond favorably to steroids (Berenholz, Eriksen, & Hirsh, 1992; Berg, Cohen, Hammerschlag, & Waltzman, 1986). Occasionally, patients with autoimmune hearing loss will also present with a rapidly progressive hearing loss. Thus, it is important to exclude such potentially treatable diseases.

A complete evaluation of patients presenting with a sudden sensorineural hearing loss should proceed forthright, to establish a diagnosis as rapidly as possible. Laboratory screening should include a complete blood count, electrolyte panel, liver function tests, syphilis screening (FTA-abs), thyroid function tests, C-reactive protein, cytoplastmic antineutrophil cytoplasmic antibody (c-ANCA), sedimentation rate, and antinuclear antibody (ANA) (Nosrati-Zarenoe et al., 2007). An MRI is also indicated to exclude a retrocochlear or central pathology such as a vestibular schwannoma. Audiometry is also a requisite portion of the evaluation, as are serial audiograms during the treatment course to monitor treatment response.

In a majority of cases, the laboratory and radiologic evaluation will be entirely normal in patients presenting with a sudden hearing loss. However, treatment should begin promptly at the first indication of a hearing loss, even while the evaluation is progressing and no clear diagnosis has been reached. Currently, the most commonly agreed upon treatment for this condition is steroids (prednisone), in dosages that are similar to those used for autoimmune inner ear disease: 1 mg/kg/day (typically 60 mg/day) for a total of 1 to 3 weeks. If a favorable response to steroids is noted by improvement on audiometry, then the treatment can be extended for an additional 1 to 2 months.

Intratympanic (IT) steroid injections such as dexamethasone have become popular to treat patients with ISSHL who have failed to recover hearing following systemic steroid therapy as salvage therapy and as primary therapy in patients in whom systemic steroid therapy is contraindicated or not tolerated (Bear & Mikulec, 2014; Hultcrantz & Nosrati-Zarenoe, 2014). As salvage therapy, intratympanic steroid injections appear to have benefit. A recent systematic review found that nearly all studies evaluating IT steroid therapy found some degree of benefit (Spear & Schwartz, 2011). Furthermore, this study performed a meta-analysis of the three highest quality studies and found a pooled pure-tone average benefit of 13.3 dB. As primary therapy, IT steroid injections appear to be at least equivalent to systemic glucocorticoid therapy. A recently completed prospective, randomized, multicenter noninferiority trial of ISSHL compared 121 patients who received oral prednisone 60 mg/day for 14 days followed by a 5-day taper to 129 patients who received four doses of intratympanic methylprednisolone 40 mg/mL over a 14-day period (Rauch et al., 2011). Overall, the study found that IT methylprednisolone was not inferior to oral prednisone treatment at 2 months posttreatment; however, two subgroups, patients with a baseline pure-tone average of at least 90 dB and dizziness—groups with worse prognosis for hearing recovery—showed a trend for better outcome with oral than IT treatment. Similarly, a recent systematic review echoed this finding, concluding that the available literature suggests that IT steroid therapy as primary treatment of ISSHL appears equivalent to high-dose oral prednisone therapy (Bear & Mikulec, 2014; Spear & Schwartz, 2011).

A number of additional therapies have been attempted in ISSHL, including hyperbaric oxygen; vasoactive and hemodilution treatments, such as pentoxyfylline, dextran, Ginko biloba, and nifedipine; magnesium; carbogen; and various vitamin and minerals. In general, with the exception of hyperbaric oxygen, no well-designed data exist to support their use (Conlin & Parnes, 2007a, 2007b). A Cochrane Review investigating hyperbaric oxygen (HBO) did find that this therapy improves hearing in patients with ISSHL who present within 2 weeks of hearing loss (Bennett, Kertesz, & Yeung, 2007). HBO leads to a 22% greater chance of hearing improvement, with a significantly increased chance in a 25% increase in pure-tone averages.

The prognosis of patients with ISSHL is variable, and depends upon several factors (Xenellis et al., 2006). Studies have shown that the more severe the initial hearing loss, the less likely hearing will recover, regardless of treatment modality used. Additionally, patients who show a predominantly low-frequency hearing loss, or up-sloping and mid-frequency losses, recover more frequently than do patients with a down-sloping or flat hearing loss. Other poor prognostic factors include the presence of vertigo, reduced speech discrimination, and a sudden hearing loss in both children and adults over the age of 40 (Byl, 1977, 1984; Mattox & Simmons, 1977; Wilson et al., 1980).

Infections

Viral

A number of viruses have been implicated in adult sensorineural hearing loss; however, the majority of evidence is inferential. Neurotropic viruses, such as herpes simplex, cytomegalovirus (CMV), and rubella are well causes of congenital hearing loss, but there is no clear causative evidence of adult hearing loss.

Ramsay Hunt Syndrome

Ramsay Hunt syndrome, or herpes zoster oticus, is a viral illness caused by infection of the geniculate ganglion of the facial nerve by the varicella zoster virus (VZV). Commonly affected nerves include cranial nerves V, VII, VIII, and IX. Infection leads to vesiculation and ulceration of the external ear, anterior two thirds of the tongue, and soft palate; ipsilateral facial palsy; otalgia; dysarthria; and gait ataxia. Otologic symptoms in adults include hearing loss (52.7%), tinnitus (24.7%), and vertigo (31.8%) (Sweeney & Gilden, 2001). Treatment includes an oral steroid, commonly a 2-week prednisone taper beginning at 1 mg/kg/day, and an antiviral medication such as acyclovir. In one large retrospective study with Ramsay Hunt syndrome, 12 patients presented with mild to moderate hearing loss and were followed with serial audiograms. Six had complete audiologic recovery, three had partial recovery, and three remained stable (Murakami et al., 1997).

Human Immunodeficiency Virus

Human immunodeficiency virus (HIV) has been implicated in sensorineural hearing loss among patient with and without AIDS. Approximately 30% to 50% of HIV-infected individuals in various stages of the disease display hearing loss as compared with 12% in a seronegative control group (Chandrasekhar et al., 2000). Affected individuals display a varying degree of hearing loss incidence, magnitude, and range of loss.

Etiologic theories of HIV-mediated hearing loss include direct infection by the HIV virus, opportunistic infection (e.g., bacterial, viral, parasitic, and fugal), central nervous system disease, malignancies (e.g., lymphoma and Kaposi sarcoma), and ototoxic therapy. Chronic and acute otitis media, otitis externa, and progression of otosyphilis are found more frequently in patients with HIV. A number of antiretroviral medications have been implicated in hearing loss and tinnitus, both sudden and gradual onset (Simdon, Watters, Bartlett, & Connick, 2001).

Otosyphilis

Otologic manifestations secondary to syphilis infection can present at any point during the time course of the illness. Patients with infections of primary or secondary syphilis typically note sudden or rapidly progressive bilateral and profound hearing loss with or without vertigo. The majority of patients, however, present with late disease with symptoms similar to Ménière's disease. Hearing loss is typically asymmetric and sensorineural with poor word recognition scores (Phillips, Gaunt, & Phillips, 2014; Yimtae, Srirompotong, & Lertsukprasert, 2007).

The diagnosis of otosyphilis can be challenging and is usually a diagnosis of exclusion based on positive serology with cochleovestibular symptoms and no other likely causes. Although the respective sensitivity and sensitivity of FTA-abs is 100% and 98%, the predictive value of a positive test in the otologic population is only 22% because of the low disease prevalence. Due to this reason and the fact that the FTA-abs assay does not distinguish between active and treated disease, the Western blot assay can eliminate the possibility of a false-positive result and evaluate for active infection.

Treatment consists of penicillin and steroids. As most patients present with latent syphilis in which spirochetes replicate more slowly, the most commonly recommended antibiotic regimen is with benzathine penicillin G 2.4 million units per week for 3 weeks. Prednisone is given 20 mg four times daily for 10 days to empirically reduce vasculitis. Long-term hearing outcomes show that less than a third of patients experience hearing improvements, but over 75% will experience at least hearing preservation (Phillips et al., 2014; Yimtae et al., 2007). Otosyphilis can present at any stage of HIV infection and should be considered in seropositive patients presenting with otologic complaints (Smith & Canalis, 1989).

Treatment of Sensorineural Hearing Loss

Regardless of the etiology, the treatment of sensorineural hearing loss is based upon the degree of hearing loss, as a measure of both the pure-tone audiometry as well as speech discrimination scores.

Patients with hearing loss not correctable by medical therapy may benefit from hearing amplification. Contemporary hearing aids are comparatively free of distortion and have been miniaturized to the point where they often may be contained entirely within the ear canal or lie inconspicuously behind the ear. To optimize the benefit, a hearing aid must be carefully selected to conform to the nature of the hearing loss. Digitally programmable hearing aids are widely available and allow optimization of speech intelligibility and improved performance in difficult listening circumstances. Aside from hearing aids, many assistive devices are available to improve comprehension in individual and group settings, to help with hearing television and radio programs, and for telephone communication.

For patients with severe to profound sensory hearing loss, the cochlear implant—an electronic device that is surgically implanted into the cochlea to stimulate the auditory nerve—offers socially beneficial auditory rehabilitation to most adults with acquired deafness and children with congenital or genetic deafness. New trends in cochlear implantation include its use for patients with only partial deafness, preserving residual hearing and allowing both acoustic and electrical hearing in the same ear, and cochlear implantation for single-sided deafness.

Tinnitus

Tinnitus is defined as the sensation of sound in the absence of an exogenous sound source. Tinnitus can accompany any form of hearing loss and its presence provides no diagnostic value in determining the cause of a hearing loss. Approximately 15% of the general population experience some type of tinnitus (Coles, 1984), with prevalence beyond 20% in aging populations (Axelsson & Ringdahl, 1987). Simple forms of tinnitus include the low-pitch tinnitus that results from the accumulation of cerumen in the external ear canal, or that due to tympanic membrane perforation. Eustachian tube insufficiency manifesting a middle-ear effusion can produce tinnitus, as can an abnormally patent Eustachian tube that produces a "blowing" tinnitus that coincides with respiration. More complex presentations of tinnitus are categorized as either objective (heard or recordable by others) or subjective (reported by the patient and not heard or recordable by others). The most common presentation of tinnitus, however, is in the setting of sensorineural hearing loss. Sensorineural hearing loss has several psychophysical correlates (Pickles, 1988). In addition to diminished hearing sensitivity, reduced frequency resolution, loudness recruitment, and tinnitus commonly associate with many forms of sensorineural hearing loss. Indeed, the vast majority of cases of clinical tinnitus are subjective and are associated with sensorineural hearing loss.

Objective Tinnitus

Objective tinnitus can be recorded objectively with a microphone or heard by another listener. Classic descriptions place the site-of-lesion of objective tinnitus peripheral to the cochlea. Contractions of soft palate and middle ear muscles can produce a clicking-like objective tinnitus. The presence of pulsatile tinnitus should prompt an evaluation of the skull base and major vessels of the head and neck, as this represents a vascular etiology. Arterial causes for pulsatile tinnitus include arteriovenous shunts (as seen with malformations, fistulae, and in Paget's disease), arterio-arterial anastomoses including rare congenital anomalies and intraluminal irregularities of the carotid system (such as atherosclerotic plaques or fibromuscular dysplasia) (Arganbright & Friedland, 2007; Baumgartner & Bogousslavsky, 2005; Chen, How, & Chern, 2007; Hafeez, Levine, & Dulli, 1999; Lerut, De Vuyst, Ghekiere, Vanopdenbosch, & Kuhweide, 2007; Magliulo, Parrotto, Sardella, Della Rocca, & Re, 2007; Sonmez, Basekim, Ozturk, Gungor, & Kizilkaya, 2007; Topal, Erbek,

Erbek, & Ozluoglu, 2007). As a result, modalities employed in the evaluation of the patient with pulsatile tinnitus should include an MRI and MR angiography and venography of the intracranial vessels or an angiogram. For arteriovenous malformations or fistulae, these can often be treated during the angiogram by the interventional radiologist (Shownkeen, Yoo, Leonetti, & Origitano, 2001). Additional modalities can include a carotid ultrasound if vascular disease in the neck is suspected.

Subjective Tinnitus

Subjective tinnitus is a common clinical disorder characterized by auditory sensations without external stimulation. The sensation can last anywhere from several seconds to a lifelong affliction. The sensation itself can be one of a high-tone ring (most commonly), buzzing, humming, crickets, whistle, hissing, or roaring. In rare cases, patients can sometimes perceive actual musical melodies. Tinnitus can affect patients in many ways, ranging from mild annoyance to severe depression and suicide (Dobie, 2003; Kaltenbach, 2006a, 2006b).

Perhaps one of the most frustrating aspects of tinnitus is our own ignorance regarding its etiology and anatomic origin (Eggermont, 2015). The best argument for a peripheral contribution from the cochlea comes from the well-documented association of tinnitus with various forms of hearing loss, including noise exposure, presbycusis, and drug-induced ototoxicity (Konig, Schaette, Kempter, & Gross, 2006). While it is widely agreed that such triggers can generate tinnitus, whether pure cochlear causes are enough to sustain tinnitus over the long term remains controversial (Baguley, 2002; Saunders, 2007). Spontaneous otoacoustic emissions are one means by which the cochlea can physically generate sound, which can be objectively recorded (Kemp, 1978). This process, representing active outer hair cell motion, has been shown to be a rare form of objective tinnitus (Plinkert, Gitter, & Zenner, 1990). However, most who have studied the phenomenon believe this is an unlikely cause of tinnitus in most patients (Penner, 1989). Loss of cochlear hair cells has also been hypothesized to contribute to tinnitus by altering the afferent auditory signal from abnormal motion of the basilar membrane (Jastreboff, 1990; Saunders, 2007). Other theories of the peripheral origin of tinnitus include loss of efferent connectivity to the outer hair cells (Chery-Croze, Truy, & Morgon, 1994), excessive release of the neurotransmitter glutamate by the inner hair cells, and reduced hair cell stiffness in response to change in its cytoskeletal structure (Saunders, 2007). Yet, if tinnitus were purely a peripheral phenomenon, then ablating cochlear function should reduce, if not abolish tinnitus. In fact, the opposite is true: Tinnitus persists following cochlear destruction or sectioning of the auditory nerve, as clinicians have known for many years (House & Brackmann, 1981). Compounding the difficulty of identifying the source of chronic, subjective tinnitus is the lack of a robust animal model for study. To date, animal models for tinnitus have been based predominantly behavioral and psychophysical paradigms (Bauer, 2003).

In the absence of compelling data that the cochlea is the site of origin of chronic tinnitus, researchers have focused upon the central auditory system and maladaptive central neuroplastic changes as underlying factor in many, if not all forms of chronic tinnitus (Eggermont, 2005, 2015). Though the initial damage may be peripheral (i.e., cochlear), a variety of evidence implicates that there are ensuing changes throughout the central auditory system, including new axonal growth, synaptogenesis and synapse degeneration, and changes in the structural properties of neurons (Saunders, 2007). Increased spontaneous neuronal activity in the dorsal cochlear nucleus (DCN) is one theory for chronic tinnitus, backed by data demonstrating cochlear damage altering outer hair cell afferent Type II fibers, which in turn leads to reduced inhibition of fusiform cells within the cochlear nucleus (Kaltenbach, 2006b). Within the DCN, increased spontaneous neuronal activation has also been attributed to bursting discharges and neuronal synchrony following ototoxic drug exposure (Baguley, 2002; Saunders, 2007).

Beyond the DCN, the auditory cortex has also been implicated in chronic tinnitus. As with the DCN, hyperactive spontaneous neuronal activity has been implicated along with neuronal synchronization (Saunders, 2007). These theories of tinnitus mirror those changes seen in the somatosensory system in response to loss of sensory input due to amputation (Rauschecker, 1999). Newer imaging modalities, including positron emission tomography (PET) and functional magnetic resonance imaging (fMRI), will hopefully shed new light onto some of the potential central causes of chronic tinnitus (Langguth et al., 2006; Saunders, 2007; Smits et al., 2007).

Tinnitus Treatment

Though tinnitus is commonly associated with hearing loss, tinnitus severity correlates poorly with the degree of hearing loss (Chung, Gannon, & Mason, 1984). Studies have shown that approximately one in seven tinnitus sufferers experience severe annoyance while 4% are severely disabled (Evered & Lawrenson, 1981).

To date, both pharmacologic and surgical approaches have shown little to no efficacy toward tinnitus. Those treatments that have demonstrated benefit relate to the use of masking external inputs, in the form of an external masking (Vernon, Griest, & Press, 1990) or cochlear implantation (Miyamoto, Wynne, McKnight, & Bichey, 1997). Behavioral adaptive methods offer benefit in enhancing tolerance of the symptom and alleviating associated anxiety.

The potential concurrence of tinnitus and depression should be assessed in all patients with subjective tinnitus. Dobie (Dobie et al., 1989) has examined the simple notion that "tinnitus causes depression" and notes that many patients believe this to be the case. However, Sullivan et al. (1988) observed that depressed subjects noted having had episodes of major depression prior to the onset of their tinnitus. In 32 patients with tinnitus and a current or past major depression, depression preceded tinnitus in 15, and the five noted simultaneous onset. Mathew et al. (Mathew, Weinman, & Mirabi, 1981) evaluated the physical symptoms of depressed patients and found that roughly half of a cohort of depressed patients complained of tinnitus, compared to 12% of controls. Does this suggest that depression causes tinnitus? Dobie notes that this seems implausible, but it is certainly likely that depression erodes the coping process that probably develops for most patients with tinnitus, thus changing a normally insignificant symptom into one with high life impact. Depression can certainly amplify disability associated with chronic medical illness (Wells & Sherbourne, 1999). For example, not all individuals with irritable bowel syndrome seek medical advice, and those who do seek medical advice differ from those who do not. Those who do seek advice demonstrate greater levels anxiety and depression after controlling for severity of gastrointestinal symptoms (Olden & Drossman, 2000).

Treatment results using antidepressants in tinnitus sufferers are encouraging. Nortriptyline, a tricyclic antidepressant administered with careful dosing adjustments, appears to improve symptoms in several domains in depressed tinnitus patients. Clearly, the largest treatment effect is a reduction in depressive symptoms. Furthermore, however, not only did patients' estimates of tinnitus severity (i.e., "How much does it bother you?") decline as depression improved, but many patients also reported quieter tinnitus. Nonetheless, the authors of this study indicate that the major therapeutic benefit was in mood and coping ability, with less change in tinnitus, per se.

Transcranial magnetic stimulation, or TMS, is newer modality being directed toward tinnitus and has shown early promise (De Ridder et al., 2007; Dornhoffer & Mennemeier, 2007; Langguth et al., 2007). Already in trials for such disorders as depression (Loo, McFarquhar, & Mitchell, 2007), TMS may offer tinnitus sufferers a nonpharmacologic, nonsurgical treatment.

References

Adkisson, G. H., & Meredith, A. P. (1990). Inner ear decompression sickness combined with a fistula of the round window. Case report [see comments]. *Annals of Otology, Rhinology, and, Laryngology, 99*(9, Pt. 1), 733–737.

American National Standards Institute. (1978). *Methods for manual pure-tone threshold audiometry* (Vol. S3.21–1978 [R1986]). New York, NY: Author.

Arganbright, J., & Friedland, D. R. (2007). Pulsatile tinnitus. *Otology & Neurootology, 29*(3), 416.

Arnold, W., Nadol, J. B., Jr., & Weidauer, H. (1981). Ultrastructural histopathology in a case of human ototoxicity due to loop diuretics. *Acta Otolaryngologica, 91*(5–6), 399–414.

Augustsson, I., Nilson, C., & Engstrand, I. (1990). The preventive value of audiometric screening of preschool and young school-children. *International Journal of Pediatric Otorhinolaryngology, 20*(1), 51–62.

Axelsson, A., & Ringdahl, A. (1987). The occurrence and severity of tinnitus: A prevalence study. In F. H (Ed.), *Proceedings of the Third International Tinnitus Seminar*. Karlsruhe, Germany: Harsch Verlag.

Bagger-Sjoback, D., Filipek, C., & Schacht, J. (1980). Characteristics and drug responses of cochlear and vestibular adenylate cyclase. *Archives of Otorhinolaryngology, 228*, 217–222.

Baguley, D. M. (2002). Mechanisms of tinnitus. *British Medical Bulletin, 63*, 195–212.

Bates, D. E., Beaumont, S. J., & Baylis, B. W. (2002). Ototoxicity induced by gentamicin and furosemide. *Annals of Pharmacotherapy, 36*(3), 446–451.

Bauch, C. D., & Olsen, W. O. (1986). The effect of 2000–4000 Hz hearing sensitivity on ABR results. *Ear and Hearing, 7*(5), 314–317.

Bauer, C. A. (2003). Animal models of tinnitus. *Otolaryngology Clinics of North America, 36*(2), 267–285, vi.

Baumgartner, R. W., & Bogousslavsky, J. (2005). Clinical manifestations of carotid dissection. *Frontiers of Neurology and Neuroscience, 20*, 70–76.

Beal, D. D., Hemenway, W. G., & Lindsay, J. R. (1967). Inner ear pathology of sudden deafness: Histopathology of acquired deafness in the adult coincident with viral infection. *Archives of Otolaryngology, 85*(6), 591–598.

Bear, Z. W., & Mikulec, A. A. (2014). Intratympanic steroid therapy for treatment of idiopathic sudden sensorineural hearing loss. *Molecular Medicine, 111*(4), 352–356.

Beaubian, A., Desjardins, S., Ormsby, E., Bayne, A., Carrier, K., Cauchy, M. J., . . . St. Pierre, A. (1990). Delay in hearing loss following drug administration: A consistent feature of amikacin ototoxicity. *Acta Otolaryngologica, 109*, 345–341.

Beiter, R. C., & Talley, J. N. (1976). High-frequency audiometry above 8000 Hz. *Audiology, 15*(3), 207–214.

Bennett, M. H., Kertesz, T., & Yeung, P. (2007). Hyperbaric oxygen for idiopathic sudden sensorineural hearing loss and tinnitus. *Cochrane Database Systematic Reviews, 1*, CD004739.

Berenholz, L. P., Eriksen, C., & Hirsh, F. A. (1992). Recovery from repeated sudden hearing loss with corticosteroid use in the presence of an acoustic neuroma. *Annals of Otology, Rhinology, and Laryngology, 101*(10), 827–831.

Berg, H. M., Cohen, N. L., Hammerschlag, P. E., & Waltzman, S. B. (1986). Acoustic neuroma presenting as sudden hearing loss with recovery. *Otolaryngology-Head and Neck Surgery, 94*(1), 15–22.

Berrettini, S., Ferri, C., Ravecca, F., LaCivita, L., Bruschini, L., Riente, L., Sellari-Franceschini, S. (1998). Progressive sensorineural hearing impairment in systemic vasculitides. *Seminars in Arthritis and Rheumatology, 27*(5), 301–318.

Billings, P. B., Keithley, E. M., & Harris, J. P. (1995). Evidence linking the 68 kilodalton antigen identified in progressive sensorineural hearing loss patient sera with heat shock protein 70. *Annals of Otology, Rhinology, and Laryngology, 104*(3), 181–188.

Bizjak, E. D., Haug, M. T., III, Schilz, R. J., Sarodia, B. D., & Dresing, J. M. (1999). Intravenous azithromycin-induced ototoxicity. *Pharmacotherapy, 19*(2), 245–248.

Blakley, B. W., Gupta, A. K., Myers, S. F., & Schwan, S. (1994). Risk factors for ototoxicity due to cisplatin. *Archives of Otolaryngology-Head & Neck Surgery, 120*(5), 541–546.

Boettcher, F. A., & Salvi, R. J. (1991). Salicylate ototoxicity: Review and synthesis. *American Journal of Otolaryngology, 12*(1), 33–47.

Bokemeyer, C., Berger, C. C., Hartmann, J. T., Kollmannsberger, C., Schmoll, H. J., Kuczyk, M. A., & Kanz, L. (1998). Analysis of risk factors for cisplatin-induced ototoxicity in patients with testicular cancer. *British Journal of Cancer, 77*(8), 1355–1362.

Bonaguri, C., Orsoni, J. G., Zavota, L., Monica, C., Russo, A., Pellistri, I., & Piazza, F. (2007). Anti-68 kDa antibodies in autoimmune sensorineural hearing loss: Are these autoantibodies really a diagnostic tool? *Autoimmunity, 40*(1), 73–78.

Bosher, S. (1979). Ethacrynic acid ototoxicity as a general model in cochlear pathology. *Advances in Otorhinolaryngology, 22*, 81–89.

Brock, P. R., Bellman, S. C., Yeomans, E. C., Pinkerton, C. R., & Pritchard, J. (1991). Cisplatin ototoxicity in children: A practical grading system. *Medical and Pediatric Oncology, 19*(4), 295–300.

Brummett, R., & Fox, K. (1989). Vancomycin and erythromycin-induced hearing loss in humans. *Antimicrobial Agents in Chemotherapy, 33*, 791–794.

Brummett, R. E. (1993). Ototoxicity of vancomycin and analogues. *Otolaryngology Clinics of North America, 26*(5), 821–828.

Brummett, R. E., Fox, K. E., Jacobs, F., Kempton, J. B., Stokes, Z., & Richmond, A. B. (1990). Augmented gentamicin ototoxicity induced by vancomycin in guinea pigs. *Archives of Otolaryngology-Head & Neck Surgery, 116*(1), 61–64.

Byl, F. M. (1977). Seventy-six cases of presumed sudden hearing loss occurring in 1973: Prognosis and incidence. *Laryngoscope, 87*(5, Pt. 1), 817–825.

Byl, F. M., Jr. (1984). Sudden hearing loss: eight years' experience and suggested prognostic table. *Laryngoscope, 94*(5, Pt. 1), 647–661.

Carhart, R. (1971). Observations on relations between thresholds for pure tones and for speech. *Journal of Speech and Hearing Disorders, 36*(4), 476–483.

Carhart, R., & Jerger, J. F. (1959). Preferred method for clinical determination of pure-tone thresholds. *Journal of Speech and Hearing Disorders, 24*, 330–345.

Cazals, Y., Li, X., Aurousseau, C., & Didier, A. (1988). Acute effects of noradrenalin-related vasoactive agents on the ototoxicity of aspirin: An experimental study in the guinea pig. *Hearing Research, 36*, 89–94.

Chandrasekhar, S. S., Brackmann, D. E., & Devgan, K. K. (1995). Utility of auditory brainstem response audiometry in diagnosis of acoustic neuromas. *American Journal of Otology, 16*(1), 63–67.

Chandrasekhar, S. S., Connelly, P. E., Brahmbhatt, S. S., Shah, C. S., Kloser, P. C., & Baredes, S. (2000). Otologic and audiologic evaluation of human immunodeficiency virus-infected patients. *American Journal of Otolaryngology, 21*(1), 1–9.

Chen, W. C., Jackson, A., Budnick, A. S., Pfister, D. G., Kraus, D. H., Hunt, M. A., . . . Wolden, S. L. (2006). Sensorineural hearing loss in combined modality treatment of nasopharyngeal carcinoma. *Cancer, 106*(4), 820–829.

Chen, Y., Huang, W. G., Zha, D. J., Qiu, J. H., Wang, J. L., Sha, S. H., & Schacht, J. (2007). Aspirin attenuates gentamicin ototoxicity: from the laboratory to the clinic. *Hearing Research, 226*(1–2), 178–182.

Chen, Y. J., How, C. K., & Chern, C. H. (2007). Cerebral dural arteriovenous fistulas presenting as pulsatile tinnitus. *Internal Medical Journal, 37*(7), 503.

Chery-Croze, S., Truy, E., & Morgon, A. (1994). Contralateral suppression of transiently evoked otoacoustic emissions and tinnitus. *British Journal of Audiology, 28*(4–5), 255–266.

Chung, D. Y., Gannon, R. P., & Mason, K. (1984). Factors affecting the prevalence of tinnitus. *Audiology, 23*(5), 441–452.

Clark, J. G. (1981). Uses and abuses of hearing loss classification. *ASHA, 23*(7), 493–500.

Coats, A. C. (1978). Human auditory nerve action potentials and brain stem evoked responses. *Archives of Otolaryngology, 104*(12), 709–717.

Cohen, S., Roland, P., Shoup, A., Lowenstein, M., Silverstein, H., Kavanaugh, A., & Harris, J. (2011). A pilot study of rituximab in immune-mediated inner ear disease. *Audiology & Neurootology, 16*(4), 214–221.

Cohen, S., Shoup, A., Weisman, M. H., & Harris, J. (2005). Etanercept treatment for autoimmune inner ear disease: Results of a pilot placebo-controlled study. *Otology & Neurootology, 26*(5), 903–907.

Coles, R. R. (1984). Epidemiology of tinnitus: (1) prevalence. *Journal of Laryngology and Otolology Supplement, 9*, 7–15.

Conlin, A. E., & Parnes, L. S. (2007a). Treatment of sudden sensorineural hearing loss: I. A systematic review. *Archives of Otolaryngology-Head & Neck Surgery, 133*(6), 573–581.

Conlin, A. E., & Parnes, L. S. (2007b). Treatment of sudden sensorineural hearing loss: II. A meta-analysis. *Archives of Otolaryngology-Head & Neck Surgery, 133*(6), 582–586.

Conway, R., Khan, S., & Foley-Nolan, D. (2011). Use of adalimumab in treatment of autoimmune sensorineural hearing loss: A word of caution. *Journal of Rheumatology, 38*(1), 176; Author reply 176.

Darmstadt, G. L., Keithley, E. M., & Harris, J. P. (1990). Effects of cyclophosphamide on the pathogenesis of cytomegalovirus-induced labyrinthitis. *Annals of Otology, Rhinology, and, Laryngology, 99*(12), 960–968.

Davis, A., Stephens, D., Rayment, A., & Thomas, K. (1992). Hearing impairments in middle age: The acceptability, benefit and cost of detection (ABCD). *British Journal of Audiology, 26*(1), 1–14.

Day, R. Graham, G. G., Bieri, D., Brown, M., Cairns. D., Harris, G., . . . Smith, J. (1989). Concentration-response relationships for salicylate-induced ototoxicity in normal volunteers. *British Journal of Clinical Pharmacology, 28*, 295–701.

de Hoog, M., van Zanten, B. A., Hop, W. C., Overbosch, E., Weisglas-Kuperus, N., & van den Anker, J. N. (2003). Newborn hearing screening: Tobramycin and vancomycin are not risk factors for hearing loss. *Journal of Pediatrics, 142*(1), 41–46.

De la Cruz, M., & Bance, M. (1998). Bilateral sudden sensorineural hearing loss following non-otologic surgery [see comments]. *Journal of Laryngology and Otology, 112*(8), 769–771.

De Ridder, D., De Mulder, G., Verstraeten, E., Seidman, M., Elisevich, K., Sunaert, S., . Møller, A. (2007). Auditory cortex stimulation for tinnitus. *Acta Neurochirurgica Supplement, 97*(Pt. 2), 451–462.

Deer, B., & Hunter-Duvar, I. (1982). Salicylate ototoxicity in the chinchilla: A behavioral and electron microscope study. *Journal of Otolaryngology, 11*, 260–271.

Department of Labor, Occupational Health and Safety Administration. (1981). Occupational noise exposure: Hearing conservation amendment. *Federal Register, 46*, 4078–4180.

Dillon, H. (1982). A quantitative examination of the sources of speech discrimination test score variability. *Ear and Hearing, 3*(2), 51–58.

Dobie, R. (1993). *Medical-legal evaluation of hearing loss*. New York, NY: Van Nostrand Reinhold.

Dobie, R., Katon, W., Sulliva, N. M., & Sakai, C. (1989). Tinnitus, depression and aging. In J. Goldstein, H. Kashima, & C. Koopmann (Eds.), *Geriatric otolaryngology* (pp. 45–49). Toronto, Canada: B.C. Decker.

Dobie, R. A. (2003). Depression and tinnitus. *Otolaryngology Clinics of North America, 36*(2), 383–388.

Dornhoffer, J. L., & Mennemeier, M. (2007). Transcranial magnetic stimulation and tinnitus: Implications for theory and practice. *Journal of Neurology, Neurosurgery, and Psychiatry, 78*(2), 113.

Eggermont, J. J. (2005). Tinnitus: Neurobiological substrates. *Drug Discovery Today, 10*(19), 1283–1290.

Eggermont, J. J. (2015). The auditory cortex and tinnitus—a review of animal and human studies. *European Journal of Neuroscience, 41*(5), 665–676.

Eggermont, J. J., Don, M., & Brackmann, D. E. (1980). Electrocochleography and auditory brainstem electric responses in patients with pontine angle tumors. *Annals of Otology, Rhinology, and Laryngology Supplement, 89*(6, Pt. 2), 1–19.

Emanuele, E., Olivieri, V., Aldeghi, A., & Minoretti, P. (2007). Topical administration of oleandrin could protect against gentamicin ototoxicity via inhibition of activator protein-1 and c-Jun N-terminal kinase. *Medical Hypotheses, 68*(3), 711.

Evered, D., & Lawrenson, G. (1981). *Tinnitus, CIBA Symposium #85*. London, UK: Pitman Medical.

Fernandez, K. A., Jeffers, P. W., Lall, K., Liberman, M. C., & Kujawa, S. G. (2015). Aging after noise exposure: Acceleration of cochlear synaptopathy in "recovered" ears. *Journal of Neuroscience, 35*(19), 7509–7520.

Festen, J. M., & Plomp, R. (1983). Relations between auditory functions in impaired hearing. *Journal of the Acoustical Society of America, 73*(2), 652–662.

Fetterman, B. L., Saunders, J. E., & Luxford, W. M. (1996). Prognosis and treatment of sudden sensorineural hearing loss. *American Journal of Otology, 17*(4), 529–536.

Finger, R. P., & Gostian, A. O. (2006). Idiopathic sudden hearing loss: Contradictory clinical evidence, placebo effects and high spontaneous recovery rate—where do we stand in assessing treatment outcomes? *Acta Otolaryngologica, 126*(11), 1124–1127.

Fisch, U. (1983). Management of sudden deafness. *Otolaryngology-Head and Neck Surgery, 91*(1), 3–8.

Fischel-Ghodsian, N. (1999). Genetic factors in aminoglycoside toxicity. *Annals of the New York Academy of Sciences, 884*, 99–109.

Fletcher, H. (1950). A method of calculating hearing loss for speech from an audiogram. *Acta Otolaryngologica Supplement, 90*, 26–37.

Fortnum, H., Davis, A., Butler, A., & Stevens, J. (1991). *Health service implications of changes in aetiology and referral patterns of hearing impaired children in Trent 1985–1993. Report to Trent Health.* Nottingham/Sheffield: MRC Institute of Hearing Research and Trent Health.

Galambos, R., & Hecox, K. E. (1978). Clinical applications of the auditory brain stem response. *Otolaryngology Clinics of North America, 11*(3), 709–722.

Gallagher, K. L., & Jones, J. K. (1979). Furosemide-induced ototoxicity. *Annals of Internal Medicine, 91*(5), 744–745.

Gates, G. A., Cooper, J. C., Kannel, W. B., & Miller, N. J. (1990). Hearing in the elderly: The Framingham cohort, 1983–1985. Part I. Basic audiometric test results. *Ear and Hearing, 11*(4), 247–256.

Goebel, J. A. (2001). The ten-minute examination of the dizzy patient. *Seminars in Neurology, 21*(4), 391–398.

Goodman, A. (1965). Reference zero levels for pure tone audiometer. *ASHA, 7*, 262–263.

Gordon, J. S., Phillips, D. S., Helt, W. J., Konrad-Martin, D., & Fausti, S. A. (2005). Evaluation of insert earphones for high-frequency bedside ototoxicity monitoring. *Journal of Rehabilitation Research and Development, 42*(3), 353–361.

Green, J. (2002). Noise-induced hearing loss. *Pediatrics, 109*(5), 987–988.

Guan, M. X., Fischel-Ghodsian, N., & Attardi, G. (2000). A biochemical basis for the inherited susceptibility to aminoglycoside ototoxicity [in process citation]. *Human Molecular Genetics, 9*(12), 1787–1793.

Haberkamp, T. J., & Tanyeri, H. M. (1999). Management of idiopathic sudden sensorineural hearing loss [see comments]. *American Journal of Otology, 20*(5), 587–592.

Hafeez, F., Levine, R. L., & Dulli, D. A. (1999). Pulsatile tinnitus in cerebrovascular arterial diseases. *Journal of Stroke and Cerebrovascular Disease, 8*(4), 217–223.

Hall, J. W. (1992). *Handbook of auditory evoked responses*. Needham Heights, MA: Allyn & Bacon.

Halmagyi, G. M., Aw, S. T., McGarvie, L. A., Todd, M. J., Bradshaw, A., Yavor, R. A., & Fagan, P. A. (2003). Superior semicircular canal dehiscence simulating otosclerosis. *Journal of Laryngology and Otology, 117*(7), 553–557.

Halmagyi, G. M., Curthoys, I. S., Colebatch, J. G., & Aw, S. T. (2005). Vestibular responses to sound. *Annals of the New York Academy of Sciences, 1039*, 54–67.

Halmagyi, G. M., Yavor, R. A., & McGarvie, L. A. (1997). Testing the vestibulo-ocular reflex. *Advances in Otorhinolaryngology, 53*, 132–154.

Harris, J. P., Moscicki, R. A., & Ryan, A. F. (1997). Immunologic disorders of the inner ear. In G. B. Hughes & M. L. Pensak (Eds.), *Clinical otology* (pp. 381–391). New York, NY: Thieme Medical.

Harris, J. P., & Ryan, A. F. (1995). Fundamental immune mechanisms of the brain and inner ear. *Otolaryngology-Head and Neck Surgery, 112*(6), 639–653.

Harris, J. P., & Sharp, P. A. (1990). Inner ear autoantibodies in patients with rapidly progressive sensorineural hearing loss. *Laryngoscope, 100*(5), 516–524.

Harris, J. P., Weisman, M. H., Derebery, J. M., Espeland, M. A., Gantz, B. J., Gulya, A. J., . . . Brookhouser, P. E. (2003). Treatment of corticosteroid-responsive autoimmune inner ear disease with methotrexate: A randomized controlled trial. *Journal of the American Medical Association, 290*(14), 1875–1883.

Hashino, E., & Shero, M. (1995). Endocytosis of aminoglycoside antibiotics in sensory hair cells. *Brain Resesearch, 704*(1), 135–140.

Havlik, R. (1986). *Aging in the eighties: Impaired senses for sound and light in persons age 65 years and over. Preliminary data from the supplement on aging to the National Health Interview Survey: United States, January–June 1984. Advance data from vital and health statistics, #125,* 86-1250). Hyattsville, MD: National Center for Health Statistics.

Hawkins, J. (1976). Drug ototoxicity. In W. Keidel & W. Neff (Eds.), *Handbook of sensory physiology: Vol. 5. Auditory system.* Berlin, Germany: Springer-Verlag.

Hawkins, J., & Engstrom, H. (1964). Effect of kanamycin on cochlear cytoarchitecture. *Acta Otolaryngologica Supplement (Stockholm), 188,* 100–112.

Henley, C., & Schacht, J. (1988). Pharmacokinetics of aminoglycoside antibiotics in inner ear fluids and their relationship to ototoxicity. *Audiology, 27,* 137–146.

Herdman, S. J., Hall, C. D., Schubert, M. C., Das, V. E., & Tusa, R. J. (2007). Recovery of dynamic visual acuity in bilateral vestibular hypofunction. *Archives of Otolaryngology-Head & Neck Surgery, 133*(4), 383–389.

Hinojosa, R., Riggs, L. C., Strauss, M., & Matz, G. J. (1995). Temporal bone histopathology of cisplatin ototoxicity. *American Journal of Otology, 16*(6), 731–740.

Hood, L. J. (1988). *Clinical applications of the auditory brainstem response* (pp. 49–63). San Diego, CA: Singular.

House, J. W., & Brackmann, D. E. (1981). Tinnitus: Surgical treatment. *Ciba Foundation Symposium, 85,* 204–216.

Huang, M., & Schacht, J. (1988). Formation of a cytotoxic metabolite from gentamicin by liver. *Biochemical Pharmacology, 40,* 11–21.

Huang, R. S., Duan, S., Shukla, S. J., Kistner, E. O., Clark, T. A., Chen, T. X., . . . Dolan, M. E. (2007). Identification of genetic variants contributing to cisplatin-induced cytotoxicity by use of a genome-wide approach. *American Journal of Human Genetics, 81*(3), 427–437.

Hughes, G. B., Barna, B. P., Kinney, S. E., Calabrese, L. H., & Nalepa, N. J. (1988). Clinical diagnosis of immune inner-ear disease. *Laryngoscope, 98*(3), 251–253.

Hughes, G. B., Freedman, M. A., Haberkamp, T. J., & Guay, M. E. (1996). Sudden sensorineural hearing loss. *Otolaryngology Clinics of North America, 29*(3), 393–405.

Hultcrantz, E., & Nosrati-Zarenoe, R. (2014). Corticosteroid treatment of idiopathic sudden sensorineural hearing loss: Analysis of an RCT and material drawn from the Swedish national database. *European Archives of Otorhinolaryngology,* 272(11), 3169–3175.

Humes, H. D. (1999). Insights into ototoxicity: Analogies to nephrotoxicity. *Annals of the New York Academy of Sciences, 884,* 15–18.

Hyppolito, M. A., de Oliveira, J. A., & Rossato, M. (2006). Cisplatin ototoxicity and otoprotection with sodium salicylate. *European Archives of Otorhinolaryngology, 263*(9), 798–803.

Ishii, T., Bernstein, J., & Bulogh, K. (1967). Distribution of tritium-labelled salicylate in the cochlea. *Annals of Otology, Rhinology, and, Laryngology, 76,* 368–372.

Jackler, R. K., & Dillon, W. P. (1988). Computed tomography and magnetic resonance imaging of the inner ear. *Otolaryngology-Head and Neck Surgery, 99*(5), 494–504.

Jackson, G., & Arcieri, G. (1971). Ototoxicity of gentamicin in man: a survey and controlled analysis of clinical experience in the United States. *Journal of Infectous Disease, 124*(Suppl.), 130–137.

Jaffe, B. F. (1973). Clinical studies in sudden deafness. *Advances in Oto-Rhino-Laryngology, 20*, 221–228.

Jastreboff, P. J. (1990). Phantom auditory perception (tinnitus): Mechanisms of generation and perception. *Neuroscience Research, 8*(4), 221–254.

Jensen, J. B., Lysaght, A. C., Liberman, M. C., Qvortrup, K., & Stankovic, K. M. (2015). Immediate and delayed cochlear neuropathy after noise exposure in pubescent mice. *PLoS One, 10*(5), e0125160.

Jerger, J., & Johnson, K. (1988). Interactions of age, gender, and sensorineural hearing loss on ABR latency. *Ear and Hearing, 9*(4), 168–176.

Jerger, J. F. (1970). Clinical experience with impedance audiometry. *Archives of Otolaryngology, 92*, 311–324.

Jewett, D. L., & Williston, J. S. (1971). Auditory-evoked far fields averaged from the scalp of humans. *Brain, 94*(4), 681–696.

Josey, A. F., Glasscock, M. E., III, & Musiek, F. E. (1988). Correlation of ABR and medical imaging in patients with cerebellopontine angle tumors. *American Journal of Otology, 9*(Suppl.), 12–16.

Juhn, S. K., Hunter, B. A., & Odland, R. M. (2001). Blood-labyrinth barrier and fluid dynamics of the inner ear. *International Tinnitus Journal, 7*(2), 72–83.

Jung, T. Park, Y. M., Miller, S. K., Rozehnal, S., Woo, H. Y., & Baer, W. (1992). Effect of exogenous arachidonic acid metabolites applied on round window membrane on hearing and their levels in the perilymph. *Acta Otolaryngologica Supplement, 493*, 171–176.

Kahlmeter, G., & Kahlager, J. (1984). Aminoglycoside toxicity and review of medical studies published between 1975 and 1982. *Journal of Antimicrobial Chemotherapy, 13*(Suppl.), 9–22.

Kaltenbach, J. A. (2006a). The dorsal cochlear nucleus as a participant in the auditory, attentional and emotional components of tinnitus. *Hearing Research, 216–217*, 224–234.

Kaltenbach, J. A. (2006b). Summary of evidence pointing to a role of the dorsal cochlear nucleus in the etiology of tinnitus. *Acta Otolaryngologica Supplement, 556*, 20–26.

Kaye, C. I., Accurso, F., La Franchi, S., Lanc, P. A., Northrup, H., Pang, S., & Schaefer, G. B. (2006). Introduction to the newborn screening fact sheets. *Pediatrics, 118*(3), 1304–1312.

Kemp, D. T. (1978). Stimulated acoustic emissions from within the human auditory system. *Journal of the Acoustical Society of America, 64*(5), 1386–1391.

Kharkheli, E., Kevanishvili, Z., Maglakelidze, T., Davitashvili, O., & Schacht, J. (2007). Does vitamin E prevent gentamicin-induced ototoxicity? *Georgian Medical News, 146*, 14–17.

Kiang, N., Liberman, M., & Levine, R. (1976). Auditory nerve activity in cats exposed to ototoxic drugs and high-intensity sounds. *Annals of Otology, Rhinology, and, Laryngology, 85*, 752–761.

Klibanov, O. M., Filicko, J. E., DeSimone, J. A., Jr., & Tice, D. S. (2003). Sensorineural hearing loss associated with intrathecal vancomycin. *Annals of Pharmacotherapy, 37*(1), 61–65.

Koegel, L., Jr. (1985). Ototoxicity: a contemporary review of aminoglycosides, loop diuretics, acetylsalicylic acid, quinine, erythromycin, and cisplatinum. *American Journal of Otology, 6*(2), 190–199.

Konig, O., Schaette, R., Kempter, R., & Gross, M. (2006). Course of hearing loss and occurrence of tinnitus. *Hearing Research, 221*(1–2), 59–64.

Krzanowski, J., & Matschinsky, F. (1971). Phosphocreatine gradient opposite to that of glycogen in the organ of Corti and the effect of salicylate on adenosine triphosphate and P-creatine in cochlear structures. *Journal of Histochemistry, 19*, 321–326.

Langguth, B., Eichhammer, P., Kreutzer, A., Maenner, P., Marienhagen, J., Kleinjung,

T., . . . Hajak, G. (2006). The impact of auditory cortex activity on characterizing and treating patients with chronic tinnitus—first results from a PET study. *Acta Otolaryngologica Supplement* (556), 84–88.

Langguth, B., Kleinjung, T., Marienhagen, J., Binder, H., Sand, P. G., Hajak, G., & Eichhammer, P. (2007). Transcranial magnetic stimulation for the treatment of tinnitus: effects on cortical excitability. *BMC Neuroscience, 8*, 45.

Lerner, S., & Matz, G. (1980). Aminoglycoside ototoxicity. *American Journal of Otology, 1*, 169–179.

Lerut, B., De Vuyst, C., Ghekiere, J., Vanopdenbosch, L., & Kuhweide, R. (2007). Post-traumatic pulsatile tinnitus: The hallmark of a direct carotico-cavernous fistula. *Journal of Laryngology and Otology*, 1–5.

Liu, Y. C., Rubin, R., & Sataloff, R. T. (2011). Treatment-refractory autoimmune sensorineural hearing loss: Response to infliximab. *Ear, Nose, and Throat Journal, 90*(1), 23–28.

Loo, C. K., McFarquhar, T. F., & Mitchell, P. B. (2007). A review of the safety of repetitive transcranial magnetic stimulation as a clinical treatment for depression. *International Journal of Neuropsychopharmacology*, 11(1), 131–147.

Luetje, C. M., & Berliner, K. I. (1997). Plasmapheresis in autoimmune inner ear disease: Long-term follow-up. *American Journal of Otology, 18*(5), 572–576.

Magliulo, G., Parrotto, D., Sardella, B., Della Rocca, C., & Re, M. (2007). Cavernous hemangioma of the tympanic membrane and external ear canal. *American Journal of Otolaryngology, 28*(3), 180–183.

Marchese-Ragona, R., Marioni, G., Marson, P., Martini, A., & Staffieri, A. (2007). The discovery of salicylate ototoxicity. *Audiology & Neurootology, 13*(1), 34–36.

Marks, S., & Schacht, J. (1983). Effects of ototoxic diuretics on cochlear Na-K-ATPase and adenylate cyclase. *Scandanavian Audiology Supplement, 14*, 131–136.

Martin, G. K., Jassir, D., Stagner, B. B., & Lonsbury-Martin, B. L. (1998). Effects of loop diuretics on the suppression tuning of distortion-product otoacoustic emissions in rabbits. *Journal of the Acoustical Society of America, 104*(2, Pt. 1), 972–983.

Mathew, R. J., Weinman, M. L., & Mirabi, M. (1981). Physical symptoms of depression. *British Journal of Psychiatry, 139*, 293–296.

Matteson, E. L., Tirzaman, O., Facer, G. W., Fabry, D. A., Kasperbauer, J., Beatty, C. W., & McDonald, T. J. (2000). Use of methotrexate for autoimmune hearing loss. *Annals of Otology, Rhinology, and, Laryngology, 109*(8, Pt. 1), 710–714.

Mattox, D. E. (1980). Medical management of sudden hearing loss. *Otolaryngology-Head and Neck Surgery, 88*(2), 111–113.

Mattox, D. E., & Lyles, C. A. (1989). Idiopathic sudden sensorineural hearing loss. *American Journal of Otology, 10*(3), 242–247.

Mattox, D. E., & Simmons, F. B. (1977). Natural history of sudden sensorineural hearing loss. *Annals of Otology, Rhinology, and, Laryngology, 86*(4, Pt. 1), 463–480.

Matz, G. (1993). Aminoglycoside cochlear ototoxicity. *Otolaryngology Clinics of North America, 26*, 705–712.

McAlpine, D., & Johnstone, B. (1990). The ototoxic mechanisms of cisplatinum. *Hearing Research 47*, 191–197.

McCabe, B. F. (1979). Autoimmune sensorineural hearing loss. *Annals of Otology, Rhinology, and, Laryngology, 88*(5, Pt. 1), 585–589.

Merchant, S. N., Rosowski, J. J., & McKenna, M. J. (2007). Superior semicircular canal dehiscence mimicking otosclerotic hearing loss. *Advances in Otorhinolaryngology, 65*, 137–145.

Millen, S. J., Toohill, R. J., & Lehman, R. H. (1982). Sudden sensorineural hearing loss: Operative complication in non-otologic surgery. *Laryngoscope, 92*(6, Pt. 1), 613–617.

Minor, L. B., Carey, J. P., Cremer, P. D., Lustig, L. R., Streubel, S. O., & Ruckenstein, M. J. (2003). Dehiscence of bone overlying the superior canal as a cause of apparent conductive hearing loss. *Otology & Neurotology, 24*(2), 270–278.

Miyamoto, R. T., Wynne, M. K., McKnight, C., & Bichey, B. (1997). Electrical suppression of tinnitus via cochlear implants. *International Tinnitus Journal, 3*(1), 35–38.

Moller, A. R., Jho, H. D., Yokota, M., & Jannetta, P. J. (1995). Contribution from crossed and uncrossed brainstem structures to the brainstem auditory evoked potentials: A study in humans. *Laryngoscope, 105*(6), 596–605.

Mom, T., Telischi, F. F., Martin, G. K., Stagner, B. B., & Lonsbury-Martin, B. L. (2000). Vasospasm of the internal auditory artery: Significance in cerebellopontine angle surgery [in process citation]. *American Journal of Otology, 21*(5), 735–742.

Montaguti, M., Bergonzoni, C., Zanetti, M. A., & Rinaldi Ceroni, A. (2007). Comparative evaluation of ABR abnormalities in patients with and without neurinoma of VIII cranial nerve. *Acta Otorhinolaryngologica Italy, 27*(2), 68–72.

Morgan, D. E., & Canalis, R. F. (1991). Auditory screening of infants. *Otolaryngology Clinics of North America, 24*(2), 277–284.

Mostafa, B. E., Tawfik, S., Hefnawi, N. G., Hassan, M. A., & Ismail, F. A. (2007). The role of deferoxamine in the prevention of gentamicin ototoxicity: A histological and audiological study in guinea pigs. *Acta Otolaryngologica, 127*(3), 234–239.

Mrena, R., Ylikoski, M., Makitie, A., Pirvola, U., & Ylikoski, J. (2007). Occupational noise-induced hearing loss reports and tinnitus in Finland. *Acta Otolaryngologica, 127*(7), 729–735.

Mulrow, C. D., & Lichtenstein, M. J. (1991). Screening for hearing impairment in the elderly: Rationale and strategy. *Journal of General Internal Medicine, 6*(3), 249–258.

Murakami, S., Hato, N., Horiuchi, J., Honda, N., Gyo, K., & Yanagihara, N. (1997). Treatment of Ramsay Hunt syndrome with acyclovir-prednisone: Significance of early diagnosis and treatment. *Annals of Neurology, 41*(3), 353–357.

Musiek, F. E. (1982). ABR in eighth-nerve and brain-stem disorders. *American Journal of Otology, 3*(3), 243–248.

Myers, E., & Bernstein, J. (1965). Salicylate ototoxicity. *Archives of Otolaryngology, 82*, 483–493.

Nedzelski, J., Schessel, D., Bryce, G., & Pfleiderer, A. (1992). Chemical labyrinthectomy: Local application of gentamicin for the treatment of unilateral Meniere's disease. *American Journal of Otology, 1992*, 13(1): 18–22.

Norris, C. H. (1988). Drugs affecting the inner ear: A review of their clinical efficacy, mechanisms of action, toxicity, and place in therapy. *Drugs, 36*(6), 754–772.

Nosrati-Zarenoe, R., Arlinger, S., & Hultcrantz, E. (2007). Idiopathic sudden sensorineural hearing loss: Results drawn from the Swedish national database. *Acta Otolaryngologica, 127*(11), 1168–1175.

Olden, K. W., & Drossman, D. A. (2000). Psychologic and psychiatric aspects of gastrointestinal disease. *Medical Clinics of North America, 84*(5), 1313–1327.

Onusko, E. (2004). Tympanometry. *American Family Physician, 70*(9), 1713–1720.

Papsin, B. C. (2005). Cochlear implantation in children with anomalous cochleovestibular anatomy. *Laryngoscope, 115*(1, Pt. 2, Suppl. 106), 1–26.

Parsons, S. K., Neault, M. W., Lehmann, L. E., Brennan, L. L., Eickhoff, C. E., Kretschmar, C. S., & Diller, L. R. (1998). Severe ototoxicity following carboplatin-containing conditioning regimen for autologous marrow transplantation for neuroblastoma. *Bone Marrow Transplant, 22*(7), 669–674.

Penner, M. J. (1989). Empirical tests demonstrating two coexisting sources of tinnitus: A case study. *Journal of Speech and Hearing Research, 32*(2), 458–462.

Penrod, J. P. (1994). Speech threshold and word recognition/discrimination testing. In J. Katz (Ed.), *Handbook of clinical audiology* (p. 174). Baltimore, MD: Williams & Wilkins.

Phillips, J. S., Gaunt, A., & Phillips, D. R. (2014). Otosyphilis: A neglected diagnosis? *Otology & Neurootology, 35*(6), 1011–1013.

Pickles, J. (1988). *An introduction to the physiology of hearing*. London, UK: Academic Press.

Plinkert, P. K., Gitter, A. H., & Zenner, H. P. (1990). Tinnitus associated spontaneous otoacoustic emissions: Active outer hair cell movements as common origin? *Acta Otolaryngologica, 110*(5–6), 342–347.

Prager, D. A., Stone, D. A., & Rose, D. N. (1987). Hearing loss screening in the neonatal intensive care unit: Auditory brain stem response versus Crib-O-Gram; a cost-effectiveness analysis. *Ear and Hearing, 8*(4), 213–216.

Quick, C., & Duvall, A. (1970). Early changes in the cochlear duct from ethacrynic acid: An electron-microscopic evaluation. *Laryngoscope, 80*, 954–963.

Ramsden, R., Rotteveel, L., Proops, D., Saeed, S., van Olphen, A., & Mylanus, E. (2007). Cochlear implantation in otosclerotic deafness. *Advances in Otorhinolaryngology, 65*, 328–334.

Rask-Andersen, H., & Stahle, J. (1980). Immunodefence of the inner ear? Lymphocyte-macrophage interaction in the endolymphatic sac. *Acta Otolaryngologica, 89*(3–4), 283–294.

Rauch, S. D., Halpin, C. F., Antonelli, P. J., Babu, S., Carey, J. P., Gantz, B. J., . . . Reda, D. J. (2011). Oral vs intratympanic corticosteroid therapy for idiopathic sudden sensorineural hearing loss: A randomized trial. *Journal of the American Medical Association, 305*(20), 2071–2079. doi:10.1001/jama.2011.679

Rauschecker, J. P. (1999). Auditory cortical plasticity: A comparison with other sensory systems. *Trends in Neuroscience*, 22(2), 74–80.

Richardson, G. P., Forge, A., Kros, C. J., Marcotti, W., Becker, D., Williams, D. S., & Steel, K. P. (1999). A missense mutation in myosin VIIA prevents aminoglycoside accumulation in early postnatal cochlear hair cells. *Annals of the New York Academyof Sciences, 884*, 110–124.

Riko, K., Hyde, M. L., & Alberti, P. W. (1985). Hearing loss in early infancy: Incidence, detection and assessment. *Laryngoscope, 95*(2), 137–145.

Rizzi, M. D., & Hirose, K. (2007). Aminoglycoside ototoxicity. *Current Opinions in Otolaryngology-Head and Neck Surgery, 15*(5), 352–357.

Rybak, L., & Matz, G. (1986). Effect of toxic agents. In C. Cummings, J. Fredrickson, L. Harker, C. Krause, & D. Schuller (Eds.), *Otolaryngology-Head and neck surgery* (2nd ed., pp. 2943–2964). St. Louis, MO: Mosby Year Book.

Rybak, L. P. (1982). Pathophysiology of furosemide ototoxicity. *Journal of Otolaryngology, 11*(2), 127–133.

Rybak, L. P. (2007). Mechanisms of cisplatin ototoxicity and progress in otoprotection. *Current Opinion in Otolaryngology-Head and Neck Surgery, 15*(5), 364–369.

Saunders, J. C. (2007). The role of central nervous system plasticity in tinnitus. *Journal of Communication Disorders, 40*(4), 313–334.

Saxena, A. K. (2007). N-acetylcysteine for preventing ototoxicity in hemodialysis patients receiving gentamicin. *Nature Clinical Practice in Nephrology, 3*(9), 478–479.

Schweitzer, V., & Olson, N. (1984). Ototoxic effect of erythromycin therapy. *Archives of Otolaryngology, 110*, 258–263.

Schweitzer, V. G. (1993). Ototoxicity of chemotherapeutic agents. *Otolaryngology Clinics of North America, 26*(5), 759–789.

Shaia, F. T., & Sheehy, J. L. (1976). Sudden sensori-neural hearing impairment: A report of 1,220 cases. *Laryngoscope, 86*(3), 389–398.

Shine, N. P., & Coates, H. (2005). Systemic ototoxicity: A review. *East African Medical Journal, 82*(10), 536–539.

Shownkeen, H., Yoo, K., Leonetti, J., & Origitano, T. C. (2001). Endovascular treatment of transverse-sigmoid sinus dural arteriovenous malformations presenting as pulsatile tinnitus. *Skull Base, 11*(1), 13–23.

Silverstein, J., Bernstein, J., & Davies, D. (1967). Salicylate ototoxicity: A biochemical and electrophysiological study. *Annals of Otology, Rhinology, and Larngology, 76*, 118–127.

Simdon, J., Watters, D., Bartlett, S., & Connick, E. (2001). Ototoxicity associated with use of nucleoside analog reverse transcriptase inhibitors: A report of 3 possible cases and review of the literature. *Clinical Infectious Diseases, 32*(11), 1623–1627.

Smith, M. E., & Canalis, R. F. (1989). Otologic manifestations of AIDS: The otosyphilis connection. *Laryngoscope, 99*(4), 365–372.

Smith, R. J., Zimmerman, B., Connolly, P. K., Jerger, S. W., & Yelich, A. (1992). Screening audiometry using the high-risk register in a level III nursery. *Archives of Otolaryngology-Head & Neck Surgery, 118*(12), 1306–1311.

Smits, M., Kovacs, S., de Ridder, D., Peeters, R. R., van Hecke, P., & Sunaert, S. (2007). Lateralization of functional magnetic resonance imaging (fMRI) activation in the auditory pathway of patients with lateralized tinnitus. *Neuroradiology, 49*(8), 669–679.

Smoorenburg, G. F., De Groot, J. C., Hamers, F. P., & Klis, S. F. (1999). Protection and spontaneous recovery from cisplatin-induced hearing loss. *Annals of the New York Academy of Sciences, 884*, 192–210.

So, H., Kim, H., Lee, J. H., Park, C., Kim, Y., Kim, E., Park, R. (2007). Cisplatin cytotoxicity of auditory cells requires secretions of proinflammatory cytokines via activation of ERK and NF-kappaB. *Journal of the Association for Research in Otolaryngology, 8*(3), 338–355.

Sonmez, G., Basekim, C. C., Ozturk, E., Gungor, A., & Kizilkaya, E. (2007). Imaging of pulsatile tinnitus: A review of 74 patients. *Clinical Imaging, 31*(2), 102–108.

Spear, S. A., & Schwartz, S. R. (2011). Intratympanic steroids for sudden sensorineural hearing loss: A systematic review. *Otolaryngology-Head and Neck Surgery, 145*(4), 534–543. doi: 10.1177/0194599811419466

Stokroos, R. J., Albers, F. W., & Tenvergert, E. M. (1998). Antiviral treatment of idiopathic sudden sensorineural hearing loss: A prospective, randomized, double-blind clinical trial. *Acta Otolaryngologica, 118*(4), 488–495.

Stypulowski, P. (1990). Mechanisms of salicylate ototoxicity. *Hearing Research, 46*, 113–126.

Sullivan, M. D., Katon, W., Dobie, R., Sakai, C., Russo, J., & Harrop-Griffiths, J. (1988). Disabling tinnitus. Association with affective disorder. *General Hospital Psychiatry, 10*(4), 285–291.

Sweeney, C. J., & Gilden, D. H. (2001). Ramsay Hunt syndrome. *Journal of Neurological Neurosurgery and Psychiatry, 71*(2), 149–154.

Tomiyama, S., & Harris, J. P. (1987). The role of the endolymphatic sac in inner ear immunity. *Acta Otolaryngologica, 103*(3–4), 182–188.

Topal, O., Erbek, S. S., Erbek, S., & Ozluoglu, L. N. (2007). Subjective pulsatile tinnitus associated with extensive pneumatization of temporal bone. *European Archives of Otorhinolaryngology, 265*(1), 123–125.

Tran Ba Huy, P., Bernard, P., & Schacht, J. (1986). Kinetics of gentamicin uptake and release in the rat: Comparison of inner ear tissues and fluids with other organs. *Journal of Clinical Investigation, 77*, 1492–1500.

Trimble, K., Blaser, S., James, A. L., & Papsin, B. C. (2007). Computed tomography and/or magnetic resonance imaging before pediatric cochlear implantation? Developing an investigative strategy. *Otology & Neurootology, 28*(3), 317–324.

Vambutas, A., Lesser, M., Mullooly, V., Pathak, S., Zahtz, G., Rosen, L., & Goldofsky, E. (2014). Early efficacy trial of anakinra in corticosteroid-resistant autoimmune inner ear disease. *Journal of Clinical Investigation, 124*(9), 4115–4122.

Van Naarden, K., Decoufle, P., & Caldwell, K. (1999). Prevalence and characteristics of children with serious hearing impairment in metropolitan Atlanta, 1991–1993. *Pediatrics, 103*(3), 570–575.

Van Prooyen-Keyzer, S., Sadik, J. C., Ulanovski, D., Parmantier, M., & Ayache, D. (2005). Study of the posterior communicating arteries of the circle of Willis in idiopathic sudden sensorineural hearing loss. *Otology & Neurootology, 26*(3), 385–386.

Van Wijk, F., Staecker, H., Keithley, E., & Lefebvre, P. P. (2006). Local perfusion of the tumor necrosis factor alpha blocker infliximab to the inner ear improves autoimmune neurosensory hearing loss. *Audiology & Neurootology, 11*(6), 357–365.

Vasama, J. P., & Linthicum, F. H., Jr. (2000). Idiopathic sudden sensorineural hearing loss: Temporal bone histopathologic study. *Annals of Otology, Rhinology, and Laryngology, 109*(6), 527–532.

Vernon, J., Griest, S., & Press, L. (1990). Attributes of tinnitus and the acceptance of masking. *American Journal of Otolaryngology, 11*(1), 44–50.

Wake, M., Tobin, S., Cone-Wesson, B., Dahl, H. H., Gillam, L., McCormick, L., . . . Williams, J. (2006). Slight/mild sensorineural hearing loss in children. *Pediatrics, 118*(5), 1842–1851.

Wallace, M. R., Miller, L. K., Nguyen, M. T., & Shields, A. R. (1994). Ototoxicity with azithromycin [Letter]. *Lancet, 343*(8891), 241.

Walsted, A., Andreassen, U. K., Berthelsen, P. G., & Olesen, A. (2000). Hearing loss after cardiopulmonary bypass surgery. *European Archives of Otorhinolaryngology, 257*(3), 124–127.

Wei, X., Zhao, L., Liu, J., Dodel, R. C., Farlow, M. R., & Du, Y. (2005). Minocycline prevents gentamicin-induced ototoxicity by inhibiting p38 MAP kinase phosphorylation and caspase 3 activation. *Neuroscience, 131*(2), 513–521.

Weiss, M., Kisiel, D., & Bhatia, P. (1990) Predictive value of brainstem evoked response in the diagnosis of acoustic neuroma. *Otolaryngology-Head and Neck Surgery, 103*, 583–585.

Wells, K. B., & Sherbourne, C. D. (1999). Functioning and utility for current health of patients with depression or chronic medical conditions in managed, primary care practices. *Archives of General Psychiatry, 56*(10), 897–904.

Williams, S., Zenner, H., & Schacht, J. (1987). Three molecular steps of aminoglycoside ototoxicity demonstrated in outer hair cells. *Hearing Research, 30*, 11–22.

Wilson, D., Hodgson, R., Gustafson, M., Hogue, S., & Mills, L. (1992). The sensitivity of auditory brainstem response testing in small acoustic neuromas. *Laryngoscope, 102*, 961–964.

Wilson, W. R., Byl, F. M., & Laird, N. (1980). The efficacy of steroids in the treatment of idiopathic sudden hearing loss: A double-blind clinical study. *Archives of Otolaryngology, 106*(12), 772–776.

Wilson, W. R., Laird, N., Moo-Young, G., Soeldner, J. S., Kavesh, D. A., & MacMeel, J. W. (1982). The relationship of idiopathic sudden hearing loss to diabetes mellitus. *Laryngoscope, 92*(2), 155–160.

Wong, L., & Bance, M. (2004). Are all cookie-bite audiograms hereditary hearing loss? *Journal of Otolaryngology, 33*(6), 390–392.

Xenellis, J., Karapatsas, I., Papadimitriou, N., Nikolopoulos, T., Maragoudakis, P., Tzagkaroulakis, M., & Ferekidis, E. (2006). Idiopathic sudden sensorineural hearing loss: Prognostic factors. *Journal of Laryngology and Otology, 120*(9), 718–724.

Yehudai, D., Shoenfeld, Y., & Toubi, E. (2006). The autoimmune characteristics of progressive or sudden sensorineural hearing loss. *Autoimmunity, 39*(2), 153–158.

Yimtae, K., Srirompotong, S., & Lertsukprasert, K. (2007). Otosyphilis: A review of 85 cases. *Otolaryngology-Head and Neck Surgery, 136*(1), 67–71.

CHAPTER 8

The Future of Cochlear Implants

Richard Tyler, Paul R. Kileny, Aniruddha K. Deshpande, Shruti Balvalli Deshpande, Camille Dunn, Marlan Hansen, and Bruce Gantz

Cochlear implants (CIs) have changed how we consider hearing loss. Over the years, selection criteria have moved from profound to severe to moderate hearing loss. Now those with good residual low-frequency hearing or only a unilateral hearing loss can benefit. Several exciting advances have been made in the last few years, pointing to new directions to improve selection criteria, fitting of the devices, increasing benefit through training, and the realistic documenting of benefit.

Preoperative Transtympanic Electric ABR

One of the challenges in cochlear implantation, in particular in the very young patient, is to determine the likelihood of efficacy in cases of significant temporal bone anomalies. These include a variety of cochlear malformations as well as hypoplastic cochlear nerves or the appearance of absent cochlear nerves. One way to manage these cases is by sophisticated high-resolution preoperative imaging studies. However, the imaging studies are not always sufficient due to resolution issues, skull orientation issues during the imaging procedure, and the fact that at times the cochlear nerve can be displaced from its normal anatomical position and may not be in the plane of the imaging study. Therefore, it is likely that in some cases of malformations giving the appearance of a completely absent cochlear nerve, there is actually a cochlear nerve that can respond to electrical stimulation, despite the fact that it is not identifiable on imaging studies. Therefore, it makes good clinical sense to have other modalities that, in addition to imaging, can help determine presence or absence of a cochlear nerve. An electrophysiologic measure that has proven to be a useful and reliable to evaluate the functional status of electrical excitability in cochlear implant candidates is the transtympanic promontory electric auditory brainstem response (TEABR). This modality borrows from the concept of promontory testing by utilizing an electrophysiologic modality to record responses to pulsatile, electrical stimuli delivered to the cochlear promontory. While not yet widespread, it is well documented that it can be effective in determining the functional status of the cochlear nerve and even provides some measure of correlation with functional outcome. In the most simplistic fashion, the presence of a response indicates the presence of an excitable cochlear nerve in these candidates. However, the actual reality is more complex bringing into play response threshold, response amplitude, consistency/replicability, and adaptation of phenomena. In a retrospective study (Kim et al., 2008), we investigated the relationship between the results of preoperative TEABR in children with inner-ear malformation with postoperative speech perception performance measures. Relatively low TEABR threshold and higher wave 5 amplitude and shorter latency were associated with better speech performance. These characteristics also coincided with the subgroup with the least complex temporal bone malformations. We concluded that preoperative TEABR is useful in determining cochlear implant candidacy and, to some extent, outcome in children with inner ear malformations. In subsequent investigations, we began looking at the phenomenon of rapid or abnormal adaptation of wave V of the TEABR. We found that in cases of hypoplastic cochlear nerves, including some patients who were categorized as auditory neuropathy spectrum disorder (ANSD), we identified an abnormal adaptation pattern characterized by a reduction of wave 5 amplitude in consecutive responses obtained at the same stimulus intensity. In some cases, responses could be obtained at a certain electrical stimulus level, but the response disappeared after one or two consecutive attempts to get a replication. We refer to this as an abnormal adaptation pattern, and many of these patients who were implanted subsequently exhibited inconsistent benefit from the cochlear implant requiring frequent reprogramming. The inconsistency in benefit could be noted during short periods such as a change to normal waking hours to a longer time from days to weeks or even to even months. So in conclusion, if used correctly, this preoperative functional test modality could help us predict the presence or absence of any electrically stimuable auditory neural elements, as well as, to some extent, level of benefit from a cochlear implant in pediatric patient populations.

Functional Near-Infrared Spectroscopy (fNIRS) Imaging of Cochlear Implant Function

Functional near-infrared spectroscopy (fNIRS) detects changes in hemodynamic response using near-infrared light and provides information about metabolic function of brain regions of interest. It is comparable to blood-oxygen-level dependent (BOLD) response in functional MRI (MRI). Using two different wavelengths, changes in the concentration of oxy- and deoxy-hemoglobin can be determined. Other optically based measurements of oxygen concentration in medicine are pulse oximetry and cerebral oximetry used intraoperatively. A single fNIRS measurement is sensitive to a volume of tissue that falls between the source of light entering the tissue and the detector receiving the light that diffuses off the tissue. fNIRS is limited by depth sensitivity, which in adult humans typically reaches 5 to 10 mm beneath the inner surface of the skull for brain activation. This can be improved with time domain methods, where depth sensitivity increases with delayed detection of the propagation of a pulse of light. Like evoked potentials or event-related potentials (ERPs), fNIRS activity needs to be averaged over the duration of the task. It is helpful to think of fNIRS as a measure reflecting percent change in hemodynamic response over time, as opposed to voltage over time for an evoked potential measurement. The optical sources and detectors are arrayed based on the 10-20 system. In our cochlear implant subjects, we concentrated on recording activity from the prefrontal and temporal cortex, inferior frontal gyrus, and superior temporal gyrus, right and left. The tasks for these measurements were a rhyming task, a tone identification task, and a passive listening task. We examined 10 adults with CIs with varying duration of experience with the implant and 10 normal-hearing controls. We had both unilateral and bilateral implantees, as well as patients with CI and contralateral hearing aid (HA) (CI+contraHA). Our preliminary data analysis indicated similarities between controls and cochlear implant subjects in terms of metabolic changes at our two detection sites, with magnitude differences: A lower concentration change has been noted in the CI subjects for identical tasks, when compared to the normal-hearing controls. Additionally, in some CI subjects, the tone identification task resulted in more frontal lobe activation than temporal cortex activation.

fNIRS has several advantages over ERP measures: no electrical interference from CI, lower cost than fMRI or positron emission tomography (PET), absence of noise interference as in MRIs, portable, tolerates movement as long the optical devices are stable, and good spatial cortical resolution. The disadvantages are the currently poorer temporal resolution compared to ERP and it cannot be carried out under any type of general anesthesia.

Automated Mapping and Telepractice in Cochlear Implants

One of the problems worldwide is that there are many underserved regions where there is no personnel for managing cochlear implants. These regions could be in remote areas of South America or Africa, but there are also such underserved regions in the United States and in Europe. Consequently, the verification, mapping, programming, and reprogramming of cochlear implants is not very accessible in these regions. There is a tendency for patients from such regions to travel to major cochlear implant centers for cochlear implant surgery and initial programming. Alternatively, it is quite common for surgeons to travel to these underserved regions and perform cochlear implant surgery. However, when the recipient returns home from the cochlear implant center, or the surgeon returns home from the remote, underserved region, there is very little help available for these patients. Many have to travel long distances to receive routine services, not to mention troubleshooting and technical updates. The solution could be telepractice, and already there are modalities for remote programming of cochlear implants via an Internet connection. Thus, an experienced cochlear implant clinician at the cochlear implant center could connect with the help of a lesser trained assistant at the remote site, to the cochlear implant of a recipient and carry out routine mapping procedures or perform software updates.

These procedures would still require some type of a behavioral response and interaction from the patient. With the availability of a video link to perform this type of programming, this could be accomplished in adults and older children. However, if the patient is very young (1–2 years old), this could be more challenging on a remote telepractice basis, and its success would depend to some extent on the skill of the assistant providing support at the other end. One solution for this could be the inclusion of some built-in neurophysiologic measures that could be used as feedback to the remote location. This can be accomplished today, to some extent through the use of electrical auditory potentials or neural response telemetry (EAPs/NRTs). To carry out any other electrophysiologic measurements, the use of an evoked potential system and the application of surface electrodes would be necessary. This would require some skill from the assistant at the recipient site. In the future, it is possible that all of these functions—stimulus generation and electrophysiologic recording—could be built into the CI itself in conjunction with the programming software. Thus, for instance, during implantation, an additional extracochlear electrode could be added to be positioned somewhere under the scalp to be used for the recording of an electric evoked potential other than the EAP/NRT. In conjunction with the external ground electrode, one could envision a single channel recording of electrically evoked auditory potentials through the CI. The signal averaging component could be provided within the software used to program the CI and thus the connection and the setup would require minimal skill at the recipient site. Also, it has been shown on a pilot basis that later auditory potentials such as the middle latency auditory evoked potential may be more accurate in determining operating levels of a CI. This application could be operated remotely via telepractice, or automated software could be developed to program each electrode based on an automated evoked potential measure (either electrically evoked auditory brainstem responses (EABR) or electrically-evoked middle latency responses (EMLR) and thus would necessitate minimal expertise and may not even necessitate a telepractice approach to this type of objective, electrophysiologic programming.

Realistic Tests for Candidacy and Evaluation

It is becoming easier to evaluate patients based on more realistic listening situations. Audiograms and word recognition scores in quiet do not adequately capture functioning of a patient when they are being considered a candidate, nor do they accurately reflect real-life benefit after receiving CIs. Much progress has been made in recent years to develop tests better representing actual benefit (including sentence recognition with different talkers) (e.g., Dunn et al., 2008; Litovsky et al., 2006).

Hearing in noise is a part of everyday life, and in most situations the speech and noise originate from different sources. Testing with speech and noise from the same loudspeaker or with speech directly in front and with noise directly behind the listener is not common in real life. In addition, presenting the same signal-to-noise ratios at both ears is not representative and eliminates important spatial hearing functions.

The focus on spatial hearing will continue to be exemplified in the future. Already there are examples of speech and noise coming from different locations on different trials (Tyler et al., 2003, 2006a, 2006b, 2007), including a preword cue from the loudspeaker that will contain the target word. Varying the stimuli on different presentations prevents the listener from using spectral cues that would not be available in real life. Such approaches have been recommended as a systematic approach to selection strategies for combinations of CIs and HAs (Perreau et al., 2007). Localization tests in the vertical plane will capture important real-life abilities, and tests of distance perception have already been explored (Tyler et al., 2002).

Music perception is much different than hearing speech. There has been a great interest in music perception in CI patients (Gfeller et al., 2000; Tyler et al., 2000). It is noteworthy that spatial hearing for music is much different than spatial hearing for speech in noise. With speech in noise, the task is to separate the stimuli. With music produced by different instruments, the task is to integrate the stimuli from different locations.

Further refinements in more questionnaires, which sample a variety of realistic listening situations, are also expected (Tyler, Perreau, & Ji, 2009). With all of the wallet-size computers available (sometimes called mobile phones), listening experiences can be documented along with time, location, and background noise.

Spatial Hearing Training

Many CI users require several months to obtain good benefit from their CIs (e.g., Chang et al., 2010). Training programs are available that can accelerate this. Furthermore, some will likely be able to obtain even higher scores with training. We have introduced the first spatial hearing training program, focused on speech perception in spatially separate noise and on localization (Tyler et al., 2010). Two loudspeakers are required, which could mean two mobile phones that are linked together. Other groups have successfully utilized music training as an aural rehabilitation procedure (e.g., Kraus & Anderson, 2014) and on lip reading (Levitt et al., 2011).

A Cochlear Implant for Tinnitus

Tinnitus can affect primary functions of thoughts and emotions, hearing, sleep, and concentration (Stouffer & Tyler, 1990; Tyler et al., 2014; Tyler & Baker, 1983). Many patients are severely handicapped. Although a variety of counseling and sound therapy procedures exist (Tyler, 2006), there is no cure.

CI recipients with preimplant tinnitus report either complete elimination or significant reduction of tinnitus after implantation (e.g., Bovo, Ciorba, & Martini, 2011; Bredberg, Walden, & Lindström, 1992; Gibson, 1992; Hazell, McKinney, & Aleksy, 1995; Kim, Shim, Kim, & Kim, 1995; Kou, Shipp, Nedzelski, 1994; Miyamoto, Wynne, McKnight, & Bichey, 1997; Perez et al., 1997; Ruckenstein, Hedgepeth, Rafter, Montes, & Bigelow, 2001; Tyler, 1994; Tyler & Kelsay, 1990). Miyamoto et al. (1997) observed an inverse relationship between the duration of tinnitus and the likelihood of tinnitus suppression postimplantation. Early evidence from laboratory trials indicated that electrical stimulation of the cochlea could reduce tinnitus in some patients (e.g., Ito & Sakakihara, 1994). Additionally, patients who received CIs for their hearing loss often reported that the reduction of their tinnitus was an even greater benefit than hearing improvement. Patients with unilateral hearing loss and tinnitus were obviously ideal candidates to receive a CI. Even if their tinnitus is not helped, there is a good chance their hearing would be improved.

We report data from two projects conducted at our lab showing what we believe represent the future direction for the use of CIs for tinnitus.

Deshpande and Tyler (2016) explored tinnitus reduction in CI recipients in a controlled laboratory setting. Participants chose up to five sounds and rated the tinnitus loudness, tinnitus annoyance, and sound acceptability before, during, and after listening to the stimuli. Data revealed (a) an ocean sound was the most preferred sound for tinnitus reduction, (b) the mean reduction in tinnitus loudness and annoyance during stimulus presentation for the most effective stimulus was 63%, and (c) the mean reduction in tinnitus loudness and annoyance after stimulus presentation for the most effective stimulus was 36% and 21%, respectively.

The other study (Tyler et al., 2015) investigated the effectiveness of auditory stimuli presented via CIs in tinnitus reduction. Six CI recipients who experienced tinnitus were recruited for this study for a total of seven data points (one subject participated twice). Stimuli that they found acceptable (such as ocean waves, spa music, amplitude, and frequency-modulated tones and noise) were added in the background of their CI. The participants tried different sounds on different days. The results indicated that (a) most participants preferred environmental sounds over tones or noise, (b) there was a 24% to 25% reduction in tinnitus loudness for the three stimuli chosen by most participants at the end of the study as compared to baseline ratings, and (c) there was a 26% to 31% reduction in tinnitus annoyance for the three stimuli chosen by most participants at the end of the study as compared to baseline ratings.

Both the above studies give us insights into the type of stimulus most likely to result in tinnitus reduction in CI users. There will soon be a CI for tinnitus. The next steps to bring this into the future include conducting large-scale trials of the use of specific auditory stimuli via CIs both in controlled laboratory settings as well as in real-life scenarios, exploring the utility of pulse trains presented via a CI in tinnitus alleviation, and investigation of pre-post tinnitus characteristics in specific populations such as individuals with single-sided deafness and recipients of the hybrid implant.

Short Electrodes to Preserve Hearing

There are many types and configurations of hearing losses that require some sort of remediation to help listeners function more successfully in their environments. For those with severe-to-profound hearing losses, a CI is often a reasonable option. However, for individuals who have a high-frequency hearing loss and mild to moderate low-frequency hearing, implanting a standard-length electrode will likely result in loss of residual hearing. Recently, the Food and Drug Administration (FDA) approved the Nucleus Hybrid L24 Cochlear Implant System (FDA, 2014). The Hybrid L24 cochlear implant is an electrode that is shorter in length than the standard electrode and is designed to stimulate the damaged high-frequency regions of the cochlear while at the same time preserving the anatomy of the inner ear in the apical region of the cochlea, increasing the chance of preserving that hearing. If hearing is preserved, the Nucleus Hybrid L24 Cochlear Implant System is designed to allow patients to hear electrically in the mid and high frequencies and acoustically in the normal to moderate low-frequency hearing loss. The approval of this device and its indications has opened up the number of patients who can benefit from electrical stimulation for hearing.

Preserving low-frequency hearing offers numerous advantages to these individuals. The research demonstrates that listeners are able to obtain improved speech perception abilities from the integration of low-frequency acoustic hearing with high-frequency electrical stimulation (Gantz & Turner, 2004, 2003; Gantz, Hanson, Turner, Oleson, Reiss, & Parkinson, 2009). It has also been shown to provide better frequency selectivity, which helps with understanding speech in environments with background noise (Turner et al., 2007). Furthermore, the preserved low-frequency hearing provides better temporal cues, which provides the benefit of recognizing melodies, giving a greater appreciation of music (Gantz, Turner, & Gfeller, 2006; Gantz, Turner, Gfeller, & Lowder 2005; Gantz & Turner, 2004, 2003; Gfeller, Olszewski, Turner, Gantz, & Oleson, 2006; Turner et al., 2007). Finally, the preservation of acoustic hearing provides patients with similar processing across ears. This has been shown to provide a clear advantage for these listeners in the realm of localization because important low-frequency timing cues are available across ears (Dunn et al., 2010; Gifford et al., 2014).

According to the Hybrid L24 clinical trial data, approximately 60% of the patients implanted with Hybrid L24 cochlear implant maintain all or some of their residual hearing (Roland et al., 2015). Furthermore, in this population of 50 subjects, significant speech perception improvements were demonstrated in the implanted ear. Specifically, averaged words in quiet improvement were 36% and 32% for sentences presented in a background of noise (Roland et al., 2015).

There seems to be a critical period between 1 and 6 months after activation of the cochlear implant where some individuals lose their low-frequency hearing. At present, we are trying to understand why this loss of hearing occurs. At the University of Iowa, we have implanted over 90 subjects with different lengths of hearing preservation cochlear implants. Our results show that the length of the device has a significant effect on amount of hearing preservation (Van Able et al., 2014). The devices that did not go as far into the cochlea had better hearing preservation. We do know, however, that for individuals who have lost their hearing, they have the capability of still doing very well with this hybrid cochlear implant. In most cases, there is little need to explant and reimplant with a longer device.

Unilateral Hearing Loss

It is estimated that approximately 60,000 people per year acquire single-sided deafness (SSD) in the United States (Hear-it.org and AudiologyOnline.com). Severe to profound sensorineural hearing loss may occur unilaterally as a result of a viral infection of the inner ear (e.g., herpetic or mumps), bacterial infection, direct trauma to the ear or head, endolymphatic hydrops, surgical complications (e.g., stapedectomy), or a variety of other etiologies. In contrast to persons with bilateral severe to profound hearing loss, individuals with unilateral severe to profound hearing loss (with much better hearing in the other ear) have not, to date, been considered cochlear implant candidates. The underlying assumption is that such individuals can function effectively with the better ear alone, even if that ear requires amplification. However, many of these hearing-impaired individuals still experience significant difficulties in understanding speech in their everyday listening environments, along with significant communication handicaps that interfere with their quality of life (Wie, Pripp, & Tvete, 2010). Furthermore, these listeners can no longer localize sound, which poses serious safety concerns. These disabilities often lead to social isolation, decreased self-esteem, and depression.

Current treatment options for those with SSD consist of a CROS (Contralateral Routing of Signals) hearing aid (Harford, 1966) and an osseointegrated (OI) device, such as a Baha (Bone-anchored hearing aid) bone conduction auditory implant. Both the CROS and Baha systems are intended to overcome the effect of the head shadow so that sounds at the poorer side are received at audible sensation levels at the better ear, but it is not true binaural input and does not provide the known benefits of binaural listening (Mueller, Cutie, & Shaw, 1990). In particular, they provide little improvement in sound localization and modest improvement in speech perception in noisy environments (Bishop et al., 2010; Desmet et al.,

2012; Hol et al., 2010; Lin et al., 2006; Linstrom, Silverman, & Yu, 2009; Nicolas et al., 2013; Niparko, Cox, & Lustig, 2003). This is in contrast to reports where a CI has been shown to provide improved ability to localize sound and understand speech in some situations with background noise compared to CROS and OI devices (Arndt et al., 2011).

The use of a cochlear implant is the only technology currently available that can restore sound to the impaired ear. Recent studies in patients with SSD who received a CI to reduce the impact of unilateral tinnitus concomitant with unilateral deafness demonstrated improved speech perception in noise benefits as well as suppression of tinnitus (Arndt et al., 2011; Buechner et al., 2010; Van de Heyning et al., 2008; Vermiere & Van de Heyning, 2009). Additionally, other studies using small groups of individuals with SSD who have received a CI reported similar findings of improved speech perception and localization abilities following cochlear implantation in the implanted ear (Arndt et al., 2011; Firszt et al., 2012; Hansen, Gantz, & Dunn, 2013 Vermeire & Van de Heyning, 2009). Yet, in all of these studies, a great deal of variability in performance has also been demonstrated. Further research is needed to better understand how this population fuses the acoustic and electric information together to take advantage of binaural benefits.

Auditory Steady-State Responses

Objective techniques to estimate hearing thresholds and to facilitate mapping in difficult-to-test cochlear implant recipients are invaluable. The auditory steady-state response (ASSR) is one such objective technique. The ASSR is an auditory evoked potential that can be elicited by a continuous, repeating sound stimulus. The ASSR can be used to (a) determine frequency-specific hearing levels, (b) interpret results based on statistical tests incorporated within the ASSR module (objective interpretation of results as opposed to subjective waveform analysis by the tester), and (c) test multiple frequencies and both ears simultaneously, thereby shortening the test duration (Korczak, Smart, Delgado, Strobel, & Bradford, 2012; Picton, John, Dimitrijevic, & Purcell, 2003; Rance, 2008).

The steady-state and continuous nature of the ASSR stimulus makes it an ideal option for being transduced through loudspeakers as well as hearing prosthetic devices (Picton et al., 1998). Picton et al. (1998); Dimitrijevic, John, and Picton (2004); and Stroebel, Swanepoel de, and Groenewald (2007) demonstrated the utility of sound-field ASSR as an objective technique for threshold estimation and/or suprathreshold word recognition in hearing aid users. Another advantage offered by the ASSR specifically related to CI assessments is that the ASSR can be recorded at higher intensities (compared to tone-burst auditory brainstem responses), thereby facilitating better accuracy in terms of the severity of hearing loss and residual hearing thresholds. This information can be a factor in the clinical decision-making process, especially for pediatric CI candidates.

The advantages offered by the ASSR make it valuable for hearing assessment, programing, and mapping of a CI. Currently, there are limited number of studies evaluating the use of the ASSR in CI recipients; however, the interest in this topic is growing. One of the technical problems of recording ASSR in CI recipients is the CI-related artifact that can contaminate the results. According to Dimitrijevic and Cone (2015):

> The CI artifact problem arises because the CI extracts the envelope of an incoming signal and uses that derived envelope to modulate electrical pulses stimulating the auditory nerve. Therefore, the stimulus modulation frequency used to elicit the ASSR becomes part of the CI electrical artifact. The algorithms used to detect the ASSR are not able to distinguish between the ASSR and the CI electrical signal. (p. 285)

The focus of the current and future research on CI-ASSR is to develop algorithms to record ASSR while controlling and minimizing artifacts.

Menard et al. (2004) were the first group of researchers to investigate the feasibility of recording the ASSR in CI users. They manipulated the CI pulse train amplitude and duration in order to study the artifacts. Sufficient charge density (pulse amplitude and duration) is required to stimulate the auditory nerve via a CI. Two assumptions were made by this group: (1) The amplitude of the artifact grows as a function of intensity and was related with the pulse amplitude, and (2) at short pulse durations, artifacts are not present. The authors were able to demonstrate that some part of the recorded response contained a physiologic response. The response grew nonlinearly as a function of duration, thereby facilitating the separation of physiologic responses from the CI artifact at supra-threshold levels. The authors demonstrated good agreement between the ASSR thresholds and behavioral thresholds in their small group of MXM Digisonic CI users.

Jeng et al. (2007) explored the possibility of recording ASSR using amplitude modulated electrical pulses in implanted guinea pig models. They used summing alternating polarity responses and spectral analysis to separate artifacts from physiologic responses. Additionally, they demonstrated that it is possible to obtain physiologic responses during ASSR by conducting three control experiments. In the first control experiment, the authors showed that a response was present when a stimulus with 100% depth of modulation was used but not when the depth was 0%, thereby implying that the response was locked to the modulation envelope of the stimulus. In the second control experiment, the researchers demonstrated that a response was detected prior to an injection with the cochleo-toxic tetrodotoxin but not after injecting the drug. In the third control experiment, Jeng et al. (2007) compared the ASSR recordings from live versus subsequently euthanized implanted guinea pigs. The response was evident in the live subjects but disappeared postmortem. In a subsequent study, Jeng et al. (2008) investigated the neural generators of the ASSR in guinea pig models by studying the latency estimates of the ASSR recordings. Results indicated that ASSRs recorded via CIs using lower modulation frequencies evoked cortical structures corresponding with a latency of approximately 22 ms. On the other hand, higher modulation frequencies evoked peripheral structures and the response latency was approximately 2 ms.

Hoffman and Wouters (2010) conducted a study on six CI participants. The purpose of the study was to investigate if ASSRs could be recorded to short-duration biphasic pulses of alternate polarities at low modulation rates in CI participants and if there was a correlation between the ASSR thresholds and behavioral thresholds. The answer to both questions was affirmative. However, as they predicted early on, the ASSRs in CI subjects were affected by artifacts. A large 100-microvolt artifact was present. The artifact was minimized by averaging separate ASSR stimuli of opposite polarities delivered directly to the CI interface. Another step in artifact reduction involved removal of the artifact by linear extrapolation of the EEG time points encompassing the artifacts. The response was reduced to 500 nanovolts and high significant correlations were obtained between ASSR and behavioral thresholds. One drawback was that the ASSR modulation rates (around 40 Hz) were low compared to clinical pulse rates that are used (e.g., 1,000 pulses per second). In order to address this issue, the authors created new stimuli-modulated high-pulse trains at 900 pulses per second and demonstrated their correlation with behavioral threshold levels (Hoffman & Wouters, 2012). They also proposed a novel statistical method based on the assumption that physiologic responses are affected by stimulus-related factors like modulation rates whereas artifacts are not. Recently, Luke, Van Deun, Hofmann, Van Wieringen, and Wouters (2015) demonstrated that parameters of the electrical ASSRs in CI subjects were significantly related to modulation detection thresholds obtained behaviorally. The study indicates that ASSR could be used as an objective measure of temporal sensitivity in CI users. The ASSR may provide an objective measure for the optimization of site-specific stimulation parameters (Luke et al., 2015) as well as a measure of speech perception in CI recipients (Fu, 2002; Won, Drennan, Nie, Jameyson, & Rubinstein, 2011). In addition to electrical ASSR, there is interest in investigating if sound-field ASSR could be a viable technique for hearing assessment in CI users. Preliminary investigations have revealed that they are contaminated by CI-related artifacts. A recent study (Deshpande, Scott, Zhang, Keith, & Dimitrijevic, 2014) investigated the characteristics of the artifacts in terms of amplitude and phase and discusses future steps for characterizing and minimizing them.

The presence of CI-related artifacts makes the acquisition of the ASSR technically challenging. However, the merits offered by this technique in terms of measuring hearing thresholds as well as supra-threshold speech perception could potentially motivate hearing scientists to develop strategies to minimize the artifacts in the future. It will be interesting to witness the possibility of the addition of another useful electrophysiologic technique like the ASSR to the CI armamentarium in the future.

Neuroimaging

Neural Plasticity and Early Implantation

A critical period for speech-language learning—during which time maximal gains in auditory-speech-language performance may be observed with optimal intervention—has long been hypothesized in young children with and without hearing impairment (Johnson & Newport, 1989; Lenneberg, 1967; Locke, 1997; Mayberry et al., 2002; Newport, 1990, 1991; Oyama, 1976; Ruben, 1997; Sharma et al., 2002, 2013; Sharma & Dorman, 2006). A high degree of neural plasticity is observed before 2½ years of age for speech and vocabulary development (Connor et al., 2006) and "at least until age 4 years" for speech perception (Tajudeen et al., 2010, p. 1259) in young CI recipients. Evidence from electrophysiologic research corroborates that the primary auditory cortex and projections leading to it are maximally sensitive to auditory inputs before 3½ years of age. Prolonged auditory deprivation beyond 7 years of age may result in irreversible changes in plasticity, and implantation after this period is likely to result in suboptimal benefits (Cardon & Sharma; 2013; Dorman et al., 2007; Gilley et al., 2008; Munivrana & Mildner, 2013; Sharma et al., 2002a, 2002b; Sharma et al., 2002, 2005, 2009, 2013; Sharma & Dorman, 2006). Subsequent research (Colletti et al., 2011; Geers, 2006; Kennedy et al., 2006; Niparko et al., 2010; Tobey et al., 2013; Yoshinaga-Itano, 2013) has indicated the importance of early intervention especially during the critical period for speech-language development.

Use of Neuroimaging Techniques With CI Candidates/Recipients

Since neural plasticity plays an especially important role in children with hearing impairment, the study of brain function has garnered attention in the last two decades. Various neuroimaging techniques have been utilized to study the brain's response to auditory stimuli. Notable among them are the following: PET (e.g., Schecklmann et al., 2013; Talavage et al., 2013), magnetoencephalography (MEG; e.g., Bavelier et al., 2001; Cansino et al., 1994; Dietrich et al., 2001; Finney, Fine, & Dobkins et al., 2001; Finney et al., 2003; Lütkenhöner et al., 1997; Pantev et al., 2006; Pelizzone et al., 1991; Rupp et al., 2004) and functional magnetic resonance imaging (fMRI; e.g., DiFrancesco et al., 2013; Leach & Holland, 2010; Patel et al., 2007; Tan et al., 2013; Weiss et al., 2012). As certain areas of the brain are activated, they consume greater oxygen, which is supplied by blood vessels. Neuroimaging techniques utilize different aspects of this blood flow to acquire images of "the brain in action." The contrasts between auditory and nonauditory conditions can be studied across groups to arrive at areas of significant importance in acquiring specific auditory and speech-language skills. For instance, fMRI has been used to study various typical and atypical auditory phenomena in children as well as adults (e.g., Anderson et al., 2001; Bilecen et al., 2000; Corina et al., 2001; Patel et al., 2007; Pfleiderer et al., 2002; Propst et al., 2010; Robson et al., 1998; Scheffler et al., 1998; Tan et al., 2013; Tschopp et al., 2000). Some drawbacks of neuroimaging techniques include scanner noise confounding with auditory paradigms, temporal resolution, and compatibility with cochlear implants. However, these drawbacks can be tackled with minor modifications to the scanning protocol (e.g., Schmithorst & Holland, 2004), use of a different neuroimaging procedure, or use of specialized CIs (discussed in the next section).

Future of Neuroimaging Techniques in CI Candidates/Recipients

If neuroimaging techniques can be used to evaluate functional organization corresponding to speech-language skills, they may also be used to predict auditory, speech, and language outcomes in different populations. Patel et al. (2007), using linear regression analysis, found that there was a significant activation in the primary auditory cortex of children with hearing loss before implantation and that the improvement in postimplant hearing thresholds correlated positively with the preimplant fMRI activation, thus establishing the possibility of fMRI as a prognostic indicator of postimplant auditory outcomes. These data can be acquired even from asleep/sedated infants (e.g., DiFrancesco et al., 2013).

Taking these findings a step further, Deshpande et al. (in press) studied the relationship between preimplant fMRI activation and postimplant speech perception and oral language outcomes in children with CIs. They identified certain areas of the brain such as the angular gyrus, supramarginal gyrus, and the cingulate gyrus to be strongly correlated with oral language performance 2 years after implantation.

Some of the future clinical applications of neuroimaging techniques in CI candidates/users include detailed presurgical mapping of the auditory and language system (e.g., Berthezene et al., 1997); understanding how the auditory and language systems vary across the life span depending on age at implantation, duration of deafness, and other influencing factors (e.g., Schmidt, Weber, & Becker, 2001); predicting regions of maximum sensitivity important for future speech, language, and auditory development (e.g., Deshpande et al., 2014; Deshpande et al., in press); observation of functional changes corresponding with changes in music perception; observation of functional brain changes in response to auditory training procedures in CI recipients (e.g., Giraud, Truy, & Frackowiak, 2000); study of compensatory strategies employed by the brain due to auditory deprivation (e.g., Scheffler et al., 1998); use of diffusion tensor imaging (DTI) for mapping white matter track differences between CI users and those with typical hearing; and the use of direct measures of parental expectation (e.g., Zaidman-Zait & Most, 2005) in conjunction with fMRI to help establish realistic expectations for parents considering a CI for their child.

Recently, Med-El's SYNCHRONY became the first CI to receive the Food and Drug Administration (FDA) approval for use in a 3.0-tesla MRI scanner (Hires, 2015). The premise behind this design is that the magnet can rotate freely within a case, thus allowing it to align itself with the external MRI magnet. With this device, the CI recipient does not need to undergo magnet removal before an MRI scan. Future neuroimaging studies will undoubtedly focus on such devices. Until other models and CI companies incorporate such features, use of neuroimaging techniques such as PET and MEG, which are CI compatible, can be adopted.

As Hall (2006) rightly pointed out, "To enhance the clinical use of auditory fMRI, standardized protocols are required that are easy and quick to use with patients and yield robust replicable results" (p. 364). Based on current data trends, it seems inevitable that, eventually, the use of neuroimaging techniques will enhance our understanding of the "CI brain" and guide us in the creation of more precise, targeted, and reliable diagnostic and rehabilitation procedures.

Totally Implanted CIs

Totally implanted cochlear implants (TICIs) offer several advantages when compared to the external devices available today (Carlson et al., 2012; Cohen, 2007). Compared to current devices, TICIs are less exposed to the environment, including moisture, sweat, dust, and water, which limit the use of current implants in certain environments. TICI devices are also less vulnerable to dislodgement and potential damage for physical activities. Finally, TICI devices are not visible, which may be desirable to many patients for social reasons (Carlson et al., 2012). However, several technical barriers exist to development of a TICI, including power source management, environmental sound detection, and management of component breakdown (Carlson et al., 2012; Cohen, 2007). A TICI will need to be powered internally, likely with the use of a rechargeable battery. Batteries will need to be able to recharge quickly, hold enough charge to power the CI for about a day, not generate significant heat, and have a very low chance of leaking potentially dangerous battery chemicals even in the event of battery failure (Carlson et al., 2012; Cohen, 2007; Briggs et al., 2008). Additionally, a TICI, placed in the subcutaneous tissue, will have limited direct access to sound sources compared to externally worn microphones of current devices. Several options for sound detection exist such as microphone placement subcutaneously in the external auditory canal or behind the ear or using the tympanic membrane and/or ossicular chain as a microphone directly (Carlson et al., 2012; Cohen, 2007; Huttenbrink et al., 2001; Maniglia et al., 1999; Zenner et al., 2000). The speech processor and related electrical components will also need to be implanted. It is likely that with the increased number of components implanted, an explantation strategy will need to be devised as component failure becomes more likely. It is also probable that TICI will need some type of external hardware for battery recharging, programming, and switching between programs. It may also be desirable to allow the TICI to be powered and stimulated using a conventional external device (Briggs et al., 2008; Cohen, 2007). Briggs et al. (2008) published a report of three patients implanted with a TICI system. They demonstrated that the devices can be safely implanted and all three subjects derived benefit when using the devices in the "invisible hearing" mode. However, each subject performed better on word scores in quiet and sentence scores in noise when using the external speech processor in conventional mode. Although technical hurdles remain, it is likely that TICI will become a device option for patients in the future.

Summary

It is interesting to look back over 25 years ago at articles that we wrote on the future of CIs (Tyler, 1990; Tyler et al.,

1992; see also Tyler, 1991). These articles emphasized careful fitting of devices, a focus on fitting strategies based on individuals (not averages; see also Tyler, 1986), the development of a completely implantable CI, implanting individuals with plentiful residual hearing, designing synergistic CIs and HAs, developing computer-assisted aural rehabilitation programs, and providing CIs to disadvantaged populations. Many of these areas have shown important advances. CIs have changed how we treat hearings loss. So many patients have benefited, and the future is bright.

References

Anderson, A. W., Marois, R., Colson, E. R., Peterson, B. S., Duncan, C. C., Ehrenkranz, R. A., . . . Ment, L. R. (2001). Neonatal auditory activation detected by functional magnetic resonance imaging. *Magnetic Resonance Imaging*, *19*(1), 1–5.

Bavelier, D., Brozinsky, C., Tomann, A., Mitchell, T., Neville, H., & Liu, G. (2001). Impact of early deafness and early exposure to sign language on the cerebral organization for motion processing. *The Journal of Neuroscience*, *21*(22), 8931–8942.

Berthezene, Y., Truy, E., Morgon, A., Giard, M. H., Hermier, M., Franconi, J. M., & Froment, J. C. (1997). Auditory cortex activation in deaf subjects during cochlear electrical stimulation: Evaluation by functional magnetic resonance imaging. *Investigative Radiology*, *32*(5), 297–301.

Bilecen, D., Seifritz, E., Radü, E. W., Schmid, N., Wetzel, S., Probst, R., & Scheffler, K. (2000). Cortical reorganization after acute unilateral hearing loss traced by fMRI. *Neurology*, *54*(3), 765–765.

Bredberg, G., Walden, J., & Lindström, B. (1992). Tinnitus after cochlear implantation. In J. M. Aran & R. Dauman (Eds.), *Proceedings of the fourth international seminar, Bordeaux* (pp. 417–422). Amsterdam, The Netherlands: Kugler.

Briggs, R. J., Eder, H. C., Seligman, P. M., Cowan, R. S., Plant, K. L., Dalton, J., . . . Patrick, J. F. (2008). Initial clinical experience with a totally implantable cochlear implant research device. *Otology & Neurotology*, *29*(2), 114–119.[

Cansino, S., Williamson, S. J., & Karron, D. (1994). Tonotopic organization of human auditory association cortex. *Brain Research*, *663*(1), 38–50.

Cardon, G., & Sharma, A. (2013). Central auditory maturation and behavioral outcome in children with auditory neuropathy spectrum disorder who use cochlear implants. *International Journal of Audiology*, *52*(9), 577–586.

Carlson, M. L., Driscoll, C. L, Gifford, R. H., & McMenomey, S. O. (2012). Cochlear implantation: Current and future device options. *Otolaryngologic Clinics of North America*, 45(1), 221–248.

Chang, S.-A., Tyler, R. S., Dunn, C. C., Ji, H., Witt, S. A., Gantz, B., & Hansen, M. (2010). Performance over time on adults with simultaneous bilateral cochlear implants. *American Academy of Audiology*, *21*(1), 35–43.

Cohen, N. (2007). The totally implantable cochlear implant. *Ear and Hearing*, 28(2, Suppl.), 100s–101s.

Colletti, L., Mandalà, M., Zoccante, L., Shannon, R. V., & Colletti, V. (2011). Infants versus older children fitted with cochlear implants: Performance over 10 years. *International Journal of Pediatric Otorhinolaryngology*, *75*(4), 504–509.

Connor, C. M., Craig, H. K., Raudenbush, S. W., Heavner, K., & Zwolan, T. A. (2006). The age at which young deaf children receive cochlear implants and their vocabulary and speech-production growth: Is there an added value for early implantation? *Ear and Hearing*, *27*(6), 628–644.

Corina, D. P., Richards, T. L., Serafini, S., Richards, A. L., Steury, K., Abbott, R. D.,

. . . Berninger, V. W. (2001). fMRI auditory language differences between dyslexic and able reading children. *Neuroreport, 12*(6), 1195–1201.

Deshpande, A. K., Tan, L., Lu, L. J., Altaye, M., & Holland, S. (2014). *fMRI predicts post-implant auditory-language outcomes as measured by parent observation reports.* Presented at ASHA, Orlando, FL.

Deshpande, A. K., Tan, L., Lu, L. J., Altaye, M., & Holland, S. (in press). fMRI as a pre-implant objective tool to predict post-implant speech perception and oral language outcomes in children with cochlear implants. *Ear and Hearing.*

Deshpande, A. K., & Tyler, R. (2016). *Laboratory trial of background sounds via cochlear implants for tinnitus reduction.* Presented at AudiologyNow!, Phoenix, AZ.

Deshpande, S. B., Scott, M. P., Zhang, F., Keith, R. W., & Dimitrijevic, A. (2014). Characterization of cochlear implant-related artifacts during sound-field recording of the auditory steady state response using an amplitude modulated stimulus: A comparison among normal hearing adults, cochlear implant recipients, and implant-in-a-box. *The Journal of the Acoustical Society of America, 136*(4), 2306.

Dietrich, V., Nieschalk, M., Stoll, W., Rajan, R., & Pantev, C. (2001). Cortical reorganization in patients with high frequency cochlear hearing loss. *Hearing Research, 158*(1), 95–101.

Dimitrijevic, A., & Cone, B. (2015). Auditory steady-state response. In J. Katz., M. Chasin, K. English, L. J. Hood, & K. L. Tillery (Eds.), *Handbook of clinical audiology* (7th ed., pp. 267–294). Philadelphia, PA: Wolters Kluwer.

Dimitrijevic, A., John, M. S., & Picton, T. W. (2004). Auditory steady-state responses and word recognition scores in normal-hearing and hearing-impaired adults. *Ear and Hearing, 25*(1), 68–84.

Dimitrijevic, A., John, M. S., Van Roon, P., Purcell, D. W., Adamonis, J., Ostroff, J., Nedzelski, J. M., & Picton, T.W. (2002). Estimating the audiogram using multiple auditory steady-state responses. *Journal of the American Academy of Audiology, 13*(4), 205–224.

Dorman, M. F., Sharma, A., Gilley, P., Martin, K., & Roland, P. (2007). Central auditory development: Evidence from CAEP measurements in children fit with cochlear implants. *Journal of Communication Disorders, 40*(4), 284–294.

Dunn, C. C., Tyler, R. S., Oakley, S. A., Gantz, B. J., & Noble, W. (2008). Comparison of speech recognition and localization performance in bilateral and unilateral cochlear implant users matched on duration of deafness and age at implantation. *Ear and Hearing, 29*(3), 352–359.

Finney, E. M., Clementz, B. A., Hickok, G., & Dobkins, K. R. (2003). Visual stimuli activate auditory cortex in deaf subjects: Evidence from MEG. *Neuroreport, 14*(11), 1425–1427.

Finney, E. M., Fine, I., & Dobkins, K. R. (2001). Visual stimuli activate auditory cortex in the deaf. *Nature Neuroscience, 4*(12), 1171–1173.

Francesco, M. W., Robertson, S. A., Karunanayaka, P., & Holland, S. K. (2013). BOLD fMRI in infants under sedation: Comparing the impact of pentobarbital and propofol on auditory and language activation. *Journal of Magnetic Resonance Imaging, 38*(5), 1184–1195.

Fu, Q. J. (2002). Temporal processing and speech recognition in cochlear implant users. *Neuroreport, 13*(13), 1635–1639.

Geers, A. E. (2006). Factors influencing spoken language outcomes in children following early cochlear implantation. *Advances in Oto-Rhino-Laryngology, 64,* 50–65.

Gfeller, K., Christ, A., Knutson, J. F., Witt, S., Murray, K. T., & Tyler, R. S. (2000). Musical backgrounds, listening habits, and aesthetic enjoyment of adult cochlear implant recipients. *Journal of the American Academy of Audiology, 11*(7), 390–406.

Gibson, W. P. R. (1992). The effect of electrical stimulation and cochlear implantation on tinnitus. In J. M. Aran & R. Dauman (Eds.) *Proceedings of the Fourth International Seminar, Bordeaux* (pp. 403–408). Amsterdam, The Netherlands: Kugler.

Gilley, P. M., Sharma, A., & Dorman, M. F. (2008). Cortical reorganization in children with cochlear implants. *Brain Research, 1239*, 56–65.

Giraud, A. L., Truy, E., & Frackowiak, R. (2000). Imaging plasticity in cochlear implant patients. *Audiology & Neuro-Otology, 6*(6), 381–393.

Hall, D. A. (2006). fMRI of the auditory cortex. In S. H. Faro & F. B. Mohamed (Eds.), *Functional MRI: Basic principles and clinical applications* (pp. 364–393). New York, NY: Springer.

Hazell, J. W. P., McKinney, C. J., & Aleksy, W. (1995). Mechanisms of tinnitus in profound deafness. *The Annals of Otology, Rhinology & Laryngology Supplement, 166*, 418–420.

Hires, E. A. N. (2015). Med-El launches new products. *The Hearing Journal, 68*(4), 41–42.

Hofmann, M., & Wouters, J. (2010). Electrically evoked auditory steady state responses in cochlear implant users. *Journal of the Association for Research in Otolaryngology, 11*(2), 267–282.

Hofmann, M., & Wouters, J. (2012). Improved electrically evoked auditory steady-state response thresholds in humans. *Journal of the Association for Research in Otolaryngology, 13*(4), 573–589.

Huttenbrink, K. B., Zahnert, T. H., Bornitz, M., Hofmann, G. (2001). Biomechanical aspects in implantable microphones and hearing aids and development of a concept with a hydroacoustical transmission. *Acta Otolaryngolica, 121*(2), 185–189.

Jeng, F. C., Abbas, P. J., Brown, C. J., Miller, C. A., Nourski, K. V., & Robinson, B. K. (2007). Electrically evoked auditory steady-state responses in guinea pigs. *Audiology & Neurotology, 12*(2), 101–112.

Jeng, F. C., Abbas, P. J., Brown, C. J., Miller, C. A., Nourski, K. V., & Robinson, B. K. (2008). Electrically evoked auditory steady-state responses in a guinea pig model: Latency estimates and effects of stimulus parameters. *Audiology & Neurotology, 13*(3), 161–171.

Johnson, J. S., & Newport, E. L. (1989). Critical period effects in second language learning: The influence of maturational state on the acquisition of English as a second language. *Cognitive Psychology, 21*(1), 60–99.

Kennedy, C. R., McCann, D. C., Campbell, M. J., Law, C. M., Mullee, M., Petrou, S., . . . Stevenson, J. (2006). Language ability after early detection of permanent childhood hearing impairment. *New England Journal of Medicine, 354*(20), 2131–2141.

Kim, H. N., Shim, Y. J., Kim, Y. M., & Kim, E. S. (1995). Effect of electrical stimulation on tinnitus in the profoundly deaf. In G. E. Reich & J. Vernon (Eds.), *Proceedings of the Fifth International Tinnitus Seminar* (pp. 508–517). Portland, OR: American Tinnitus Association.

Korczak, P., Smart, J., Delgado, R., Strobel, T., & Bradford, C. (2012). Auditory steady-state responses. *Journal of the American Academy of Audiology, 23*(3), 146–170.

Kou, B. S., Shipp, D. B., & Nedzelski, J. M. (1994). Subjective benefits reported by adult Nucleus 22-channel cochlear implant users. *Journal of Otolaryngology, 23*(1), 8–14.

Kraus, N., & Anderson, S. (2014). Better speech processing in smaller amplitudes. *Hearing Journal, 67*(3), 48.

Leach, J. L., & Holland, S. K. (2010). Functional MRI in children: Clinical and research applications. *Pediatric Radiology, 40*(1), 31–49.

Lenneberg, E. H. (1967). *Biological foundations of language*. New York, NY: Wiley.

Levitt, H., Oden, C., Simon, H., Noack, C., & Lotze, A. (2011). Entertainment overcomes barriers of auditory training. *Hearing Journal, 64*(8), 40, 42.

Litovsky, R., Parkinson, A., Arcaroli, J., & Sammeth, C. (2006). Simultaneous bilateral cochlear implantation in adults: A multicenter clinical study. *Ear and Hearing, 27*(6), 714–731.

Locke, J. L. (1997). A theory of neurolinguistic development. *Brain and Language, 58*(2), 265–326.

Luke, R., Van Deun, L., Hofmann, M., Van Wieringen, A., & Wouters, J. (2015). Assessing temporal modulation sensitivity using electrically evoked auditory steady state responses. *Hearing Research, 324*, 37–45.

Lütkenhöner, B., & Steinsträter, O. (1997). High-precision neuromagnetic study of the functional organization of the human auditory cortex. *Audiology & Neuro-Otology, 3*(2–3), 191–213.

Maniglia, A. J., Abbass, H., Azar, T., Kane, M., Amantic, P., Garverick S., Ko, WH., Frenz, W., Falk, T., (1999). The middle ear bioelectronic microphone for a totally implantable cochlear hearing de-vice for profound and total hearing loss. *American Journal of Otology, 20*(5), 602–611.

Mayberry, R. I., Lock, E., & Kazmi, H. (2002). Linguistic ability and early language exposure. *Nature, 417*(6884), 38.

Menard, M., Gallego, S., Truy, E., Berger-Vachon, C., Durrant, J. D., & Collet, L. (2004). Auditory steady-state response evaluation of auditory thresholds in cochlear implant patients. *International Journal of Audiology, 43*(Suppl. 1), S39–S43.

Munivrana, B., & Mildner, V. (2013). Cortical auditory evoked potentials in unsuccessful cochlear implant users. *Clinical Linguistics & Phonetics, 27*(6–7), 472–483.

Newport, E. L. (1990). Maturational constraints on language learning. *Cognitive Science, 14*(1), 11–28.

Newport, E. L. (1991). Contrasting conceptions of the critical period for language. In S. Carey & R. Gelman (Eds.), *The epigenesis of mind: Essays on biology and cognition* (pp. 111–130). Hillsdale, NJ: Erlbaum.

Niparko, J. K., Tobey, E. A., Thal, D. J., Eisenberg, L. S., Wang, N. Y., Quittner, A. L., . . . CDaCI Investigative Team. (2010). Spoken language development in children following cochlear implantation. *JAMA, 303*(15), 1498–1506.

Osaki, Y., Nishimura, H., Takasawa, M., Imaizumi, M., Kawashima, T., Iwaki, T., . . . Kubo, T. (2005). Neural mechanism of residual inhibition of tinnitus in cochlear implant users. *NeuroReport, 16*(15), 1625–1628. http://doi.org/10.1097/01.wnr.0000183899.85277.08

Oyama, S. (1976). A sensitive period for the acquisition of a nonnative phonological system. *Journal of Psycholinguistic Research, 5*(3), 261–283.

Pantev, C., Dinnesen, A., Ross, B., Wollbrink, A., & Knief, A. (2006). Dynamics of auditory plasticity after cochlear implantation: A longitudinal study. *Cerebral Cortex, 16*(1), 31–36.

Patel, A. M., Cahill, L. D., Ret, J., Schmithorst, V., Choo, D., & Holland, S. (2007). Functional magnetic resonance imaging of hearing-impaired children under sedation before cochlear implantation. *Archives of Otolaryngology-Head & Neck Surgery, 133*(7), 677–683.

Pelizzone, M., Kasper, A., Hari, R., Karhu, J., & Montandon, P. (1991). Bilateral electrical stimulation of a congenitally-deaf ear and of an acquired-deaf ear. *Acta Oto-Laryngologica, 111*(2), 263–268.

Perez, R., Shaul, C., Vardi, M., Muhanna, N., Kileny, P. R., & Sichel, J. Y. (2015). Multiple electrostimulation treatments to the promontory for tinnitus. *Otology & Neurotology 36*(2), 366–372.

Perreau, A., Tyler, R. S., Witt, S., & Dunn, C. (2007). Selection strategies for binaural and monaural cochlear implantation. *American Journal of Audiology, 16*(2), 85–93.

Pfleiderer, B., Ostermann, J., Michael, N., & Heindel, W. (2002). Visualization of auditory habituation by fMRI. *Neuroimage, 17*(4), 1705–1710.

Picton, T. W., Durieux-Smith, A., Champagne, S. C., Whittingham, J., Moran, L. M., Giguère, C., & Beauregard, Y. (1998). Objective evaluation of aided thresholds using auditory steady-state responses. *Journal of the American Academy of Audiology, 9*(5), 315–331.

Picton, T. W., John, M. S., Dimitrijevic, A., & Purcell, D. (2003). Human auditory steady-state responses: Respuestas auditivas de estado estable en humanos. *International Journal of Audiology, 42*(4), 177–219.

Propst, E. J., Greinwald, J. H., & Schmithorst, V. (2010). Neuroanatomic differences in children with unilateral sensorineural hearing loss detected using functional magnetic resonance imaging. *Archives of Otolaryngology-Head & Neck Surgery, 136*(1), 22–26.

Rance, G. (Ed.). (2008). *The auditory steady-state response.* San Diego, CA: Plural.

René, H., Gifford, D., Grantham, W., Sheffield, S. W., Davis, T. J., Dwyer, R., & Dorman, M. F. (2014). Localization and interaural time difference (ITD) thresholds for cochlear implant recipients with preserved acoustic hearing in the implanted ear. *Hearing Research, 312,* 28–37.

Robson, M. D., Dorosz, J. L., & Gore, J. C. (1998). Measurements of the temporal fMRI response of the human auditory cortex to trains of tones. *Neuroimage, 7*(3), 185–198.

Ruben, R. J. (1997). A time frame of critical/sensitive periods of language development. *Acta Oto-Laryngologica, 117*(2), 202–205.

Rupp, A., Gutschalk, A., Uppenkamp, S., & Scherg, M. (2004). Middle latency auditory-evoked fields reflect psychoacoustic gap detection thresholds in human listeners. *Journal of Neurophysiology, 92*(4), 2239–2247.

Schecklmann, M., Landgrebe, M., Poeppl, T. B., Kreuzer, P., Männer, P., Marienhagen, J., . . . Langguth, B. (2013). Neural correlates of tinnitus duration and distress: a positron emission tomography study. *Human Brain Mapping, 34*(1), 233–240.

Scheffler, K., Bilecen, D., Schmid, N., Tschopp, K., & Seelig, J. (1998). Auditory cortical responses in hearing subjects and unilateral deaf patients as detected by functional magnetic resonance imaging. *Cerebral Cortex, 8*(2), 156–163.

Schmidt, A. M., Weber, B. P., & Becker, H. (2001). Functional magnetic resonance imaging of the auditory cortex as a diagnostic tool in cochlear implant candidates. *Neuroimaging Clinics of North America, 11*(2), 297–304.

Schmithorst, V. J., & Holland, S. K. (2004). Event-related fMRI technique for auditory processing with hemodynamics unrelated to acoustic gradient noise. *Magnetic Resonance in Medicine, 51*(2), 399–402.

Sharma, A., & Dorman, M. F. (2006). Central auditory development in children with cochlear implants: Clinical implications. *Advances in Oto-Rhino-Laryngology, 64,* 66–88.

Sharma, A., Dorman, M. F., & Spahr, A. J. (2002a). A sensitive period for the development of the central auditory system in children with cochlear implants: Implications for age of implantation. *Ear and Hearing, 23*(6), 532–539.

Sharma, A., Dorman, M. F., & Spahr, A. J. (2002b). Rapid development of cortical auditory evoked potentials after early cochlear implantation. *Neuroreport, 13*(10), 1365–1368.

Sharma, A., Dorman, M., Spahr, A., & Todd, N. W. (2002). Early cochlear implantation in children allows normal development of central auditory pathways. *The Annals of Otology, Rhinology & Laryngology Supplement, 189,* 38–41.

Sharma, A., Glick, H., Campbell, J., & Biever, A. (2013). Central auditory development in children with hearing impairment: Clinical relevance of the P1 CAEP biomarker in children with multiple disabilities. *Hearing, Balance and Communication, 11*(3), 110–120.

Sharma, A., Martin, K., Roland, P., Bauer, P., Sweeney, M. H., Gilley, P., & Dorman, M. (2005). P1 latency as a biomarker for central auditory development in children with hearing impairment. *Journal of the American Academy of Audiology, 16*(8), 564–573.

Sharma, A., Nash, A. A., & Dorman, M. (2009). Cortical development, plasticity and re-organization in children with cochlear implants. *Journal of Communication Disorders, 42*(4), 272–279.

Stroebel, D., Swanepoel de, W., & Groenewald, E. (2007). Aided auditory steady-state responses in infants. *International Journal of Audiology, 46*(6), 287–292.

Tajudeen, B. A., Waltzman, S. B., Jethanamest, D., & Svirsky, M. A. (2010). Speech perception in congenitally deaf children receiving cochlear implants in the first year of life. *Otology & Neurotology, 31*(8), 1254.

Talavage, T. M., Gonzalez-Castillo, J., & Scott, S. K. (2014). Auditory neuroimaging with fMRI and PET. *Hearing Research, 307*, 4–15.

Tan, L., Chen, Y., Maloney, T. C., Caré, M. M., Holland, S. K., & Lu, L. J. (2013). Combined analysis of sMRI and fMRI imaging data provides accurate disease markers for hearing impairment. *NeuroImage: Clinical, 3*, 416–428.

Tobey, E. A., Thal, D., Niparko, J. K., Eisenberg, L. S., Quittner, A. L., & Wang, N. Y. (2013). Influence of implantation age on school-age language performance in pediatric cochlear implant users. *International Journal of Audiology, 52*(4), 219–229.

Tschopp, K., Schillinger, C., Schmid, N., Rausch, M., Bilecen, D., & Scheffler, K. (2000). Detection of central auditory compensation in unilateral deafness with functional magnetic resonance tomography. *Laryngo-Rhino-Otologie, 79*(12), 753–757.

Tyler, R., Ji, H., Perreau, H., Witt, S., Noble, W., & Coelho, C. (2014). Development and validation of the Tinnitus Primary Function Questionnaire. *American Journal of Audiology, 23*, 260–272.

Tyler, R., Preece, J., Wilson, B., Rubinstein, J., Parkinson, A., Wolaver, A., & Gantz, B. (2002). Distance, localization and speech perception pilot studies with bilateral cochlear implants. In T. Kubo, Y. Takahashi, & T. Iwaki (Eds.), *Cochlear implants—An update* (pp. 517–522). Amsterdam, The Netherlands: Kugler Publications.

Tyler, R. S. (1979). Measuring hearing loss in the future. *British Journal of Audiology, 13*(Suppl. 2), 29–40.

Tyler, R. S. (1986). Adjusting a hearing aid to amplify speech to the MCL. *The Hearing Journal, 39*(8), 24–27.

Tyler, R. S. (1990). What should be implemented in future cochlear implants? *Acta Oto-Laryngologica* (Stockholm) Supplement, 469, 268–275.

Tyler, R. S. (1991). What can we learn about hearing aids from cochlear implants? *Ear and Hearing, 12*(6, Suppl.), 177S–186S.

Tyler, R. S. (1994). Advantages and disadvantages expected and reported by cochlear implant patients. *American Journal of Otolaryngology, 15*, 523–531.

Tyler, R. S., & Baker, L. J. (1983). Difficulties experienced by tinnitus sufferers. *Journal of Speech and Hearing Disorders, 48*(2), 150–154.

Tyler, R. S., Dunn, C. C., Witt, S. A., & Noble, W. G. (2007). Speech perception and localization with adults with bilateral sequential cochlear implants. *Ear and Hearing, 28*(2), 86S–90S.

Tyler, R. S., Dunn, C. C., Witt, S. A., Noble, W., Gantz, B. J., Rubinstein, J. T., Parkinson, A. J., & Branin, S. C. (2006a). Soundfield hearing for patients with cochlear implants and hearing aids. In H. R. Cooper & L.C. Craddock (Eds.), *Cochlear implants: A practical guide* (pp. 338–365). London, UK: Whur Publishers.

Tyler, R. S., Dunn, C. C., Witt, S. A., & Preece, J. P. (2003). Update on bilateral cochlear implantation. *Current Opinion in Otolaryngology, 11*(5), 388–393.

Tyler, R. S., Gfeller, K., & Mehr, M. A. (2000). A preliminary investigation comparing one and eight channels at fast and slow rates on music appraisal in adults with cochlear implants. *Cochlear Implants International, 1*(2), 82–87.

Tyler, R. S., Keiner, A. J., Walker, K., Deshpande, A. K., Witt, S., Killian, M., ... Gantz, B. (2015). A series of case studies of tinnitus suppression with mixed background stimuli in a cochlear implant. *American Journal of Audiology, 24*(3), 398–410.

Tyler, R. S., & Kelsay, D. (1990). Advantages and disadvantages reported by some of the better cochlear implant patients. *American Journal of Otolaryngology, 11*(4), 282–289.

Tyler, R. S., Noble, W., Dunn, C., & Witt, S. (2006b). Some benefits and limitations of binaural cochlear implants and our ability to measure them. *International Journal of Audiology, 45*(Suppl. 1), S113–S119.

Tyler, R. S., Opie, J. M., Fryauf-Bertschy, H., & Gantz, B. J. (1992). Future directions for cochlear implants. *Journal of Speech-Language-Pathology and Audiology, 16*(2), 151–164.

Tyler, R. S., Perreau, A., & Ji, H. (2009). Validation of the Spatial Hearing Questionnaire. *Ear and Hearing, 30*(4), 419–430.

Tyler, R. S., Witt, S. A., Dunn, C. C., & Wang, W. (2010). Initial development of a spatially separated speech-in-noise and localization training program. *Journal of the American Academy of Audiology, 21*(6), 390–403.

Weiss, J. P., Bernal, B., Balkany, T. J., Altman, N., Jethanamest, D., & Andersson, E. (2012). fMRI evaluation of cochlear implant candidacy in diffuse cortical cytomegalovirus disease. *The Laryngoscope, 122*(9), 2064–2066.

Won, J. H., Drennan, W. R., Nie, K., Jameyson, E. M., & Rubinstein, J. T. (2011). Acoustic temporal modulation detection and speech perception in cochlear implant listeners. *The Journal of the Acoustical Society of America, 130*(1), 376–388.

Yoshinaga-Itano, C. (2013). Principles and guidelines for early intervention after confirmation that a child is deaf or hard of hearing. *Journal of Deaf Studies and Deaf Education, 19*(2), 143–175.

Zaidman-Zait, A., & Most, T. (2005) Cochlear implants in children with hearing loss: Maternal expectations and impact on the family. *Volta Review, 105*, 129–150.

Zenner, H. P., Leysieffer, H., Maassen, M., Lehner, R., Lenarz, T., Baumann, J., . . . McElveen, Jr., J. T. (2000). Human studies of a piezoelectric transducer and a microphone for a totally implantable electronic hearing device. *American Journal of Otology, 21*(2), 196–204.

CHAPTER 9

Novel Approaches for Protection and Restoration of Hearing

Min Young Lee and Yehoash Raphael

Sensorineural hearing loss (SNHL) is usually irreversible, primarily because lost hair cells and auditory neurons cannot be replaced and mutations affecting the cochlea cannot be cured. Some cases of sudden hearing loss that respond to steroids, along with fluctuating hearing such as seen in Ménière's disease, are perhaps the only exceptions. Due to the lack of biological treatments for SNHL, clinical management and rehabilitation rely on amplification with hearing aids or a cochlear implant (CI) when applicable. Although the CI is extremely helpful in cases of severe-to-profound hearing loss, it does not replicate the same quality of audiologic perception the normal cochlea has. Therefore, novel approaches for protection and restoration of hearing are needed. Recent progress in understanding the biological basis of genetic deafness, along with gene therapy approaches and availability of improved animal models replicating human disease, facilitate the quest for novel therapies to provide protection and regeneration of the inner ear.

This review chapter is focused on future novel treatments based on gene therapy: the delivery of genetic material to target (diseased) cells. In some cases, especially when the therapeutic gene product is secreted, it is possible to deliver genes to cells in the vicinity of target cells. This is usually accomplished by use of modified viruses called viral vectors, which can enter cells and deliver their load (the exogenous gene) into the cell. In other cases, gene expression is manipulated for overexpression or inhibition by inserting small interfering RNA (siRNA) or small molecules. The cochlea is a difficult site to reach for delivery of reagents, but the fluid-filled spaces help spread the injected reagents away from the site of injection. There are various delivery routes for gene therapy of the inner ear: transtympanic (injection into middle ear), trans–round window (injection into perilymph), and canalostomy or cochleostomy (injection into either perilymph or endolymph) (Fukui et al., 2013). In most cases, delivery routes are determined by the target cells and the ability of the viral vector to reach them. In this chapter, four common aims for gene therapy of the inner ear will be introduced, and for each we discuss recent animal studies and future directions. Although invaluable progress has been made in the design and testing of novel therapies, it is important to emphasize that none of them are currently available in the otology clinic.

Protection

Background

Progress in biomedical sciences is facilitating and enhancing novel therapies for preserving structure and function of the inner ear. Several cell types are of specific interest for protection because they are known to degenerate due to a variety of insults. The most common targets for novel protective measures are hair cells, neurons, synapses, and cells of the stria vascularis. Of these targets, the hair cells are probably the most commonly addressed, because of their small number, critical role in hearing, and the fact that cochlear (mammalian) hair cells do not regenerate spontaneously (Izumikawa et al., 2008; Roberson et al., 1994). Among the insults that could lead to hair cell loss are acoustic overstimulation, cisplatin (common chemotherapeutic), aminoglycosides (antimicrobial therapy), and infections. Given the lack of regenerative capacity, methods to minimize hair cell loss and other types of cochlear cells are needed. Currently pursued approaches for protection against each of these categories of insults are listed below.

Animal Experiments

Aminoglycosides, developed in the 1940s, are one of the oldest classes of antibiotics still in use today (Chen et al., 2009). One of these is gentamicin, which is currently widely used for the treatment of certain infectious conditions secondary to genetic disorders (Chen et al., 2014), including cystic fibrosis (Xue et al., 2014) and Rett syndrome (Brendel et al., 2009; Brendel et al., 2011; Popescu et al., 2010). Use of aminoglycosides poses risks of toxic side effects, especially in the kidney and the inner ear. In the ear, these side effects include acute loss of hair cells and, in some cases, degeneration of auditory nerve cells (spiral ganglion neurons, SGNs) (Hawkins et al., 1977; Johnsson et al., 1981; Schacht, 1999). Degeneration of SGNs may be secondary to the loss of inner hair cells or possibly directly due to aminoglycoside exposure. Animal studies have been performed to test reagents that would reduce or eliminate the ototoxicity of aminoglycosides for both hair cells and SGNs.

One family of molecules experimentally used for protecting against ototoxicity are growth factors. One of the first molecules to be tested was glial cell line–derived neurotrophic factor (GDNF). GDNF transgene overexpression mediated by adenovirus vector prevented hair cell degeneration by aminoglycoside ototoxicity when the vector was given alone (Yagi et al., 1999) or in combination with transforming growth factor–β1 (TGF-β1) (Ben-Yosef et al., 2003). Similar results were obtained when GDNF levels were elevated using another type of viral vector, adeno-associated virus vector (AAV; Liu et al., 2008). Other growth factors used for protecting against aminoglycoside ototoxicity are members of the neurotrophin family. Neurotrophins are members of a family of proteins critical to the development and maintenance of neural systems, including the inner ear and its innervation. Within the auditory system, brain-derived neurotrophic factor (BDNF) and neurotrophin-3 (NT3) are the two neurotrophins essential for normal auditory neural development (Pirvola et al., 1992). These neurotrophins have been shown to protect both hair cells and SGNs (Atkinson et al., 2012; Miller et al., 1997; Nakaizumi et al., 2004; Ruan et al., 1999; Yagi et al., 2000).

To introduce elevated levels of growth factors in the cochlea, it is necessary to infuse them with mini-osmotic pumps, inject viral vectors carrying the neurotrophin gene, or use alginates or other materials presoaked with the factors (Noushi et al., 2005). The limitations of mini-osmotic pumps are the risk of infections where the cannula is inserted, the need to reload the pump for continued expression, and the potential degradation of the therapeutic molecule when it is in body temperature for days or weeks. The risks and limitation of viral vectors are described elsewhere in this review (Side Effects of Viruses). None of these methods are immediately clinically applicable, but the data are useful in selecting the molecules for protective use in the future. In addition to elevating levels of the growth factor itself, attempts to influence levels of expression of the receptor for neurotrophins also have shown a protective effect (Yu et al., 2013).

Other families of molecules have been used in experiments for assessing protection against aminoglycoside ototoxicity. One approach has been to stop cell death by overexpression of bcl-2, a molecule that protects against apoptosis. Genes delivered via an adenoviral vector to the inner ear showed protection against transtympanic aminoglycosides for both the cochlea and vestibular organs (Pfannenstiel et al., 2009). Another approach is to use heat-shock protein (Hsp) overexpression. Hsp70 was found to be protective agent against ototoxic insults (Baker et al., 2015; Takada et al., 2015; Taleb et al., 2009). Using compounds with antioxidant activity is an especially attractive way to protect against ototoxicity because some of these medications can be given orally and are available over the counter. Antioxidants have been shown to protect against ototoxicity in several animal studies (Kawamoto et al., 2004; Ojano-Dirain et al., 2014; Song et al., 1996) and in clinical trials in humans (Chen et al., 2007).

Cisplatin is a chemotherapeutic agent used commonly to treat squamous cell cancers of the head and neck (Goncalves et al., 2013). One major side effect of this pharmacologic agent is SNHL that is symmetrical and starts in the high-frequency region and progresses to lower frequencies in a dose-related and cumulative fashion (Chirtes et al., 2014). The average incidence of SNHL among cisplatin-treated patients is relatively high (62%; Marshak et al., 2014). Despite this high frequency of cisplatin-induced hearing loss, there are no definite preventive measures or alternative nonototoxic chemotherapeutic agents (Chirtes et al., 2014). One molecule that has been tested for its protective ability is X-linked inhibitor of apoptosis protein, which inhibits apoptosis effectors such as caspases and other cell death pathways (Deveraux et al., 1997). This protein was upregulated by AAV2-mediated gene therapy in rats, resulting in protection of hair cells from cisplatin ototoxicity (Cooper et al., 2006). Protection by the same proteins was also accomplished using a less invasive delivery technique, based on AAV transfer across the round window membrane (Jie et al., 2015). Neurotrophins also have been shown to protect against cisplatin ototoxicity (Bowers et al., 2002), as was siRNA that was used to block expression of transient receptor potential vanilloid1 (TRPV1) and STAT (Mukherjea et al., 2010; Mukherjea et al., 2008).

Acoustic trauma is one of the most common etiologies causing hearing losses. Exposure sources include occupational (industrial, military, etc.), recreational (shooting firearms, loud music), or accidental (loud alarms, sirens, etc.; Clifford et al., 2009). The National Institutes of Health (NIH) estimated that about 26 million U.S. adult citizens have noise-induced hearing loss (NIHL) (Clifford et al., 2009), and its economic and social impact are suspected to be immeasurable. Therefore, there are ongoing efforts to minimize and compensate NIHL (Folmer et al., 2002; Kopke et al., 2007; Le Prell et al., 2007; Seidman et al., 2013; Verbeek et al., 2009). The main site of permanent pathology due to acoustic trauma is hair cell loss, although the ribbon synapse connecting the inner hair cell with the auditory nerve can be the first site-of-lesion, and progressive degeneration of auditory neurons may also be involved (Kujawa et al., 2015).

Similar to treatments for lesions caused by aminoglycosides or cisplatin, experimental routes for protecting against acoustic trauma usually require surgical interventions that would limit the clinical applicability of the procedure. However, and as noted previously, results are useful in identifying molecular pathways that can be used for protection. Neurotrophins, antioxidants, antiapoptotic reagents, and other molecules have been used to reduce the outcome of acoustic trauma (Bielefeld, 2013; Coleman et al., 2007; Du et al., 2011; Shoji et al., 2000; Yamashita et al., 2005). In most cases, the protective agents were administered before the exposure, which is suboptimal for clinical use.

Next Steps

Treatments that cause more damage than benefit are not clinically useful. There are several hurdles that need to be traversed to make preventive treatments applicable. One important task is to deliver the protective agent to the cochlea reliably with minimal manipulation. The other task is to identify or develop agents that will rescue cochlear cells and potentially hearing even when applied after the insult has occurred, be it ototoxicity or exposure to loud acoustic signals (noise or sound). Until these are accomplished, it is important to remind ourselves and our patients that at least for acoustic trauma, prevention is an important goal. Thus, whenever possible, it is advisable to use passive attenuation devices like foam ear plugs or ear muffs to reduce the impact of the acoustic exposure, or try to avoid these situations all together.

Regeneration

Background

The term regeneration is used differently when referring to neurons or epithelial cells. For neurons, regeneration traditionally means the regrowth of axons or dendrites to reconnect with a target, although more recently it became clear that only in a few systems can neurons be replaced once lost (Verma et al., 2015). This is not the case in the spiral ganglion, where lost neurons are not spontaneously replaced. For epithelial cells, the term regeneration usually means generation of a new cell to replace a missing or lost cell. Both types of biological regeneration are needed in the cochlea, and research is under way to induce or enhance regenerative capabilities. At the level of the neuron, one goal of regeneration is to restore the ribbon synapse after acoustic trauma (discussed in this section) while another need is to induce sprouting of nerve fibers to the proximity of the CI (discussed with the CI, below). For epithelial tissues, the main need is to induce hair cell regeneration. Most epithelial cells turn over during life and are capable of accelerated regeneration when injured. The one clear exception is the population of cochlear hair cells in the mammalian organ of Corti. For this reason, SNHL due to hair cell degeneration is permanent. Therefore, studies on ways to induce hair cell regeneration are of high significance.

Animal Experiments

Several theoretical and practical approaches have been formulated for inducing hair cell regeneration in the cochlea. One approach is to induce nonsensory cell proliferation within the cochlea in an attempt to grow new hair cells. This approach was enthusiastically motivated by the observation that supporting cells that keep dividing in transgenic mice with P27 deficiency eventually become new hair cells (Chen et al., 1999; Kanzaki et al., 2006; Lowenheim et al., 1999). Continued proliferation in the organ of Corti was also observed in another transgenic mouse, where the Rb1 gene was depleted (Sage et al., 2005). In all these mice, hearing was disrupted and new hair cells appeared to degenerate. Another related finding was that nonsensory cells that remain in the auditory epithelium after hair cell loss can proliferate spontaneously in the guinea pig, yet new hair cells were never identified in this experimental protocol (Kim et al., 2007). It appears that proliferation alone does not lead to formation of stable hair cells in mammalian cochlea. Nevertheless, when combined with other approaches for generating new hair cells, the ability to induce proliferation and increase the number of cells in the auditory epithelium will likely be essential because an adequate number of cells are needed, including supporting cells, which have their own important roles in hearing. For now, a way to efficiently induce proliferation in the auditory epithelium of wild-type animals (by means other than transgenesis) has not been presented.

In parallel, other studies have been pursuing treatments aimed at inducing transdifferentiation, namely, converting the phenotype of supporting cells to new hair cells. The concept is "borrowed" from the few other systems where transdifferentiation is known to occur spontaneously, most notably, the basilar papilla of birds (Corwin et al., 1988; Ryals et al., 1988), the vestibular sensory epithelium in mammals (Forge et al., 1993), and in some pathologic conditions such as Barrett's metaplasia (Slack et al., 2010). In addition to the concept itself, the molecular approach to achieving transdifferentiation also has been influenced by those same systems. Practically, the approach is to first identify molecules and signaling pathways that are known to regulate development as well as transdifferentiation in systems where they spontaneously occur. Once identified, these pathways can be manipulated (blocked or enhanced) as in the mammalian auditory system. Below are some examples of initial successes in inducing regeneration of auditory hair cells.

The first successes in generating new hair cells in vivo were accomplished using viral-mediated overexpression of *Atoh1*, a gene necessary for developmental differentiation of hair cells. When injected into the scala media of guinea pig ears, *Atoh1* induced formation of a small number of ectopic hair cells (Kawamoto et al., 2003). The number of new hair cells was increased by overexpressing, in parallel, a gene (Skp2) that removes the inhibition on cell division in the auditory epithelium (Minoda et al., 2007). An alternative to overexpression of hair cell genes in supporting cells is the inhibition of supporting cell-encoding genes. This approach was tested in wild-type mice that were exposed to noise to eliminate many hair cells and then received injection of a γ-secretase inhibitor to block notch signaling in the supporting cells (Mizutari et al., 2013). The procedure resulted in formation of new hair cells and improvement of thresholds. These results are very exciting because a small molecule like a γ-secretase inhibitor should be easily introduced into the cochlea once the approach is optimized. Combination therapies, where more than one pathway is manipulated, will likely enhance the outcome of attempts to regenerate hair cells, as already demonstrated by some experiments (Atkinson et al., 2014). Together, these results suggest that manipulating expression levels of specific genes and pathways can induce hair cell regeneration, possibly paving the way to clinically feasible approaches.

The other attractive target for regeneration is the ribbon synapse, which is damaged early after acoustic trauma. Studies in transgenic animals identified NT-3 as a molecule that can induce synaptic regeneration as long as the hair cells survive (Wan et al., 2014), leading to restoration of thresholds after acoustic trauma. Next, it will be necessary to design therapies that induce NT-3 elevation in the proximity of the synapse without using transgenesis, to mimic the results obtained in transgenic mice.

Next Steps

Over the last few years, the progress in identifying pathways that can lead to regeneration has been substantial, and as a result, we are now closer to reversing the permanent nature of cochlear hair cell loss. However, much more needs to be accomplished to bring the technology closer to human applicability. The main task is to identify a reliable and consistent way to induce efficient regeneration of hair cells in the correct place. To accomplish this task, it is necessary to identify additional genes involved in the process and determine how to manipulate gene expression levels for inducing regeneration. Part of this task is being addressed by bioinformatics work comparing gene expression between different conditions, species, and ages (Fujioka et al., 2015; Ku et al., 2014; Maass et al., 2015; Thomas et al., 2015). In addition, it is necessary to design practical and safe means for delivering the therapeutics into the cochlea.

Treatment of Genetic Hearing Loss

Background

Genetic disease may contribute to more than half of the cases of hearing loss that require clinical intervention. In addition, some cases of environmentally induced hearing loss may also be influenced by hereditary predisposition. With this high prevalence of genetic inner ear diseases, diagnostics and treatments have high priority. Regretfully, many of the therapies that are being developed for environmental causes of auditory lesions, including most of the approaches for protection and regeneration, would not be applicable for genetic disease. For instance, hair cell regeneration based on transdifferentiation of supporting cells may place new hair cells in a cochlea that suffered acoustic trauma or an ototoxic lesion and improve hearing in these ears, but, when the cause of hair cell death is a mutation (affecting the hair cell or another cell in the cochlea), any new hair cells would still be susceptible to the same morbidity. Therefore, approaches that treat the underlying genetic basis are needed before regeneration can be contemplated. The most likely routes for such genetic treatment are insertion of the wild-type gene into the cells where the function is needed or treatment with a downstream target, thereby bypassing the defective gene product. In cases where mutations are dominant negative, treatment can be aimed at reducing or antagonizing the mutated gene product. In cases when the affected cells are absent due to degeneration or lack of development, cell replacement therapy may also become relevant.

Therapies for genetic inner ear disease may need to be developed specifically for each mutation. This is different from environmental disease, where the focus is usually on the lost cell (such as hair cell, neuron, or both), and the approach is to replace the missing cells. In environmental deafness, it may not make a difference why a hair cell was lost (overstimulation, medication side effects, or aging), and the treatment would be the same for all, namely, hair cell regeneration. In contrast, when designing biological treatments for genetic disease, it will be necessary to consider each mutation individually.

Several factors need to be addressed. A prerequisite for designing treatments for genetic inner ear disease is that the gene involved in the pathology be clearly identified. It is likely that most of the prevalent mutations are among the ~100 already identified, but additional rarer disease states, not yet linked with a specific gene, may need to be characterized (Atik et al., 2015). Once the gene is identified, it is essential to know what cells need the gene product to survive and function. It is also important to know the function of the gene and whether it has a role in development or in the maintenance and function of the mature ear, or both. It may not be necessary to replace the gene or the cell if function in the mature ear can be substituted, for instance, by providing a gene product downstream of the affected gene. It is necessary to understand whether these cells are present in the ear of the affected individual and realize that in some cases, the cells fail to develop or degenerate early in life, further complicating potential treatment options. Clearly, mutations where cells are present but dysfunctional would be first targets for feasible therapies, whereas diseases in which cells fail to develop present more complex challenges.

The next task is to design a vector (carrier) that will insert the wild-type gene (transgene) into the cell that needs it, which is the original goal of gene therapy (Crystal, 2014; Lundstrom, 2004; Warnock et al., 2011). The task involves multiple considerations and challenges. It is necessary to identify a carrier that will have selective ability to deliver the transgene into the target cell. Once there, duration of gene expression and some control on the level of gene expression are needed. It is necessary to determine that side effects from expressing the transgene in cells other than the target are not worrisome. Each of these challenging and complex considerations must be addressed for each of the known mutations. The levels of complexity are demonstrated for the following examples.

Gene Carrier Vectors

One important consideration is the gene carrier. The most efficient way to introduce a gene into cells is with a viral vector, which is a virus engineered to deliver the transgene without causing disease (the transgene is inserted and viral genes essential to replicate the virus in the host cell or that are recognized by the host immune system are deleted). Viruses are adapted to enter cells, but they may not be able to enter the optimal target cell with high efficiency or specificity. For instance, adenovirus vectors will express transgenes in most cells, but their affinity to hair cells is low (Excoffon et al., 2006; Ishimoto et al., 2002; Kawamoto et al., 2003; Venail et al., 2007). They usually infect nonsensory cells in the organ of Corti, and even for that goal, they need to be introduced into the scala media (endolymph), which is a difficult task in the human ear.

Gene Size

Another major challenge in the development of gene transfer technology is gene size. Some genes (e.g., myosins) are very large, and their size exceeds the space available for accepting a transgene in the virus. This difficulty is now being addressed by improved viral vector technology, where genes are spliced into two (or more) parts, each inserted into a viral vector and the transgenes are designed to fuse in the cell into a functional protein once they are synthesized by the transgene (Koo et al., 2014).

Side Effects of Viruses

A third challenge is to minimize side effects of using viruses. Experimental animals often are raised in pathogen-free environments, and their immune systems are not sensitized to viruses. Experiments using two sequential treatments of an adenovirus in guinea pig have shown that the second application induces a severe immune response (Ishimoto et al., 2003). Because humans usually have had exposure to adenoviruses, the risk of a severe immune response, even if the viral vector cannot replicate, should be considered. That being said, there is reassuring evidence from other systems that viral vectors, when injected into small and relatively isolated body areas such as the eye, do not elicit a detrimental immune response (Zhang et al., 2015).

Other limitations for use of viruses for gene delivery are related to the inability to regulate duration or levels of gene expression. Duration may be too long for some applications and too short for others. When long-term gene expression is needed, there is a risk that the transduced cells may die (due to turnover or the presence of the virus), leading to premature cessation of gene expression.

These are just a few examples of the complex issues involved in designing and attempting phenotypic rescue. While these appear monumental, the availability of sequenced genes and the mouse models, as well as the ability to generate those not yet available, are instrumental in testing ways to advance therapies for hereditary inner ear disease. Indeed, with the knowledge of the underlying molecular genetic mechanisms that cause inner ear disease, the possibility for designing and implementing novel therapeutics is becoming increasingly more likely. A few seminal papers have been published in recent years showing first steps toward clinically feasible therapies. These exciting data are summarized below.

Animal Models and Progress in Treatments

Mice are the animal of choice for modeling human hereditary inner ear diseases. In some cases, a mutation appears spontaneously in a mouse colony and later is identified as a mutation in a gene that also causes human phenotypes (Mochizuki et al., 2010). In other cases, mice with a specific mutation are deliberately produced in a lab (Hardisty-Hughes et al., 2010). In most cases, the mutant mice exhibit some features of the human phenotype, but not all. Although the homologous gene is targeted, several factors make it unlikely that the mutant mouse will be a perfect model for the human disease, even in cases of loss of function mutations. Differences between mice and humans may include the severity of the disease, the need for homozygosity to lead to a phenotype, and the structural and functional manifestations of the disease. Nevertheless, progress has been made using mice that model human hereditary inner ear disease to develop gene-based interventions for alleviating the disease.

The first set of examples involves the gene that is the most commonly cause of human hereditary deafness, *GJB2*, which encodes for the gap junction protein connexin26 (Cx26). Several mouse models for *GJB2* have been generated for research (Cohen-Salmon et al., 2002; Crispino et al., 2011; Kamiya et al., 2014; Takada et al., 2014; Wang et al., 2009). Most of them appear to have a cochlear phenotype that is more severe than in most human patients, with progressive hair cell degeneration as well as loss of auditory neurons. In contrast, in the majority of the human patients, hearing is impaired but stable, and the auditory neurons survive, making these ears excellent candidates for cochlear implant therapy. Vestibular involvement in these mouse mutants appears to be minimal or absent, similar to the condition in humans (Lee et al., 2015).

In some *GJB2* mutants, there were attempts to accomplish a phenotypic rescue. In one early study, a mouse with a dominant-negative mutation of the *GJB2* gene was treated with siRNA to downregulate production of the abnormal protein that underlies the disease. This procedure partially improved hearing in these mouse ears (Maeda et al., 2005). More recent studies have attempted to use gene therapy in Cx26 mutant mice. In one study, the model mouse featured a deletion of Cx26 mediated by Foxg1 cre (Foxg1-cCx26KO). An AAV vector with the wild-type gene was injected into the scala media of early postnatal conditional *Gjb2* knockout, resulting in restoration of the gap junction network in the organ of Corti and rescue of the hair cells and the neurons (Yu et al., 2014). Thresholds, however, were not improved by the procedure. The next tasks should be to improve the procedure to also accomplish functional restoration of hearing with gene transfer and to have similar results in mature ears.

Using another mouse mutant with deleted Cx26 gene (mice were designed to express Cre recombinase under the control of the P0 promoter, leading to deletion of Cx26 in a large number of supporting cells: Cx26fl/flP0-Cre), similar gene transfer procedure rescued cells, restored gap junctions, and also improved hearing thresholds. These results were obtained in neonatal (immature) mice, which are much less developed than newborn humans. For clinically feasible approaches, this rescue needs to be accomplished in mature ears, a result not yet demonstrated. Nevertheless, this study again demonstrated that progress in gene therapy for hereditary deafness is advancing but not ready as yet for clinical use (Iizuka et al., 2015).

Other sets of experiments have dealt with less common mutations. *Vglut3* encodes a vesicular glutamate transporter, which is needed for the transport of the neurotransmitter glutamate into secretory vesicles in inner hair cells (Takamori et al., 2002). When VGLUT3 in the inner hair cells is absent due to a mutation, glutamate is not released by inner hair cells, and auditory neurons do not depolarize in response to sound, leading to severe-to-profound hearing loss (Seal et al., 2008). Phenotypic rescue was attempted for this mouse model by using a viral vector (AAV1) to insert the wild-type mouse *Vglut3* gene into cochlear inner hair cells (Akil et al., 2012). When this procedure was performed in neonatal mice, the ears demonstrated nearly complete rescue of ABR thresholds, and the morphology of the synaptic region, also affected by the mutation, was partly restored. Performing the same treatment on mature mice is the next major challenge to be met to make this therapy applicable to human ears.

Mutations in the transmembrane channel-like 1 (TMC1) protein, a component of the transduction channel in hair cells, cause DFNB7/11 and DFNA36, which are autosomal recessive and dominant deafness, respectively (Kurima et al., 2002; Scott et al., 1998; Vreugde et al., 2002). Mice that model these mutations exhibit severe hearing loss. Inserting the wild-type gene into hair cells of these mice with AAV vectors at a stage when the ear is developing has resulted in hearing improvement (Askew et al., 2015). Once this therapy advances past the next challenge, accomplishing this task in mature ears, it can also be considered a treatment for hearing restoration patients who carry TMC1 mutations.

The effects of the mutations mentioned are localized mostly to the sensory epithelium (organ of Corti) and to specific cell types in that region (supporting cells for GJB2; hair cells for the other mutations listed above). Gene therapy for a mutation affecting the stria vascularis also has recently been applied to mice model of Jervell and Lange-Nielsen (JLN) syndrome (Tranebjaerg et al., 1993). JLN patients are deaf due to a mutation in the potassium channel subunit; the mouse model (Kcnq1(−/−) mice) exhibits a similar deafness phenotype. Inserting the wild-type gene using an AAV vector into the marginal cells of the stria vascularis cells of developing mice resulted in correction of structural pathologies and improvement of ABR thresholds in these animals (Chang et al., 2015). In addition to its specific importance for JLN syndrome patients, these data demonstrate that other mutations affecting the stria vascularis can be addressed therapeutically with similar gene transfer approaches.

Next Steps

The main hurdle in translating these therapies from mice to patients is to understand what additional steps are necessary to make these approaches successful in mature ears. In addition, it is also necessary to verify the long-term effects, the lack of side effects, and the safety of the viral vectors. While these tasks are not trivial, the progress so far accomplished is a proof of principle that phenotypic rescue for hereditary hearing loss is feasible, especially in ears where the cellular components required for hearing are present despite the mutation. In other types of genetic deafness, where cells degenerate, cell replacement therapies will also be needed. These genetic diseases are likely to need more complex therapy than replacement of a single gene, but progress is being made.

At the technical level, it would be desirable to have transgene expression restricted to the target cell. That could potentially be accomplished by driving expression of the vector with promoter that is active in the target cell and not elsewhere. Another option is to design vectors that transduce the target cells selectively. Progress in gene transfer vector design may avail novel and advanced options to enhance therapeutic work with transgenes in the inner ear and elsewhere (O'Reilly et al., 2014).

Augmentation of Cochlear Implantation (CI)

Background

The CI prosthesis is widely and successfully used around the globe to treat severe-to-profound hearing loss at all ages. The different components of the prosthesis have been improved by extensive engineering over many years. More recently, as tissue engineering procedures are gaining momentum and feasibility, attempts are also being made to enhance the biology or health of the receiving cochlea in order to improve the outcome of hearing with the prosthesis. It is hypothesized that cochlear "health" can be improved by procedures that address specific areas or subtissues in the cochlea (Pfingst, Zhou, et al., 2015). Examples for these cochlear elements include hair cells, SGNs, peripheral fibers and myelination of neurons, and connective tissue in the cochlear fluid spaces. Experiments have demonstrated that some of these elements correlate with performance with the implant (Landry et al., 2013; Pfingst, Hughes, et al., 2015). The extent of surviving hair cells (and residual hearing) usually correlates directly with performance (Miranda et al., 2014). Evidence for a correlation between SGN density and performance has been less conclusive, with suggestions that a minimum "threshold" number of neurons is clearly needed, but above this threshold, the differences in performance are not significant (Fayad et al., 2006). It can be hypothesized that less connective tissue and better preservation of myelin will contribute positively to the performance with the implant and that the presence of nerve fibers in proximity to the electrodes will similarly improve outcomes.

In recognition of the influence of these elements of cochlear health on CI outcomes, efforts have been invested in "hearing preservation" associated with CI (Huarte et al., 2014; Mowry et al., 2012). Parameters that have been experimented with include length of the implant electrode (depth of insertion into the scala tympani), flexibility of the material, treatment with steroid, use of lubricating Healon to prevent cochlear trauma while inserting the electrode, and more (Chang et al., 2009; Ramos et al., 2015). In parallel, recent advances in tissue engineering, such as gene-transfer approaches, electrodes that elute reagents, or mini-osmotic pumps are presenting opportunities for additional measures to enhance the health of the cochlea. Such biologically based therapies may further improve implant function. Here, we review several approaches that are being developed on animal models for enhancing the conditions in the receiving ear.

Animal Experiments

Most experiments for enhancing CI outcomes in deaf ears have been aimed at maximizing survival of SGNs and improving their functional status. These experiments utilized several animal models. The most common small animal for combining CI, stimulation, and measurements is the guinea pig (Chouard, 2015; Pfingst et al., 2011), although experiments in cats (Beitel et al., 2000) and other animals have also been performed. When it is necessary to address augmentation of cochlear health in hereditary deafness models for mutations that are commonly treated with CIs, mice need to be used. While they provide the best possible model, the models are not a perfect match for the human disease, and the use of a full-scale cochlear implant is excluded due to the small size of the mouse ear. Nevertheless, a short electrode with one stimulating site has been successfully placed in a mouse ear (Mistry et al., 2014).

Enhancing survival of SGNs in ears that are depleted of hair cells has been among the more frequently studied areas for improving the health of the deaf cochlea. The animal models are suboptimal because in many human ears, neurons do not degenerate following hair cell loss, whereas the procedures for eliminating hair cells in animals usually lead to nerve degeneration as an unwanted consequence. However, these models do allow for testing protocols aimed at enhancing SGN density in deafened ears. The most commonly used molecules are the neurotrophins (Budenz et al., 2012). Early experiments utilized mini-osmotic pumps to infuse neurotrophins into the cochlea and showed that both hair cells and neurons can be rescued (Miller et al., 1997). When gene vectors became available, experimentation has shifted to gene therapy approaches, which were also efficacious in preserving neurons. Both BDNF and NT3 were shown to have positive effects (Budenz et al., 2015; Nakaizumi et al., 2004; Shibata et al., 2010; Wise et al., 2010).

In addition to preserving SGN cell bodies, it also has been theorized that providing a nerve ending closer to the site of stimulation would reduce impedance and current spread and lead to better performance with the implant. To that end, neurotrophins have been used to attract auditory nerve fibers to sprout and grow toward the site of the implant (Budenz et al., 2015; Shibata et al., 2010; Wise et al., 2010; Wise et al., 2011). Psychophysical and electrophysiologic measurements have shown some benefits of this procedure, but a direct casuality between the presence of regenerated peripheral fibers and the functional responses has not been conclusively demonstrated. Because the fibers that grow into the basilar membrane area are not myelinated, it is possible that the speed of conductivity of neural signals is too slow to be relevant. Future approaches will need to attempt to induce myelination of these fibers, also.

Neurotrophins can also help preserve hair cells and potentially may be used for hearing preservation. Other approaches for preserving hair cells while implanting an ear include the use of dexamethasone, a protocol that is already a clinical routine of many surgeons and serves more than one goal (Douchement et al., 2015; Eshraghi et al., 2011). The ability of neurotrophins to preserve neurons and induce peripheral fiber sprouting also has been demonstrated in models of genetic deafness (Fukui et al., 2012; Takada et al., 2014).

Before these techniques can be used in the clinic, it will be necessary to ascertain the lack of side effect of the viral vector, the duration of positive effects, and the actual functional benefit derived from the treatment. Combining neurotrophin therapy with other means to enhance myelination and preserve hair cells may prove to provide critical enhancement of function with the cochlear implant. In addition, it may be needed to reduce the extent of connective tissue growth in the scala tympani in deaf ears that receive a cochlear implant. Experiments along these lines will likely address components of the immune system, which has been linked to this response (Aftab et al., 2010).

Next Steps

Many of the innovative techniques discussed may be clinically relevant in concert with CI therapy because a surgery is being performed, already, and can be combined with additional therapeutic measures. However, before applying these protocols to enhance cochlear health in the clinic, it will be necessary to better characterize if and how each of the protocols improves outcomes of the prosthesis. Prevention of unwanted side effects and improvements in technology for gene delivery will also be necessary. Once efficacy for enhancing cochlear health is proven, performing these therapies in tandem with CI insertion is somewhat more applicable than some of the other procedures described in this chapter, because a surgery is being performed anyway, providing access and opportunity to insert reagents into the cochlea.

Present and Future Implications in the Audiology Clinic

Because some of the animal studies described above showed favorable outcomes of gene therapy in treating hearing loss, the following steps should be preparing for applying these therapies to humans. However, there are a few important issues to be considered before beginning clinical trials. First, side effects of the therapy need to be evaluated. Potential side effects could be from the vector (actions by it or reactions to it) or the surgical method to deliver the vector. Expressing the transgene in the wrong cells can also be detrimental. Growth factors and agents inducing proliferation pose a risk of tumor formation. Immune response to the viral vector is always a concern. Duration of gene expression and regulation of quantity are also important, and hard to regulate with current generation vectors. Currently, the most effective gene transfection in the cochlea requires invasive surgical methods that disrupt the bony capsule or round window; however, there are reports of some success in transfecting the sensory epithelium with noninvasive approaches (Jero et al., 2001; Konishi et al., 2008; Mukherjea et al., 2010). This may seem to be a long list of serious considerations, but successes to this point provide hope for their eventual resolution, which would provide us new and promising avenues for treating hearing loss.

Acknowledgments. We thank Drs. Gregory Basura, Takaomi Kurioka, and Donald Swiderski for helpful comments on the manuscript. Our work is supported by the R. Jamison and Betty Williams Professorship, the Organogenesis Research Team Program from the Center for Organogenesis at the University of Michigan, and NIH/NIDCD grants R01-DC010412, R01 DC009410, T32-DC005356, and P30-DC05188.

References

Aftab, S., Semaan, M. T., Murray, G. S., & Megerian, C. A. (2010). Cochlear implantation outcomes in patients with autoimmune and immune-mediated inner ear disease. *Otology & Neurootology, 31,* 1337–1342.

Akil, O., Seal, R. P., Burke, K., Wang, C., Alemi, A., During, M., & Lustig, L. R. (2012). Restoration of hearing in the VGLUT3 knockout mouse using virally mediated gene therapy. *Neuron, 75,* 283–293.

Askew, C., Rochat, C., Pan, B., Asai, Y., Ahmed, H., Child, E., & Holt, J. R. (2015). Tmc gene therapy restores auditory function in deaf mice. *Science Translational Medicine, 7,* 295ra108.

Atik, T., Bademci, G., Diaz-Horta, O., Blanton, S. H., & Tekin, M. (2015). Whole-exome sequencing and its impact in hereditary hearing loss. *Genetics Research, 97,* e4.

Atkinson, P. J., Wise, A. K., Flynn, B. O., Nayagam, B. A., Hume, C. R., O'Leary, S. J., & Richardson, R.T. (2012). Neurotrophin gene therapy for sustained neural preservation after deafness. *PLoS One, 7,* e52338.

Baker, T. G., Roy, S., Brandon, C. S., Kramarenko, I. K., Francis, S. P., Taleb, M., Cunningham, L. L. (2015). Heat shock protein-mediated protection against cisplatin-induced hair cell death. *Journal of the Association for Research in Otolaryngology, 16,* 67–80.

Atkinson, P. J., Wise, A .K., Flynn, B. O., Nayagam, B. A., & Richardson, R.T. (2014). Hair cell regeneration after ATOH1 gene therapy in the cochlea of profoundly deaf adult guinea pigs. *PLoS One, 9,* e102077.

Beitel, R. E., Snyder, R. L., Schreiner, C. E., Raggio, M. W., & Leake, P. A. (2000). Electrical cochlear stimulation in the deaf cat: Comparisons between psychophysical and central auditory neuronal thresholds. *Journal of Neurophysiology, 83,* 2145–2162.

Ben-Yosef, T., Belyantseva, I. A., Saunders, T. L., Hughes, E. D., Kawamoto, K., Van Itallie, C. M., & Friedman, T. B. (2003). Claudin 14 knockout mice, a model for autosomal recessive deafness DFNB29, are deaf due to cochlear hair cell degeneration. *Human Molecular Genetics, 12,* 2049–2061.

Bielefeld, E. C. (2013). Reduction in impulse noise-induced permanent threshold shift with intracochlear application of an NADPH oxidase inhibitor. *Journal of the American Academy of Audiology, 24,* 461–473.

Bowers, W. J., Chen, X., Guo, H., Frisina, D. R., Federoff, H. J., & Frisina, R. D. (2002). Neurotrophin-3 transduction attenuates cisplatin spiral ganglion neuron ototoxicity in the cochlea. *Molecular Therapy, 6,* 12–18.

Brendel, C., Belakhov, V., Werner, H., Wegener, E., Gartner, J., Nudelman, I., Baasov, T., & Huppke, P. (2011). Readthrough of nonsense mutations in Rett syndrome: Evaluation of novel aminoglycosides and generation of a new mouse model. *Journal of Molecular Medicine (Berlin, Germany), 89,* 389–398.

Brendel, C., Klahold, E., Gartner, J., & Huppke, P. (2009). Suppression of nonsense mutations in Rett syndrome by aminoglycoside antibiotics. *Pediatric Research, 65,* 520–523.

Budenz, C. L., Pfingst, B. E., & Raphael, Y. (2012). The use of neurotrophin therapy in the inner ear to augment cochlear implantation outcomes. *Anatomical Record (Hoboken), 295,* 1896–1908.

Budenz, C. L., Wong, H. T., Swiderski, D. L., Shibata, S. B., Pfingst, B. E., & Raphael, Y. (2015). Differential effects of AAV.BDNF and AAV.Ntf3 in the deafened adult guinea pig ear. *Scientific Reports, 5,* 8619.

Chang, A., Eastwood, H., Sly, D., James, D., Richardson, R., & O'Leary, S. (2009). Factors influencing the efficacy of round window dexamethasone protection of residual hearing post-cochlear implant surgery. *Hearing Research, 255,* 67–72.

Chang, Q., Wang, J., Li, Q., Kim, Y., Zhou, B., Wang, Y., Li, H., & Lin, X. (2015). Virally mediated Kcnq1 gene replacement therapy in the immature scala media restores hearing in a mouse model of human Jervell and Lange-Nielsen deafness syndrome. *EMBO Molecular Medicine, 7,* 1077–1086.

Chen, C., Chen, Y., Wu, P., & Chen, B. (2014). Update on new medicinal applications of gentamicin: evidence-based review. *Journal of the Formosan Medical Association, 113,* 72–82.

Chen, L. F., & Kaye, D. (2009). Current use for old antibacterial agents: Polymyxins, rifamycins, and aminoglycosides. *Infectious Disease Clinics of North America, 23,* 1053–1075, x.

Chen, P., & Segil, N. (1999). p27(Kip1) links cell proliferation to morphogenesis in the developing organ of Corti. *Development, 126,* 1581–1590.

Chen, Y., Huang, W. G., Zha, D. J., Qiu, J. H., Wang, J. L., Sha, S. H., & Schacht, J. (2007). Aspirin attenuates gentamicin ototoxicity: from the laboratory to the clinic. *Hearing Research, 226,* 178–182.

Chirtes, F., & Albu, S. (2014). Prevention and restoration of hearing loss associated with the use of cisplatin. *BioMed Research International, 2014,* 925485.

Chouard, C. H. (2015). The early days of the multi channel cochlear implant: Efforts and achievement in France. *Hearing Research, 322,* 47–51.

Clifford, R. E., & Rogers, R. A. (2009). Impulse noise: Theoretical solutions to the quandary of cochlear protection. *Annals of Otology, Rhinology, and Laryngology, 118,* 417–427.

Cohen-Salmon, M., Ott, T., Michel, V., & Hardelin, J. P., Perfettini, I., Eybalin, M., Petit, C. (2002). Targeted ablation of connexin26 in the inner ear epithelial gap junction network causes hearing impairment and cell death. *Current Biology, 12,* 1106–1111.

Coleman, J. K., Littlesunday, C., Jackson, R., & Meyer, T. (2007). AM-111 protects against permanent hearing loss from impulse noise trauma. *Hearing Research, 226,* 70–78.

Cooper, L. B., Chan, D. K., Roediger, F. C., Shaffer, B. R., Fraser, J. F., Musatov, S., Selesnick, S. H., & Kaplitt, M. G. (2006). AAV-mediated delivery of the caspase inhibitor XIAP protects against cisplatin ototoxicity. *Otology & Neurootology, 27,* 484–490.

Corwin, J. T., & Cotanche, D. A. (1988). Regeneration of sensory hair cells after acoustic trauma. *Science, 240,* 1772–1774.

Crispino, G., Di Pasquale, G., Scimemi, P., Rodriguez, L., Galindo Ramirez, F., De Siati, R. D., & Mammano, F. (2011). BAAV mediated GJB2 gene transfer restores gap junction coupling in cochlear organotypic cultures from deaf Cx26Sox10Cre mice. *PLoS One, 6,* e23279.

Crystal, R. G. (2014). Adenovirus: The first effective in vivo gene delivery vector. *Human Gene Therapy, 25,* 3–11.

Deveraux, Q. L., Takahashi, R., Salvesen, G. S., & Reed, J. C. (1997). X-linked IAP is a direct inhibitor of cell-death proteases. *Nature, 388,* 300–304.

Douchement, D., Terranti, A., Lamblin, J., Salleron, J., Siepmann, F., Siepmann, J., & Vincent, C. (2015). Dexamethasone eluting electrodes for cochlear implantation: Effect on residual hearing. *Cochlear Implants International, 16,* 195–200.

Du, X., Choi, C. H., Chen, K., Cheng, W., Floyd, R. A., & Kopke, R. D. (2011). Reduced formation of oxidative stress biomarkers and migration of mononuclear phagocytes in the cochleae of chinchilla after antioxidant treatment in acute acoustic trauma. *International Journal of Otolaryngology, 2011,* 612690.

Eshraghi, A. A., Dinh, C. T., Bohorquez, J., Angeli, S., Abi-Hachem, R., & Van De Water, T. R. (2011). Local drug delivery to conserve hearing: Mechanisms of action of eluted dexamethasone within the cochlea. *Cochlear Implants International, 12*(Suppl. 1), S51–S53.

Excoffon, K. J., Avenarius, M. R., Hansen, M. R., Kimberling, W. J., Najmabadi, H., Smith, R. J., & Zabner, J. (2006). The coxsackievirus and adenovirus receptor: A new adhesion protein in cochlear development. *Hearing Research, 215,* 1–9.

Fayad, J. N., & Linthicum, F. H., Jr. (2006). Multichannel cochlear implants: Relation of histopathology to performance. *Laryngoscope, 116,* 1310–1320.

Folmer, R. L., Griest, S. E., & Martin, W. H. (2002). Hearing conservation education programs for children: A review. *Journal of School Health, 72,* 51–57.

Forge, A., Li, L., Corwin, J. T., & Nevill, G. (1993). Ultrastructural evidence for hair cell regeneration in the mammalian inner ear. *Science, 259,* 1616–1619.

Fujioka, M., Okano, H., & Edge, A. S. (2015). Manipulating cell fate in the cochlea: A feasible therapy for hearing loss. *Trends in Neuroscience, 38,* 139–144.

Fukui, H., & Raphael, Y. (2013). Gene therapy for the inner ear. *Hearing Research, 297,* 99–105.

Fukui, H., Wong, H. T., Beyer, L. A., Case, B. G., Swiderski, D. L., Di Polo, A., Ryan, A. F., & Raphael, Y. (2012). BDNF gene therapy induces auditory nerve survival and fiber sprouting in deaf Pou4f3 mutant mice. *Scientific Reports, 2,* 838.

Goncalves, M. S., Silveira, A. F., Teixeira, A. R., & Hyppolito, M. A. (2013). Mechanisms of cisplatin ototoxicity: Theoretical review. *Journal of Laryngology and Otology, 127,* 536–541.

Hardisty-Hughes, R. E., Parker, A., & Brown, S. D. (2010). A hearing and vestibular phenotyping pipeline to identify mouse mutants with hearing impairment. *Nature Protocols, 5,* 177–190.

Hawkins, J. E., Jr., Stebbins, W. C., Johnsson, L. G., Moody, D. B., & Muraski, A. (1977). The patas monkey as a model for dihydrostreptomycin ototoxicity. *Acta Otolaryngologica (Stockholm), 83,* 123–129.

Huarte, R. M., & Roland, J. T., Jr. (2014). Toward hearing preservation in cochlear implant surgery. *Current Opinion in Otolaryngology & Head and Neck Surgery, 22,* 349–352.

Iizuka, T., Kamiya, K., Gotoh, S., Sugitani, Y., Suzuki, M., Noda, T., & Ikeda, K. (2015). Perinatal Gjb2 gene transfer rescues hearing in a mouse model of hereditary deafness. *Human Molecular Genetics, 24,* 3651–3661.

Ishimoto, S., Kawamoto, K., Kanzaki, S., & Raphael, Y. (2002). Gene transfer into supporting cells of the organ of Corti. *Hearing Research, 173,* 187–197.

Ishimoto, S., Kawamoto, K., Stover, T., Kanzaki, S., Yamasoba, T., & Raphael, Y. (2003). A glucocorticoid reduces adverse effects of adenovirus vectors in the cochlea. *Audiology & Neurootology, 8,* 70–79.

Izumikawa, M., Batts, S. A., Miyazawa, T., Swiderski, D. L., & Raphael, Y. (2008). Response of the flat cochlear epithelium to forced expression of Atoh1. *Hearing Research, 240,* 52–56.

Jero, J., Coling, D. E., & Lalwani, A. K. (2001). The use of Preyer's reflex in evaluation of hearing in mice. *Acta Otolaryngologica, 121,* 585–589.

Jie, H., Tao, S., Liu, L., Xia, L., Charko, A., Yu, Z., & Wang, J. (2015). Cochlear protection against cisplatin by viral transfection of X-linked inhibitor of apoptosis protein across round window membrane. *Gene Therapy, 22,* 546–552.

Johnsson, L. G., Hawkins, J. E., Jr., Kingsley, T. C., Black, F. O., & Matz, G. J. (1981). Aminoglycoside-induced cochlear pathology in man. *Acta Otolaryngologica Supplement, 383,* 1–19.

Kamiya, K., Yum, S. W., Kurebayashi, N., Muraki, M., Ogawa, K., Karasawa, K., &

Ikeda, K. (2014). Assembly of the cochlear gap junction macromolecular complex requires connexin 26. *Journal of Clinical Investigation, 124,* 1598–1607.

Kanzaki, S., Beyer, L. A., Swiderski, D. L., Izumikawa, M., Stover, T., Kawamoto, K., & Raphael, Y. (2006). p27(Kip1) deficiency causes organ of Corti pathology and hearing loss. *Hearing Research, 214,* 28–36.

Karlsson, K. K., Flock, B., & Flock, A. (1991). Ultrastructural changes in the outer hair cells of the guinea pig cochlea after exposure to quinine. *Acta Otolaryngologica (Stockholm)., 111,* 500–505.

Kawamoto, K., Ishimoto, S., Minoda, R., Brough, D. E., & Raphael, Y. (2003). Math1 gene transfer generates new cochlear hair cells in mature guinea pigs in vivo. *Journal of Neuroscience, 23,* 4395–4400.

Kawamoto, K., Sha, S. H., Minoda, R., Izumikawa, M., Kuriyama, H., Schacht, J., & Raphael, Y. (2004). Antioxidant gene therapy can protect hearing and hair cells from ototoxicity. *Molecular Therapy, 9,* 173–181.

Kim, Y. H., & Raphael, Y. (2007). Cell division and maintenance of epithelial integrity in the deafened auditory epithelium. *Cell Cycle, 6,* 612–619.

Konishi, M., Kawamoto, K., Izumikawa, M., Kuriyama, H., & Yamashita, T. (2008). Gene transfer into guinea pig cochlea using adeno-associated virus vectors. *The Journal of Gene Medicine, 10,* 610–618.

Koo, T., Popplewell, L., Athanasopoulos, T., & Dickson, G. (2014). Triple trans-splicing adeno-associated virus vectors capable of transferring the coding sequence for full-length dystrophin protein into dystrophic mice. *Human Gene Therapy, 25,* 98–108.

Kopke, R. D., Jackson, R. L., Coleman, J. K., Liu, J., Bielefeld, E. C., & Balough, B. J. (2007). NAC for noise: from the bench top to the clinic. *Hearing Research, 226,* 114–125.

Ku, Y. C., Renaud, N. A., Veile, R. A., Helms, C., Voelker, C. C., Warchol, M. E., & Lovett, M. (2014). The transcriptome of utricle hair cell regeneration in the avian inner ear. *Journal of Neuroscience, 34,* 3523–3535.

Kujawa, S. G., & Liberman, M. C. (2015). Synaptopathy in the noise-exposed and aging cochlea: Primary neural degeneration in acquired sensorineural hearing loss. *Hearing Research, 330,* 191–199.

Kurima, K., Peters, L. M., Yang, Y., Riazuddin, S., Ahmed, Z. M., Naz, S., & Griffith, A. J. (2002). Dominant and recessive deafness caused by mutations of a novel gene, TMC1, required for cochlear hair-cell function. *Nature Genetics, 30,* 277–284.

Landry, T. G., Fallon, J. B., Wise, A. K., & Shepherd, R. K. (2013). Chronic neurotrophin delivery promotes ectopic neurite growth from the spiral ganglion of deafened cochleae without compromising the spatial selectivity of cochlear implants. *Journal of Comparative Neurology, 521,* 2818–2832.

Le Prell, C. G., Hughes, L. F., & Miller, J. M. (2007). Free radical scavengers vitamins A, C, and E plus magnesium reduce noise trauma. *Free Radical Biology & Medicine, 42,* 1454–1463.

Lee, M. Y., Takada, T., Takada, Y., Kappy, & M. D., Beyer, L. A., Swiderski, D. L., Raphael, Y. (2015). Mice with conditional deletion of Cx26 exhibit no vestibular phenotype despite secondary loss of Cx30 in the vestibular end organs. *Hearing Research, 328,* 102–112.

Liu, Y., Okada, T., Shimazaki, K., Sheykholeslami, K., Nomoto, T., Muramatsu, S., & Ozawa, K. (2008). Protection against aminoglycoside-induced ototoxicity by regulated AAV vector-mediated GDNF gene transfer into the cochlea. *Molecular Therapy, 16,* 474–480.

Lowenheim, H., Furness, D. N., Kil, J., Zinn, C., Gultig, K., Fero, M. L., & Zenner, H. P. (1999). Gene disruption of p27(Kip1) allows cell proliferation in the postnatal and adult organ of Corti. *Proceedings of the National Academy of Sciences U. S. A., 96,* 4084–4088.

Lundstrom, K. (2004). Gene therapy applications of viral vectors. *Technology in Cancer Research & Treatment, 3,* 467–477.

Maass, J. C., Gu, R., Basch, M. L., Waldhaus, J., Lopez, E. M., Xia, A., & Groves, A. K. (2015). Changes in the regulation of the Notch signaling pathway are temporally correlated with regenerative failure in the mouse cochlea. *Frontiers in Cellular Neuroscience, 9,* 110.

Maeda, Y., Fukushima, K., Nishizaki, K., & Smith, R. J. (2005). In vitro and in vivo suppression of GJB2 expression by RNA interference. *Human Molecular Genetics, 14,* 1641-1650.

Marshak, T., Steiner, M., Kaminer, M., Levy, L., & Shupak, A. (2014). Prevention of cisplatin-induced hearing loss by intratympanic dexamethasone: A randomized controlled study. *Otolaryngology-Head and Neck Surgery, 150,* 983–990.

Miller, J. M., Chi, D. H., O'Keeffe, L. J., Kruszka, P., Raphael, Y., & Altschuler, R. A. (1997). Neurotrophins can enhance spiral ganglion cell survival after inner hair cell loss. *International Journal of Developmental Neuroscience, 15,* 631–643.

Minoda, R., Izumikawa, M., Kawamoto, K., Zhang, H., & Raphael, Y. (2007). Manipulating cell cycle regulation in the mature cochlea. *Hearing Research, 232,* 44–51.

Miranda, P. C., Sampaio, A. L., Lopes, R. A., Ramos Venosa, A., & de Oliveira, C. A. (2014). Hearing preservation in cochlear implant surgery. *International Journal of Otolaryngology, 2014,* ID 468515, 1–6.

Mistry, N., Nolan, L. S., Saeed, S. R., Forge, A., & Taylor, R. R. (2014). Cochlear implantation in the mouse via the round window: Effects of array insertion. *Hearing Research, 312,* 81–90.

Mizutari, K., Fujioka, M., Hosoya, M., Bramhall, N., Okano, H. J., Okano, H., & Edge, A. S. (2013). Notch inhibition induces cochlear hair cell regeneration and recovery of hearing after acoustic trauma. *Neuron, 77,* 58–69.

Mochizuki, E., Okumura, K., Ishikawa, M., Yoshimoto, S., Yamaguchi, J., Seki, Y., & Kikkawa, Y. (2010). Phenotypic and expression analysis of a novel spontaneous myosin VI null mutant mouse. *Experimental Animals, 59,* 57–71.

Mowry, S. E., Woodson, E., & Gantz, B. J. (2012). New frontiers in cochlear implantation: acoustic plus electric hearing, hearing preservation, and more. *Otolaryngology Clinics of North America, 45,* 187–203.

Mukherjea, D., Jajoo, S., Kaur, T., Sheehan, K. E., Ramkumar, V., & Rybak, L. P. (2010). Transtympanic administration of short interfering (si)RNA for the NOX3 isoform of NADPH oxidase protects against cisplatin-induced hearing loss in the rat. *Antioxidant & Redox Signaling, 13,* 589–598.

Mukherjea, D., Jajoo, S., Whitworth, C., Bunch, J. R., Turner, J. G., Rybak, L. P., & Ramkumar, V. (2008). Short interfering RNA against transient receptor potential vanilloid 1 attenuates cisplatin-induced hearing loss in the rat. *Journal of Neuroscience, 28,* 13056–13065.

Nakaizumi, T., Kawamoto, K., Minoda, R., & Raphael, Y. (2004). Adenovirus-mediated expression of brain-derived neurotrophic factor protects spiral ganglion neurons from ototoxic damage. *Audiology & Neurootology, 9,* 135–143.

Noushi, F., Richardson, R. T., Hardman, J., Clark, G., & O'Leary, S. (2005). Delivery of neurotrophin-3 to the cochlea using alginate beads. *Otology & Neurootology, 26,* 528–533.

O'Reilly, M., Federoff, H. J., Fong, Y., Kohn, D. B., Patterson, A. P., Ahmed, N., & Corrigan-Curay, J. (2014). Gene therapy: Charting a future course—summary of a National Institutes of Health Workshop, April 12, 2013. *Human Gene Therapy, 25,* 488–497.

Ojano-Dirain, C. P., Antonelli, P. J., & Le Prell, C. G. (2014). Mitochondria-targeted antioxidant MitoQ reduces gentamicin-induced ototoxicity. *Otology & Neurootology, 35,* 533–539.

Pfannenstiel, S. C., Praetorius, M., Plinkert, P. K., Brough, D. E., & Staecker, H. (2009).

Bcl-2 gene therapy prevents aminoglycoside-induced degeneration of auditory and vestibular hair cells. *Audiology & Neurootology, 14,* 254–266.

Pfingst, B. E., Hughes, A. P., Colesa, D. J., Watts, M. M., Strahl, S. B., & Raphael, Y. (2015). Insertion trauma and recovery of function after cochlear implantation: Evidence from objective functional measures. *Hearing Research, 330 (Part A),* 98–105.

Pfingst, B. E., Bowling, S. A., Colesa, D. J., Garadat, S. N., Raphael, Y., Shibata, S. B., & Zhou, N. (2011). Cochlear infrastructure for electrical hearing. *Hearing Research, 28,* 65–73.

Pfingst, B. E., Zhou, N., Colesa, D. J., Watts, M. M., Strahl, S. B., Garadat, S. N., & Zwolan, T. A. (2015). Importance of cochlear health for implant function. *Hearing Research, 322,* 77–88.

Pirvola, U., Ylikoski, J., Palgi, J., Lehtonen, E., Arumäe, U., & Saarma, M. (1992). Brain-derived neurotrophic factor and neurotrophin 3 mRNAs in the peripheral target fields of developing inner ear ganglia. *Proceedings of the National Academy of Sciences, U. S. A., 89,* 9915–9919.

Popescu, A. C., Sidorova, E., Zhang, G., & Eubanks, J. H. (2010). Aminoglycoside-mediated partial suppression of MECP2 nonsense mutations responsible for Rett syndrome in vitro. *Journal of Neuroscience Research, 88,* 2316–2324.

Ramos, B. F., Tsuji, R. K., Bento, R. F., Goffi-Gomez, M. V., Ramos, H. F., Samuel, P. A., & Brito, R. (2015). Hearing preservation using topical dexamethasone alone and associated with hyaluronic acid in cochlear implantation. *Acta Otolaryngologica, 135,* 473–477.

Roberson, D. W., & Rubel, E. W. (1994). Cell division in the gerbil cochlea after acoustic trauma. *American Journal of Otology, 15,* 28–34.

Ruan, R. S., Leong, S. K., Mark, I., & Yeoh, K. H. (1999). Effects of BDNF and NT-3 on hair cell survival in guinea pig cochlea damaged by kanamycin treatment. *NeuroReport, 10,* 2067–2071.

Ryals, B. M., & Rubel, E. W. (1988). Hair cell regeneration after acoustic trauma in adult Coturnix quail. *Science, 240,* 1774–1776.

Sage, C., Huang, M., Karimi, K., Gutierrez, G., Vollrath, M. A., Zhang, D. S., & Chen, Z. Y. (2005). Proliferation of functional hair cells in vivo in the absence of the retinoblastoma protein. *Science, 307,* 1114–1118.

Schacht, J. (1999). Biochemistry and pharmacology of aminoglycoside-induced hearing loss. *Acta Physiologica, Pharmacologica, et Therapeutica Latinoamericana, 49,* 251–256.

Scott, D. A., Greinwald, J. H., Jr., Marietta, J. R., Drury, S., Swiderski, R. E., Vinas, A., & Sheffield, V. C. (1998). Identification and mutation analysis of a cochlear-expressed, zinc finger protein gene at the DFNB7/11 and dn hearing-loss loci on human chromosome 9q and mouse chromosome 19. *Gene, 215,* 461–469.

Seal, R. P., Akil, O., Yi, E., Weber, C. M., Grant, L., Yoo, J., & Edwards, R. H. (2008). Sensorineural deafness and seizures in mice lacking vesicular glutamate transporter 3. *Neuron, 57,* 263–275.

Seidman, M. D., Tang, W., Bai, V. U., Ahmad, N., Jiang, H., Media, J., & Standring, R. T. (2013). Resveratrol decreases noise-induced cyclooxygenase-2 expression in the rat cochlea. *Otolaryngology-Head and Neck Surgery, 148,* 827–833.

Shibata, S. B., Cortez, S. R., Beyer, L. A., Wiler, J. A., Di Polo, A., Pfingst, B. E., & Raphael, Y. (2010). Transgenic BDNF induces nerve fiber regrowth into the auditory epithelium in deaf cochleae. *Experimental Neurology, 223,* 464–472.

Shoji, F., Miller, A. L., Mitchell, A., Yamasoba, T., Altschuler, R. A., & Miller, J. M. (2000). Differential protective effects of neurotrophins in the attenuation of noise-induced hair cell loss. *Hearing Research, 146,* 134–142.

Slack, J. M., Colleypriest, B. J., Quinlan, J. M., Yu, W. Y., Farrant, M. J., & Tosh, D. (2010). Barrett's metaplasia: molecular mechanisms and nutritional influences. *Biochemical Society Transactions, 38,* 313–319.

Song, B. B., & Schacht, J. (1996). Variable efficacy of radical scavengers and iron chelators to attenuate gentamicin ototoxicity in guinea pig in vivo. *Hearing Research, 94,* 87–93.

TTakada, Y., Beyer, L. A., Swiderski, D. L., O'Neal, A. L., Prieskorn, D. M., Shivatzki, S., & Raphael, Y. (2014). Connexin 26 null mice exhibit spiral ganglion degeneration that can be blocked by BDNF gene therapy. *Hearing Research, 309,* 124–135.

akada, Y., Takada, T., Lee, M. Y., Swiderski, D. L., Kabara, L. L., Dolan, D. F., & Raphael, Y. (2015). Ototoxicity-induced loss of hearing and inner hair cells is attenuated by HSP70 gene transfer. Molecular therapy. *Methods & Clinical Development, 2,* 15019.

Takamori, S., Malherbe, P., Broger, C., & Jahn, R. (2002). Molecular cloning and functional characterization of human vesicular glutamate transporter 3. *EMBO Reports, 3,* 798–803.

Taleb, M., Brandon, C. S., Lee, F. S., Harris, K. C., Dillmann, W. H., & Cunningham, L. L. (2009). Hsp70 inhibits aminoglycoside-induced hearing loss and cochlear hair cell death. *Cell Stress Chaperones, 14,* 427–437.

Thomas, E. D., Cruz, I. A., Hailey, D. W., & Raible, D. W. (2015). There and back again: development and regeneration of the zebrafish lateral line system. Wiley interdisciplinary reviews. *Developmental Biology, 4,* 1–16.

Tranebjaerg, L., Samson, R. A., & Green, G. E. (1993). Jervell and Lange-Nielsen syndrome. In R. A. Pagon, M. P. Adam, H. H. Ardinger, S. E. Wallace, A. Amemiya, L. J. H. Bean, . . . K. Stephens (Eds.), *Gene reviews.* Seattle, WA: University of Washington.

Venail, F., Wang, J., Ruel, J., Ballana, E., Rebillard, G., Eybalin, M., & Puel, J. L. (2007). Coxsackie adenovirus receptor and alpha nu beta3/alpha nu beta5 integrins in adenovirus gene transfer of rat cochlea. *Gene Therapy, 14,* 30–37.

Verbeek, J. H., Kateman, E., Morata, T. C., Dreschler, W., & Sorgdrager, B. (2009). Interventions to prevent occupational noise induced hearing loss. *Cochrane Database Systematic Reviews,* CD006396, 10, 1–109.

Verma, V., Samanthapudi, K., & Raviprakash, R. (2015). Classic studies on the potential of stem cell neuroregeneration. *Journal of the History of the Neurosciences, 26,* 1–19.

Vreugde, S., Erven, A., Kros, C. J., Marcotti, W., Fuchs, H., Kurima, K., & Steel, K. P. (2002). Beethoven, a mouse model for dominant, progressive hearing loss DFNA36. *Nature Genetics, 30,* 257–258.

Wan, G., Gomez-Casati, M. E., Gigliello, A. R., Liberman, M. C., & Corfas, G. (2014). Neurotrophin-3 regulates ribbon synapse density in the cochlea and induces synapse regeneration after acoustic trauma. *eLife 3,* e03564, 1–18.

Wang, Y., Chang, Q., Tang, W., Sun, Y., Zhou, B., Li, H., & Lin, X. (2009). Targeted connexin26 ablation arrests postnatal development of the organ of Corti. *Biochemical and Biophysical Research Communications, 385,* 33–37.

Warnock, J. N., Daigre, C., & Al-Rubeai, M. (2011). Introduction to viral vectors. *Methods in Molecular Biology, 737,* 1–25.

Wise, A. K., Hume, C. R., Flynn, B. O., Jeelall, Y. S., Suhr, C. L., Sgro, B. E., & Richardson, R.T. (2010). Effects of localized neurotrophin gene expression on spiral ganglion neuron resprouting in the deafened cochlea. *Molecular Therapy, 18,* 1111–1122.

Wise, A. K., Tu, T., Atkinson, P. J., Flynn, B. O., Sgro, B. E., Hume, C., & Richardson, R. T. (2011). The effect of deafness duration on neurotrophin gene therapy for spiral ganglion neuron protection. *Hearing Research, 278,* 69–76.

Xue, X., Mutyam, V., Tang, L., Biswas, S., Du, M., Jackson, L. A., & Rowe, S. M.

(2014). Synthetic aminoglycosides efficiently suppress cystic fibrosis transmembrane conductance regulator nonsense mutations and are enhanced by ivacaftor. *American Journal Respiratory Cell and Molecular Biology, 50,* 805–816.

Yagi, M., Kanzaki, S., Kawamoto, K., Shin, B., Shah, P. P., Magal, E., Raphael, Y. (2000). Spiral ganglion neurons are protected from degeneration by GDNF gene therapy. *Journal of the Association for Research in Otolaryngology, 1,* 315–325.

Yagi, M., Magal, E., Sheng, Z., Ang, K. A., & Raphael, Y. (1999). Hair cell protection from aminoglycoside ototoxicity by adenovirus-mediated overexpression of glial cell line-derived neurotrophic factor. *Human Gene Therapy, 10,* 813–823.

Yamashita, D., Jiang, H. Y., Le Prell, C. G., Schacht, J., & Miller, J. M. (2005). Post-exposure treatment attenuates noise-induced hearing loss. *Neuroscience, 134,* 633–642.

Yu, Q., Chang, Q., Liu, X., Wang, Y., Li, H., Gong, S., Ye, K., & Lin, X. (2013). Protection of spiral ganglion neurons from degeneration using small-molecule TrkB receptor agonists. *Journal of Neuroscience, 33,* 13042–13052.

Yu, Q., Wang, Y., Chang, Q., Wang, J., Gong, S., Li, H., & Lin, X. (2014). Virally expressed connexin26 restores gap junction function in the cochlea of conditional Gjb2 knockout mice. *Gene Therapy, 21,* 71–80.

Zhang, J. X., Wang, N. L., & Lu, Q. J. (2015). Development of gene and stem cell therapy for ocular neurodegeneration. *International Journal of Ophthalmology, 8,* 622–630.

CHAPTER 10

The Olivocochlear System

A Current Understanding of Its Molecular Biology and Functional Roles in Development and Noise-Induced Hearing Loss

Douglas E. Vetter

A Historical Introduction to Control Over Inflow of Sensory Nerve Activity

Sensory processing is a remarkably dynamic process, a fact that is underappreciated. To most, the dynamics of all incoming sensory nerve impulses is "supposed" to represent, with high fidelity, the interaction of our sensory receptors with the external world. This information should faithfully represent what is happening outside according to the sensitivity of the receptors, and be passed along to higher brain centers that then "decide" what to do with such information. It seems almost to be an epiphany to learn of the complex underpinnings of sensory information processing. The true nature of all sensory systems is that the brain does not simply collect sensory information as unprocessed activity, but rather molds the sensory code all along the pathway from peripheral receptor to the highest levels of the cortex. Of particular importance, and seemingly least intuitive, is the realization that the brain interacts with the peripheral sensory system to alter not only the sensory nerve fiber activity but also the peripheral sensory elements themselves. The brain thereby both indirectly, through changes to the receptors, or directly, at the level of the transmitted "code" sent along afferent nerve fibers, shapes how we experience the outside world via a feedback system of nerve fibers. Thus, the idea of sensory encoding is not a simple, linear process, but is highly dynamic—one does not simply get an output that is always in some straightforward way equal to the input. At a fundamental level then, when feedback from the brain impinges upon the peripheral sensory apparatus, sensory processing is not encoding the outside world faithfully. This leads to basic questions that include what benefit accrues from this strategy, under what circumstances does this occur, etc. With respect to the auditory system, an early (1956) report on the role of attention in decreasing auditory activity at the level of the cochlear nucleus went so far as suggest that an understanding of the mechanisms underlying such modulation of the afferent auditory activity may play an important role in the "selective exclusion of sensory messages along their passage toward mechanisms of perception and consciousness" (Hernandez-Peon et al., 1956). Other early forays into trying to understand the significance and implications of this kind of modulation of sensory systems in general expressed the idea that "if we could decide where and how the divergence [of activity under conditions of selective attention and without such attention] arises, we should be nearer to understanding how the level of consciousness is reached" (Adrian, 1954). Interesting that the mystery of consciousness haunted even early neurobiologists so long ago! Thus, while the phenomenon of modulation over sensory nerve activity was not clearly understood, early investigators nevertheless began to ascribe high significance to the process.

It should be understood that much of the impact exerted by these types of "descending" inputs back to the peripheral sensory elements is typically classified as modulatory, yet one must also recognize that the modulation of sensory input to the brain sculpts our representation of the external environment to create what we consider to be a normal sensory experience. Our current understanding of hearing makes room for such sculpting of information flow into the brain. This occurs via a set of descending nerve fibers constituting the olivocochlear (OC) pathway, oftentimes simply referred to as the "efferent system." While the efferent system can be considered fairly well understood in terms of its role in altering hair cell physiology and cochlear mechanics, the relatively fundamental cellular processes underlying hearing, there are many aspects of this system that remain clouded in uncertainty. What are the impacts of either the loss or diminution of efferent feedback on the more general aspects of hearing? Such uncertainty may stem from an incomplete understanding of its role in hearing as a fully integrated neurobiologic concept, and is perhaps due to the fact that without descending modulation by the efferent system, hearing is not normal, but nonetheless does occur, albeit with modest changes. Added to this is the fact that the structure, and therefore to some extent also the function, of the efferent system can be considered almost as unique between species, thus engendering a need for exercising caution when extrapolating results from animal models back to the role of the efferent system in human audition. Perhaps most surprising to students beginning to explore the auditory/vestibular sciences, however, is that some of the largest gaps in our knowledge of the OC system lie not where one may expect, in the basic science of synaptic innervation, neurochemistry, etc., but fundamentally in what role the OC system plays in the human hearing experience. Yet, the very facts that an efferent system can be recognized back to the earliest forms of life using hair cell systems for orientation and balance before hearing emerged (Roberts & Meredith, 1992), and that the structure of the OC system seems to be modified depending on the auditory environment of the subject under study, should be viewed as a potential boon, providing clues not only to the ultimate role of normal efferent function in hearing, but in how hearing occurs at all.

How representations of our auditory world are modified at the periphery is an interesting story that draws from the anatomic, physiologic, molecular, and developmental fields of neuroscience, and begins before neurobiology emerged out of the more general medical fields of anatomy, neurology and pathology. Interestingly, while we believe we have a firm grasp on the molecular underpinnings of the efferent system, at least in the cochlea, the very route that efferent fibers take into the human inner ear is still unclear (does it progress though the anastomosis of Oort?). We will consider, in turn, the initial characterization of the efferent system, aspects of the neurochemistry of the efferent fibers and their pattern of innervation to cochlear hair cells, and molecular aspects of the system. Gaps in our knowledge and potential experiments that would increase our fundamental understanding of the role of this system will be described when appropriate.

Early History of Investigations Into the Olivocochlear System

Work on the OC system has a relatively long history. The first discussion of neural innervation of the inner ear can be traced back to Arnold in his dissertation of 1826, and further explored in 1851. Additionally, Held reported connections to the auditory periphery in 1897 (see Ross, 1981, for discussion of these references). But it is normally acknowledged that the OC system, as we now understand it in anatomical terms, was discovered and reported first by Grant Rasmussen, although much earlier descriptions of a centrally located fascicle of nerve fibers that ultimately were found to be centrifugal in nature (Bischoff, 1899; Leidler, 1914; Papez, 1930) exist. In a series of manuscripts, first in abstract form in 1942, and then as full papers in 1946 and 1953, Rasmussen used the Marchi technique to selectively visualize and trace anterograde degenerating myelin and axons resulting from lesions made in the superior olivary complex (Rasmussen, 1942, 1946, 1953). For the next 30 years following Rasmussen's reports, however, the function, if not the existence itself, of the efferent system was hotly debated. This was due mainly to two problems not settled in the earliest descriptions of the efferent system. First, the Marchi degeneration method becomes less informative as fibers lose their myelin sheath and become thin. The ability to follow the degenerating fibers in the light microscope ceases to exist under these conditions. Second, while the gross region of the brainstem known to give rise to the efferent fibers, and even the sidedness of the organization of fibers, was understood, the exact location of the cell bodies of origin was unknown. The alternate viewpoint to the description of the efferent system as a sensory feedback pathway to the inner ear was that the system was autonomic in nature (e.g., Ross, 1981). Indeed, the prominent investigators supporting a neural role for the efferent system were suggested by those in opposition to such a viewpoint to exemplify the definition of a tragedy as quoted by T. H. Huxley (1904) (and paraphrased here): The idea of a tragedy is a deduction (read here as the neural feedback nature of the OC system) killed by a fact. Which *fact* was evident that should dissuade one from the neural feedback nature of the OC system was less clear than perhaps the proponents of the autonomic nature of the OC system themselves recognized. Nonetheless, investigations continued into the neural nature of the system. With the advent of electron microscopy, at least the idea that hair cells of the cochlea were indeed the targets of neural innervation (despite the continued paucity of information of the origins of these fibers) could be proven and finally accepted. Both the idea of a potential feedback system to the sensory periphery and early hesitation to the idea of the efferent fibers as a neural feedback control system is perhaps less surprising when one considers the general state of neurobiologic knowledge at the time the OC system was described. Dawson (1947a, 1947b, 1958) was reporting on the possibility of peripheral feedback processes related to the motor system, while others were examining the activity of the muscle spindle in 1954 (Hagbarth & Kerr, 1954), now a classic example of feedback in a reflex arc. These ideas continued the slow unraveling, begun much earlier by Head and Holmes, of the relatively long-held idea that the brain (specifically the cortex) received incoming volleys of nerve activity essentially unaltered (Head & Holmes, 1911).

In attempts to more precisely localize the cell bodies of origin for the OC system, Rossi and Cortesina (1963), and later Brown and Howlett (1972), followed the logical assumption that since the OC fibers could be differentiated from afferent cochlear fibers by their high acetylcholinesterase (AChE) levels (Rossi, 1960), the cell bodies should also be heavily AChE positive. This approach helped narrow down those nuclei of the superior olivary complex that most likely contained OC neurons, but did not unambiguously delineate the cells of origin. Revealing the precise localization of the cell bodies giving rise to the fibers and terminals of the OC system awaited the evolution of new neuroanatomic techniques.

The Olivocochlear System and Developments in Modern Tract Tracing Techniques

The development of horseradish peroxidase (HRP) as a tract tracer (Kristensson & Olsson, 1971a, 1971b, 1973; Kristensson et al., 1971; LaVail, 1975; LaVail & LaVail, 1972, 1974, 1975) allowed investigators to unambiguously trace neural connections using the process of retrograde axonal transport of injected substances from the site of injection. This technique finally enabled researchers to map with confidence the location of cell bodies giving rise to the OC fibers (Warr, 1975), and to begin recognizing that the OC system was in fact composed of a number of different classes of neurons. Soon after, it became apparent that the neuroanatomy—the location, number, and type of neurons making up the OC system—varied between species. Arguably, one of the most significant steps forward in understanding the OC system made possible by neuronal tract tracing came with the realization that the OC system can generally be divided into two zones (Figure 10–1) based on the position of the cell bodies with respect to the medial superior olive (MSO) (Guinan et al., 1983, 1984). The Lateral Olivocochlear System (LOCS) nomenclature was adopted for OC cells located lateral to the MSO. The LOCS was found to be predominantly an ipsilateral system (with some relatively small contralateral contribution, depending on species) of small neurons located either in (rodents) or directly around and in the hilus (cat) of the lateral superior olive (LSO). The cell bodies of origin for the LOCS were found to terminate under the IHCs, directly on the spiral ganglion neuron (SGN) dendrites and, in adults, to a far lesser degree also on the IHC itself. Ipsilaterally, the location of LOCS neurons within the LSO covers the full frequency map of the LSO, while those smaller numbers found in the contralateral LSO in some species seem to be limited in their location along the characteristic frequency map to low frequencies (Robertson, 1985; Robertson et al., 1987; Warr, 1975). The Medial Olivocochlear System (MOCS), as defined by Warr, is composed of those OC neurons located bilaterally and medial to MSO. Thus, there are elements of MOCS that produce a crossed system of fibers (the majority of the MOCS fibers in rodents), and another portion of MOCS fibers that remains uncrossed. Thus, a portion of the MOCS neurons represents Rasmussen's original crossed, bilateral system. The cell bodies of

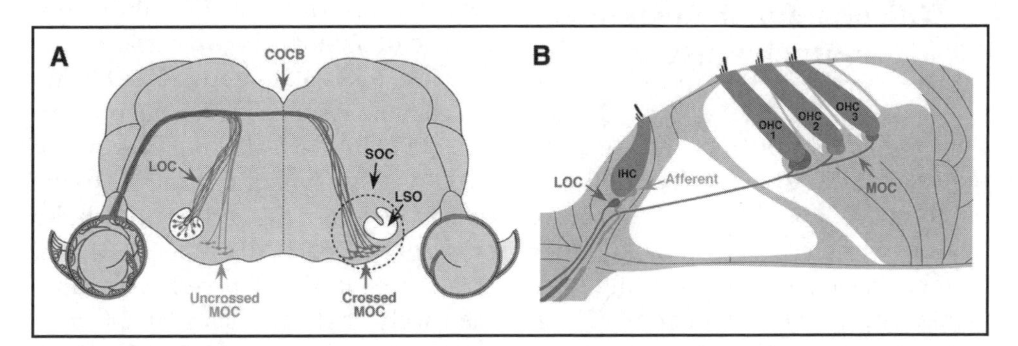

Figure 10–1. Schematic illustrations of the brainstem origins (**A**) and peripheral targets (**B**) of the olivocochlear system of rodents. **A.** Cross section of the brainstem at the level of the LSO reveals the location of the LOCS cell bodies in the body of the ipsilateral LSO, while the MOCS cell bodies are located bilaterally medial to LSO. Note that Shell OC neurons are not included in the schematic. **B.** Cross section through the organ of Corti demonstrates the terminations of LOCS fibers on the dendrites of the SGNs, and the terminations of the MOCS fibers directly onto the cell body of the outer hair cells. Reproduced from Maison et al. (2013) with permission of the Society for Neuroscience.

origin for the MOCS are located in the ventral nucleus of the trapezoid body (VNTB) and terminate directly on the OHCs. This system seems to be fixed in its location among mammals except for the bat species *Rhinolophus rouxi* (the horseshoe bat). *R. rouxi* possess a single, ipsilateral, AChE-positive group of neurons projecting to the cochlea that are situated between the LSO and the VNTB (Aschoff & Ostwald, 1987). This nucleus was termed the nucleus olivocochlearis, and matched well with the fact that *R. rouxi* is the only mammal that does not possess OC innervation of its OHCs. Thus, anatomical studies on the OC system demonstrated it to be a complex system with respect to the location of cells of origin and the origins of the synaptic terminations on their target. Additionally, and not surprisingly, the structure of the OC system seemed to be influenced to some degree by the auditory environment the subject was required to attend to.

Olivocochlear System Neurochemistry

With the advent of tract tracing techniques, the OC system was shown to be composed of different classes of neurons situated in various superior olivary nuclei, thereby leading to a deeper understanding of the OC system of mammals. Once the cell bodies of the OC system were mapped, attention naturally turned to elucidating the neurochemistry of the system. The earliest investigations into the neurochemical nature of the OC system used the thiocholine method for demonstrating the presence of cholinesterase, and thus an indirect method for establishing acetylcholine pathways or nuclei. Indeed, all motor nuclei and previously recognized efferent pathways were found to be cholinesterase positive. Rossi and Cortesina (1963) and later Brown and Howlett (1972) used the cholinesterase technique to establish the OC system as a cholinergic pathway. Yet, evidence soon emerged from biochemical experiments that within the OC system was some small element of a GABAergic transmission system (Fex & Wenthold, 1976). Based on the low GAD activity measured (approximately 1/10th that of choline acetyltransferase, ChAT), however, this early report suggested that GABA was not likely to be a neurotransmitter of the OC system. This was a curious finding given that efferent activity of the OC system inhibited cochlear function and elevated auditory thresholds. Perhaps more confounding, evidence existed that the cholinergic neurotransmission itself may underlie the inhibition (Fex, 1967; Fex & Adams, 1978). Ultimately, the use of an anti-glutamic acid decarboxylase (GAD) antibody (GAD being the enzyme responsible for the production of GABA) revealed the presence of GAD-positive synaptic terminals in the cochlea of guinea pig and rat (Altschuler et al., 1989; Fex & Altschuler, 1984; Fex et al., 1986). This proved that more than acetylcholine exists as an OC neurotransmitter. Functionality of the GABAergic system in the cochlea remains a topic of debate to this day, although data demonstrate some developmental/maintenance-like roles as well as a role in contributing to normal auditory thresholds (Maison et al., 2006).

A Neurochemical Map of the OC System

Based on the findings that GAD immunoreactive terminals were observed under both the IHCs and the OHCs, one would expect GAD-positive OC neurons in both the LOCS and MOCS. However, no such direct evidence had previously been found for localization of non-ACh neurons of the OC system. A confluence of abilities to trace neural connections and the availability of good antibodies useful for immunohistochemistry made it possible to begin constructing a high-quality neurochemical map of the OC system. By injecting HRP into the cochlea, followed by processing for HRP to localize the OC neurons, and sequentially immunostaining for GAD and choline acetyltransferase (ChAT), the first neurochemical map of OC neurons was produced (Vetter et al., 1991). While these data clearly indicated that the MOCS was exclusively cholinergic, it also showed that the LOCS was composed of two distinctly different cell populations: a ChAT-positive (presumably cholinergic) and a separate GAD-positive (presumably GABAergic) neuron pool. Furthermore, we also showed that the ChAT-positive cell bodies of origin of the LOCS expressed calcitonin gene-related peptide (CGRP), a potential neurotransmitter/modulator previously linked with cholinergic neurotransmission (New & Mudge, 1986) and shown to be expressed and functional in hair cell systems (Adams et al., 1987). Just as in the case of GAD, CGRP was exclusively part of the LOCS, and never part of the MOCS neuronal pool.

It had previously been shown that in the cochlea, GAD-positive terminals were situated below both the IHC region as well as the OHCs. Our own immunostaining data on the rat cochlea agreed with these prior studies. However, these data went on to demonstrate that CGRP-positive terminals in the cochlea were also present under both the IHCs and OHCs. Thus, GAD-positive and CGRP-positive OC neurons were located only in the LOCS cell bodies of the OC system, yet sent terminals to the OHC region. This suggests that the LOCS/MOCS dichotomy perhaps does not hold for the terminal fields of the OC cells, and is useful only for neuronal positioning within the CNS. In an attempt to further examine (and perhaps reconcile) this discrepancy, Maison and coworkers examined the neurochemical nature of the MOCS using surgical transections of the OC fiber bundle coupled with immunohistochemical localization of vesicular ACh transporter (VAT), GAD, and CGRP. Their work demonstrated extensive co-localization of VAT with GAD and CGRP in *all* OC terminals under both inner and outer hair cells, and further, that the OC fiber system under outer hair cells arises exclusively from the MOCS pool of neurons (Maison et al., 2003). Yet, these data do not fully lay to rest the initial findings described in our original mapping efforts—specifically, that no GAD- or CGRP-positive OC neurons are found in the MOCS pool of cells. Thus, further work is required to more completely reconcile the data generated by each group. Related to this line of investigation, we published findings that urocortin (Ucn), a neuropeptide related to corticotropin releasing factor (CRF, also termed corticotropin releasing hormone, CRH), is expressed by the LOCS cell bodies, and sends CRF fibers and terminals only to the IHC (Vetter et al., 2002). These findings seem to uphold the original ideas of peripheral terminal fields of LOCS and MOCS being associated exclusively with different pools of OC neurons. Our data were based on protein (immunohistochemical) localization, but previous work by the Vale lab (Vaughan et al., 1995) also demonstrated that Ucn mRNA is found only in the LSO. This strongly suggests Ucn protein is not expressed in regions occupied by the MOCS cell bodies, even at potentially very low levels.

In addition to the OC neurons labeled in the expected domains of the superior olive, we found various HRP-labeled neurons scattered around the dorsal and ventral aspects of the LSO of rat. While these had been described previously, the numbers of neurons we had managed to label from the cochlea were more significant than one might expect given the previously published descriptions. The extra-LSO OC neurons were consistently immunolabeled only by anti-ChAT antibodies, and were significantly larger than the intrinsic LSO OC neurons. While of some interest, owing to the extra-LSO location of these (apparent) LOCS neurons, and their size (seemingly more similar to MOCS rather than LOCS neurons), their numbers were still unimpressive, engendering the opinion that these were aberrantly located MOCS neurons. However, with the continued evolution of tract tracing substances, this view changed dramatically.

Use of cholera toxin B subunit was proving to be a sensitive retrograde tract tracer at the time (Ericson & Blomqvist, 1988). Using a conjugate of CTB with HRP (CTB-HRP), it proved possible to reveal the neurons of the OC system with Golgi-like detail (Vetter & Mugnaini, 1992). While the location of the main components of the OC system was replicated, the CTB-HRP technique revealed a much clearer picture of the OC neurons previously observed around the LSO in the rat material. The HRP-based techniques revealed approximately 20 of these OC neurons surrounding the LSO, but the CTB-HRP tracing experiments revealed these neurons to be part of a more extensive network of more than 100 neurons. Sectioning the material in all three planes further revealed that these neurons form a shell around the LSO (Figure 10–2), thus giving rise to the nomenclature of Shell OC neurons.

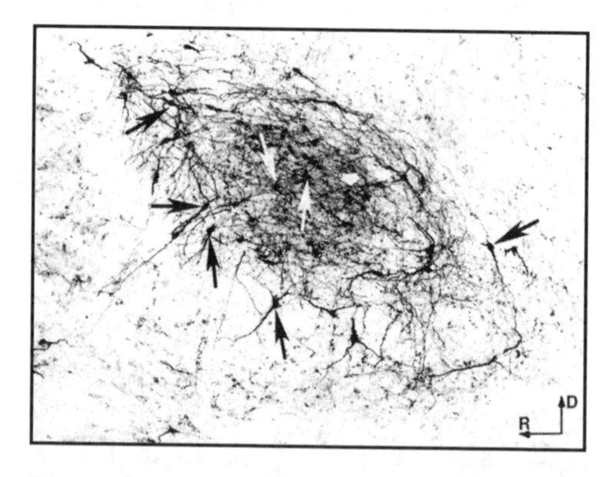

Figure 10–2. Cytology and location of Shell OC neurons is demonstrated following retrograde labeling of CTB-HRP from the cochlea. Shell OC neurons, indicated by large black arrows, completely surround the body of the LSO in this sagittal section. Some of their dendrites (*white arrowhead*) penetrate the LSO. Small LOCS neurons within the body of the LSO are indicated by white arrows. Reproduced from Vetter and Mugnaini (1992) with permission of Springer.

These data began to firmly establish the Shell OC neurons as a third distinct population of OC neurons in rodents.

Projections to the Cochlea of OC Neurons and Their Terminal Fields

LOCS terminations. One of the major differentiators of the OC system of neurons lies in their peripheral projections to the cochlea, as described briefly above. Axons of the LOCS neurons are thin, unmyelinated fibers. Axonal labeling demonstrates that these fibers make up the vast majority of the inner spiral bundle (Guinan et al., 1983; Liberman et al., 1990) in the vicinity of the inner hair cells. Despite the apparently homogeneous nature of the cell type composing the classic LOCS, early examinations of terminal arbors of LOCS neurons suggested that two types of fibers could be differentiated within the cochlea (Brown, 1987). The LOCS fibers were classified as either unidirectional or bidirectional, with the greatest majority being of the unidirectional variety. The directionality nomenclature refers to the direction the fibers spiral within the cochlea. Bidirectional fibers were found to bifurcate upon entering the inner hair cell region, and to send long branches, 1 mm or longer in length, both basally and apically. At the time, no information describing differences with respect to the cell bodies of origin in the LOCS compartment explained the different fiber types. Using fine stereologically guided injections of tracers in and around the LSO, Warr showed (Figure 10–3) that it is the Shell OC neurons that give rise to the bidirectional efferent fibers, while the classic (LSO intrinsic) LOCS neurons give rise to the unidirectional fibers (Warr et al., 1997). This makes sense given the numbers of Shell OC neurons versus intrinsic LSO OC neurons, and the relative abundance of each fiber type.

MOCS terminations. The MOCS terminations are distinct from those of the LOCS fibers. MOCS fibers terminate directly on the basal aspect of outer hair cells (see Figure 10–1). While the LOCS terminal field can be thought of as generally homogeneous, the MOCS fiber terminal pattern is more complex. As a basic organizational plan, those fibers emanating from the crossed MOCS cell bodies (thus, the majority of MOCS fibers) innervate the cochlea most densely in the middle of the cochlear span (Maison et al., 2003), while those fibers from the uncrossed system of MOCS cell bodies tend to be distributed more evenly across the apical-basal span of the cochlea (Guinan et al., 1984). MOCS innervation is also complicated by the demonstration that numerous en passant–like swellings are located below the outer hair cells, in contact with the Type II afferent SGN fibers innervating the outer hair cells (Thiers et al., 2002). Finally, recent evidence also suggests that efferent innervation may not be confined to fibers with origins in the brain. Strong evidence now exists demonstrating that all Type II SGN afferent fibers make reciprocal synapses with outer hair cells (Thiers et al., 2008). The local circuit between afferent fibers and outer hair cells is further modulated by traditional MOCS fiber innervation on the Type II fibers. The significance of this complex local synaptic circuitry is yet to be discerned, but it is intriguing to think that afferent

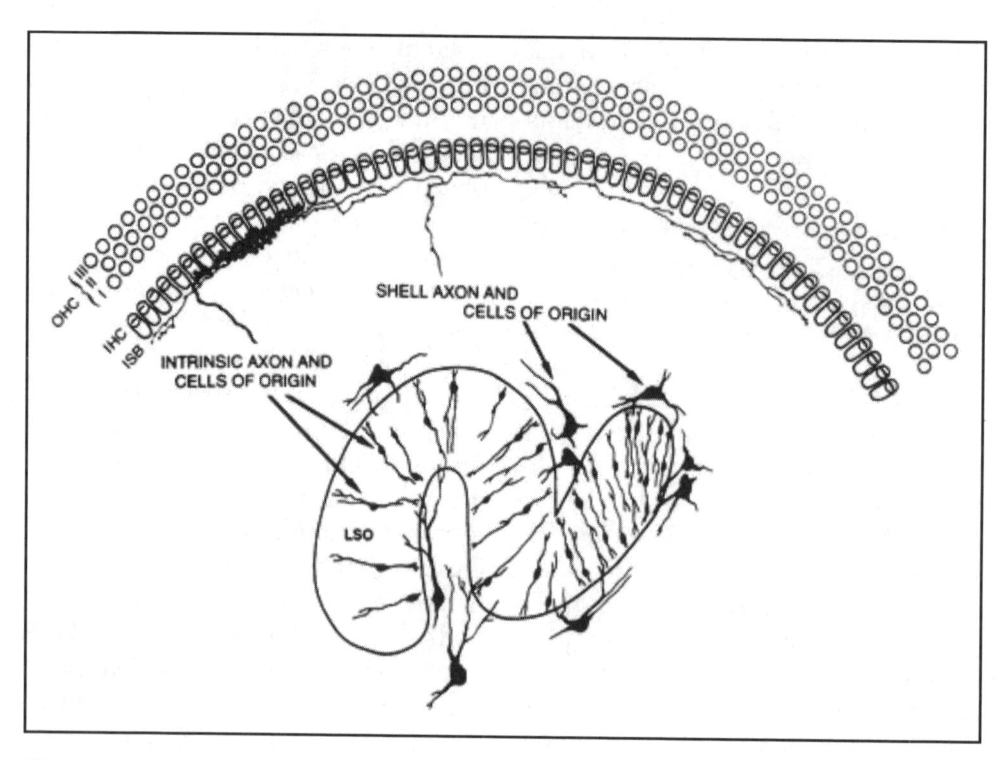

Figure 10–3. Summary diagram of the locations and typical projections of the Shell and intrinsic LOCS neurons. Location of the cell bodies was revealed with CTB-HRP, and ramifications of axons within the cochlea were revealed with BDA injections into regions of the superior olive adjacent to, or within, the LSO (OHC, outer hair cell; IHC, inner hair cell; ISB, inner spiral bundle). Reproduced from Warr et al. (1997) with permission of Elsevier.

fibers may be involved in modulating normal outer hair cell function. Because of the role of outer hair cells in cochlear amplification, this reciprocal feedback may be found to take on greater significance in future experiments.

Tracing of OC fibers by intracellular/fiber fillings continue to indicate that OC innervation to the different target zones is accomplished by distinct and separate populations of fibers. Based on tract tracing data, no evidence to date has demonstrated the existence of OC neurons that project to both the inner and outer hair cells regions. Thus, the origins of the CGRP and GAD-positive terminals on the outer hair cells seems not to be resolved by the possibility of a population of LOCS cells sending collaterals to the outer hair cells. Yet, centrally, no cell bodies labeled by tract tracing from the cochlea have been shown to express CGRP or GAD. This long-standing mystery seems to involve a basic property of the OC system innervation that is in need of resolution.

Nonclassical Efferent-Like Innervation of the Cochlea

Terminal fields of efferent-like fibers target structures other than hair cell related elements. For example, significant innervation of Hensen's and Deiters' cells has been demonstrated (Burgess et al., 1997; Liberman et al., 1990; Wright & Preston, 1976). The neurochemical identity of these fibers is apparently complex, with at least some being revealed as GABAergic in nature (Eybalin et al., 1988). More recently, the origins of these fibers have been revealed to be a mixture of a small number of classic MOCS fibers and a greater number of Type II SGN afferent fibers (Fechner et al., 1998; Fechner et al., 2001). The vast majority of support cell innervation occurs in the apical regions of the cochlea, perhaps indicating a difference in the mechanisms or metabolic sensory processing requirements between the basal and the apical (high and low frequency) portions of the cochlea. The functional significance of these types of efferent-like innervations remains to be revealed, but it is curious to note that it is the apex, at least in rodent models, that is often most resistant to ototoxic and noise-induced damage.

The Molecular Biology of the Olivocochlear System

Early work demonstrated that activation of the efferent system produces an inhibition of cochlear activity (Fex, 1967). It was somewhat surprising then to find that the early neurochemical analysis of the efferent system suggested that this pathway was cholinergic. Even when GABA was later shown to be involved in the efferent projections, the number of fibers and the location of those fibers that were stimulated to demonstrate the inhibitory nature of OC activation seemed to suggest that the cholinergic OC system was inhibitory. Additionally, the speed of response ruled out muscarinic ACh receptors, but at the time, all known nicotinic acetylcholine receptors (nAChRs) generated only excitatory responses. Ultimately, the first clues in understanding the early physiological data came from the work of Paul Fuchs and colleagues, who showed that the synaptic response of hair cells to efferent release of ACh was more complicated than originally thought. It was demonstrated that a massive hyperpolarization of hair cells in response to ACh was preceded by an initial depolarization of the hair cell (Fuchs & Murrow, 1992). Thus, a typical excitatory nicotinic receptor could be involved in OC function if it was also coupled to another channel, such as a potassium channel, whose activity ultimately hyperpolarized the cell. Soon after this landmark work was published, new nAChR subunit genes were cloned (Elgoyhen et al., 1994; Elgoyhen et al., 2001). The alpha 9 gene was cloned first, and using heterologous expression physiology techniques, it was shown to be sensitive to ACh. From a molecular (DNA sequence) perspective, this new clone was clearly of the nicotinic variety of ACh receptor subunits. But its pharmacologic profile was unusual for most nAChRs. It was highly sensitive to block by strychnine, a property shared by only one other mammalian nAChR, the alpha 7 receptor subunit. This was significant because the early physiology of the efferent system was shown to be blocked by strychnine, but an alpha 7 function in efferent activity could be ruled out (at that time) because these receptors were only known to be excitatory. In situ hybridization (Figure 10–4) using a partial clone of the alpha 9 gene unambiguously demonstrated its expression in cochlear hair cells (Elgoyhen et al., 1994), thereby making a strong case that it was the alpha 9 subunit that produced the nAChR encoding efferent activity at the outer hair cell. Alpha 9 expression has never been observed in the CNS. Expression of alpha 9 in adult inner hair cells was somewhat surprising given that direct efferent innervation to inner hair cells is shed just prior to the onset of hearing in rodents. But it was later revealed that expression levels in inner hair cells was high during the time of direct efferent innervation of the inner hair cells, and that expression levels drop following rearrangement of the efferent innervation (Luo et al., 1998). The adult expression levels therefore seemed to be a leftover of the maturation process. Yet, questions persisted following the cloning of the alpha 9 gene. Activity in *Xenopus* oocytes was woefully small. More important, numerous biophysical properties of nAChRs composed of homomeric alpha 9 subunits, including the current-voltage relationships, calcium sensitivity, and especially the desensitization properties did not match the "hair cell efferent nAChR." Using the alpha 9 DNA sequence, the GenBank-expressed sequence tags database was probed, and from that information, the alpha 10 gene was discovered and cloned (Elgoyhen et al., 2001). While the alpha 10 gene did not produce an ACh-inducible response when expressed on its own in heterologous expression systems, or even when combined with numerous other alpha subunit genes, coexpression of alpha 9 with alpha 10 fully recapitulated the native hair cell nAChR physiologic profile.

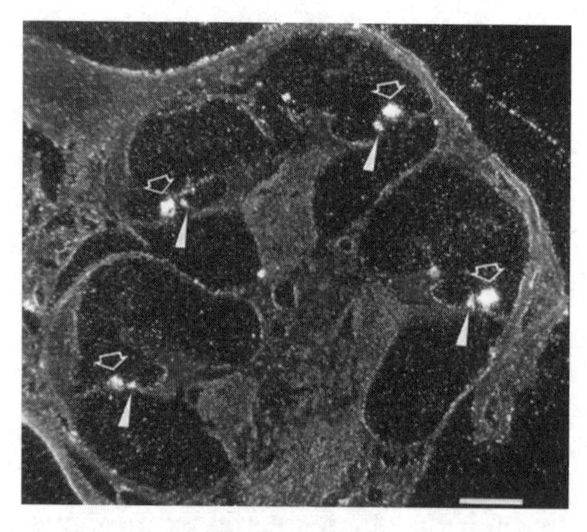

Figure 10–4. Dark-field microscopy of a mid-modiolar section of the rat cochlea hybridized with an alpha 9 cRNA probe reveals intense labeling within the outer hair cell region (*open arrowheads*) and, to a lesser extent, within the inner hair cell region (*closed arrowheads* in all turns of the cochlea; scale, 1 mm). Reproduced from Elgoyhen et al. (1994) with permission of Elsevier.

While the cloning of alpha 9 and alpha 10 genes seemed to solve the long-standing riddle of the identity of the hair cell nAChR, the fact that these proteins generated an ACh receptor that was highly permeable to calcium (Katz et al., 2000) did nothing to directly solve the puzzle of efferent based inhibition of cochlear processing. However, it did, perhaps, limit the kinds of processes that could be envisioned to explain the hyperpolarization of hair cells. First, hyperpolarization would most easily be explained by the activation of potassium channels. Second, if this was the case, the potassium channels involved should be activated, at least in part, by calcium influx, and ideally also be located close to the efferent synapse. The puzzle was finally solved by the demonstration that the sK2 channel, an apamin-sensitive, small-conductance calcium-activated potassium channel originally cloned by the Adelman group (Bond et al., 1999; Köhler et al., 1996), is expressed in hair cells (Dulon et al., 1998). Thus, the cholinergic receptor subserving neurotransmission events between the descending OC axons and the hair cells was shown to be composed of the highly calcium permeable alpha 9/10 nAChRs functionally linked to the calcium-activated sK2 potassium channel. Together, this seemingly unusual two-channel system represents a fast inhibitory mechanism that modulates OHC function, and indirectly controls auditory sensitivity via its effects on basilar membrane mechanics (among other potential mechanisms). In fact, this dual-channel system is anything but unusual, and is conserved from fish (and perhaps even "lower") all the way up to mammals. A clear selective pressure seems to be at work to maintain this system across species. Thus, this dual-channel mechanism underlying hyperpolarization of hair cells following efferent activity should more correctly be viewed as a specialized, rather than an unusual, mechanism of hair cell modulation. The cloning of the alpha 9, alpha 10, and sK2 genes was performed at a time when genetic manipulation techniques began to be employed in neurobiology. Understanding the role of each of the major components of OC system synaptic function thus seemed to be at hand, with the requirement of simple genetic manipulations to solve the deeper questions of the efferent system from a molecular level.

Use of Genomic Manipulation Techniques in Attempts to Answer Questions Pertaining to the Role of the Olivocochlear System

The alpha 9 gene was the first of the tripartite molecular ensemble responsible for producing the hair cell response to efferent stimulation that was targeted for genetic manipulation. Because the alpha 10 gene had not been cloned at the time the alpha 9 gene knockout mouse was produced, it was naively assumed that ablation of the alpha 9 gene should fully prevent efferent activity, and perhaps allow for a molecular dissection of the roles played by the various neurotransmitters expressed in the efferent terminals. It is noteworthy that the alpha 10 gene, once cloned, proved to not be functional on its own, and so interpretation of data from the alpha 9 null mouse can be interpreted as a functional knockout of nAChR-mediated responses. Keeping in mind that the major neurotransmitter of the efferent system has always been assumed to be ACh, it then was a formality that examination of the alpha 9 null mice indeed demonstrated no discernable efferent function (Vetter et al., 1999). The lack of efferent activation of hair cells was demonstrated indirectly by examining the modulation of sound-evoked amplitudes of the 2f1–f2 distortion product otoacoustic emissions (DPOAEs) and the compound action potential (CAP) of the afferent nerve during delivery of electrical shocks in the floor of the fourth ventricle, where the crossed MOCS fibers decussate and also meet with the uncrossed MOCS fibers. In normal animals, these shocks diminish both the sound-evoked DPOAE and CAP amplitudes. In the alpha 9 null mice, shocks to the MOCS fibers did not diminish these responses to sound (Figure 10–5). These results therefore demonstrated that all of the previously recognized neural functions normally impacted by efferent fiber activity do so via the alpha 9 containing nAChRs. Interestingly, with the development and characterization of the alpha 10 null mouse line (Vetter et al., 2007), very similar results were obtained. This suggested that expression of alpha 9 homomers by hair cells (i.e., without alpha 10) is not sufficient for normal efferent functionality (quantitative PCR [qPCR]) showed that alpha 9 gene expression was slightly downregulated in the alpha 10 null mice, but was still 75% of the baseline expression level). Finally, in both the alpha 9 and alpha 10 null mice, baseline auditory functionality was unaltered. Auditory brainstem responses (ABRs) revealed a normal threshold across the frequency spectrum tested, and amplitudes of DPOAEs were also normal. This reinforces the idea that efferent activity *modulates* peripheral auditory processing, but is not responsible for setting normal thresholds. Finally, it should be recognized that the analyses of function in the absence of alpha 9 or alpha 10 subunits suggest that the GABAergic OC system is not intimately involved in the "classic" OC function, since no changes to DP amplitudes, for example, were invoked following OC fiber shocks in the null mice.

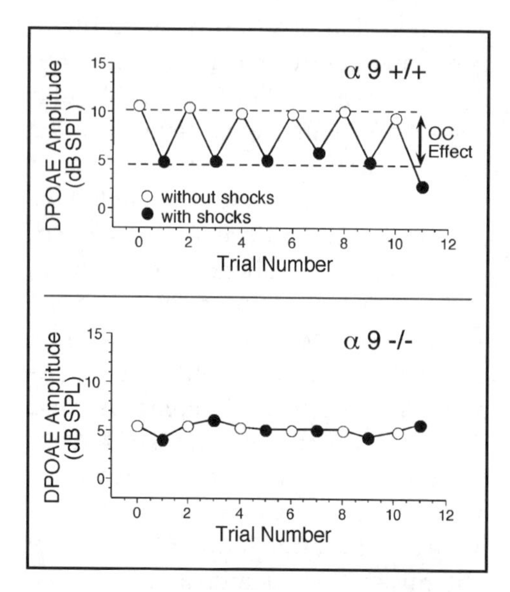

Figure 10–5. DPOAE-based assay of OC effects in wild type (+/+) and null (−/−) mice. Amplitudes of the 2f1-f2 DPOAE were monitored without (*open circles*) or during the presence of (*solid circles*) OC fiber electrical stimulation at the floor of the fourth ventricle. Note that the approximately 5-dB difference in the no-shock baseline between wild type and null mice represent normal interanimal variability. Reproduced from Vetter et al. (1999) with permission of Elsevier.

The alpha 9 and alpha 10 null mice contained constitutively ablated genes because this was the state of the art of gene manipulation at the time these mouse lines were created. However, two circumstances made further manipulations of the alpha 9 gene with refined techniques attractive. First, a large literature already existed with respect to understanding the biophysical mechanisms of nAChR activation and gating (e.g., Miyazawa et al., 2003). This body of work pinpointed critical amino acids for a normally functioning nAChR. Mutation of one of these amino acids, a leucine at the 9′ position

(L9′) of the M2 domain (lining the pore) was critically involved in establishing the normal permeation and desensitization kinetics of the channel. Substitution mutation of the 9′ leucine led to slower desensitization, increased spontaneous activity of the channel, and greater activation by ACh (Filatov & White, 1995; Kosolapov et al., 2000; Labarca et al., 1995). The role of this highly conserved leucine was therefore hypothesized to be involved in setting the mean open time of the channel, and that this occurs through interactions between the leucine and other polar regions of the receptor when the channel opens following ACh binding. It was further hypothesized, and later proven, that a substitution of the leucine with a more polar amino acid such as threonine (L9′T) would result in stronger, stabilized open confirmations of the channel that also exhibited spontaneous opening (Kosolapov et al., 2000). Critically, the L9′T mutation of the alpha 9 gene was shown to produce the same responses observed when the same mutation was carried out in other nAChRs (Plazas et al., 2005). This led directly to the idea that manipulating the 9′ leucine of the M2 (pore lining) domain of the alpha 9 gene to produce a new mutant mouse line could be informative with respect to efferent functionality and normal peripheral auditory processing. With evolving techniques for genetic manipulation and production of point mutant mouse lines, it became possible to pursue such a goal. Examination of the L9′T point mutant (knock-in) mouse line revealed the expected results: application of ACh led to hypersensitive, slowly desensitizing hair cell ACh-evoked currents, ABR and DPOAE thresholds were slightly raised (due to the spontaneous opening of the alpha 9 channels, effectively mimicking OC fiber activity), and brief OC fiber bundle shocks (2 minutes of 150 ms, duration, 200 shocks/s) led to a long-lasting (approximately 10 minutes) decrease in DPOAE amplitudes (Figure 10–6). These results were completely blocked by strychnine, a potent blocker of alpha 9 activity (Taranda et al., 2009).

Finally, two groups (working simultaneously, but independently) used the sK2 null mice (Bond et al., 2004) to probe the role of the calcium activated potassium channel in peripheral auditory processing (Kong et al., 2008; Murthy, Maison, et al., 2009). One of the lures for these experiments was the realization that if the nAChR normally depolarizes the hair cell prior to activation of the sK2 channel that then hyperpolarizes the hair cell, it should be possible to "reverse the sign" of the efferent synapse by ablating the sK2 gene. Loss of sK2 expression should leave in place the depolarization induced by the calcium entry via the alpha 9/10 heteromeric receptor. Surprisingly, however, loss of sK2 was shown to also result in loss of *all* cholinergic synaptic currents (despite apparent normal expression of alpha 9, and elevated expression of alpha 10 genes, assessed via qPCR) and a complete loss of the OC-mediated suppression of the DPOAE amplitudes. Like the alpha 9 nulls, however, ABR thresholds appeared normal. Thus, the synaptic reversal did not materialize. Further, it seems that the system has some kind of physiologic break on it that disallows excitatory OC activity. This has yet to be fully explained.

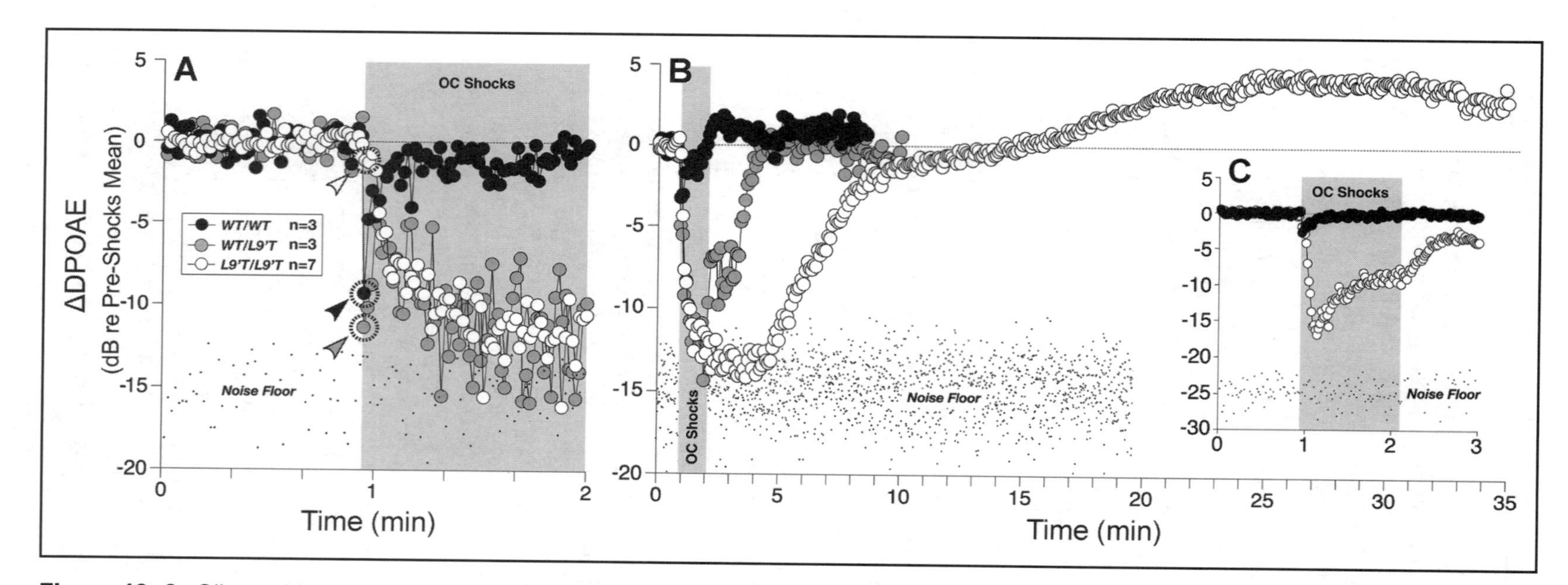

Figure 10–6. Olivocochlear-mediated suppression of DPOAE amplitudes in the alpha 9 L9′T point mutant mouse. **A.** DPOAE amplitudes measured before, during, and after a 70-second shock train to the OC bundle (*gray boxes*) are shown on two different time scales to emphasize the onset effects (**A**) and the offset effects (**B**) of OC activation. DPOAE amplitudes from each experiment are normalized to the average preshock value and then averaged across ears within a genotype (group sizes are shown in the key in **A**). Arrowheads in **A** indicate the first point after shock-train onset for each genotype. Data acquisition times are approximately 1 second/point. For **A** and **B**, the f2 primary was at 22.6 kHz, and primary levels were adjusted to produce a DPOAE approximately 15 dB above the noise floor. Suppression is so strong in mutant ears that the DPOAEs are driven into the noise. (WT, wild type). **C.** To reveal full suppression magnitude, we raised primary levels until preshock DPOAEs were 25 dB above the noise floor. Peak suppression in alpha 9 L9′T/L9′T reached approximately 17 dB, whereas for alpha 9 wild type, effects were less than approximately 5 dB at these higher stimulus levels. Reproduced from Taranda et al. (2009) with permission of authors (open source publication).

The Role of Early Efferent Activity in Formation/Maintenance of Efferent Synapses

With the exception of the lack of ACh-inducible activity in the sK2 null mice, the cellular and peripheral auditory physiologic findings coming from the mutant mouse lines were to be expected. The results proved the hypotheses that cholinergic signaling is the main mechanism by which efferent fiber activity impacts hair cell function, and by extension, cochlear mechanics. However, each of the null and the knock-in mouse lines also delivered a surprise in the form of altered synaptic morphology following gene ablation. Ablation of either the alpha 9 or alpha 10 gene resulted in hypertrophied efferent synaptic terminals under outer hair cells (Figure 10–7). Additionally, while outer hair cells are typically contacted by three to five synaptic terminals in

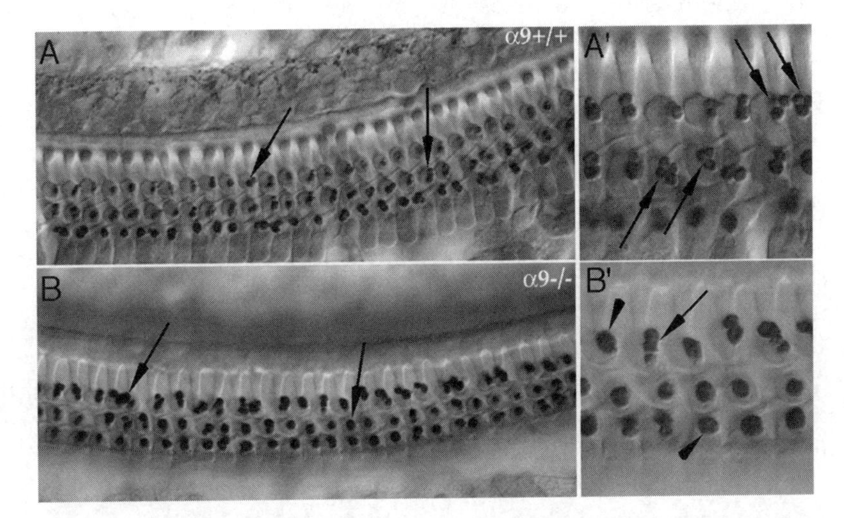

Figure 10–7. Vesicular acetylcholine transporter (VAT) immunohistochemistry reveals abnormal cholinergic efferent innervation to the outer hair cell region of alpha 9 null mice. **A.** Small efferent terminals under OHCs of wild type mice contact all rows of outer hair cells. **A'.** High magnification reveals normal morphologic features of efferent terminals under outer hair cells in wild type mice. Arrows point to polyinnervated hair cells, one of which is contacted by four efferent terminals. The third row of hair cells is out of the plane of focus in this picture. **B.** VAT immunostaining of cochleae from alpha 9 knockout mice illustrate that cholinergic innervation persists in the knockout, but that the efferent terminals under outer hair cells are abnormal in their morphology. Three rows of large immunostained terminals can be seen under the rows of outer hair cells. **B'.** High magnification better reveals the abnormal VAT-positive efferent terminals of the alpha 9 null mice. While some hair cells received primarily normal complements of terminals (*arrow*), most hair cells were contacted by a single hypertrophied terminal (*arrowheads*). Reproduced from Vetter et al. (1999) with permission of Elsevier.

the wild type mouse, the alpha 9 and alpha 10 null mice generally possessed only one or two hypertrophied synaptic terminals. This suggested that activity at the nicotinic receptor was important for establishing normal synaptic contacts between OC fibers and hair cells. It was later established that OC activity through the alpha 9 containing receptor complex played a significant role in modulating expression of cell adhesion molecules (Murthy, Taranda, et al., 2009). The molecular mechanism by which such modulation occurs remains unknown. Without normal expression of adhesion molecules, synapses could be envisioned as free to explore and expand beyond their normal bounds, effectively giving the first fibers to arrive at OHCs a competitive edge in establishing synaptic territory, perhaps thereby also explaining the decrease in numbers of synaptic boutons observed under the outer hair cells.

While loss of alpha 9 and alpha 10 gene expression resulted in decreased numbers of synaptic boutons contacting outer hair cells, the L9′T knock-in mouse, which exhibited increased ACh-mediated responses, possessed a hyperinnervation (Figure 10–8) of efferent boutons onto the outer hair cells (Murthy, Taranda, et al., 2009). Finally, work on the sK2 null mouse line revealed a degeneration of efferent innervation to outer hair cells (see Figure 10–8) that was hastened by the compounded loss of the alpha 10 gene as assessed in double sK2, nAChR alpha 10 null mice (Murthy, Maison, et al., 2009). Importantly, very young sK2 null mice exhibited a normal MOCS innervation to outer hair cells, suggesting that normal efferent activity is not necessary for generating the initial contact between hair cell and efferent fiber, but is critical for maintenance of such contacts. These data in aggregate revealed the unexpected finding that early efferent activity seems to be critically important for establishing normal MOCS innervation to the outer hair cells. While there were examples of effects on the LOCS innervation pattern under inner hair cells in the alpha 9 and alpha 10 mutant lines, the effects were generally small. This stands in stark contrast, however, to the degeneration of even the LOCS complement of efferent fibers in the sK2 null mice (Murthy, Maison, et al., 2009).

The Role of Early Efferent Activity in Maturation of Peripheral and Central Auditory Structures

Invasion of the inner ear sensory epithelium by efferent fibers begins as early as embryonic day (E)12 in mice (Fritzsch & Nichols, 1993). At this stage, the inner ear is more correctly identified as an otocyst that has only recently closed. At E12, hair cells have only begun to differentiate the day before (in mice), and are still a day away from the most robust proliferation (Ruben, 1967). Why should efferent fibers invade the future sensory epithelium so early?

All developing sensory systems, and numerous other CNS structures, share the feature that during their development, they exhibit spontaneous activity (for expanded discussions, see Blankenship & Feller, 2010; Wang & Bergles, 2015), defined as activity induced intrinsically without input from external stimuli. The role(s) such early spontaneous activity plays in the developing

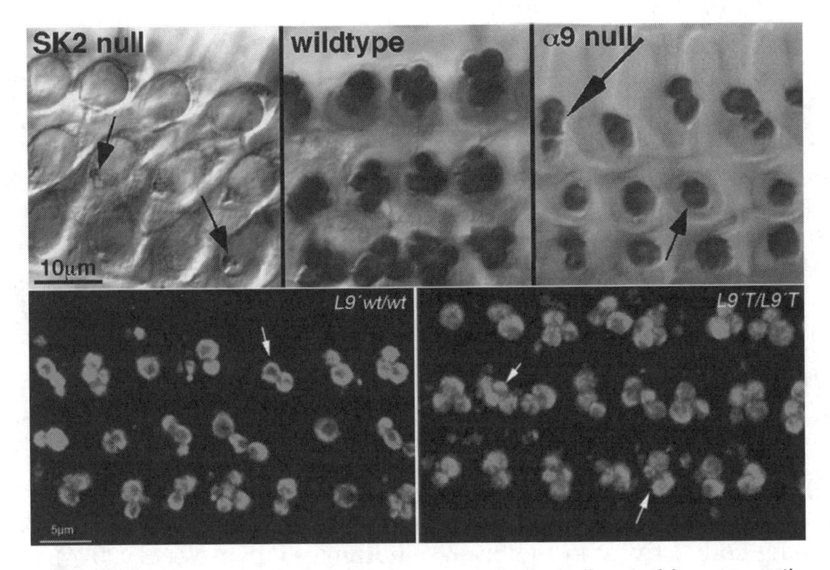

Figure 10–8. Top three panels demonstrate that olivocochlear synaptic boutons degenerate in SK2 null mice. Synaptophysin immunostaining reveals olivocochlear (efferent) synaptic terminals under outer hair cells. In wild type mice (*middle panel*), outer hair cells are contacted by 2–5 synaptic boutons (*arrows*). In alpha 9 null mice (*right panel*), olivocochlear terminals are most often less numerous (1–2 per outer hair cell) but larger (hypertrophied, *small arrow*), although occasionally hair cells are contacted by 2–3 boutons (*large arrow*). In SK2 null mice (*left panel*), terminals (*arrows*) progressively degenerate (*scale bar applies to top three panels*). The bottom two panels demonstrate that the alpha 9 L9′T point mutation results in profound innervation changes to OHCs. The left bottom figure shows that wild type mice exhibit the expected innervation pattern of medial olivocochlear fibers to the OHCs. **B.** In the alpha 9 point mutant mice, each OHC is contacted by a larger than normal contingent of olivocochlear terminals (*arrows*).

nervous system has been the topic of great interest. Spontaneous activity has been implicated in differentiation, proliferation, and migration of neurons, and has also been shown to exert significant influence over structural maturation of axonal projections and dendritic arborization (Friauf & Lohmann, 1999; Shatz, 1990, 1996; Wang & Bergles, 2015). It therefore seems possible that the efferent system might invade the developing otocyst because this projection is somehow involved in developmental or maturational aspects not only of the inner ear, but also perhaps of central auditory and vestibular pathway circuitry as well via interactions with early spontaneous activity of the periphery.

Spontaneous activity within the developing cochlea is different from the spontaneous activity observed in hearing-competent subjects. After hearing onset, spontaneous activity of auditory nerve fibers consists of a continuous neural activity that can range from a few

Hz to 100 Hz or more (Liberman, 1978). The spontaneous activity observed during development, however, takes the form of brief, high-frequency bursts of activity interspersed with long silent periods (Khazipov et al., 2004; Meister et al., 1991; Tritsch & Bergles, 2010) that slowly changes over to adult-like activity with age (Romand, 1984; Walsh & McGee, 1987). The early spontaneous activity was shown to propagate into the CNS along central auditory pathways (Tritsch et al., 2010). This is significant because while genetic programming plays a major part in establishing connectivity in the brain, refinement of these connections seems to be driven by activity (Huberman et al., 2008; O'Leary et al., 2007). If the efferent fibers are active, it could therefore be that this early efferent innervation of the developing inner ear plays a role in the development not only of the inner ear itself, but perhaps in the larger scheme, along CNS auditory pathways as well. The nAChR alpha 9 null mice (which are "functionally de-efferented" in that no efferent-mediated responses can be elicited from these mice, as described above) were used to examine these possibilities. It was demonstrated that loss of alpha 9 expression results in alteration of prehearing activity in the medial nucleus of the trapezoid body (MNTB). Activity bursts in the alpha 9 null mice were found to be shorter in duration, with shorter interspike but longer interburst intervals, and the firing rate within bursts was longer in the null mice (Clause et al., 2014). Additionally, it was shown that the topographic refinement of MNTB-LSO connectivity was altered in the alpha 9 null mice. Electrophysiologic experiments using techniques involving the uncaging of glutamate demonstrated that the input area capable of eliciting responses of principal MNTB neurons does not undergo the same (normally occurring) decrease, suggestive of synaptic pruning as the system matures that is observed in wild type mice. These functional assessments suggest that the input area remained larger in the alpha 9 nulls compared to wild types. Anatomic reconstructions of MNTB axons invading the LSO after hearing onset confirmed that pruning does not occur normally in the absence of efferent activity, that is, in the alpha 9 null mice (Clause et al., 2014).

Defects in inner hair cell exocytotic mechanisms were also found using the alpha 9 null mice, further defining a role for efferent innervation during the perinatal maturation period. During the prehearing phase of activity, inner hair cells release neurotransmitter with a high degree of calcium cooperativity. This changes dramatically by the time hearing onset occurs, and neurotransmitter release becomes linearly dependent on, and more sensitive to, calcium entry. This has been suggested to underlie a broadening of the hair cell dynamic range, which is important for encoding continuous and finely graded signals (Brandt et al., 2003; Brandt et al., 2005; Johnson et al., 2005; Johnson et al., 2008; Matthews, 1996; Matthews & Fuchs, 2010). This normal linearization process was shown to be dependent on the typical spontaneous activity found just before the onset of hearing (Johnson, Kuhn, et al., 2013). Thus, any interference with the spontaneous activity of the inner ear could in theory also alter the process of hair cell maturation with respect to normal afferent neurotransmission. Data from the alpha 9 null mice demonstrated that the transition from nonlinear to a more linear calcium dependency of neurotransmission (at inner hair cells) did not occur in the absence of efferent activity (Johnson, Wedemeyer, et al., 2013). This can be correlated with the previous findings that the maximal sensitivity of inner hair cells to ACh occurs over the second postnatal week, suggestive of maximal sensitivity to efferent input, and, as described above, that the second postnatal week represents a critical period for the normal developmental linearization of exocytotic calcium dependency (Johnson, Kuhn, et al., 2013).

Finally, a gene microarray study of the alpha 9 null cochlea (Turcan et al., 2010) examined the postnatal maturational period during which dynamic rearrangement of OC innervation to hair cells occurs (Knipper et al., 1997). Two main findings came from this approach. First, the alpha 9 null mice possessed a cochlea that, up until P7, was indistinguishable from the wild type cochlea based on a global gene expression profile. However, upon exiting the period of dynamic rearrangement, the cochleae of the alpha 9 null mice were significantly different from wild type cochleae, and appeared to cluster with the wild type cochleae of much younger ages. Second, loss of the alpha 9 gene was accompanied by the increased expression of genes encoding GABA receptor subunits ($GABA_A1$ and $GABA_A2$) as well as GAD, the GABA synthetic enzyme targeted by the anti-GAD antibody described above.

Together, these data demonstrate a significant role for efferent innervation in developmental/maturational processes beginning in the periphery, but cascading throughout the central auditory pathways. The kinds of defects observed in the absence of cholinergic efferent activity take on even greater significance when one considers that auditory processing is so intimately tied to temporal aspects of stimuli. Given that ABRs were apparently normal in the alpha 9 null mice, future experiments examining the consequences of abnormal efferent activity will likely require sophisticated experimental designs that allow for a deeper probing of function, including single-fiber electrophysiologic analysis under different, perhaps complex, acoustic environments.

A GABAergic Element to the Olivocochlear System?

In the mature brain, GABA is the major inhibitory neurotransmitter and is considered to balance excitation produced by the major excitatory neurotransmitter, glutamate. Tipping the balance of excitation and inhibition in any way results in CNS dysfunction/disease. It is also well known that during development of the brain, GABA neurotransmission is excitatory, rather than inhibitory. This is due to the balance in expression of two cation-chloride transporters, the inwardly directed Na^+-K^+-Cl^- cotransporter (NKCC1) and the outwardly directed Cl^- extruding K^+-Cl^- cotransporter, KCC2. The combined activity of these transporters determines the intracellular chloride concentration, which then dictates the direction of ion movement when chloride channels open. When NKCC1 is highly expressed, cells will respond to GABA by depolarizing due to a loss of the negative chloride ion from the cell. NKCC1 is highly expressed in embryonic and early postnatal neurons. A "GABA shift" occurs during maturation of the brain, however, led by increasing postnatal expression levels and activity of KCC2, which extrudes chloride from the cell, progressively lowering the internal chloride concentration and thereby also decreasing the E_{GABA}. The sign of the GABA synapse then becomes inhibitory.

GABA has been shown to play significant roles in early neural development, before KCC2 produces the mature GABA response. In general, GABA has been shown to have trophic effects on the neural proliferative zones of the brain, affecting the proliferation of neural progenitor cells (Wang & Kriegstein, 2009). Early excitatory GABA activity has also been shown to be important for normal morphologic development of neurons. When KCC2 is prematurely expressed in the brain, the length of neurites and dendrites are significantly stunted (Cancedda et al., 2007).

GABA binds to two main types of receptor, the ionotropic $GABA_A$ and the metabotropic $GABA_B$ receptors. $GABA_A$ receptors are multimeric receptors composed of α, β, γ, and δ subunits that form a hetropentameric structure. There are numerous subtypes of the α, β, and γ subunits, and depending on the subunit composition of the receptor, the GABA receptor can be localized either in the synaptic cleft (those receptors containing $\alpha1$ and $\gamma2$ subunits) or extrasynaptically (those containing $\alpha4$, $\alpha5$, $\alpha6$, and δ subunits). The $GABA_A$ receptor subunits are part of the same larger family of receptors, the Cys-Loop receptors that include alpha 9 and alpha 10 nAChRs subunits. The cloning of $GABA_B$ receptors (Kaupmann et al., 1997) revealed similarities between those receptors the metabotropic glutamate receptors. and $GABA_B$ receptors are G-protein coupled ($G_{i/o}$), and are composed of two subunits that dimerize, the $GABA_{B1}$ and $GABA_{B2}$ subunits. $GABA_B$ receptors have been localized both pre- and postsynaptically and are coupled to activation of postsynaptic potassium channels and inhibition of presynaptic calcium channels.

As described above, GABAergic synaptic OC terminals have been demonstrated in the cochlea by glutamic acid decarboxylase immunohistochemistry. Various $GABA_A$ subunits have also been localized in the cochlea, specifically in spiral ganglion cells, outer hair cells, and within the region of LOCS innervation below the inner hair cells (Plinkert et al., 1989; Plinkert et al., 1993; Yamamoto et al., 2002). Interestingly, the alpha 9 and alpha 10 null mice did not reveal a typical neuronal-like role for GABA in OC function, based on the fact that OC fiber shocks of alpha 9 and alpha 10 null mice did not produce *any* effects on DPOAEs or afferent CAPs. Few investigations have been carried out on GABAergic functionality in the cochlea/OC system. This may be due to the more complex nature of the receptor, which is composed of numerous different subunits that could, in principle, substitute for each other in the case of generating a constitutive null mouse line. However, seven $GABA_A$ subunit null mouse lines have been generated by different labs and were examined for cochlear-based structural or physiologic phenotypes (Maison et al.,

2006). Constitutive ablation of the α1, α2, α6, and δ subunits did not generate a phenotype identifiable at the level of the cochlea. Null ablation of α5, β2, and β3 produced ABR threshold elevations. Because DPOAE thresholds were also raised in these animals, a significant portion of the ABR threshold shifts observed likely stem from outer hair cell dysfunction. Both the α5 and β2 null mice showed an increasing ABR shift relative to the DP shift, suggesting that with aging, a neuropathy is also taking place in the cochleae of these animals. Additionally, in the α5, β2, and β3 null mice, efferent innervation was significantly reduced. Perhaps most surprisingly, at 24 weeks of age, all inner and outer hair cells in the basal 15% of the cochlea were lost in the α5, β2, and β3 null mice. Thus, the phenotypes produced by ablating the various $GABA_A$ subunits are complex, and may involve either metabolic or neurotrophic support of the system. Because of the constitutive nature of the gene ablations performed, it remains to be seen whether some of the phenotypes are the product of abnormal development, but this will have to await conditional null mouse lines.

Cochlear expression of the $GABA_B$ subunit has been examined using transgenic mice expressing a green fluorescent protein (GFP)–$GABA_{B1}$ fusion protein (Schuler et al., 2001). The $GABA_{B1}$ subunit was localized to Type I and Type II SGNs and their afferent terminals in adult cochleae (Maison et al., 2009). No expression was observed in hair cells. $GABA_{B1}$ null mice were used to examine the role of this subunit in cochlear physiology. Results indicated the presence of an approximately 10-dB ABR threshold elevation that was matched by elevation of the DPOAEs (Maison et al., 2009). This suggests that the ABR threshold elevation is the product of outer hair cell dysfunction and diminution of the cochlear amplifier (assuming, as most do, that defects in the cochlear amplifier are translated in a 1:1 relation to altered ABR thrsesholds). OC suppression of DPOAE amplitudes was unaffected by ablation of the $GABA_{B1}$ gene (Maison et al., 2009).

The $GABA_{B1}$ null mice were also used to examine the electrophysiologic state of the OC synapse (Wedemeyer et al., 2013). Using the same "reporter" mouse line and null mouse line as described above, $GABA_{B1}$ was localized to OC nerve terminals in P9-14 mice. Physiologically, while GABA failed to evoke inhibitory currents in hair cells, electrical stimulation of OC fibers activated presynaptic $GABA_B$ receptors that in turn downregulated the amount of ACh released at the OC fiber-hair cell synapse. This inhibition was shown to occur via the P/Q-type voltage-gated calcium channels of the presynaptic terminal. Because these experiments must be done in relatively young (often prehearing) animals, interpretations of the electrophysiology must be limited to this age, especially given the discordant findings of localization of the $GABA_{B1}$ protein with age. However, what these experiments definitively show is that OC fibers modulate their own synaptic activity during critical developmental times, and may be able to perform in a similar way in adults. This also indicates that the effects generated by OC activity during the early, spontaneous activity stages of auditory system development are likely to be fine-tuned and more complex than originally believed.

The Olivocochlear System, Noise-Induced Hearing Loss, and Aging

To say that the role of the OC system in hearing, and potential role in protection against noise-induced hearing loss, remains incomplete is probably to understate the degree to which this question remains contentious. What seems to be clear is that the lab setting and experimental design used in studying the OC system may give a seemingly clear indication of function that is in reality somewhat less certain. Arguments against an actual protective effect served by the OC system (other than it producing a lab-setting induced epiphenomenon) is further explored elsewhere (Kirk & Smith, 2003; Smith & Keil, 2015). Simply put, the arguments are concerned with what the OC system can be forced to do under experimental conditions, versus what it likely does under more natural settings, and what the function is that the OC system has evolved to do. The evolutionary arguments made for the actual role of the OC system are compelling, but further experimental analyses are required to settle the issue over the protective role of the OC system. Here, we will briefly introduce some of the experiments that have been used to make the case for a protective role of the OC system in order to illustrate some of the concerns raised in the arguments against such a role.

The prevalent theory concerning mechanisms of cochlear protection is that olivocochlear (OC) system activation is protective against NIHL (Rajan & Johnstone, 1988). The exact mechanism by which protection is generated is complex, being a combination of both neural and mechanical consequences, although it is assumed that the modulation of basilar membrane mechanical properties, via OC driven modulation of outer hair cell electromotility, is the major effector. This model has not changed substantially over the past several decades. Early studies, however, initially did not find an effect on sensitivity to acoustic trauma following OC fiber transection (Trahiotis & Elliott, 1970), but later investigations (Handrock & Zeisberg, 1982) began to demonstrate that loss of OC fiber innervation (specifically MOCS innervation to outer hair cells) leads to a greater susceptibility to noise-induced hearing loss (NIHL). Electrical activation of the OC system was also shown to reduce temporary threshold shifts when applied simultaneously with acoustic overexposure (Rajan, 1988; Reiter & Liberman, 1995). While evidence exists that OC stimulation can provide protection against NIHL under specific conditions, other studies employing electrical or acoustic stimulation of the OC bundle often fail to reveal protection against acoustic injury (Liberman, 1991). Additionally, not all potentially damaging sounds efficiently activate the OC system (Sridhar et al., 1995; Wiederhold, 1970; Wiederhold & Kiang, 1970). Discrepancies over the intensity levels normally encountered in the environment versus those required for OC-based protection in the lab (typically greater than 100 dB) suggest that the OC system may not protect against most noise exposures outside of the laboratory setting (Kirk & Smith, 2003). Slow effects from the medial OC (MOC) system, which are normally assumed to underlie the protection (Liberman & Gao, 1995), are maximal only at high frequencies (Sridhar et al., 1995) and require nonphysiologically intense long-duration OC stimulation for its generation (Sridhar et al., 1995). Unilateral surgically de-efferented ears show no difference in threshold shifts compared to nonlesioned ears following *moderately* intense noise exposure, suggesting that the outcome is no different in the absence or presence of the "protective" MOC (Liberman & Gao, 1995; Zheng et al., 1997). Human data seem to also raise concerns over the possibility of a protective role of the OC system. A recent clinical report found no significant correlation between measures of efferent suppression of distortion product otoacoustic emissions (DPOAEs, a readout of outer hair cell activity and indicative of OC system activation and effects on the outer hair cells) and protection against TTS (Hannah et al., 2014). Yet, other recent experiments designed to illustrate the protective nature of the OC system have reported protective effects against moderate intensity noise (Maison et al., 2013). Significantly, de-efferentation resulted in a greater loss of hearing and afferent synapses associated with the inner hair cells. The experimental design, however, used a relatively nonphysiologic stimulus paradigm (1 week of continuous noise exposures), limiting interpretation of the results, but nonetheless demonstrating what the OC system could be pushed to do. Because in this study the efferent innervation was completely lost via surgical transection of the OC bundle, it is unclear whether the results are a product of losing classical OC-mediated effects (via the nicotinic receptors) or whether the results are produced secondary to loss of some other component of efferent-based action beyond the typical mechanical modulation associated with classic medial OC function (see further discussion below). These and other reports (Kirk & Smith, 2003) spanning almost three decades of research underscore the debates and lingering concerns regarding the relative efficacy, and especially the primacy, of the protective nature of the OC system. A more complete analysis of experiments investigating the role of the OC system in protection against acoustic trauma has recently been published (Fuente, 2015), to which the reader is directed.

Various groups have used mutant mouse lines to probe the role of the OC system with respect to a potential role in providing protection against NIHL. While the most obvious mutant mouse line to use in investigating these issues would seem to be the nAChR alpha 9 null mice, the background strain (129SvEv) used to produce these mice unexpectedly also exhibited a robust resistance to NIHL as a basal phenotype (Yoshida & Liberman, 2000). Thus, testing the role of the alpha 9 gene on protection in this background was not possible. However, a mouse line that overexpresses the alpha 9 gene was used to show that extra copies of the alpha 9 gene conveyed resistance to NIHL without any changes detected in basal hearing thresholds (Maison et al., 2002). Similarly, because of the degeneration of the efferent system in the sK2 null mouse line, these animals proved to not be a good model for examining the role of the sK2 gene in NIHL. But a mouse line overexpressing the sK2 gene was used to demonstrate that extra copies of the sK2 gene produced a more robust suppression of DPOAE amplitudes during OC fiber electrical stimulation. Interestingly, the sK2 overexpresser line did not demonstrate enhanced resistance to NIHL as one may expect (Maison et al., 2007). This seemed to suggest that there is a cap on the degree of resistance the system can express against NIHL. However, the alpha 9 L9′T point mutant mouse line was used to demonstrate that the efferent system could be forced to produce significantly greater suppression levels of DPOAE amplitudes with OC fiber bundle electrical stimulation compared to the wild type state (see Figure 10–6). In addition to the effect on DPOAE amplitudes being larger, the kinetics of suppression of the DPOAE amplitudes was shown to be slower and longer lasting as well. Importantly, it was shown that the L9′T mutation produces significantly greater protection against NIHL and that the degree of protection depends on the number of copies of point mutant subunits expressed, since heterozygote mice were intermediate in level of protection compared to homozygous L9′T mice (Figure 10–9). Interestingly, this protection occurred for permanent threshold changes but was ineffective at blocking temporary threshold elevations (Taranda et al., 2009). Finally, work on the $GABA_{B1}$ null mouse line demonstrated that loss of $GABA_{B1}$ gene expression produced resistance to NIHL, but specifically only to permanent threshold shifts and not to temporary shifts (Maison et al., 2009). The segregation of protection only to permanent threshold shifts demonstrated by the $GABA_{B1}$ and alpha 9 L9′T point mutant mice could hint at different mechanisms underlying protection against the different types of hearing loss. This idea is strengthened by the demonstration that different surgical approaches to ablating the OC fibers, resulting in full OC fiber loss or only in loss of the crossed component of the OC system, produce different kinds of protection. Loss of all OC system fibers produces a significant rise in noise-induced permanent thresholds, but transection only of the crossed OC bundle did not (Kujawa & Liberman, 1997).

Finally, data from surgically manipulated wild type mice indicate a role for the OC system in long-term maintenance of cochlear function. Surgical ablation of the OC bundle, without further exposure to high intensity sound, produces an age-accelerated reduction of ABR neural responses and a related loss of afferent synaptic connectivity between SGN fibers and hair cells (Liberman et al., 2014). Interestingly, these kinds of defects are also detectable following exposure to "moderate"-intensity noise that has been shown recently to not produce permanent temporary threshold shifts, but to nonetheless induce suprathreshold response deficits (Kujawa & Liberman, 2009). Because suprathreshold hearing deficits are typically associated with the aged human population, and because these experiments begin to relate OC system functionality to protection of suprathreshold responses, it may be that a significant element of human presbycusis occurs only after an initial lack of protection normally afforded by the OC system.

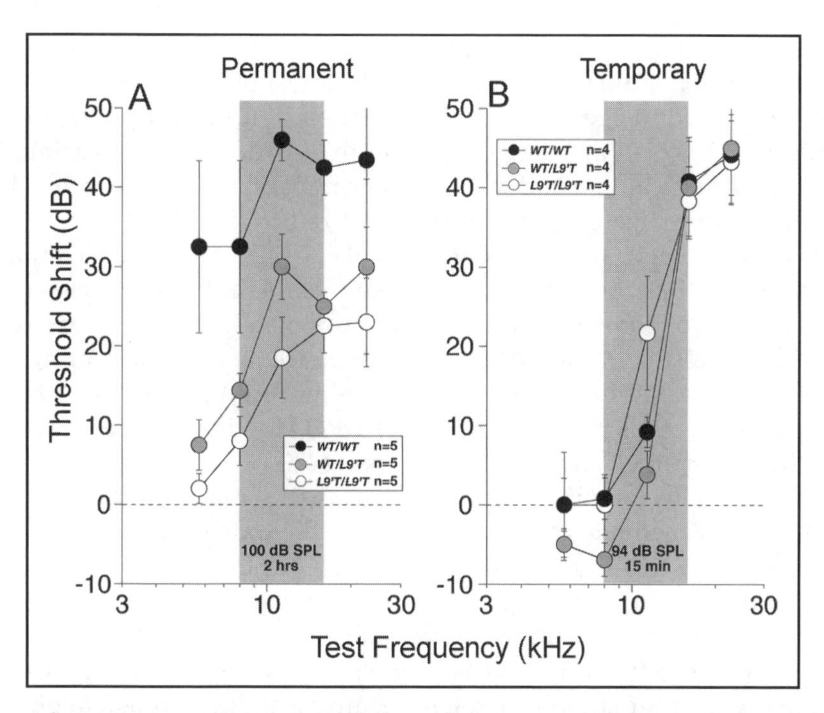

Figure 10–9. Alpha 9 L9′T point mutant mice are resistant to permanent acoustic injury. **A.** Permanent noise-induced threshold shift measured 1 week after exposure to an 8- to 16-kHz noiseband at 100 dB for 2 hours. **B.** Temporary noise-induced threshold shifts measured 6 hours after exposure to the same noise band at 94 dB for 15 minutes. Each panel shows mean thresholds ± SEM for the group, with group sizes indicated in the symbol key (WT, wild type). Reproduced from Taranda et al. (2009) with permission of authors (open source publication).

The Question of Experimental Approaches to Studying the Olivocochlear Problem and Why It Is Significant

An example of the pitfalls in interpretation of experiments induced by experimental design probably best serves to define the problem in understanding the role of the OC system. As evidenced above from the kinds of experiments performed to date on the OC system, there has been a long-standing interest in better understanding the role(s) of the efferent innervation to cochlear function. One of the more critical aspects of OC function that has captured many investigators' interest is the role of the OC system in setting up normal hearing. Thus, Walsh and colleagues set out to answer this question using a relatively straightforward, well-defined strategy that included OC bundle transection (Walsh et al., 1998). This experiment employed young, neonatal cats (most between P1 and P5), which are unresponsive to sound, and in which the cochlear epithelium is incompletely developed. Following OC bundle transection at this age, the animals were allowed to age for between 2 months and 2 years (an average age used for postsurgical analysis was 1 year old). The investigators performed numerous physiologic analyses of the auditory nerve (thus the use of larger animals instead of rodents). In short, those with successful OC bundle transections (whether midline or side cuts of the OC bundle, thus transecting the crossed MOCS or all OC fibers respectively) demonstrated significant (average 20 dB, but sometimes greater than 60 dB) elevations of auditory nerve fiber thresholds. At frequencies greater than 2 kHz, spontaneous activity of the auditory nerve fibers was generally decreased. The pathophysiology described is similar to that observed following either acoustic or drug-induced injury, and the full complement of physiologic changes observed suggested primarily an effect on the outer hair cell cochlear amplifier system. Thus, the data strongly support a role for the OC system in the normal development of cochlear physiology via impacts on the outer hair cell component.

Soon after this report, our own lab published the first results obtained with the alpha 9 null mice (Vetter et al., 1999). Because the alpha 9 nAChR is expressed in hair cells, it was assumed that the constitutive loss of this gene would generate results similar to those described above. However, while auditory nerve fiber physiology was not examined, baseline ABR/CAP-based auditory thresholds were completely normal in the null mice, seemingly in opposition to the data developed with neonatal OC bundle transection. This apparent discrepancy stood for approximately 3 years until the publication of ABR threshold changes following null ablation of urocortin (Ucn), a member of the corticotropin-releasing factor (hormone) family of peptides that we localized to the LOCS of mice (Vetter et al., 2002). Briefly, the Ucn null mice exhibited ABR threshold elevations of approximately 15 dB at best characteristic frequencies of hearing, which were mirrored by threshold elevation of DPOAEs. These mice recapitulated the results of the Walsh group.

The difference in the approaches were, obviously, surgical transection of the OC fiber bundle that ablated *all* neurotransmitters/modulators normally present in the OC synaptic terminals versus the more refined loss of one neurotransmitter receptor (alpha 9–containing nAChRs) or one neurotransmitter (Ucn). The lesson is that the OC system is neurochemically heterogeneous, and interpretations of effects produced by experimental manipulations of the OC system must be made with caution and in light of the full consequence of the manipulation used. In the end, of course, both sets of data from the different groups are valid, but each must be interpreted with respect to the experimental design used in the study.

Conclusion

In this review, historical and modern neurobiologic approaches to understanding the OC system have been discussed. Because of the long history of the system and the varied approaches taken to its study, it is impossible to fully recount all of the important contributions made by investigators over the years. It is hoped that interested readers will use this review as a springboard to further delve into those aspects of OC system functionality that most interests them. As always, any incorrect attributes to any study described are solely the responsibility of the author, and if such problems are encountered, it is hoped that the reader is not too confused over such issues. The investigation of the OC system remains a robust endeavor to this day, with new and exciting data coming at what seems to be an increasingly rapid rate. It is hoped that this review will not be rendered obsolete too soon!

Acknowledgments. This work is dedicated to my past students, who did significant work to establish our small place within the larger historic landscape of the continuing story of the olivocochlear system. The author's own work on the olivocochlear system as described herein was made possible through funding from the NIH (R01DC006258, R21DC015124) and funds from the Univ. Mississippi Medical Center Office of Research. The author declares no financial or scientific conflicts of interest in preparing this manuscript.

References

Adams, J. C., Mroz, E. A., & Sewell, W. F. (1987). A possible neurotransmitter role for CGRP in a hair-cell sensory organ. *Brain Research, 419,* 347–351.

Altschuler, R. A., Sheridan, C. E., Horn, J. W., & Wenthold, R. J. (1989). Immunocytochemical localization of glutamate immunoreactivity in the guinea pig cochlea. *Hearing Research, 42,* 167–173.

Aschoff, A., & Ostwald, J. (1987). Different origins of cochlear efferents in some bat species, rats, and guinea pigs. *The Journal of Comparative Neurology, 264,* 56–72.

Bischoff, E. (1899). Ueber den intermedullaren Verlauf des Facialis. *Neurologisches Zentralblatt, 18,* 1014–1016.

Blankenship, A. G., & Feller, M. B. (2010). Mechanisms underlying spontaneous patterned activity in developing neural circuits. *Nature Reviews, Neuroscience, 11,* 18–29.

Bond, C., Herson, P., Strassmaier, T., Hammond, R., Stackman, R., Maylie, J., & Adelman, J. (2004). Small conductance Ca2+-activated K+ channel knock-out mice reveal the identity of calcium-dependent afterhyperpolarization currents. *Journal of Neuroscience, 24,* 5301–5306.

Bond, C. T., Maylie, J., & Adelman, J. P. (1999). Small-conductance calcium-activated potassium channels. *Annals of the New York Academy of Sciences, 868,* 370–378.

Brandt, A., Khimich, D., & Moser, T. (2005). Few CaV1.3 channels regulate the exocytosis of a synaptic vesicle at the hair cell ribbon synapse. *Journal of Neuroscience, 25,* 11577–11585.

Brandt, A., Striessnig, J., & Moser, T. (2003). CaV1.3 channels are essential for development and presynaptic activity of cochlear inner hair cells. *Journal of Neuroscience, 23,* 10832–10840.

Brown, J. C., & Howlett, B. (1972). The olivocochlear tract in the rat and its bearing on the homologies of some constituent cell groups of the mammalian superior olivary complex: A thiocholine study. *Acta Anatomica. (Basel), 83,* 505–526.

Brown, M. C. (1987). Morphology of labeled afferent fibers in the guinea pig cochlea. *The Journal of Comparative Neurology, 260,* 591–604.

Burgess, B. J., Adams, J. C., & Nadol, J. B., Jr. (1997). Morphologic evidence for innervation of Deiters' and Hensen's cells in the guinea pig. *Hearing Research, 108,* 74–82.

Cancedda, L., Fiumelli, H., Chen, K., & Poo, M. M. (2007). Excitatory GABA action is essential for morphological maturation of cortical neurons in vivo. *Journal of Neuroscience, 27,* 5224–5235.

Clause, A., Kim, G., Sonntag, M., Weisz, C. J., Vetter, D. E., Rubsamen, R., & Kandler, K. (2014). The precise temporal pattern of prehearing spontaneous activity is necessary for tonotopic map refinement. *Neuron, 82,* 822–835.

Dawson, G. D. (1947a). Investigations on a patient subject to myoclonic seizures after sensory stimulation. *Journal of Neurology, Neurosurgery, and Psychiatry, 10,* 141–162.

Dawson, G. D. (1947b). Cerebral responses to electrical stimulation of peripheral nerve in man. *Journal of Neurology, Neurosurgery, and Psychiatry, 10,* 134–140.

Dawson, G. D. (1958). The central control of sensory inflow. *Proceedings of the Royal Society of Medicine, 51,* 531–535.

Dulon, D., Luo, L., Zhang, C., & Ryan, A. (1998). Expression of small-conductance calcium-activated potassium channels (SK) in outer hair cells of the rat cochlea. *European Journal of Neuroscience, 10,* 907–915.

Elgoyhen, A., Vetter, D., Katz, E., Rothlin, C., Heinemann, S., & Boulter, J. (2001). alpha10: A determinant of nicotinic cholinergic receptor function in mammalian vestibular and cochlear mechanosensory hair cells. *Proceedings of the National Academy of Sciences USA, 98,* 3501–3506.

Elgoyhen, A. B., Johnson, D. S., Boulter, J., Vetter, D. E., & Heinemann, S. (1994). Alpha 9: An acetylcholine receptor with novel pharmacological properties expressed in rat cochlear hair cells. *Cell, 79,* 705–715.

Ericson, H., & Blomqvist, A. (1988). Tracing of neuronal connections with cholera toxin subunit B: Light and electron microscopic immunohistochemistry using monoclonal antibodies. *Journal of Neuroscience Methods, 24,* 225–235.

Eybalin, M., Parnaud, C., Geffard, M., & Pujol R. (1988). Immunoelectron microscopy identifies several types of GABA-containing efferent synapses in the guinea-pig organ of Corti. *Neuroscience, 24,* 29–38.

Fechner, F. P., Burgess, B. J., Adams, J. C., Liberman, M. C., & Nadol, J. B., Jr. (1998). Dense innervation of Deiters' and Hensen's cells persists after chronic deefferentation of guinea pig cochleas. *The Journal of Comparative Neurology, 400,* 299–309.

Fechner, F. P., Nadol, J. J., Burgess, B. J., & Brown, M. C. (2001). Innervation of supporting cells in the apical turns of the guinea pig cochlea is from type II afferent fibers. *The Journal of Comparative Neurology, 429,* 289–298.

Fex, J. (1967). Efferent inhibition in the cochlea related to hair-cell dc activity: Study of postsynaptic activity of the crossed olivocochlear fibres in the cat. *Journal of the Acoustical Society of America, 41,* 666–675.

Fex, J., & Adams, J. C. (1978). alpha-Bungarotoxin blocks reversibly cholinergic inhibition in the cochlea. *Brain Research, 159,* 440–444.

Fex, J., & Altschuler, R. A. (1984). Glutamic acid decarboxylase immunoreactivity of olivocochlear neurons in the organ of Corti of guinea pig and rat. *Hearing Research, 15,* 123–131.

Fex, J., Altschuler, R. A., Kachar, B., Wenthold, R. J., & Zempel, J. M. (1986). GABA visualized by immunocytochemistry in the guinea pig cochlea in axons and endings of efferent neurons. *Brain Research, 366,* 106–117.

Fex, J., & Wenthold, R. J. (1976). Choline acetyltransferase, glutamate decarboxylase and tyrosine hydroxylase in the cochlea and cochlear nucleus of the guinea pig. *Brain Research, 109,* 575–585.

Filatov, G. N., & White, M. M. (1995). The role of conserved leucines in the M2 domain of the acetylcholine receptor in channel gating. *Molecular Pharmacology, 48,* 379–384.

Friauf, E., & Lohmann, C. (1999). Development of auditory brainstem circuitry.

Activity-dependent and activity-independent processes. *Cell Tissue Research, 297,* 187–195.

Fritzsch, B., & Nichols, D. H. (1993). DiI reveals a prenatal arrival of efferents at the differentiating otocyst of mice. *Hearing Research, 65,* 51–60.

Fuchs, P., & Murrow, B. (1992). A novel cholinergic receptor mediates inhibition of chick cochlear hair cells. *Proceedings of the Royal Society B: Biological Sciences, 248,* 35–40.

Fuente, A. (2015). The olivocochlear system and protection from acoustic trauma: A mini literature review. *Frontiers in Systems Neuroscience, 9,* 94.

Guinan, J. J., Jr., Warr, W. B., & Norris, B. E. (1983). Differential olivocochlear projections from lateral versus medial zones of the superior olivary complex. *The Journal of Comparative Neurology, 221,* 358–370.

Guinan, J. J., Jr., Warr, W. B., & Norris, B. E. (1984). Topographic organization of the olivocochlear projections from the lateral and medial zones of the superior olivary complex. *The Journal of Comparative Neurology, 226,* 21–27.

Hagbarth, K. E., & Kerr, D. I. (1954). Central influences on spinal afferent conduction. *Journal of Neurophysiology, 17,* 295–307.

Handrock, M., & Zeisberg, J. (1982). The influence of the effect system on adaptation, temporary and permanent threshold shift. *Archives of Otorhinolaryngology, 234,* 191–195.

Hannah, K., Ingeborg, D., Leen, M., Annelies, B., Birgit, P., Freya, S., & Bart V. (2014). Evaluation of the olivocochlear efferent reflex strength in the susceptibility to temporary hearing deterioration after music exposure in young adults. *Noise Health, 16,* 108–115.

Head, H., & Holmes, G. (1911). Sensory disturbances from cerebral lesions. *Brain, 34,* 102–254.

Hernandez-Peon, R., Scherrer, H., & Jouvet, M. (1956). Modification of electric activity in cochlear nucleus during attention in unanesthetized cats. *Science, 123,* 331–332.

Huberman, A. D., Feller, M. B., & Chapman, B. (2008). Mechanisms underlying development of visual maps and receptive fields. *Annual Review of Neuroscience, 31,* 479–509.

Huxley, T. H. (1904). Discourses, biological and geological. In *Collected essays* (Vol. 8, p. 229). London, UK: McMillan and Co..

Johnson, S. L., Kuhn, S., Franz, C., Ingham, N., Furness, D. N., Knipper, M., . . . Marcotti, W. (2013). Presynaptic maturation in auditory hair cells requires a critical period of sensory-independent spiking activity. *Proceedings of the National Academy of Sciences U S A, 110,* 8720–8725.

Johnson, S. L., Forge, A., Knipper, M., Munkner, S., & Marcotti, W. (2008). Tonotopic variation in the calcium dependence of neurotransmitter release and vesicle pool replenishment at mammalian auditory ribbon synapses. *Journal of Neuroscience, 28,* 7670–7678.

Johnson, S. L., Marcotti, W., & Kros, C. J. (2005). Increase in efficiency and reduction in Ca2+ dependence of exocytosis during development of mouse inner hair cells. *The Journal of Physiology, 563,* 177–191.

Johnson, S. L., Wedemeyer, C., Vetter, D. E., Adachi, R., Holley, M. C., Elgoyhen, A. B., & Marcotti, W. (2013). Cholinergic efferent synaptic transmission regulates the maturation of auditory hair cell ribbon synapses. *Open Biology, 3,* 130163.

Katz, E., Verbitsky, M., Rothlin, C. V., Vetter, D. E., Heinemann, S. F., & Elgoyhen, A. B. (2000). High calcium permeability and calcium block of the alpha9 nicotinic acetylcholine receptor. *Hearing Research, 141,* 117–128.

Kaupmann, K., Huggel, K., Heid, J., Flor, P. J., Bischoff, S., Mickel, S. J., . . . Bettler, B. (1997). Expression cloning of GABA(B) receptors uncovers similarity to metabotropic glutamate receptors. *Nature, 386,* 239–246.

Khazipov, R., Sirota, A., Leinekugel, X., Holmes, G. L., Ben-Ari, Y., & Buzsaki, G. (2004). Early motor activity drives spindle bursts in the developing somatosensory cortex. *Nature, 432,* 758–761.

Kirk, E. C., & Smith, D. W. (2003). Protection from acoustic trauma is not a primary function of the medial olivocochlear efferent system. *Journal of the Association of Research in Otolaryngology, 4,* 445–465.

Knipper, M., Kopschall, I., Rohbock, K., Kopke, A. K., Bonk, I., Zimmermann, U., & Zenner, H. (1997). Transient expression of NMDA receptors during rearrangement of AMPA-receptor-expressing fibers in the developing inner ear. *Cell Tissue Research, 287,* 23–41.

Köhler, M., Hirschberg, B., Bond, C., Kinzie, J., Marrion, N., Maylie, J., & Adelman, J. (1996). Small-conductance, calcium-activated potassium channels from mammalian brain. *Science, 273,* 1709–1714.

Kong, J. H., Adelman, J. P., & Fuchs, P. A. (2008). Expression of the SK2 calcium-activated potassium channel is required for cholinergic function in mouse cochlear hair cells. *The Journal of Physiology, 586,* 5471–5485.

Kosolapov, A. V., Filatov, G. N., & White, M. M. (2000). Acetylcholine receptor gating is influenced by the polarity of amino acids at position 9′ in the M2 domain. *The Journal of Membrane Biology, 174,* 191–197.

Kristensson, K., & Olsson, Y. (1971a). Uptake and retrograde axonal transport of peroxidase in hypoglossal neurons. Electron microscopical localization in the neuronal perikaryon. *Acta Neuropathologica, 19,* 1–9.

Kristensson, K., & Olsson, Y. (1971b). Retrograde axonal transport of protein. *Brain Research, 29,* 363–365.

Kristensson, K., & Olsson, Y. (1973). Diffusion pathways and retrograde axonal transport of protein tracers in peripheral nerves. *Progress in Neurobiology, 1,* 87–109.

Kristensson, K., Olsson, Y., & Sjostrand, J. (1971). Axonal uptake and retrograde transport of exogenous proteins in the hypoglossal nerve. *Brain Research, 32,* 399–406.

Kujawa, S. G., & Liberman, M. C. (1997). Conditioning-related protection from acoustic injury: effects of chronic deefferentation and sham surgery. *Journal of Neurophysiology, 78,* 3095–3106.

Kujawa, S. G., & Liberman, M. C. (2009). Adding insult to injury: cochlear nerve degeneration after "temporary" noise-induced hearing loss. *Journal of Neuroscience, 29,* 14077–14085.

Labarca, C., Nowak M. W., Zhang, H., Tang, L., Deshpande, P., & Lester, H. A. (1995). Channel gating governed symmetrically by conserved leucine residues in the M2 domain of nicotinic receptors. *Nature, 376,* 514–516.

LaVail, J. H. (1975). The retrograde transport method. *Federation Proceedings, 34,* 1618–1624.

LaVail, J. H., & LaVail, M. M. (1972). Retrograde axonal transport in the central nervous system. *Science, 176,* 1416–1417.

LaVail, J. H., & LaVail, M. M. (1974). The retrograde intraaxonal transport of horseradish peroxidase in the chick visual system: A light and electron microscopic study. *The Journal of Comparative Neurology, 157,* 303–357.

LaVail, M. M., & LaVail, J. H. (1975). Retrograde intraaxonal transport of horseradish peroxidase in retinal ganglion cells of the chick. *Brain Research, 85,* 273–280.

Leidler, R. (1914). Experimentelle Untersuchungen uber das Endigungsgebeit des Nervus vestibularis. 2. Mitteilung. *Arbeiten aus dem Neurologischen Institute an der Wiener Universitat, 21,* 151–212.

Liberman, M. C. (1978). Auditory-nerve response from cats raised in a low-noise chamber. *Journal of the Acoustical Society of America, 63,* 442–455.

Liberman, M. C. (1991). The olivocochlear efferent bundle and susceptibility of the inner ear to acoustic injury. *Journal of Neurophysiology, 65,* 123–132.

Liberman, M. C., Dodds, L. W., & Pierce, S. (1990). Afferent and efferent innervation of the cat cochlea: quantitative analysis with light and electron microscopy. *The*

Journal of Comparative Neurology, 301, 443–460.

Liberman, M. C., & Gao, W. Y. (1995). Chronic cochlear de-efferentation and susceptibility to permanent acoustic injury. *Hearing Research, 90,* 158–168.

Liberman, M. C., Liberman, L. D., & Maison, S. F. (2014). Efferent feedback slows cochlear aging. *Journal of Neuroscience, 34,* 4599–4607.

Luo, L., Bennett, T., Jung, H., & Ryan, A. (1998). Developmental expression of alpha 9 acetylcholine receptor mRNA in the rat cochlea and vestibular inner ear. *The Journal of Comparative Neurology, 393,* 320–331.

Maison, S. F., Adams, J. C., & Liberman, M. C. (2003). Olivocochlear innervation in the mouse: immunocytochemical maps, crossed versus uncrossed contributions, and transmitter colocalization. *The Journal of Comparative Neurology, 455,* 406–416.

Maison, S. F., Casanova, E., Holstein, G. R., Bettler, B., & Liberman, M. C. (2009). Loss of GABAB receptors in cochlear neurons: Threshold elevation suggests modulation of outer hair cell function by type II afferent fibers. *Journal of the Association for Research in Otolaryngology, 10,* 50–63.

Maison, S. F., Luebke, A. E., Liberman, M. C., & Zuo, J. (2002). Efferent protection from acoustic injury is mediated via alpha9 nicotinic acetylcholine receptors on outer hair cells. *Journal of Neuroscience, 22,* 10838–10846.

Maison, S. F., Parker, L. L., Young, L., Adelman, J. P., Zuo, J., & Liberman, M. C. (2007). Overexpression of SK2 channels enhances efferent suppression of cochlear responses without enhancing noise resistance. *Journal of Neurophysiology, 97,* 2930–2936.

Maison, S. F., Rosahl, T. W., Homanics, G. E., & Liberman, M. C. (2006). Functional role of GABAergic innervation of the cochlea: Phenotypic analysis of mice lacking GABA(A) receptor subunits alpha 1, alpha 2, alpha 5, alpha 6, beta 2, beta 3, or delta. *Journal of Neuroscience, 26,* 10315–10326.

Maison, S. F., Usubuchi, H., & Liberman, M. C. (2013). Efferent feedback minimizes cochlear neuropathy from moderate noise exposure. *Journal of Neuroscience, 33,* 5542–5552.

Matthews, G., & Fuchs, P. (2010). The diverse roles of ribbon synapses in sensory neurotransmission. *Nature Reviews, Neuroscience, 11,* 812–822.

Matthews, G. H. (1996). Calcium dependence of neurostransmitter release. *Seminars in the Neurosciences, 8,* 329–334.

Meister, M., Wong, R. O., Baylor, D. A., & Shatz, C. J. (1991). Synchronous bursts of action potentials in ganglion cells of the developing mammalian retina. *Science, 252,* 939–943.

Miyazawa, A., Fujiyoshi, Y., & Unwin, N. (2003). Structure and gating mechanism of the acetylcholine receptor pore. *Nature, 423,* 949–955.

Murthy, V., Maison, S. F., Taranda, J., Haque, N., Bond, C. T., Elgoyhen, A. B., . . . Vetter, D. E. (2009). SK2 channels are required for function and long-term survival of efferent synapses on mammalian outer hair cells. *Molecular and Cellular Neuroscience, 40,* 39–49.

Murthy, V., Taranda, J., Elgoyhen, A. B., & Vetter, D. E. (2009). Activity of nAChRs containing alpha9 subunits modulates synapse stabilization via bidirectional signaling programs. *Developmental Neurobiology, 69,* 931–949.

New, H. V., & Mudge, A. W. (1986). Calcitonin gene-related peptide regulates muscle acetylcholine receptor synthesis. *Nature, 323,* 809–811.

O'Leary, D. D., Chou, S. J., & Sahara, S. (2007). Area patterning of the mammalian cortex. *Neuron, 56,* 252–269.

Papez, J. W. (1930). Superior olivary nucleus —Its fiber connections. *Archives of Neurology and Psychiatry, 24,* 1–20.

Plazas, P. V., De Rosa, M. J., Gomez-Casati, M. E., Verbitsky, M., Weisstaub, N., Katz, E., . . . Elgoyhen, A. B. (2005). Key roles of hydrophobic rings of TM2 in gating of the alpha9alpha10 nicotinic cholinergic receptor. *British Journal of Pharmacology, 145,* 963–974.

Plinkert, P. K., Gitter, A. H., Mohler, H., & Zenner, H. P. (1993). Structure, pharmacology and function of GABA-A receptors in cochlear outer hair cells. *European Archives of Otorhinolaryngology, 250,* 351–357.

Plinkert, P. K., Mohler, H., & Zenner, H. P. (1989). A subpopulation of outer hair cells possessing GABA receptors with tonotopic organization. *Archives of Otorhinolaryngology, 246,* 417–422.

Rajan, R. (1988). Effect of electrical stimulation of the crossed olivocochlear bundle on temporary threshold shifts in auditory sensitivity. I. Dependence on electrical stimulation parameters. *Journal of Neurophysiology, 60,* 549–568.

Rajan, R., & Johnstone, B. M. (1988). Binaural acoustic stimulation exercises protective effects at the cochlea that mimic the effects of electrical stimulation of an auditory efferent pathway. *Brain Research, 459,* 241–255.

Rasmussen, G. L. (1942). An efferent cochlear bundle. *Anatomical Record, 82,* 441.

Rasmussen, G. L. (1946). The olivary peduncle and other fiber projections of the superior olivary complex. *Journal of Comparative Neurology, 84,* 141–219.

Rasmussen, G. L. (1953). Further observations of the efferent cochlear bundle. *Journal of Comparative Neurology, 99,* 61–74.

Reiter, E. R., & Liberman, M. C. (1995). Efferent-mediated protection from acoustic overexposure: relation to slow effects of olivocochlear stimulation. *Journal of Neurophysiology, 73,* 506–514.

Roberts, B. L., & Meredith, G. E. (1992). The efferent innervation of the ear: Variations on an enigma. In D. B. Webster, R. R. Fay, & A. N. Popper (Eds.), *The evolutionary biology of hearing* (pp. 185–210). New York, NY: Springer-Verlag.

Robertson, D. (1985). Brainstem location of efferent neurones projecting to the guinea pig cochlea. *Hearing Research, 20,* 79–84.

Robertson, D., Cole, K. S., & Corbett, K. (1987). Quantitative estimate of bilaterally projecting medial olivocochlear neurones in the guinea pig brainstem. *Hearing Research, 27,* 177–181.

Romand, R. (1984). Functional properties of auditory-nerve fibers during postnatal development in the kitten. *Experimental Brain Research, 56,* 395–402.

Ross, M. (1981). Centrally originating efferent terminals on hair cells: Fact or fancy? In T. Gualtierotti (Ed.), *The vestibular system: Function and morphology* (pp. 160–183). New York, NY: Springer-Verlag.

Rossi, G. (1960). Acetylcholinersterase in the ganglionic cells of the VIIIth nerve. *Italian General Review, Otorhinolaryngology, 2,* 587–596.

Rossi, G., & Cortesina, G. (1963). Research on the efferent innervation of the inner ear. *Journal of Laryngology, 77,* 202–233.

Ruben, R. J. (1967). Development of the inner ear of the mouse: A radioautographic study of terminal mitoses. *Acta Oto-Laryngologica Supplement, 220,* 221–244.

Schuler, V., Lüscher, C., Blanchet, C., Klix, N., Sansig, G., Klebs, K., . . . Bettler, B. (2001). Epilepsy, hyperalgesia, impaired memory, and loss of pre- and postsynaptic GABA(B) responses in mice lacking GABA(B(1)). *Neuron, 31,* 47–58.

Shatz, C. (1990). Impulse activity and the patterning of connections during CNS development. *Neuron, 5,* 745–756.

Shatz, C. (1996). Emergence of order in visual system development. *Proceedings of the National Academy of Sciences USA, 93,* 602–608.

Smith, D. W., & Keil, A. (2015). The biological role of the medial olivocochlear efferents in hearing: separating evolved function from exaptation. *Frontiers in Systems Neuroscience, 9,* 12.

Sridhar, T. S., Liberman, M. C., Brown, M. C., & Sewell, W. F. (1995). A novel cholinergic "slow effect" of efferent stimulation on cochlear potentials in the guinea pig. *Journal of Neuroscience, 15,* 3667–3678.

Taranda, J., Maison, S. F., Ballestero, J. A., Katz, E., Savino, J., Vetter, D. E., . . . Elgoyhen, A. B. (2009). A point mutation in the hair cell nicotinic cholinergic receptor prolongs cochlear inhibition and enhances noise protection. *PLoS Biology, 7,* e18.

Thiers, F. A., Burgess, B. J., & Nadol, J. B., Jr. (2002). Axodendritic and dendrodendritic synapses within outer spiral bundles in a human. *Hearing Research, 164,* 97–104.

Thiers, F. A., Nadol, J. B., Jr., & Liberman, M. C. (2008). Reciprocal synapses between outer hair cells and their afferent terminals: Evidence for a local neural network in the mammalian cochlea. *Journal of the Association for Research in Otolaryngology, 9,* 477–489.

Trahiotis, C., & Elliott, D. N. (1970). Behavioral investigation of some possible effects of sectioning the crossed olivocochlear bundle. *Journal of the Acoustical Society of America, 47,* 592–596.

Tritsch, N. X., & Bergles, D. E. (2010). Developmental regulation of spontaneous activity in the mammalian cochlea. *Journal of Neuroscience, 30,* 1539–1550.

Tritsch, N. X., Rodríguez-Contreras, A., Crins, T. T. H., Wang, H. C., Borst, J. G. G., & Bergles, D. E. (2010). Calcium action potentials in hair cells pattern auditory neuron activity before hearing onset. *Nature Neuroscience, 13,* 1050–1052.

Turcan, S., Slonim, D. K., & Vetter, D. E. (2010). Lack of nAChR activity depresses cochlear maturation and up-regulates GABA system components: Temporal profiling of gene expression in α9 null mice. *PLoS One, 5,* e9058.

Vaughan, J., Donaldson, C., Bittencourt, J., Perrin, M., Lewis, K., Sutton, S., . . . Rivier, C. (1995). Urocortin, a mammalian neuropeptide related to fish urotensin I and to corticotropin-releasing factor. *Nature, 378,* 287–292.

Vetter, D. E., Adams, J. C., & Mugnaini, E. (1991). Chemically distinct rat olivocochlear neurons. *Synapse, 7,* 21–43.

Vetter, D. E., Katz, E., Maison, S. F., Taranda, J., Turcan, S., Ballestero, J., . . . Boulter, J. (2007). The α10 nicotinic acetylcholine receptor subunit is required for normal synaptic function and integrity of the olivocochlear system. *Proceedings of the National Academy of Sciences USA, 104,* 20594–20599.

Vetter, D. E., Li, C., Zhao, L., Contarino, A., Liberman, M. C., Smith, G. W., . . . Lee, K.-F. (2002). Urocortin-deficient mice show hearing impairment and increased anxiety-like behavior. *Nature Genetics, 31,* 363–369.

Vetter, D. E., Liberman, M. C., Mann, J., Barhanin, J., Boulter, J., Brown, M. C., . . . Elgoyhen, A. B. (1999). Role of alpha9 nicotinic ACh receptor subunits in the development and function of cochlear efferent innervation. *Neuron, 23,* 93–103.

Vetter, D. E., & Mugnaini, E. (1992). Distribution and dendritic features of three groups of rat olivocochlear neurons. A study with two retrograde cholera toxin tracers. *Anatomy and Embryology (Berlin), 185,* 1–16.

Walsh, E. J., & McGee, J. (1987). Postnatal development of auditory nerve and cochlear nucleus neuronal responses in kittens. *Hearing Research, 28,* 97–116.

Walsh, E., McGee, J., McFadden, S., & Liberman, M. (1998). Long-term effects of sectioning the olivocochlear bundle in neonatal cats. *Journal of Neuroscience, 18,* 3859–3869.

Wang, D. D., & Kriegstein, A. R. (2009). Defining the role of GABA in cortical development. *The Journal of Physiology, 587,* 1873–1879.

Wang, H. C., & Bergles, D. E. (2015). Spontaneous activity in the developing auditory system. *Cell and Tissue Research, 361,* 65–75.

Warr, W. B. (1975). Olivocochlear and vestibular efferent neurons of the feline brain stem: their location, morphology and number determined by retrograde axonal transport and acetylcholinesterase histochemistry. *The Journal of Comparative Neurology, 161,* 159–181.

Warr, W. B., Boche, J. B., & Neely, S. T. (1997). Efferent innervation of the inner hair cell region: origins and terminations of two lateral olivocochlear systems. *Hearing Research, 108,* 89–111.

Wedemeyer, C., Zorrilla de San Martin, J., Ballestero, J., Gomez-Casati, M. E., Torbidoni, A. V., Fuchs, P. A., & Katz, E. (2013). Activation of presynaptic GABA(B(1a,2)) receptors inhibits synaptic transmission at mammalian inhibitory cholinergic olivocochlear-hair cell synapses. *Journal of Neuroscience, 33,* 15477–15487.

Wiederhold, M. L. (1970). Variations in the effects of electric stimulation of the crossed olivocochlear bundle on cat single auditory-nerve-fiber responses to tone bursts. *Journal of the Acoustical Society of America, 48,* 966–977.

Wiederhold, M. L., & Kiang, N. Y. (1970). Effects of electric stimulation of the crossed olivocochlear bundle on single auditory-nerve fibers in the cat. *Journal of the Acoustical Society of America, 48,* 950–965.

Wright, C. G., & Preston, R. E. (1976). Efferent nerve fibers associated with the outermost supporting cells of the organ of Corti in the guinea pig. *Acta Oto-Laryngologica, 82,* 41–47.

Yamamoto, Y., Matsubara, A., Ishii, K., Makinae, K., Sasaki, A., & Shinkawa H. (2002). Localization of gamma-aminobutyric acid A receptor subunits in the rat spiral ganglion and organ of Corti. *Acta Oto-Laryngologica, 122,* 709–714.

Yoshida, N., & Liberman, M. C. (2000). Sound conditioning reduces noise-induced permanent threshold shift in mice. *Hearing Research, 148,* 213–219.

Zheng, X. Y., Henderson, D., McFadden, S. L., & Hu, B. H. (1997). The role of the cochlear efferent system in acquired resistance to noise-induced hearing loss. *Hearing Research, 104,* 191–203.

CHAPTER 11

Current Progress With Auditory Midbrain Implants*

Hubert H. Lim, James F. Patrick, and Thomas Lenarz

Abbreviations

A1: primary auditory cortex

ABI: auditory brainstem implant

ACC: core/primary auditory cortex regions

AM: amplitude modulation

AMI: auditory midbrain implant

CI: cochlear implant

CT: computed tomography (imaging)

DRNL: dual-resonance nonlinear (model)

DSS: dual-site stimulation (within an ICC lamina)

IC: inferior colliculus

ICC: central nucleus of inferior colliculus

LFP: local field potential

MGV: ventral division of medial geniculate nucleus

MRI: magnetic resonance imaging

NF2: neurofibromatosis type 2

PABI: penetrating auditory brainstem implant

PSTH: poststimulus time histogram

R: correlation coefficient

SC: superior colliculus

SSS: single-site stimulation (within an ICC lamina)

**Note.* This book chapter was derived from the text and figures presented in a previously published review paper (Lim, H. H., & Lenarz, T. (2015). Auditory midbrain implant: Research and development towards a second clinical trial. *Hearing Research, 322,* 212–223).

There are hundreds of thousands of individuals implanted with a neural device for restoring sensory, motor, or autonomic function as well as for treating neurologic and psychiatric disorders (Johnson et al., 2013; Konrad & Shanks, 2010; Navarro et al., 2005). These devices interface with the peripheral or central nervous system, and can be fully implanted into the body or head with wireless capabilities. A few successful examples include implants that have restored useful hearing, vision, movement, and bladder control as well as those that have suppressed pain, seizures, and tinnitus. The field of neuroengineering has experienced tremendous growth in technological, scientific, and clinical advances over the past decade. Even greater developments and research in neural interfacing are expected in the upcoming decade, propelled forward by international efforts and investments in mapping, modeling, and understanding the human brain relevant for clinical applications. Two major examples of recent international support are the Brain Research through Advancing Innovative Neurotechnologies (BRAIN) Initiative in the United States and the Human Brain Project (HBP) in Europe to fund brain and neuroengineering research with a projected budget of several billions of dollars over 10 years. The fact that a significant component of these funds is directed toward neural prosthetic research is a revealing sign that neural stimulators are becoming more widely considered as potential treatment options for various brain disorders.

One neural prostheses that particularly stands out among these different devices mentioned above is known as the cochlear implant (CI), which is designed for implantation into the cochlea for electrically stimulating nearby auditory nerve fibers for hearing restoration (Figure 11–1) (Wilson & Dorman, 2008; Zeng, Rebscher, Harrison, Sun, & Feng, 2008). Over 350,000 patients have received a CI, with many of these individuals capable of speech perception and even the ability to converse over the telephone. Children, including infants younger than 1 year of age, have been implanted with a CI and have been able to integrate into mainstream schools. Therefore, the CI has been remarkably successful in restoring hearing to many deaf individuals, which in turn has guided the development of other neural prostheses for sensory or motor restoration, such as the visual prosthesis or a neural-controlled prosthetic limb (Weber, Friesen, & Miller, 2012; Weiland, Cho, & Humayun, 2011). The monumental achievements of the CI are attributed to the continuous efforts of several visionaries, including André Djourno, William House, Blair Simmons, Graeme Clark, and Ingeborg Hochmair (Eisenberg, 2015; Lenarz, 1998; Mudry & Mills, 2013).

In thinking about the future of auditory prostheses, the question arises as to how hearing performance can be further improved beyond what is possible with current devices, not only for those who are implanted with a CI but also for those who do not have a functional auditory nerve or implantable cochlea. There are exciting efforts toward improving the design of CIs (e.g., new electrode arrays and binaural or bimodal implants) and activation of the auditory nerve (e.g., current steering techniques, direct nerve stimulation, and optical activation methods) for

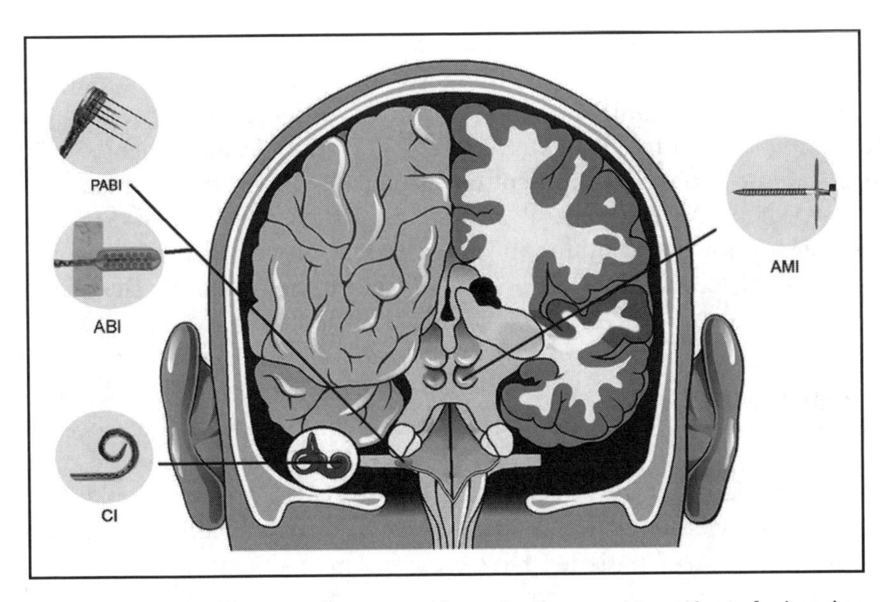

Figure 11–1. Different auditory neural prosthetics used in patients for hearing restoration. CI: cochlear implant, which consists of an electrode array that is implanted into the cochlea and used for auditory nerve stimulation. ABI: auditory brainstem implant, which is used for surface stimulation of the cochlear nucleus. PABI: penetrating auditory brainstem implant, which is used for penetrating stimulation of the cochlear nucleus. AMI: auditory midbrain implant, which is used for penetrating stimulation of the auditory midbrain (i.e., the inferior colliculus). There are several companies that build these types of implant devices. The examples shown in this figure are developed by Cochlear Limited (Australia). Reproduced from Lenarz et al. (2006), with permission of Wolters Kluwer Health, Inc.

achieving better performance in noisy environments and with more complex inputs such as music, tonal languages, and multiple talkers. Various technological, modeling, signal processing, physiology, and psychophysics research to achieve these improvements have been pursued in the CI field. The focus of this chapter is to present the development and translation of devices for stimulation beyond the auditory nerve within more central auditory structures, particularly the inferior colliculus (IC). Central auditory implants can provide an alternative hearing option for those who cannot benefit from a CI. Furthermore, a major limitation in achieving higher performance with CIs appears to be the limited number of independent information channels available through cochlear stimulation (Friesen, Shannon, Baskent, & Wang, 2001). The CI sends current through a bony modiolar wall of the cochlea with scattered flow of electrical charge to a variable distribution and reduced number of auditory neurons associated with deafness. Central auditory prostheses may provide a way for achieving more specific activation of a greater number of frequency channels of information than is currently possible with CIs.

This chapter presents the rationale for the AMI and the results of the first clinical trial using a multisite single-shank array. The animal and human studies leading to the development of a new two-shank AMI array will then be presented followed by an update on the second clinical trial.

Rationale for the AMI

The CI can provide high levels of speech understanding, at least in quiet environments, for many deaf patients. However, the CI is designed for electrically activating the auditory nerve. For those patients without a functional auditory nerve (e.g., due to a head injury or tumor removal surgery, or being born without a nerve) or without an implantable cochlea to enable array insertion (e.g., due to ossification or head trauma), then the only hearing option is a central auditory implant. The first device, known as the ABI, was implanted as early as 1979 at the House Ear Institute in Los Angeles, California, by William Hitselberger and William House. It consisted of two ball electrodes with a fabric backing that was built in collaboration with Douglas McCreery from the Huntington Medical Research Institutes in Pasadena, California. The ABI was positioned onto the surface of the cochlear nucleus. Further details of the development of the first ABIs are provided in (Schwartz, Otto, Shannon, Hitselberger, & Brackmann, 2008; Sennaroglu & Ziyal, 2012). The ABI was initially designed and justified for patients with a genetic disease known as neurofibromatosis type 2 (NF2), which is usually associated with bilateral acoustic neuromas. Due to removal of these tumors and complete damage of the auditory nerves, the patients became bilaterally deaf and unable to benefit from CIs. Since the cochlear nucleus was already approached during tumor removal, it was then possible to place the electrodes on its surface with minimal added surgical risk. A total of 25 patients were implanted with an ABI by 1992 (Schwartz et al., 2008). Since 1992, the single channel ABI has been developed into a multisite surface array (see Figure 11–1) by several implant companies (e.g., Advanced Bionics Corporation, USA; Cochlear Limited, Australia; Med-El Company, Austria; MXM Digisonic, France) and implanted in over 1,200 patients worldwide with etiologies no longer limited to NF2 (e.g., those with nerve aplasia/avulsion or cochlear ossification).

The current status of the ABI is that it can achieve high levels of hearing performance in some patients (Behr et al., 2014; Colletti, Shannon, & Colletti, 2014; Colletti, Shannon, Carner, Veronese, & Colletti, 2009; Matthies et al., 2014). There appears to be certain types of deaf patients who achieve good hearing performance with an ABI. For example, one study by (V. Colletti et al., 2009) showed that over half of the 48 nontumor (i.e., non-NF2) adult patients implanted with the ABI achieved reasonable speech perception with a few reaching levels comparable to the top CI patients. These nontumor patients obtained an average score of 59% on an open-set speech test compared to an average score of 10% across 32 NF2 adult patients. Considering that similar implants, stimulation strategies, and surgical approaches were used for both patient groups in the same clinic, these findings suggested that the limited performance observed in NF2 patients may be related to tumor damage, including surgical damage, of the cochlear nucleus (Behr et al., 2014; Colletti & Shannon, 2005). Even within the nontumor group, it appeared that those with cochlear ossification or who lost their auditory nerve due to head trauma performed better than those who had cochlear malformations or auditory neuropathy (Colletti et al., 2009). Similar trends have also been observed in children with ABIs in which those with cochlear damage due to ossification or head trauma achieved better performance over other groups (Colletti et al., 2014).

The fact that the ABI can provide sufficient speech understanding in some patients demonstrates that artificial electrical stimulation even within the brain can restore sufficient hearing function. The question now arises as to how we can further improve central auditory prostheses so that a majority of implanted patients can achieve sufficient hearing performance, especially those with NF2 tumors. There are recent reports indicating that a few NF2 ABI patients are able to achieve speech understanding comparable to typical CI patients (Behr et al., 2014; Colletti, Shannon, & Colletti, 2012; Matthies et al., 2014). One proposed reason for these encouraging results is that the surgeons were able to minimize damage to the brainstem during tumor removal surgery and/or array implantation. In over 1,000 NF2 patients with ABIs, however, only a few of them have achieved high levels of speech perception, revealing the difficulties in minimizing brainstem damage and/or accurately placing the array onto the cochlear nucleus (Colletti & Shannon, 2005; Lenarz et al., 2002; Schwartz et al., 2008; Sennaroglu & Ziyal, 2012), assuming those are the main reasons for the limited hearing performance. Figure 11–2 shows data for the speech perception performance across multiple patients implanted with the ABI and evaluated at Hannover Medical School. ABI patients typically achieve an average of about 5 to 10 words/min without lip-reading cues (audio-only condition) even after several years of implant use. They can achieve about 30 to 40 words/min with lip-reading cues (audiovisual condition). Normal hearing performance is about 85 to 100 words/min whereas CI patients in the same Hannover clinic achieve an average of about 40 words/min without lip-reading cues (n = 864; (Krueger et al., 2008; Strauss-Schier, Battmer, Rost, Allum-Mecklenburg, & Lenarz, 1995), which is much higher than what is possible by typical ABI NF2 patients.

Considering the factors described above, the authors of this chapter seek to improve central auditory prostheses by stimulating within the inferior colliculus (IC), particularly its central nucleus (ICC), and initially targeting those with NF2. Unlike the brainstem, the midbrain is directly visible during surgery (images are shown later) and is not surrounded by distorted or damaged brain structures caused by a NF2 tumor and/or its removal (Samii et al., 2007; Vince et al., 2010). Surrounding the brainstem, there are also caudal cranial nerves involved with critical functions such as breathing and swallowing that may not be easily visible during surgery. The trochlear nerve is the only nerve near the midbrain and is directly visible during surgery. In terms

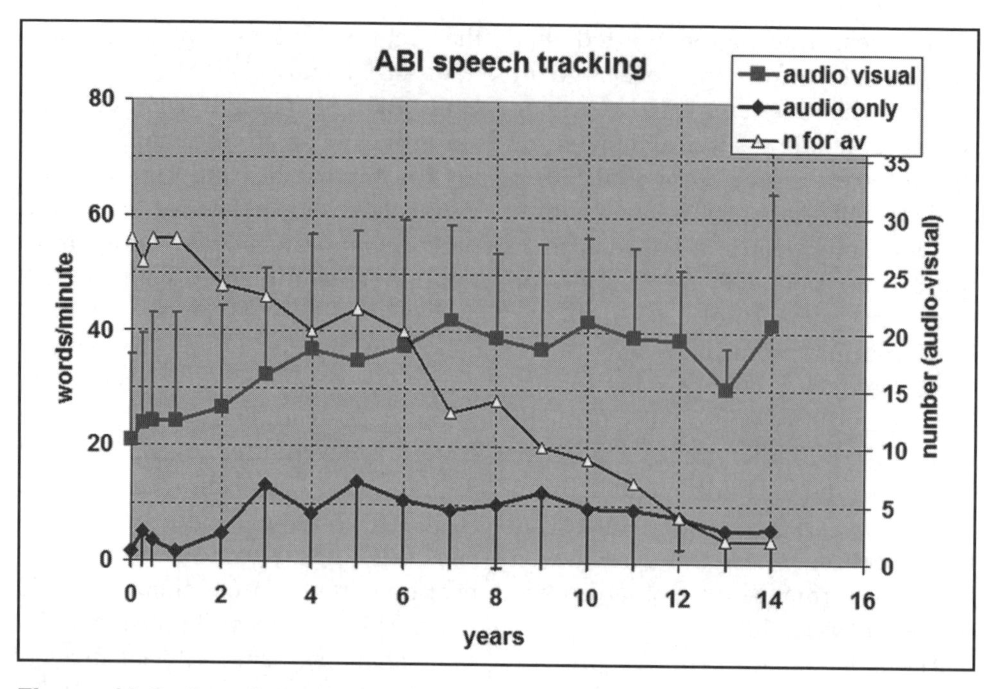

Figure 11–2. Speech tracking results over time in ABI patients at Medical University Hannover. Speech tracking (modified open set, chance level of 0%, in words per minute labeled on the left) involved reading a story to the patient who was asked to repeat the words of the cited sentences. The number of correct words in 5 minutes was obtained and divided by 5 to obtain the correct number of words per minute; "audio visual" consists of the ABI and lip-reading, while "audio only" is only the ABI. "n for av" is the total number of subjects included in calculating the mean and standard deviation bars for the other two curves shown in the plot and is labeled on the right. Further details on the speech tracking method are presented in Lim et al. (2007).

of function, the ICC has a well-defined tonotopic organization (De Martino et al., 2013; Lim, Lenarz, Joseph, & Lenarz, 2013; Oliver, 2005; Ress & Chandrasekaran, 2013; Schreiner & Langner, 1997), which is favorable for implementing an auditory prosthesis (Shannon, Fu, & Galvin, 2004; Xu & Pfingst, 2008). The IC is also the initial converging center of the central auditory system (Casseday, Fremouw, & Covey, 2002; Ehret, 1997). Once the sound information is transmitted from the auditory nerve to the brain, it gets processed across multiple structures within the brainstem through several diverging pathways (Cant & Benson, 2003). The ascending sound information and pathways then converge, for the most part, into the ICC en route to the thalamus and cortex. In other words, whichever neural pathways through the brainstem are involved with transmission of speech information to higher perceptual centers, it should be possible to implant electrode sites within specific regions of the ICC to access and stimulate those pathways. Whether artificial electrical stimulation of those pathways can restore sufficient speech perception needs to be assessed in future AMI patients.

There is some concern about the surgical risks associated with implanting an electrode array into the midbrain. However, several exciting developments in the field of central neural prostheses provide a positive perspective on this topic. No one could have imagined 35 years ago that the ABI would be considered safe enough to be implanted into children as young as 1 year old (Sennaroglu et al., 2011); the FDA recently approved children as young as 18 months in the United States. Continuous improvements in the safety of the surgical approach and implant technology have made this a reality. Significant progress is also occurring for the use of deep brain stimulation (DBS) to treat various neurologic and psychiatric conditions, with more than 100,000 patients now implanted with a penetrating DBS array (Johnson et al., 2013). There are surgical risks with DBS surgery but it is not far-fetched to assume that in the future, innovative solutions will bring these complications to nearly zero. It is important to note that the standard DBS array is approximately 20 times greater in volume than an AMI shank (Figure 11–3A) and penetrates through several centimeters of cortical tissue to reach subcortical structures versus the several millimeters of tissue penetration of the AMI; thus, it has significantly more risk than the AMI array yet is being implanted in an increasing number of adults and children for various brain disorders. A recent innovative technology has pushed the field of central neural prostheses even further. A 96-site, three-dimensional penetrating array was implanted into the motor cortex in people with tetraplegia to record neural signals and control assistive devices (Hochberg et al., 2012), demonstrating that micro-machined, high-density arrays can be safely implanted into the brain. These major achievements in the neural implant field provide increasing confidence that the surgical risks of the AMI will be reduced down to nearly zero in the future and the AMI can eventually be used in a broader clinical population beyond NF2 patients.

Findings From the First Clinical Trial

Overview

The motivation for the first AMI clinical trial was to provide an alternative hearing option to the ABI in NF2 patients. AMI research and development began around 2000. Thomas Lenarz and Minoo Lenarz (currently at University Hospital of Berlin-Charité) initiated AMI developments at Hannover Medical School with James Patrick and his team from Cochlear Limited, developing an electrode array for use in humans (Figure 11–3 A and B). The AMI array consists of a single shank with 20 linearly spaced sites and was designed to be aligned along the tonotopic gradient of the ICC. They collaborated with Hubert Lim and David Anderson at the University of Michigan to validate this technology in animal studies, eventually obtaining sufficient evidence and approvals to begin the first clinical trial in 2006–2008 in which five adult NF2 patients were implanted with the device. Prior to the clinical trial, these researchers and clinicians had shown that ICC stimulation achieves low-threshold and frequency-specific auditory activation in animals that was

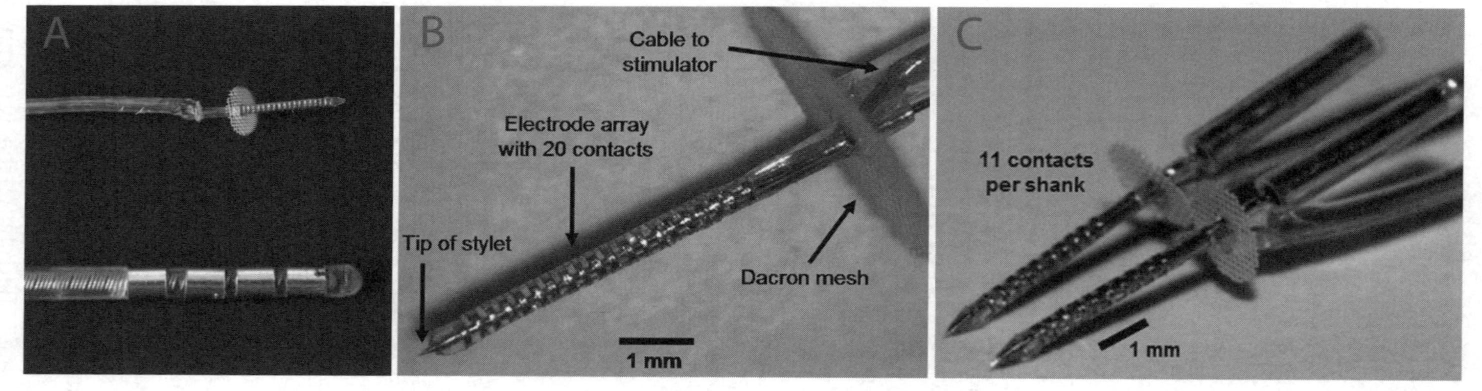

Figure 11–3. AMI arrays developed by Cochlear Limited (Australia). **A** (*top*) and **B** show the AMI array currently implanted into humans with 20 ring sites (200 µm spacing, 200 µm thickness, 400 µm diameter) along a silicone carrier. Dacron mesh prevents overinsertion of the array into the IC and tethers it to the brain. The AMI array is much smaller than current deep brain stimulation arrays used for various neurologic and psychiatric disorders in which **A** (*bottom*) shows an example array developed by Medtronic (USA). **C** shows the new two-shank AMI array that will be implanted into deaf patients in a second clinical trial. Each shank consists of 11 ring sites along a silicone carrier (300-µm site spacing except for one site positioned closer to the Dacron mesh for tinnitus treatment; see the text for further details). Reproduced from Lenarz et al. (2006) and Samii et al. (2007), with permission of Wolters Kluwer Health, Inc.

better or comparable to CI stimulation (Lenarz, Lim, Patrick, Anderson, & Lenarz, 2006; Lim & Anderson, 2006). They also showed in a cat model that long-term implantation and stimulation of the AMI device was safe without any major side effects and induced minimal tissue damage that was comparable to other clinically approved brain implants (Lenarz et al., 2007). In terms of sound coding, multiple studies have shown that ICC neurons are capable of following the temporal modulations of acoustic stimuli up to or beyond 100 Hz and the ICC has a well-defined tonotopic organization (Geniec & Morest, 1971; Joris, Schreiner, & Rees, 2004; Langner, Albert, & Briede, 2002; Oliver, 2005; Rees & Langner, 2005; Schreiner & Langner, 1997). Considering that speech perception, at least in quiet backgrounds, is possible with temporal modulations as low as ~50 Hz with just 4 to 8 frequency channels (Friesen et al., 2001; Shannon, Zeng, Kamath, Wygonski, & Ekelid, 1995; Zeng, 2004), they envisioned that the AMI would be able to restore reasonable speech perception using a CI-based strategy. In particular, each electrode site in a specific frequency region would be presented with an amplitude modulated pulse train following the bandpass-filtered envelope extracted for the corresponding frequency channel.

After obtaining the necessary approvals, five patients were implanted with the AMI and provided with a CI-based strategy. Encouragingly, the AMI has proven to be safe in all five patients for over 7 years and has provided improvements in lip-reading capabilities and environmental awareness with some speech perception, comparable to the range of performance achieved by most ABI NF2 patients (H. H. Lim, Lenarz, & Lenarz, 2009, 2011; Schwartz et al., 2008; Sennaroglu & Ziyal, 2012). These clinical results demonstrate that useful hearing can be provided by IC stimulation. However, the patients have not yet achieved sufficient speech perception without lip-reading cues. Therefore, there is still a critical need to improve the AMI if it is going to be considered as an alternative to the ABI.

Surgical Limitations

One major limitation in the first clinical trial was related to the difficulties in accurately placing the AMI array into the ICC (Figure 11–4). Out of five patients, only one (patient AMI-3) was implanted across the tonotopic gradient of the ICC. All other patients were implanted predominantly in other regions, including the dorsal and rostral IC, brachium of IC, and lateral lemniscus. As expected, AMI-3 exhibited the best hearing performance and a clear pitch organization across the sites (Lim et al., 2013; Lim et al., 2009) consistent with the tonotopy that is expected from animal and human studies (De Martino et al., 2013; Geniec & Morest, 1971; Malmierca et al., 2008; Oliver, 2005; Ress & Chandrasekaran, 2013; Schreiner & Langner, 1997).

Details of the surgical approach for the first AMI trial are provided in (Samii et al., 2007), with its limitations described in (Lim et al., 2009). Briefly, the array implantation was performed after removing the NF2 tumor at the brainstem level using a modified lateral suboccipital approach in a semi-sitting position. The cerebellum was retracted

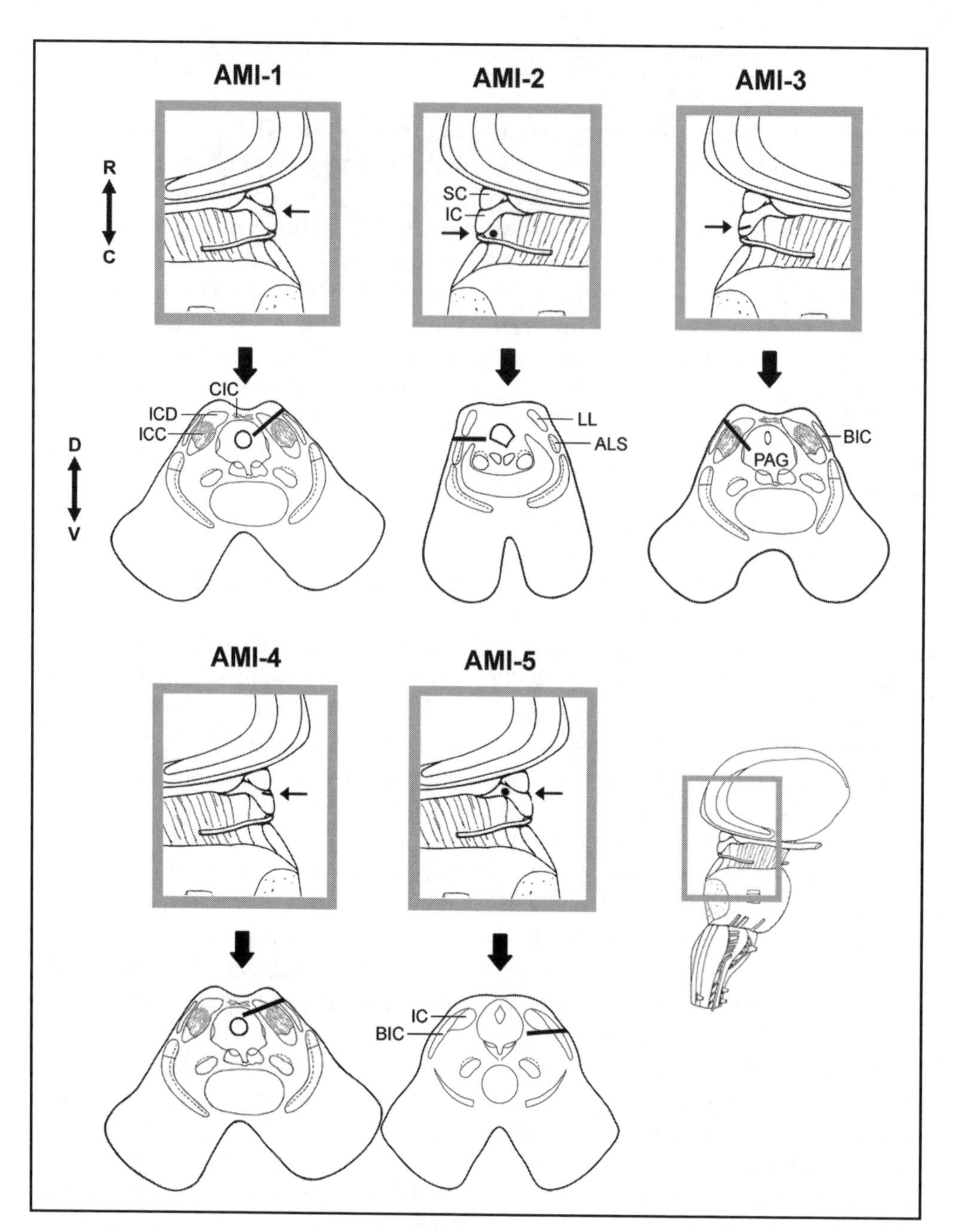

Figure 11–4. Array placement across patients in the first AMI clinical trial. For each patient, the parasagittal (*top, gray box*) and axial (*bottom, below gray box*) sections show the location and orientation of the array within the midbrain. Arrow in parasagittal section points to the caudorostral location of the array and the corresponding axial section below. The black line (or dot for AMI-2 and AMI-5) representing the array in each section corresponds to the trajectory of the array across several superimposed CT-MRI slices. ALS, anterolateral system; BIC, brachium of inferior colliculus; CIC, commissure of inferior colliculus; IC, inferior colliculus; ICC, inferior colliculus central nucleus; ICD, inferior colliculus dorsal nucleus; LL, lateral lemniscus; PAG, periaqueductal gray; SC, superior colliculus. Anatomical directions: C, caudal; D, dorsal; R, rostral; V, ventral. Further details of the reconstruction technique, anatomy of the midbrain, and AMI surgery are presented in (Lim et al., 2007; Samii et al., 2007). Only AMI-3 was properly implanted into the target region of the ICC with the array aligned along its tonotopic axis. Adapted from Lim et al. (2007).

medially to expose the tumor. After the tumor was removed, the cerebellum was allowed to drop downward due to gravity, and the IC surface could be directly viewed through the same skull opening. The main surgical limitation in the first clinical trial was the use of a small craniotomy, which made it difficult to view several key anatomic landmarks defining the outer borders of the IC and to determine the orientation of the array relative to the surface of the IC during insertion. These landmarks include the rostral border of the IC with the superior colliculus (SC), midline between both ICs, and caudal IC edge corresponding to the exit point of the trochlear nerve. The array needs to be aligned along the tonotopic gradient of the ICC, which requires an angle of insertion of about 40° relative to the sagittal plane (see Figure 11–4 for the location and orientation of the frequency laminae of the ICC; (Geniec & Morest, 1971; Kretschmann & Weinrich, 1992)). A small craniotomy was initially used to minimize surgical risks. As described later, an improved surgical approach has been developed with a larger exposure up to the midline that can still access the NF2 tumor more laterally and then approach the IC more medially with complete visibility of the landmarks mentioned above.

Frequency-Specific Activation but Limited Temporal Coding Abilities

For the one patient implanted into the ICC (AMI-3), a systematic pitch organization from low to high was observed for stimulation of superficial to deeper AMI sites as expected from the tonotopic organization observed in animal and human studies (De Martino et al., 2013; Malmierca et al., 2008; Ress & Chandrasekaran, 2013; Schreiner & Langner, 1997). However, poor temporal coding abilities were observed for AMI-3 (and the other AMI patients). Speech performance depends on both spectral and temporal cues (Nie, Barco, & Zeng, 2006; Shannon, 2002; Xu & Pfingst, 2008), and thus transmission of degraded temporal information may be limiting speech perception performance in the AMI patients. In particular, AMI-3 (and the other AMI patients) exhibited poor temporal modulation detection (i.e., ability to detect small changes in amplitude modulation, AM) and temporal resolution (i.e., ability to detect small temporal changes) compared to CI patients (Lim et al., 2008; McKay, Lim, & Lenarz, 2013). CI users can achieve reliable AM detection beyond 150 to 300 Hz (Fraser & McKay, 2012), whereas the best AMI patient exhibited degraded capabilities even at 20 to 50 Hz (Figure 11–5) (McKay et al., 2013). What was surprising was the drastic difference between CI and AMI stimulation for shorter interval pulse trains. CI patients exhibit lower thresholds (and louder percepts) as the pulse rate increases (Figure 11–6) (Kreft, Donaldson, & Nelson, 2004; McKay & McDermott, 1998; Shannon, 1989). This is attributed to a short-term integrator that sums the incoming activity within a short window (~5 ms) to track the fast temporal features that can contribute to speech understanding (McKay & McDermott, 1998; Viemeister, 1979; Viemeister & Wakefield, 1991). AMI stimulation does not exhibit this short-term integration (see Figure 11–6) (Lim et al., 2008; McKay et al., 2013).

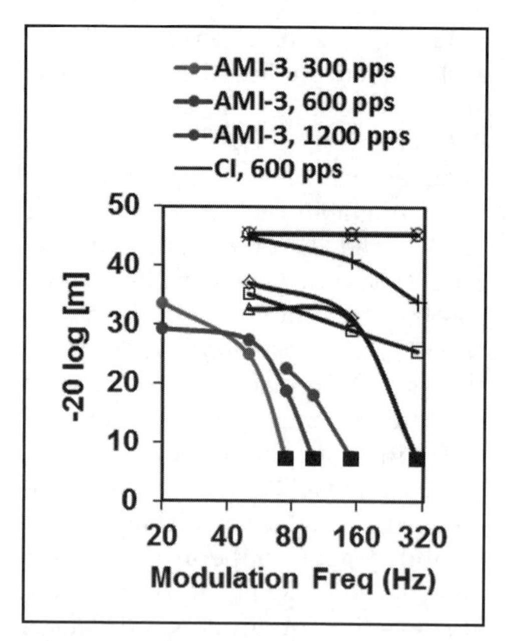

Figure 11–5. AM detection ability is lower for AMI-3 than for six typical CI users. A three-interval task with one modulated and two nonmodulated pulse trains was presented in a randomized sequence, and the subject selected the modulated interval. The modulation depth (m; higher ordinate value means less depth and better detection) for 70% correct was identified for each modulation frequency (Hz) and carrier rate (pps: pulses per second). Black square: maximum depth used. Adapted from McKay et al. (2013). Further details on the methods and results are presented in that publication.

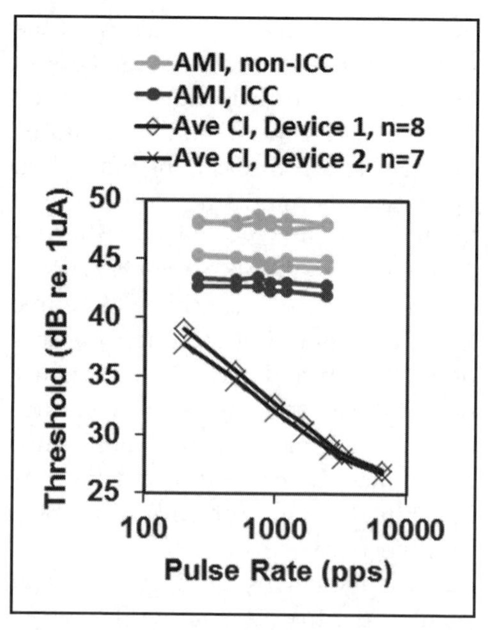

Figure 11–6. Detection thresholds do not decrease with higher pulse train rates for AMI as in CIs. AMI data are for individually stimulated sites within and outside of ICC. Adapted from Lim et al. (2008). CI data are averages across subjects for two different devices. Adapted from Kreft et al. (2004). Further details on the methods and results are presented in those corresponding publications. pps: pulses per second.

Animal and Human Studies Toward a Second Clinical Trial

Improving Neural Activation and Possibly Temporal Coding

To better understand what may be limiting short-term integration and AM detection abilities in the first AMI patients, a previous study performed ICC stimulation experiments in six ketamine-anesthetized guinea pigs (Calixto et al., 2012). Two single-shank AMI arrays were implanted parallel to each other (1.5 mm apart) with sites aligned along the tonotopic axis of the ICC. Two electrical biphasic pulses (200 µs/phase), either on one site or between two sites with varying interpulse delays (0–100 ms), were presented. The two sites were positioned into a similar ICC lamina to assess how stimulation of one versus two regions along an isofrequency lamina affected auditory cortical activity. The neural activity was recorded in the primary auditory cortex (A1) in a similar frequency region as the stimulated ICC lamina. The study discovered that stimulation of a single site with two pulses elicits strong refractory effects for shorter interpulse intervals approaching full refractory below an interval of 2 ms (Figure 11–7, SSS). In other words, AMI stimulation using multiple pulses with short intervals contributes little or no additional A1 activity than that of a single pulse, consistent with Figure 11–6 in which there was no decrease in threshold (or increase in loudness) with increasing pulse rates using the AMI. This is in contrast to CI stimulation that achieves increased activity, lower thresholds, and louder percepts for shorter interpulse intervals or higher pulse rates (McKay & McDermott, 1998; Middlebrooks, 2004; Shannon, 1985) (e.g., lower thresholds are shown in Figure 11–6 for CI stimulation).

The ICC is a three-dimensional structure with two-dimensional isofrequency laminae that have shown

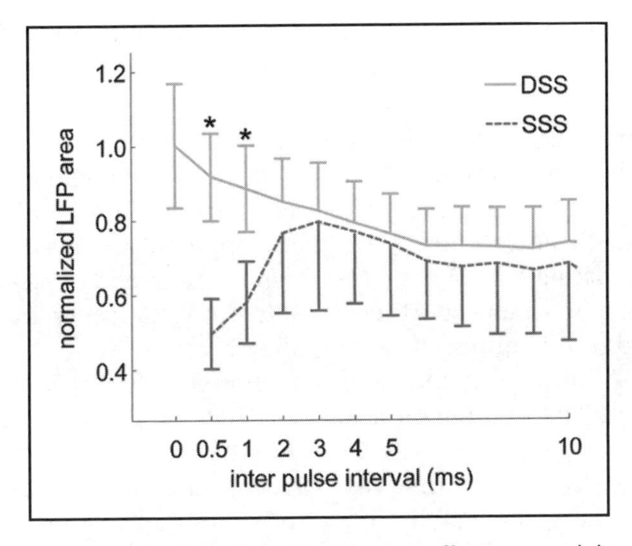

Figure 11–7. Strong suppressive effects are minimized with a two-shank AMI array. Local field potential (LFP) recordings from primary auditory cortex in response to two-pulse ICC stimulation with varying interpulse intervals either on one site (SSS) or across two sites (DSS, in same lamina). Ordinate is LFP activity to two pulses divided by the linear sum of activity to each pulse. For SSS, 0.5 means no contribution of activity for the second pulse (full refractory). Mean and standard deviation bars are shown for data across six animals (SSS: $n = 41$, DSS: $n = 72$; asterisks: $p < 0.0001$). Two single-shank AMI arrays (shown in Figure 11–3B) separated by 1.5 mm were used for these experiments. Reproduced from Calixto et al. (2012).

to code for varying temporal features of sound across different neurons (Ehret, 1997; Langner et al., 2002; Rees & Langner, 2005). Unlike stimulation of the cochlea, stimulation of a single site within a given frequency region in the ICC may not sufficiently activate higher centers with repeated electrical pulses. Instead, multisite stimulation within a lamina may be needed to achieve improved temporal activation. Encouragingly, the study by (Calixto et al., 2012) showed that stimulation of two sites within an ICC lamina elicits enhanced A1 activity with shorter intervals and overcomes the strong refractory effects observed for SSS (see Figure 11–7, DSS). This type of enhanced activity cannot be simply achieved by activating more sites across different frequency laminae but requires activation of multiple sites within the same lamina (Straka, Schendel, & Lim, 2013). Therefore, stimulating at least two sites along a lamina may restore short-term integration and could improve hearing performance. Additionally, the same study discovered that stimulation of only a single site in a lamina elicits strong suppressive effects that last beyond 100 ms in which activity to a second pulse is significantly reduced due to the activity to the first pulse. In fact, activity to the second pulse could be completely suppressed even beyond 100 ms (i.e., 4%–100% suppression at 100 ms, $n = 41$). In contrast, stimulation of two sites could exhibit full recovery and even enhanced activity to the second pulse by 100 ms (i.e., 77% suppression up to 214% enhancement, $n = 72$), exhibiting patterns closer to what is observed for two-click acoustic stimulation with varying delays (Brosch, Schulz, & Scheich, 1999; Eggermont & Smith, 1995; Wehr & Zador, 2005). The significant suppressive effects exceeding 100 ms for single-site stimulation within an ICC lamina is likely limiting AM detection abilities in which activated neurons cannot sufficiently follow the envelope fluctuations. These findings suggest that AMI stimulation of at least two sites along an ICC lamina could greatly improve temporal coding abilities on a short (<5 ms) and long (beyond 100 ms) scale, which in turn could improve speech understanding.

Can a CI-Based Strategy Still Work in the ICC?

For neural prostheses, it is challenging to develop completely new hardware and software since considerable testing and approvals are needed before using them in humans. Instead, it is favorable to use components and algorithms already approved for human use, such as those in CIs (Patrick, Busby, & Gibson, 2006). Several studies have shown that ICC neurons can follow envelope modulations of simple and natural stimuli up to a few hundred hertz (Krishna & Semple, 2000; Langner et al., 2002; Rees & Moller, 1987; Suta, Kvasnak, Popelar, & Syka, 2003; Woolley, Gill, & Theunissen, 2006). One particular study performed experiments in 10 ketamine-anesthetized guinea pigs to further assess if a CI type of strategy could potentially be effective for the AMI (Rode et al., 2013). Natural vocalizations (i.e., guinea pig speech; Figure 11–8), which exhibit components with similar temporal and spectral patterns to human speech, were presented

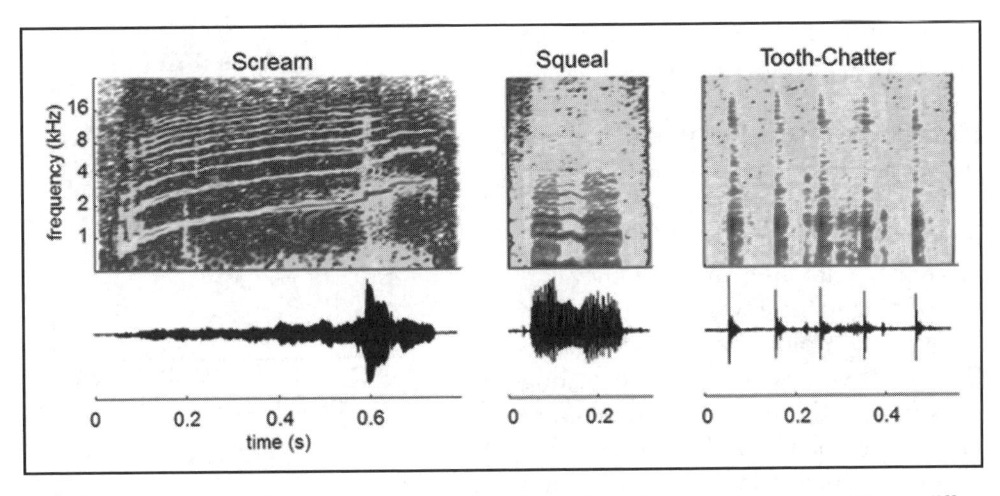

Figure 11–8. Top: Spectrograms and associated time domain waveforms. Three different types of vocalizations (replotted in gray scale) from color figure taken from Rode et al. (2013).

to the left ear of the animals and neural spiking activity was recorded across the right ICC using 32-site arrays in multiple locations per animal.

The unique aspect of that study was that a peripheral ear model was used to obtain an estimate for the true envelope pattern of the sound stimuli that reaches the basilar membrane of the cochlea. Sound travels through the eardrum and middle ear components to reach inside the fluid-filled cochlea. The fluid vibrations then cause the basilar membrane within the cochlea to fluctuate, which in turn activates hair cells and the corresponding auditory nerve fibers going to the brain. Previous studies have not typically accounted for this preprocessing that occurs from the eardrum to the basilar membrane when characterizing the effects of speech sounds on different neurons within the ICC. Fortunately, there is already a reasonably accurate mathematical model of the peripheral ear in the guinea pig, which is known as the dual-resonance nonlinear (DRNL) model (Meddis, O'Mard, & Lopez-Poveda, 2001; Sumner, O'Mard, Lopez-Poveda, & Meddis, 2003). The vocalizations were inputted into the DRNL model to obtain the output that is observed at a given frequency region along the basilar membrane. More specifically, the envelope of the output signal was extracted (up to ~100 Hz) since speech perception has been strongly correlated with the ability to transmit sufficient envelope cues to the brain (Shannon et al., 1995). What is important about this preprocessing is that it resembles the type of preprocessing already implemented in CI stimulation strategies, which electrically stimulate each electrode site with the envelope pattern of the bandpass-filtered components of the original inputted sound signal. The advantage of using the DRNL model is that it is extracting the frequency components using what is believed to be more natural processing steps compared to arbitrary bandpass filters as used in previous studies. Cross-correlation analysis can then be performed between each of those envelope signals and the temporal spiking pattern of ICC neurons (i.e., smoothed poststimulus time histograms [PSTHs]) located in a region with the same frequency corresponding to the envelope signal.

Figure 11–9 plots the correlation coefficient (*R*) values across all recording sites in the ICC from 10 animals and different frequency laminae. There were multiple neurons that exhibited high correlation values close to 1 for all three vocalizations, and thus for a wide range of spectral and temporal sound patterns. Based on visual inspection of all the raw data, *R* values ≥0.85 corresponded to neurons that accurately followed the stimulus envelope (for further examples and justifications of this criterion, see (Rode et al., 2013)). The high 0.85 criterion was achieved by 15%, 60%, and 58% of neurons for scream, squeal, and tooth-chatter, respectively. It can also be seen from Figure 11–9 that the majority of cases still had a moderately high *R* value above ~0.70. These results demonstrate that ICC neurons can follow the envelope structure of natural stimuli across different frequencies. This is an important finding because it indicates that a CI-based strategy may potentially be used for the AMI to restore sufficient speech perception as long as the right neurons are being activated (further discussed in the next section). Combined with the findings from the previous section, these results suggest that improved activation of the auditory system may be achieved with the AMI by using a CI-based strategy except that the pulse patterns for each frequency channel would be presented in an alternating or time-varying

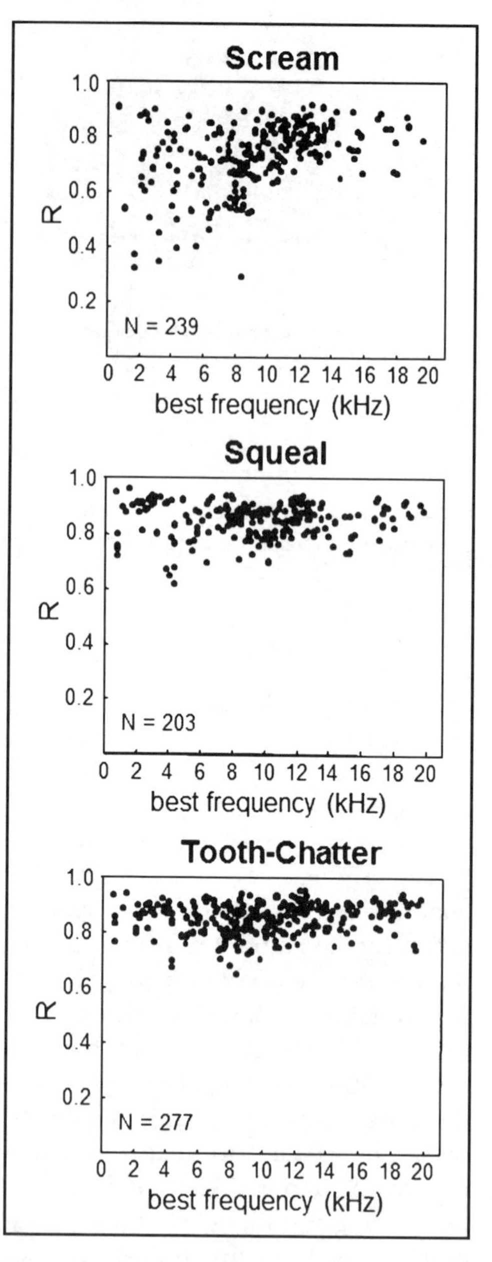

Figure 11–9. Envelope correlation coefficient (*R*) values for ICC neurons located in different frequency regions. *R* is calculated between the neuron's spiking pattern and the envelope of the vocalization for the best-matched frequency component outputted from the DRNL model. Further details on the DRNL model and correlation analysis are provided in the text. *N*: total number of neurons (i.e., multiunit sites) per vocalization. Reproduced from Rode et al. (2013).

sequence across two sites (i.e., using a two-shank array) instead of just one site in each ICC lamina. Although it may be possible to insert more than two shanks into the ICC, there would be greater surgical risks and significant technological challenges in making an AMI array with smaller dimensions and a higher density of sites and implanting multiple shanks into the ICC.

A Specific Midbrain Pathway That May Improve Speech Perception With the AMI

One caveat to the findings presented in the previous section is that not all ICC neurons had high *R* values greater than 0.85. If the neurons with the highest *R* values are located in a specific region within each ICC lamina, then it may be possible to insert a two-shank AMI array into that region (i.e., the two shanks cross each lamina to position two electrodes in that region) and systematically activate those neurons with a CI-based strategy. However, if the high *R* values are scattered throughout the ICC, then positioning only two shanks within the ICC may not sufficiently access enough of those high *R*-valued neurons. A study systematically investigating how the *R* values change as a function of location across an isofrequency lamina of the ICC still needs to be performed to answer that question. However, there are a few studies suggesting that there may be a better region within the ICC for AMI stimulation. One recent study performed experiments in 12 ketamine-anesthetized guinea pigs in which multisite arrays were used to position sites fully across a given isofrequency lamina of the ICC. Details of the methods and results are provided in (Straka, Schmitz, & Lim, 2014). This study discovered that along a given ICC lamina, there exists two subregions: a rostral-lateral region and a caudal-medial region (Figure 11–10A). The rostral-lateral ICC exhibited more precise temporal firing, shorter latencies, stronger activity, lower thresholds, and greater spatial synchrony across neurons in response to acoustic stimuli compared to the caudal-medial ICC, as listed in Figure 11–10B. In other words, there appears to exist a dual lemniscal organization within the ICC in which one pathway may be designed for more robustly transmitting sound cues to higher centers.

The concept of a dual lemniscal organization was first revealed in the 1980s (Morel & Imig, 1987; Rodrigues-Dagaeff et al., 1989), specifically for projections from the ventral division of the medial geniculate nucleus (MGV) up to the core/primary auditory cortex regions (ACC) in a cat model. The dual lemniscal pathway hypothesis was further expanded in 2006 to 2007 to include pathways from the brainstem up through the ICC, MGV, and ACC across several species, including gerbil, rat, and guinea pig (Cant & Benson, 2006, 2007; H. H. Lim & Anderson, 2007; Polley, Read, Storace, & Merzenich, 2007). Together, these studies revealed two segregated anatomical and functional pathways through the ICC (caudal-medial versus rostral-lateral regions), MGV (caudal versus rostral regions), and ACC (A1 versus core regions outside of A1). Figure 11–10A provides a simplified schematic summarizing the dual lemniscal pathways. The differences in coding properties between the rostral versus caudal MGV, demonstrated by Rodrigues-Dagaeff (1989) in

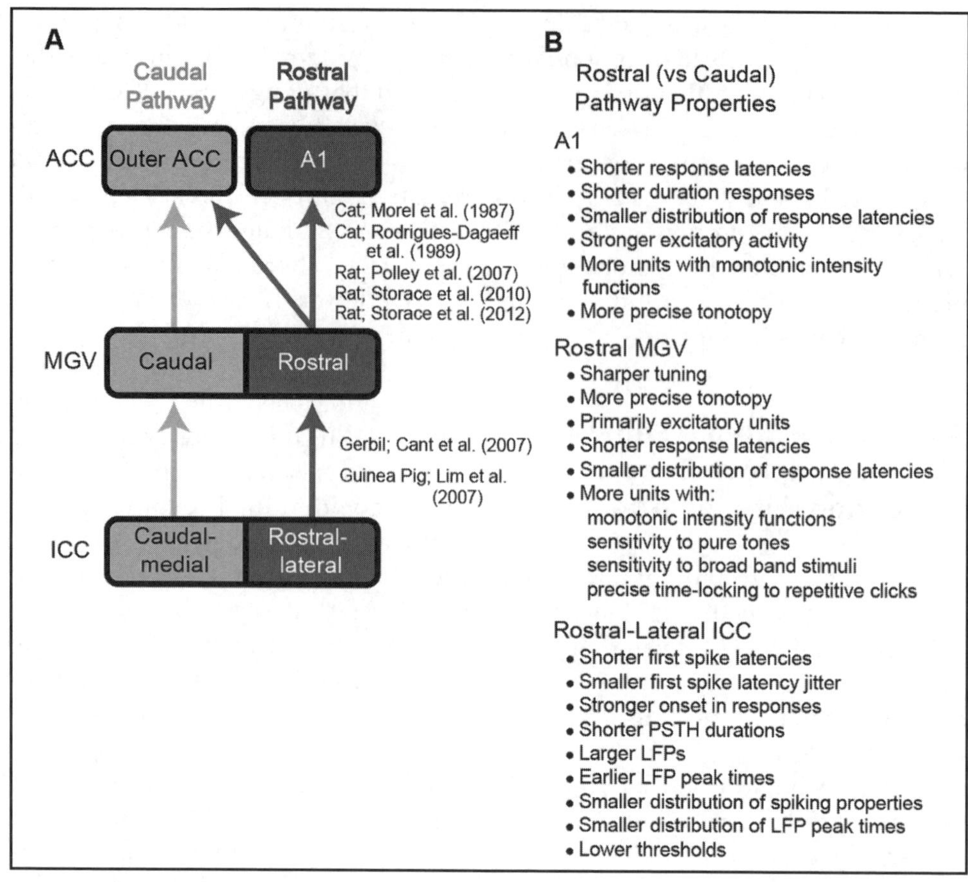

Figure 11–10. Schematic of anatomic projections and physiologic responses for the dual lemniscal pathways hypothesis. **A.** The rostral and caudal ascending pathways show spatially segregated anatomic projections from the ICC up to ACC. Overlapping projections between the two pathways are not shown. **B.** In contrast to the caudal pathway, the rostral pathway also shows different responses to acoustic stimuli in A1 (Phillips et al., 1995; Polley et al., 2007; Storace et al., 2012; Wallace et al., 2000), the rostral MGV (Rodrigues-Dagaeff et al., 1989), and the rostral-lateral ICC (Straka, Schmitz, et al., 2014). Reproduced from Straka et al. (2014).

cat and listed in Figure 11–10B, suggest that the rostral pathway is designed for stronger excitatory activation and more temporally and spectrally precise transmission of information up to higher centers. Many of these differences in coding properties between the dual pathways have also been shown in ACC (Phillips, Semple, & Kitzes, 1995; Polley et al., 2007; C. Schreiner, Froemke, & Atencio, 2011; Storace, Higgins, Chikar, Oliver, & Read, 2012; Wallace, Rutkowski, & Palmer, 2000) and more recently in ICC (Straka, Schmitz, et al., 2014), as listed in Figure 11–10B.

The high *R* values discussed in the previous section may correspond to neurons within this "rostral" pathway (i.e., rostral-lateral portion of a given isofrequency lamina of the ICC) and if targeted with the AMI, could enable high levels of speech perception. This is not to claim that speech information is only coded in this pathway but to suggest that artificial stimulation with modulated electrical pulse trains may somehow activate this pathway to provide sufficient speech understanding. Previous studies in guinea pigs have already shown that electrical stimulation of the rostral-lateral versus caudal-medial ICC achieves lower activation thresholds, stronger responses, smaller discriminable level steps, shorter response latencies, and more temporally precise firing within A1 (H. H. Lim & Anderson, 2007; Neuheiser et al., 2010). Stimulation of this rostral-lateral ICC region with two sites within a given lamina can also minimize or overcome the strong suppressive effects described previously, which is not typically or sufficiently achieved for stimulation of more caudal-medial ICC locations (Straka, McMahon, Markovitz, & Lim, 2014). In future AMI patients and by targeting the rostral-lateral ICC, there will be a unique opportunity to test if these findings in animals also occur in humans and if they relate to better speech understanding.

Improving the Surgical Approach for Array Implantation

As described previously, the main surgical limitation in the first clinical trial was the use of a small craniotomy, through which it was not possible to clearly identify key anatomical landmarks surrounding the IC. An improved surgical approach for the second clinical trial was developed based on cadaver studies. This new approach still uses a modified lateral suboccipital exposure (Samii et al., 2007), except that the skull opening is extended up to the midline (Figure 11–11A). This type of paramedian exposure has been safely used in the neurosurgical field for operating on lesions in the IC, SC, superior and middle cerebellar peduncles, and quadrangular lobules of the cerebellum (Ogata & Yonekawa, 1997), and can be used for

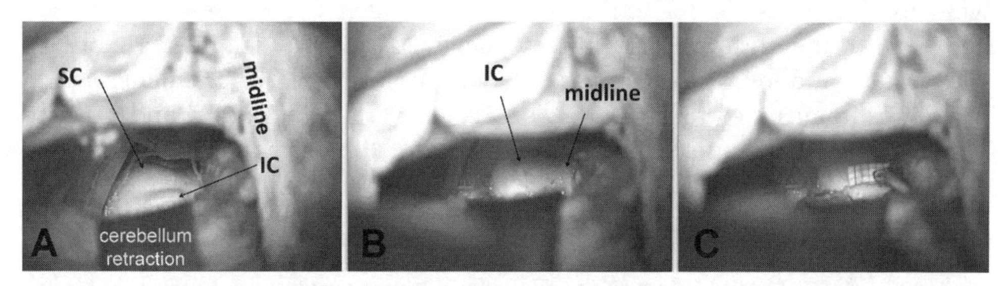

Figure 11–11. A refined surgical approach for AMI array implantation. **A.** Modified lateral suboccipital approach (*left side of the head in this image*) with the craniotomy extended to the midline. The tentorium is located immediately above the skull opening. The cerebellum is retracted downward to expose the midbrain. The IC and SC are clearly visible through this exposure. **B.** The midline and caudal edge of the IC (at the exit point of the trochlea nerve; not shown) can also be identified through this exposure. **C.** Measurements can be made along the IC surface relative to the different anatomic landmarks to identify the location for inserting the AMI array. Further details on the surgical approach and AMI implantation are provided in the text.

AMI implantation (Vince et al., 2010). With the expanded exposure, the NF2 tumor can still be accessed more laterally and then the IC can be approached more medially. In patients not requiring tumor removal, a traditional paramedian approach without the lateral exposure can be used to minimize the opening of the skull (V. Colletti et al., 2007; Ogata & Yonekawa, 1997; Vince et al., 2010). Once the dura is opened along the sinuses and the cerebellum is retracted downward, the IC and SC surfaces can be seen after pushing aside the overlying arachnoid and blood vessels (see Figure 11–11A). The viewed structures are confirmed to be the IC and SC using CT-MRI guided brain navigation (using the systems from Brainlab AG, Germany and Fiagon GmbH, Germany). Then the IC-SC border (rostral edge of IC), midline (medial edge of IC), and exit point of the trochlear nerve (caudal edge of IC; not shown) can be identified through the expanded craniotomy (Figure 11–11A and B), which is in contrast to the previous surgical approach in which these landmarks were not clearly visible due to the limited view of the midbrain. The direct view of the midline of the brain also provides a frame of reference for determining the sagittal plane and inserting the AMI array into the IC at an angle of 40° relative to that plane to align it along the tonotopic gradient of the ICC.

A major advantage of this new surgical approach is that the distances relative to these anatomic landmarks can be measured during surgery to identify the locations along the IC surface for inserting each shank of the AMI array (Figure 11–11C). Based on ICC stimulation studies in animals (H. H. Lim & Anderson, 2007; Neuheiser et al., 2010; Straka, McMahon, et al., 2014), anatomical and functional data of the IC in humans (De Martino et al., 2013; Geniec & Morest, 1971; H. H. Lim et al., 2013; H. H. Lim et al., 2009; Ress & Chandrasekaran, 2013), and IC surgical studies in cadavers (unpublished observations), the authors of this chapter have determined coordinates for inserting a two-shank AMI array into the ICC, particularly its rostral-lateral portion based on the findings presented in the previous section. The first shank will be inserted at a position of 0.25 caudally from the IC-SC border (rostral-to-caudal location normalized to the total distance between the IC-SC border and the exit point of trochlear nerve). Using a normalized location minimizes errors associated with variations in brain size across patients. Since there is no visible lateral landmark, the first shank will be inserted at a position of 7 mm from the midline. During surgery, electrical stimulation with a bipolar electrode along the surface of the IC and noninvasive neural recordings of the corresponding auditory cortical activity (i.e., middle-latency responses) may provide a way to identify a lateral landmark for more accurate array placement for each patient. Preliminary data and descriptions for this intraoperative technique are described in (H. H. Lim et al., 2009) and will be further explored in future AMI surgeries. The second shank will then be inserted about 1.5 mm diagonally toward the caudal and medial direction relative to the first shank. These new steps for positioning the AMI array into the IC are expected to improve placement of the electrode sites along the tonotopic gradient of the ICC compared to what was possible in the first clinical trial. Furthermore, these steps should enable placement of the arrays into the rostral-lateral portion of the ICC.

Two-Shank AMI Clinical Trial

Based on the encouraging animal and human findings described above, the authors of this chapter collaborated with Cochlear Limited (led by James Patrick) to design a new AMI device that consists of two shanks with 11 sites along each shank (see Figure 11–3C; note that 22 sites is the channel limit of the stimulator developed by Cochlear Limited). The shanks will be individually inserted into the ICC as described in the previous section. The previous single-shank array was able to obtain a reasonable range of pitches with 11 sites (i.e., ~2-mm spatial span with a site spacing of 200 μm; (H. H. Lim et al., 2013)). To sufficiently span the tonotopic axis of the ICC with the new two-shank AMI array, each shank was designed with a site spacing of 300 μm for 10 of the 11 sites (i.e., ~2.7-mm spatial span). The site spacing may be slightly larger than the ~200-μm thickness of each ICC lamina (Geniec & Morest, 1971; Oliver, 2005). However, the current level can be increased on each site to access adjacent laminae. This site spacing across the frequency dimension is considerably finer than what is possible with the CI, which still achieves high performance levels (Shannon, 2002; Wilson & Dorman, 2008). As shown in Figure 11–3C, the 11th site on each shank is positioned closer to the Dacron mesh. Since some of the patients implanted with the AMI will also have tinnitus, activation of the outer regions of the IC with those superficial sites may provide a way to suppress tinnitus using stimulation strategies derived from animal experiments (Offutt, Ryan, Konop, & Lim, 2014).

The second AMI clinical trial is currently under way and is funded by the National Institutes of Health (grant number U01DC013030). The clinical trial will be performed at Hannover Medical School in collaboration with Cochlear Limited and University of Minnesota for implanting the two-shank AMI device in five adult NF2 patients who cannot sufficiently benefit from a CI or an ABI. The clinical study is expected to begin in 2016. The primary objectives of this study are to demonstrate the safety and reliability of the new two-shank AMI array and the ability to consistently position the array into the ICC across patients. The secondary objective is to show that the two-shank AMI can achieve hearing performance greater than what is typically achievable with the ABI devices used in NF2 deaf patients. Success with these initial patients will open up opportunities for expanding the use of the AMI to a larger patient population and in clinics within different countries, including the United States.

Conclusions

The first AMI clinical study demonstrated that implantation and stimulation of a single-shank electrode array within the midbrain can be safe and provide useful hearing on a daily basis. However, there were difficulties in accurately placing the array into the IC in which only one out of five patients had sites properly aligned along the tonotopic gradient of the ICC. Based on psychophysical testing in the implanted AMI patients and experiments in animals, stimulation of individual sites on the single-shank array produces strong refractory and suppressive effects within the auditory pathway, which likely contributes to the poor temporal coding abilities and limited speech perception observed for the AMI patients. To address these two issues, animal experiments were performed to identify ways to minimize the refractory and suppressive effects, and cadaver studies were performed to improve the surgical approach for implanting the AMI array into the ICC. At least in animals, it appears that stimulation of two sites within each ICC lamina can sufficiently overcome these refractory and suppressive effects, especially when stimulating within the rostral-lateral portion of each lamina. Using a modified lateral suboccipital approach that is extended to the midline and identifying several key anatomic landmarks, it also appears that the AMI array can be consistently inserted into the rostral-lateral portion of the ICC.

Considering the encouraging findings described above, a new two-shank AMI array was developed in collaboration with Cochlear Limited that will target the rostral-lateral region of the ICC in a second clinical trial funded by the National Institutes of Health. This new array design will have two shanks aligned along the tonotopic gradient of the ICC with two sites positioned within each isofrequency lamina. Based on additional animal studies, a CI-based stimulation strategy within the ICC will initially be used in the patients except that the pulse pattern for each frequency channel will be distributed across two sites in each ICC lamina with varying delays between the pulses. Speech performance tests and various psychophysical measurements will be performed in the implanted AMI patients to evaluate this CI-based strategy while also investigating other types of stimulation patterns for improving hearing performance. Demonstrating the safety and reliability of the AMI in this second clinical trial as well as achieving better speech perception performance with the AMI compared to current ABI devices will revive research interests and discussions in using penetrating electrode arrays for central auditory prostheses.

One major limitation in achieving significant improvements in hearing performance with current ABI and CI devices appears to be the limited number of independent information channels possible with these implants (Friesen et al., 2001; Kuchta, Otto, Shannon, Hitselberger, & Brackmann, 2004). The CI sends current through a bony modiolar wall of the cochlea with scattered flow of electrical charge to a variable distribution and reduced number of auditory neurons associated with deafness. The ABI is placed on the surface of the brainstem, resulting in high stimulation levels and broad current spread to activate the appropriate neurons within deeper regions. Therefore, development of new types of central auditory prostheses such as the AMI, penetrating auditory brainstem implant (see Figure 11–1; McCreery, 2008), or auditory thalamic implant (Atencio, Shih, Schreiner, & Cheung, 2014) may eventually lead to innovative solutions for achieving hearing performance beyond what is possible with current technologies.

Acknowledgments. Minoo Lenarz who was previously at Hannover Medical School and later at University Hospital of Berlin-Charité (Germany) played a critical part in the initiation and progress of the AMI research and clinical study. David Anderson from University of Michigan (USA) played a critical role in the initiation and progress of the AMI animal research. Gert Joseph, Urte Rost, Joerg Pesch, Nicole Neben, Thilo Rode, and Rolf-Dieter Battmer contributed to the fitting and testing of the AMI patients at Hannover Medical School (Germany). Thilo Rode, Roger Calixto, Anke Neuheiser, Tanja Hartmann, Günter Reuter, Uta Reich, Gerrit Paasche, Verena Scheper, and Andrej Kral contributed to the AMI animal studies at Hannover Medical School, while Malgorzata Straka and Sarah Offutt performed AMI animal studies at University of Minnesota (United States). The AMI surgeries were led by and performed together with Madjid Samii and Amir Samii at the International Neuroscience Institute (Germany). The improved surgical approach in cadaver studies was developed together with Amir Samii, Omid Majdani, Markus Pietsch, Peter Erfurt, and Sven Balster at Hannover Medical School. The engineers and scientists at Cochlear Limited, including Frank Risi, Jason Leavens, Godofredo (JR) Timbol, Shahram Manoucherhi, Adrian Cryer, Peter Gibson, and Brett Swanson developed the AMI devices and software. Colette McKay from Bionics Institute (Australia) and Robert Shannon from University of Southern California (United States) helped with AMI and ABI psychophysical studies. Ray Meddis (University of Essex, UK) as well as Christian Sumner and Mark Steadman (Nottingham University, UK) provided and helped with the code for the DRNL model. Funding was provided by Cochlear Limited, German Research Foundation (SFB 599, Cluster of Excellence Hearing4All), Germany Ministry of Research and Education (01GQ0816), funds from University of Minnesota, and National Institutes of Health (P41EB2030, R03DC011589, U01DC013030).

References

Atencio, C. A., Shih, J. Y., Schreiner, C. E., & Cheung, S. W. (2014). Primary auditory cortical responses to electrical stimulation of the thalamus. *Journal of Neurophysiology, 111*(5), 1077–1087. doi:10.1152/jn.00749.2012

Behr, R., Colletti, V., Matthies, C., Morita, A., Nakatomi, H., Dominique, L., . . . Skarzynski, H. (2014). New outcomes with auditory brainstem implants in NF2 Patients. *Otology & Neurotology, 35*(10), 1844–1851. doi:10.1097/MAO.0000000000000584

Brosch, M., Schulz, A., & Scheich, H. (1999). Processing of sound sequences in macaque auditory cortex: response enhancement. *Journal of Neurophysiology, 82*(3), 1542–1559. Retrieved from http://www.ncbi.nlm.nih.gov/pubmed/10482768

Calixto, R., Lenarz, M., Neuheiser, A., Scheper, V., Lenarz, T., & Lim, H. H. (2012). Coactivation of different neurons within an isofrequency lamina of the inferior colliculus elicits enhanced auditory cortical activation. *Journal of Neurophysiology, 108*(4), 1199–1210. doi:10.1152/jn.00111.2012

Cant, N. B., & Benson, C. G. (2003). Parallel auditory pathways: Projection patterns of the different neuronal populations in the dorsal and ventral cochlear nuclei.

Brain Research Bulletin, 60(5–6), 457–474. Retrieved from http://www.ncbi.nlm.nih.gov/entrez/query.fcgi?cmd=Retrieve&db=PubMed&dopt=Citation&list_uids=12787867

Cant, N. B., & Benson, C. G. (2006). Organization of the inferior colliculus of the gerbil (Meriones unguiculatus): differences in distribution of projections from the cochlear nuclei and the superior olivary complex. *Journal of Comparative Neurology, 495*(5), 511–528. Retrieved from http://www.ncbi.nlm.nih.gov/entrez/query.fcgi?cmd=Retrieve&db=PubMed&dopt=Citation&list_uids=16498677

Cant, N. B., & Benson, C. G. (2007). Multiple topographically organized projections connect the central nucleus of the inferior colliculus to the ventral division of the medial geniculate nucleus in the gerbil, Meriones unguiculatus. *Journal of Comparative Neurology, 503*(3), 432–453. Retrieved from http://www.ncbi.nlm.nih.gov/entrez/query.fcgi?cmd=Retrieve&db=PubMed&dopt=Citation&list_uids=17503483

Casseday, J. H., Fremouw, T., & Covey, E. (2002). The inferior colliculus: A hub for the central auditory system. In D. Oertel, R. R. Fay, & A. N. Popper (Eds.), *Springer handbook of auditory research: Integrative functions in the mammalian auditory pathway* (Vol. 15, pp. 238–318). New York, NY: Springer-Verlag.

Colletti, L., Shannon, R., & Colletti, V. (2012). Auditory brainstem implants for neurofibromatosis type 2. *Current Opinion in Otolaryngology & Head and Neck Surgery, 20*(5), 353–357. doi:10.1097/MOO.0b013e328357613d

Colletti, L., Shannon, R. V., & Colletti, V. (2014). The Development of Auditory Perception in Children after Auditory Brainstem Implantation. *Audiology & Neurotology, 19*(6), 386–394. doi:10.1159/000363684

Colletti, V., Shannon, R., Carner, M., Sacchetto, L., Turazzi, S., Masotto, B., & Colletti, L. (2007). The first successful case of hearing produced by electrical stimulation of the human midbrain. *Otology & Neurotology, 28*(1), 39–43. Retrieved from http://www.ncbi.nlm.nih.gov/entrez/query.fcgi?cmd=Retrieve&db=PubMed&dopt=Citation&list_uids=17195744

Colletti, V., Shannon, R., Carner, M., Veronese, S., & Colletti, L. (2009). Outcomes in nontumor adults fitted with the auditory brainstem implant: 10 years' experience. *Otology & Neurotology, 30*, 614–618. Retrieved from http://www.ncbi.nlm.nih.gov/entrez/query.fcgi?cmd=Retrieve&db=PubMed&dopt=Citation&list_uids=19546832

Colletti, V., & Shannon, R. V. (2005). Open set speech perception with auditory brainstem implant? *Laryngoscope, 115*(11), 1974–1978. Retrieved from http://www.ncbi.nlm.nih.gov/entrez/query.fcgi?cmd=Retrieve&db=PubMed&dopt=Citation&list_uids=16319608

De Martino, F., Moerel, M., van de Moortele, P. F., Ugurbil, K., Goebel, R., Yacoub, E., & Formisano, E. (2013). Spatial organization of frequency preference and selectivity in the human inferior colliculus. *Nature Communications, 4*, 1386. doi:10.1038/ncomms2379

Eggermont, J. J., & Smith, G. M. (1995). Synchrony between single-unit activity and local field potentials in relation to periodicity coding in primary auditory cortex. *Journal of Neurophysiology, 73*(1), 227–245. Retrieved from http://www.ncbi.nlm.nih.gov/pubmed/7714568

Ehret, G. (1997). The auditory midbrain, a "shunting yard" of acoustical information processing. In G. Ehret & R. Romand (Eds.), *The central auditory system* (pp. 259–316). New York, NY: Oxford University Press.

Eisenberg, L. S. (2015). The contributions of William F. House to the field of implantable auditory devices. *Hearing Research, 322*, 52–56. doi:10.1016/j.heares.2014.08.003

Fraser, M., & McKay, C. M. (2012). Temporal modulation transfer functions in cochlear implantees using a method that limits overall loudness cues. *Hearing Research, 283*(1–2), 59–69. doi:10.1016/j.heares.2011.11.009

Friesen, L. M., Shannon, R. V., Baskent, D., & Wang, X. (2001). Speech recognition in noise as a function of the number of spectral channels: Comparison of acoustic hearing and cochlear implants. *The Journal of the Acoustical Society of America, 110*(2), 1150–1163. Retrieved from http://www.ncbi.nlm.nih.gov/entrez/query.fcgi?cmd=Retrieve&db=PubMed&dopt=Citation&list_uids=11519582

Geniec, P., & Morest, D. K. (1971). The neuronal architecture of the human posterior colliculus. A study with the Golgi method. *Acta Otolaryngologica Supplement, 295,* 1–33. Retrieved from http://www.ncbi.nlm.nih.gov/entrez/query.fcgi?cmd=Retrieve&db=PubMed&dopt=Citation&list_uids=4117000

Hochberg, L. R., Bacher, D., Jarosiewicz, B., Masse, N. Y., Simeral, J. D., Vogel, J., . . . Donoghue, J. P. (2012). Reach and grasp by people with tetraplegia using a neurally controlled robotic arm. *Nature, 485*(7398), 372–375. doi:10.1038/nature11076

Johnson, M. D., Lim, H. H., Netoff, T. I., Connolly, A. T., Johnson, N., Roy, A., . . . He, B. (2013). Neuromodulation for brain disorders: challenges and opportunities. *IEEE Transactions on Biomedical Engineering, 60*(3), 610-624. doi:10.1109/TBME.2013.2244890

Joris, P. X., Schreiner, C. E., & Rees, A. (2004). Neural processing of amplitude-modulated sounds. *Physiological Reviews, 84*(2), 541–577. Retrieved from http://www.ncbi.nlm.nih.gov/entrez/query.fcgi?cmd=Retrieve&db=PubMed&dopt=Citation&list_uids=15044682

Konrad, P., & Shanks, T. (2010). Implantable brain computer interface: challenges to neurotechnology translation. *Neurobiology of Disease, 38*(3), 369–375. doi:10.1016/j.nbd.2009.12.007

Kreft, H. A., Donaldson, G. S., & Nelson, D. A. (2004). Effects of pulse rate on threshold and dynamic range in Clarion cochlear-implant users. *The Journal of the Acoustical Society of America, 115*(5 Pt. 1), 1885–1888. Retrieved from http://www.ncbi.nlm.nih.gov/entrez/query.fcgi?cmd=Retrieve&db=PubMed&dopt=Citation&list_uids=15139595

Kretschmann, H. J., & Weinrich, W. (1992). *Cranial neuroimaging and clinical neuroanatomy: Magnetic resonance imaging and computed tomography* (2nd ed.). New York, NY: Thieme Medical.

Krishna, B. S., & Semple, M. N. (2000). Auditory temporal processing: responses to sinusoidally amplitude-modulated tones in the inferior colliculus. *Journal of Neurophysiology, 84*(1), 255–273. Retrieved from http://www.ncbi.nlm.nih.gov/entrez/query.fcgi?cmd=Retrieve&db=PubMed&dopt=Citation&list_uids=10899201

Krueger, B., Joseph, G., Rost, U., Strauss-Schier, A., Lenarz, T., & Buechner, A. (2008). Performance groups in adult cochlear implant users: Speech perception results from 1984 until today. *Otology & Neurotology, 29*(4), 509–512. doi:10.1097/MAO.0b013e318171972f

Kuchta, J., Otto, S. R., Shannon, R. V., Hitselberger, W. E., & Brackmann, D. E. (2004). The multichannel auditory brainstem implant: How many electrodes make sense? *Journal of Neurosurgery, 100*(1), 16–23. Retrieved from http://www.ncbi.nlm.nih.gov/entrez/query.fcgi?cmd=Retrieve&db=PubMed&dopt=Citation&list_uids=14743907

Langner, G., Albert, M., & Briede, T. (2002). Temporal and spatial coding of periodicity information in the inferior colliculus of awake chinchilla (Chinchilla laniger). *Hearing Research, 168*(1–2), 110–130. Retrieved from http://www.ncbi.nlm.nih.gov/entrez/query.fcgi?cmd=Retrieve&db=PubMed&dopt=Citation&list_uids=12117514

Lenarz, M., Lim, H. H., Lenarz, T., Reich, U., Marquardt, N., Klingberg, M. N., . . . Stan, A. C. (2007). Auditory midbrain implant: Histomorphologic effects of long-term implantation and electric stimulation of a new deep brain stimulation array. *Otology & Neurotology, 28*(8), 1045–1052. Retrieved from http://www.ncbi.nlm.nih.gov/entrez/query.fcgi?cmd=Retrieve&db=PubMed&dopt=Citation&list_uids=18043431

Lenarz, M., Lim, H. H., Patrick, J. F., Anderson, D. J., & Lenarz, T. (2006). Electrophysiological validation of a human prototype auditory midbrain implant in a guinea pig model. *Journal of the Association for Research in Otolaryngology, 7*, 383–398. Retrieved from http://www.ncbi.nlm.nih.gov/entrez/query.fcgi?cmd=Retrieve&db=PubMed&dopt=Citation&list_uids=17075701

Lenarz, M., Matthies, C., Lesinski-Schiedat, A., Frohne, C., Rost, U., Illg, A., . . . Lenarz, T. (2002). Auditory brainstem implant part II: Subjective assessment of functional outcome. *Otology & Neurotology, 23*(5), 694–697. Retrieved from http://www.ncbi.nlm.nih.gov/entrez/query.fcgi?cmd=Retrieve&db=PubMed&dopt=Citation&list_uids=12218621

Lenarz, T. (Ed.) (1998). *Cochlea-implantat*. Berlin Heidelberg, Germany: Springer.

Lim, H. H., & Anderson, D. J. (2006). Auditory cortical responses to electrical stimulation of the inferior colliculus: Implications for an auditory midbrain implant. *Journal of Neurophysiology, 96*(3), 975–988. Retrieved from http://www.ncbi.nlm.nih.gov/entrez/query.fcgi?cmd=Retrieve&db=PubMed&dopt=Citation&list_uids=16723413

Lim, H. H., & Anderson, D. J. (2007). Spatially distinct functional output regions within the central nucleus of the inferior colliculus: Implications for an auditory midbrain implant. *Journal of Neuroscience, 27*(32), 8733–8743. Retrieved from http://www.ncbi.nlm.nih.gov/entrez/query.fcgi?cmd=Retrieve&db=PubMed&dopt=Citation&list_uids=17687050

Lim, H. H., Lenarz, M., Joseph, G., & Lenarz, T. (2013). Frequency representation within the human brain: Stability versus plasticity. *Scientific Reports, 3*, 1474. doi:10.1038/srep01474

Lim, H. H., Lenarz, M., & Lenarz, T. (2009). Auditory midbrain implant: A review. *Trends in Amplification, 13*(3), 149–180. Retrieved from http://www.ncbi.nlm.nih.gov/entrez/query.fcgi?cmd=Retrieve&db=PubMed&dopt=Citation&list_uids=19762428

Lim, H. H., Lenarz, M., & Lenarz, T. (2011). Midbrain Auditory Prostheses. In F. G. Zeng, R. R. Fay, & A. N. Popper (Eds.), *Auditory prostheses: New horizons* (Vol. 39, pp. 207–232). New York, NY: Springer Science+Business Media.

Lim, H. H., Lenarz, T., Joseph, G., Battmer, R. D., Patrick, J. F., & Lenarz, M. (2008). Effects of phase duration and pulse rate on loudness and pitch percepts in the first auditory midbrain implant patients: Comparison to cochlear implant and auditory brainstem implant results. *Neuroscience, 154*(1), 370–380. Retrieved from http://www.ncbi.nlm.nih.gov/entrez/query.fcgi?cmd=Retrieve&db=PubMed&dopt=Citation&list_uids=18384971

Lim, H. H., Lenarz, T., Joseph, G., Battmer, R. D., Samii, A., Samii, M., . . . Lenarz, M. (2007). Electrical stimulation of the midbrain for hearing restoration: Insight into the functional organization of the human central auditory system. *Journal of Neuroscience, 27*(49), 13541–13551. Retrieved from http://www.ncbi.nlm.nih.gov/entrez/query.fcgi?cmd=Retrieve&db=PubMed&dopt=Citation&list_uids=18057212

Malmierca, M. S., Izquierdo, M. A., Cristaudo, S., Hernandez, O., Perez-Gonzalez, D., Covey, E., & Oliver, D. L. (2008). A discontinuous tonotopic organization in the inferior colliculus of the rat. *Journal of Neuroscience, 28*(18), 4767–4776. doi:10.1523/JNEUROSCI.0238-08.2008

Matthies, C., Brill, S., Varallyay, C., Solymosi, L., Gelbrich, G., Roosen, K., . . . Muller, J. (2014). Auditory brainstem implants in neurofibromatosis Type 2: Is open speech perception feasible? *Journal of Neurosurgery, 120*(2), 546–558. doi:10.3171/2013.9.JNS12686

McCreery, D. B. (2008). Cochlear nucleus auditory prostheses. *Hearing Research, 242*(1–2), 64–73. Retrieved from http://www.ncbi.nlm.nih.gov/entrez/query.fcgi?cmd=Retrieve&db=PubMed&dopt=Citation&list_uids=18207678

McKay, C. M., Lim, H. H., & Lenarz, T. (2013). Temporal processing in the auditory system: Insights from cochlear and auditory midbrain implantees. *Journal of the Association for Research in Otolaryngology, 14*(1), 103–124. doi:10.1007/s10162-012-0354-z

McKay, C. M., & McDermott, H. J. (1998). Loudness perception with pulsatile electrical stimulation: the effect of interpulse intervals. *The Journal of the Acoustical Society of America, 104*(2 Pt. 1), 1061–1074. Retrieved from http://www.ncbi.nlm.nih.gov/entrez/query.fcgi?cmd=Retrieve&db=PubMed&dopt=Citation&list_uids=9714925

Meddis, R., O'Mard, L. P., & Lopez-Poveda, E. A. (2001). A computational algorithm for computing nonlinear auditory frequency selectivity. *The Journal of the Acoustical Society of America, 109*(6), 2852–2861. Retrieved from http://www.ncbi.nlm.nih.gov/pubmed/11425128

Middlebrooks, J. C. (2004). Effects of cochlear-implant pulse rate and inter-channel timing on channel interactions and thresholds. *The Journal of the Acoustical Society of America, 116*(1), 452–468. Retrieved from http://www.ncbi.nlm.nih.gov/entrez/query.fcgi?cmd=Retrieve&db=PubMed&dopt=Citation&list_uids=15296005

Morel, A., & Imig, T. J. (1987). Thalamic projections to fields A, AI, P, and VP in the cat auditory cortex. *Journal of Comparative Neurology, 265*(1), 119–144. doi:10.1002/cne.902650109

Mudry, A., & Mills, M. (2013). The early history of the cochlear implant: A retrospective. *JAMA Otolaryngology-Head & Neck Surgery, 139*(5), 446–453. doi:10.1001/jamaoto.2013.293

Navarro, X., Krueger, T. B., Lago, N., Micera, S., Stieglitz, T., & Dario, P. (2005). A critical review of interfaces with the peripheral nervous system for the control of neuroprostheses and hybrid bionic systems. *Journal of the Peripheral Nervous System, 10*(3), 229–258. doi:10.1111/j.1085-9489.2005.10303.x

Neuheiser, A., Lenarz, M., Reuter, G., Calixto, R., Nolte, I., Lenarz, T., & Lim, H. H. (2010). Effects of pulse phase duration and location of stimulation within the inferior colliculus on auditory cortical evoked potentials in a guinea pig model. *Journal of the Association for Research in Otolaryngology, 11*(4), 689–708. doi:10.1007/s10162-010-0229-0

Nie, K., Barco, A., & Zeng, F. G. (2006). Spectral and temporal cues in cochlear implant speech perception. *Ear and Hearing, 27*(2), 208–217. doi:10.1097/01.aud.0000202312.31837.25

Offutt, S. J., Ryan, K. J., Konop, A. E., & Lim, H. H. (2014). Suppression and facilitation of auditory neurons through coordinated acoustic and midbrain stimulation: Investigating a deep brain stimulator for tinnitus. *Journal of Neural Engineering, 11*(6), 066001. doi:10.1088/1741-2560/11/6/066001

Ogata, N., & Yonekawa, Y. (1997). Paramedian supracerebellar approach to the upper brain stem and peduncular lesions. *Neurosurgery, 40*(1), 101–104; discussion 104–105. Retrieved from http://www.ncbi.nlm.nih.gov/entrez/query.fcgi?cmd=Retrieve&db=PubMed&dopt=Citation&list_uids=8971831

Oliver, D. L. (2005). Neuronal organization in the inferior colliculus. In J. A. Winer & C. E. Schreiner (Eds.), *The inferior colliculus* (pp. 69–114). New York, NY: Springer Science+Business Media.

Patrick, J. F., Busby, P. A., & Gibson, P. J. (2006). The development of the Nucleus

Freedom Cochlear implant system. *Trends in Amplification, 10*(4), 175–200. Retrieved from http://www.ncbi.nlm.nih.gov/entrez/query.fcgi?cmd=Retrieve&db=PubMed&dopt=Citation&list_uids=17172547

Phillips, D. P., Semple, M. N., & Kitzes, L. M. (1995). Factors shaping the tone level sensitivity of single neurons in posterior field of cat auditory cortex. *Journal of Neurophysiology, 73*(2), 674–686. Retrieved from http://www.ncbi.nlm.nih.gov/pubmed/7760126

Polley, D. B., Read, H. L., Storace, D. A., & Merzenich, M. M. (2007). Multiparametric auditory receptive field organization across five cortical fields in the albino rat. *Journal of Neurophysiology, 97*(5), 3621–3638. doi:10.1152/jn.01298.2006

Rees, A., & Langner, G. (2005). Temporal coding in the auditory midbrain. In J. A. Winer & C. E. Schreiner (Eds.), *The inferior colliculus* (pp. 346–376). New York, NY: Springer Science+Business Media.

Rees, A., & Moller, A. R. (1987). Stimulus properties influencing the responses of inferior colliculus neurons to amplitude-modulated sounds. *Hearing Research, 27*(2), 129–143. Retrieved from http://www.ncbi.nlm.nih.gov/entrez/query.fcgi?cmd=Retrieve&db=PubMed&dopt=Citation&list_uids=3610842

Ress, D., & Chandrasekaran, B. (2013). Tonotopic organization in the depth of human inferior colliculus. *Frontiers in Human Neuroscience, 7*, 586. doi:10.3389/fnhum.2013.00586

Rode, T., Hartmann, T., Hubka, P., Scheper, V., Lenarz, M., Lenarz, T., . . . Lim, H. H. (2013). Neural representation in the auditory midbrain of the envelope of vocalizations based on a peripheral ear model. *Frontiers in Neural Circuits, 7*, 166. doi:10.3389/fncir.2013.00166

Rodrigues-Dagaeff, C., Simm, G., De Ribaupierre, Y., Villa, A., De Ribaupierre, F., & Rouiller, E. M. (1989). Functional organization of the ventral division of the medial geniculate body of the cat: Evidence for a rostro-caudal gradient of response properties and cortical projections. *Hearing Research, 39*(1–2), 103–125. Retrieved from http://www.ncbi.nlm.nih.gov/entrez/query.fcgi?cmd=Retrieve&db=PubMed&dopt=Citation&list_uids=2737959

Samii, A., Lenarz, M., Majdani, O., Lim, H. H., Samii, M., & Lenarz, T. (2007). Auditory midbrain implant: A combined approach for vestibular schwannoma surgery and device implantation. *Otology & Neurotology, 28*(1), 31–38. Retrieved from http://www.ncbi.nlm.nih.gov/entrez/query.fcgi?cmd=Retrieve&db=PubMed&dopt=Citation&list_uids=17195743

Schreiner, C., Froemke, R., & Atencio, C. (2011). Spectral processing in auditory cortex. In J. A. Winer & C. E. Schreiner (Eds.), *The auditory cortex* (pp. 275–308). New York, NY: Springer.

Schreiner, C. E., & Langner, G. (1997). Laminar fine structure of frequency organization in auditory midbrain. *Nature, 388*(6640), 383–386. Retrieved from http://www.ncbi.nlm.nih.gov/entrez/query.fcgi?cmd=Retrieve&db=PubMed&dopt=Citation&list_uids=9237756

Schwartz, M. S., Otto, S. R., Shannon, R. V., Hitselberger, W. E., & Brackmann, D. E. (2008). Auditory brainstem implants. *Neurotherapeutics, 5*(1), 128–136.

Sennaroglu, L., Colletti, V., Manrique, M., Laszig, R., Offeciers, E., Saeed, S., . . . Konradsson, K. (2011). Auditory brainstem implantation in children and non-neurofibromatosis type 2 patients: A consensus statement. *Otology & Neurotology, 32*(2), 187–191. doi:10.1097/MAO.0b013e318206fc1e

Sennaroglu, L., & Ziyal, I. (2012). Auditory brainstem implantation. *Auris Nasus Larynx, 39*(5), 439–450. doi:10.1016/j.anl.2011.10.013

Shannon, R. V. (1985). Threshold and loudness functions for pulsatile stimulation of cochlear implants. *Hearing Research,*

18(2), 135–143. Retrieved from http://www.ncbi.nlm.nih.gov/entrez/query.fcgi?cmd=Retrieve&db=PubMed&dopt=Citation&list_uids=3840159

Shannon, R. V. (1989). A model of threshold for pulsatile electrical stimulation of cochlear implants. *Hearing Research, 40*(3), 197–204. Retrieved from http://www.ncbi.nlm.nih.gov/entrez/query.fcgi?cmd=Retrieve&db=PubMed&dopt=Citation&list_uids=2793602

Shannon, R. V. (2002). The relative importance of amplitude, temporal, and spectral cues for cochlear implant processor design. *American Journal of Audiology, 11*(2), 124–127. Retrieved from http://www.ncbi.nlm.nih.gov/entrez/query.fcgi?cmd=Retrieve&db=PubMed&dopt=Citation&list_uids=12691223

Shannon, R. V., Fu, Q. J., & Galvin, J., 3rd. (2004). The number of spectral channels required for speech recognition depends on the difficulty of the listening situation. *Acta Otolaryngologica Supplement* (552), 50–54. Retrieved from http://www.ncbi.nlm.nih.gov/entrez/query.fcgi?cmd=Retrieve&db=PubMed&dopt=Citation&list_uids=15219048

Shannon, R. V., Zeng, F. G., Kamath, V., Wygonski, J., & Ekelid, M. (1995). Speech recognition with primarily temporal cues. *Science, 270*(5234), 303–304. Retrieved from http://www.ncbi.nlm.nih.gov/entrez/query.fcgi?cmd=Retrieve&db=PubMed&dopt=Citation&list_uids=7569981

Storace, D. A., Higgins, N. C., Chikar, J. A., Oliver, D. L., & Read, H. L. (2012). Gene expression identifies distinct ascending glutamatergic pathways to frequency-organized auditory cortex in the rat brain. *Journal of Neuroscience, 32*(45), 15759–15768. doi:10.1523/JNEUROSCI.1310-12.2012

Straka, M. M., McMahon, M., Markovitz, C. D., & Lim, H. H. (2014). Effects of location and timing of co-activated neurons in the auditory midbrain on cortical activity: Implications for a new central auditory prosthesis. *Journal of Neural Engineering, 11*(4), 046021. doi:10.1088/1741-2560/11/4/046021

Straka, M. M., Schendel, D., & Lim, H. H. (2013). Neural integration and enhancement from the inferior colliculus up to different layers of auditory cortex. *Journal of Neurophysiology, 110*(4), 1009–1020. doi:10.1152/jn.00022.2013

Straka, M. M., Schmitz, S., & Lim, H. H. (2014). Response features across the auditory midbrain reveal an organization consistent with a dual lemniscal pathway. *Journal of Neurophysiology, 112*(4), 981–998. doi:10.1152/jn.00008.2014

Strauss-Schier, A., Battmer, R. D., Rost, U., Allum-Mecklenburg, D. J., & Lenarz, T. (1995). Speech-tracking results for adults. *Annals of Otology Rhinology and Laryngology Supplement, 166*, 88–91. Retrieved from http://www.ncbi.nlm.nih.gov/entrez/query.fcgi?cmd=Retrieve&db=PubMed&dopt=Citation&list_uids=7668770

Sumner, C. J., O'Mard, L. P., Lopez-Poveda, E. A., & Meddis, R. (2003). A nonlinear filter-bank model of the guinea-pig cochlear nerve: rate responses. *The Journal of the Acoustical Society of America, 113*(6), 3264–3274. Retrieved from http://www.ncbi.nlm.nih.gov/pubmed/12822799

Suta, D., Kvasnak, E., Popelar, J., & Syka, J. (2003). Representation of species-specific vocalizations in the inferior colliculus of the guinea pig. *Journal of Neurophysiology, 90*(6), 3794–3808. Retrieved from http://www.ncbi.nlm.nih.gov/entrez/query.fcgi?cmd=Retrieve&db=PubMed&dopt=Citation&list_uids=12944528

Viemeister, N. F. (1979). Temporal modulation transfer functions based upon modulation thresholds. *The Journal of the Acoustical Society of America, 66*(5), 1364–1380. Retrieved from http://www.ncbi.nlm.nih.gov/entrez/query.fcgi?cmd=Retrieve&db=PubMed&dopt=Citation&list_uids=500975

Viemeister, N. F., & Wakefield, G. H. (1991). Temporal integration and multiple looks.

The Journal of the Acoustical Society of America, 90(2 Pt. 1), 858–865. Retrieved from http://www.ncbi.nlm.nih.gov/entrez/query.fcgi?cmd=Retrieve&db=PubMed&dopt=Citation&list_uids=1939890

Vince, G. H., Herbold, C., Coburger, J., Westermaier, T., Drenckhahn, D., Schuetz, A., . . . Matthies, C. (2010). An anatomical assessment of the supracerebellar midline and paramedian approaches to the inferior colliculus for auditory midbrain implants using a neuronavigation model on cadaveric specimens. *Journal of Clinical Neuroscience, 17*(1), 107–112. doi:10.1016/j.jocn.2009.06.034

Wallace, M. N., Rutkowski, R. G., & Palmer, A. R. (2000). Identification and localisation of auditory areas in guinea pig cortex. *Experimental Brain Research, 132*(4), 445–456. Retrieved from http://www.ncbi.nlm.nih.gov/entrez/query.fcgi?cmd=Retrieve&db=PubMed&dopt=Citation&list_uids=10912825

Weber, D. J., Friesen, R., & Miller, L. E. (2012). Interfacing the somatosensory system to restore touch and proprioception: Essential considerations. *Journal of Motor Behavior, 44*(6), 403–418. doi:10.1080/00222895.2012.735283

Wehr, M., & Zador, A. M. (2005). Synaptic mechanisms of forward suppression in rat auditory cortex. *Neuron, 47*(3), 437–445. doi:10.1016/j.neuron.2005.06.009

Weiland, J. D., Cho, A. K., & Humayun, M. S. (2011). Retinal prostheses: Current clinical results and future needs. *Ophthalmology, 118*(11), 2227–2237. doi:10.1016/j.ophtha.2011.08.042

Wilson, B. S., & Dorman, M. F. (2008). Cochlear implants: A remarkable past and a brilliant future. *Hearing Research, 242*(1–2), 3–21.

Woolley, S. M., Gill, P. R., & Theunissen, F. E. (2006). Stimulus-dependent auditory tuning results in synchronous population coding of vocalizations in the songbird midbrain. *Journal of Neuroscience, 26*(9), 2499–2512. doi:10.1523/JNEUROSCI.3731-05.2006

Xu, L., & Pfingst, B. E. (2008). Spectral and temporal cues for speech recognition: implications for auditory prostheses. *Hearing Research, 242*(1–2), 132–140. doi:10.1016/j.heares.2007.12.010

Zeng, F. G. (2004). Trends in cochlear implants. *Trends in Amplification, 8*(1), 1–34. Retrieved from http://www.ncbi.nlm.nih.gov/entrez/query.fcgi?cmd=Retrieve&db=PubMed&dopt=Citation&list_uids=15247993

Zeng, F. G., Rebscher, S., Harrison, W., Sun, X., & Feng, H. (2008). Cochlear implants: system design, integration, and evaluation. *IEEE Reviews in Biomedical Engineering, 1*, 115–142. doi:10.1109/RBME.2008.2008250

CHAPTER 12

Perception and Psychoacoustics of Speech in Cochlear Implant Users

Deniz Başkent, Etienne Gaudrain,
Terrin Nichole Tamati, and Anita Wagner

Cochlear implants (CIs) are prosthetic devices that restore hearing in deaf individuals via electric stimulation of the auditory nerve through an electrode array inserted in the cochlea. The device consists of a microphone and an externally worn speech processor that converts acoustic signals into electric signals. The speech signal coded with electric pulses is then delivered via a wireless transmission system through the scalp to an electrode array, implanted in the cochlea, traditionally in the scala tympani (Grayden & Clark, 2006).

While research with CIs dates back to the 1950s (Djourno & Eyriès, 1957), the Food and Drug Administration (FDA) has approved CI use in adults in 1984, in children 2 years and older in 1989, and in children 12 months and older in 2000. The FDA reports that approximately 324,000 people worldwide had received CIs by 2012.

Despite the relatively long history of the implant, the speech signal transmitted via the modern CIs is still inherently degraded in fine spectrotemporal details (e.g., Loizou, 1998; Rubinstein, 2004). The device mainly delivers slow-varying amplitude envelopes of speech modulating (usually) fixed-rate digital pulses, delivered at a small number of contact points (electrodes). This degraded signal is recognized and reinterpreted by the brain as speech (Fu, 2002; Shannon, Zeng, Kamath, Wygonski, & Ekelid, 1995). One of the main forms of degradation is the reduced spectral resolution (Friesen, Shannon, Başkent, & Wang, 2001; Fu & Nogaki, 2005; Henry, Turner, & Behrens, 2005). This reduction does not come from the small number of electrodes per se but instead from the channel interactions caused by the spatial overlap of the broad stimulation from individual electrodes (e.g., Shannon, 1983; Stickney et al., 2006). Coding of temporal fine structure is also limited, mainly caused by the characteristics of the electric stimulation of the auditory nerve (Rubinstein & Hong, 2003). Other factors that can affect the quality of the speech signal delivered via the CI include the position of the electrode array, such as the insertion depth (Başkent & Shannon, 2005; Dorman, Loizou, & Rainey, 1997; Hochmair, Hochmair, Nopp, Waller, & Jolly, 2015; Skinner et al., 2002) or the proximity to the spiral ganglia (Holden et al., 2013), potential mismatch in the frequency-place mapping (Başkent & Shannon, 2004; Siciliano, Faulkner, Rosen, & Mair, 2010; Venail et al., 2015), limited dynamic range of electric hearing (Zeng et al., 2002), presence of acoustic low-frequency hearing in the implanted or contralateral ear (Cullington & Zeng, 2010; Gantz, Turner, Gfeller, & Lowder, 2005; Gifford, Dorman, McKarns, & Spahr, 2007), robustness of the electrode-nerve interface (Bierer & Faulkner, 2010), neural survival patterns (Khan et al., 2005) and potential dead regions (Kasturi, Loizou, Dorman, & Spahr, 2002; Shannon, Galvin, & Başkent, 2002), cochlear abnormalities and surgical factors (Finley & Skinner, 2008; Sennaroğlu, 2010), and device-related factors such as sound-processing strategy (Wilson et al., 1991), electrode design, configuration, and stimulation mode (Stickney et al., 2006; Zwolan, Kileny, Ashbaugh, & Telian, 1996) and stimulation rate (Friesen, Shannon, & Cruz, 2005; Vandali, Whitford, Plant, & Clark, 2000).

After a postimplantation adaptation period (Lazard, Innes-Brown, & Barone, 2014), many CI users achieve a good level of speech understanding in quiet, ideal listening conditions (Rouger et al., 2007). However, performance across individual CI users is still highly variable (Blamey et al., 2013, also see Figure 12–1). Additionally, comprehension of speech is further degraded by other, external factors, for example, due to interfering sounds or poor room acoustics, remains to be a challenge for these individuals (Friesen et al., 2001; Fu & Nogaki, 2005; Nelson, Jin, Carney, & Nelson, 2003; Stickney, Zeng, Litovsky, & Assmann, 2004).

Historical Perspective

Speech perception research with CIs dates back to the single-channel devices in the 1970s and 1980s (Danley & Fretz, 1982; Douek, Fourcin, Moore, & Clark, 1977; Hochmair & Hochmair-Desoyer, 1983; Merzenich, Michelson, Schindler, Pettit, & Reid, 1973). These devices were only able to deliver the slow-varying amplitude envelope of the broadband speech signal, with a severely limited dynamic range and spectral information (Millar, Tong, & Clark, 1984). Due to using a single point of stimulation in the cochlea, the implant could not evoke any place pitch percept (i.e., the pitch percept evoked by stimulating different locations of the tonotopically organized cochlear partition). Some temporal pitch percept could be achieved via the rate of stimulation. Yet, this was thought to be limited by the nerve refractory period, allowing only partial transmission of voice pitch (defined by fundamental frequency, F0), and first formant (F1) information (Dorman & Spahr, 2006).

Despite these limitations, these devices succeeded in providing some seg-

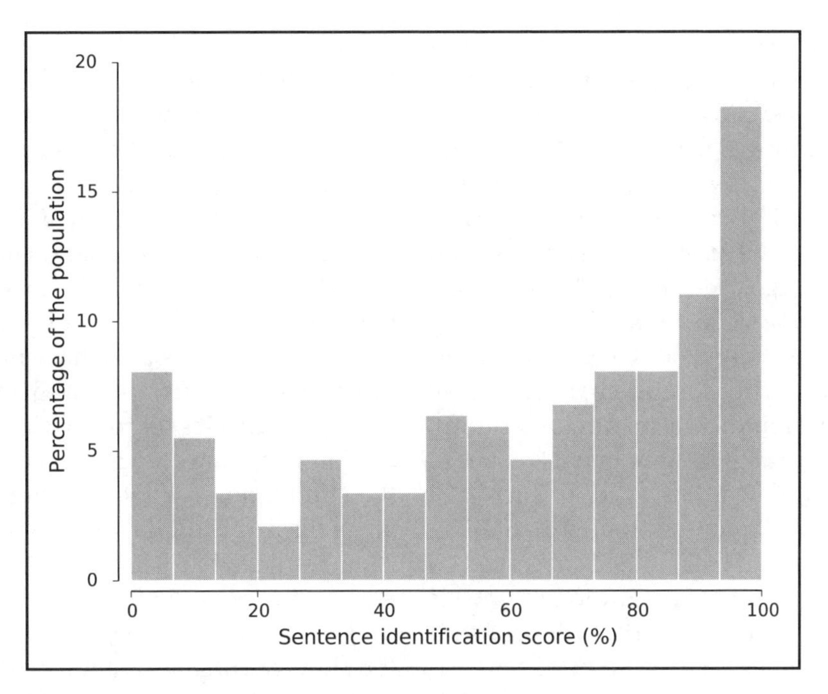

Figure 12–1. Distribution of percent correct scores for postimplantation sentence identification in the CI user population. Adapted from Blamey et al. (2013).

mental and supra-segmental speech cues. Overall intensity fluctuations can already provide segmental information and stress patterns (Rosen, Walliker, Brimacombe, & Edgerton, 1989). Voice pitch, even if partial, can help with perceiving information on voicing (i.e., discrimination of voiced consonants from unvoiced ones, such as /z/ vs. /s/) and manner (i.e., discrimination of stop consonants, such as /p/, from fricatives, such as /s/). Pitch fluctuations can provide sentence prosody (i.e., discrimination of a question from a statement). Through F1, some vowel identification can be achieved (Dorman & Spahr, 2006). However, lack of spectral resolution means place of articulation cues are lost (i.e., consonants such as /p, t, k/ that mostly differ in their place cue could not be discriminated). Lack of higher formant cues leads to confusions in discriminating different vowels from each other (e.g., /i/ vs. /u/) and also in discriminating fricatives from each other (e.g., /s/ vs. /ʃ/). With such limited transmission of speech cues, as a result, while some CI users showed open-set speech recognition abilities with auditory input only (Berliner, Tonokawa, Dye, & House, 1989), most CI users could only derive useful speech perception benefit in closed-set phoneme or word discrimination, or in combination with visual speech cues (Rosen et al., 1989).

In modern multichannel CIs, the cochlear tonotopic organization is taken into account (Greenwood, 1990); low-frequency components of speech are delivered to apical electrodes while high-frequency components are delivered to basal electrodes. Multichannel CIs thus present a significant improvement over single-channel devices in transmitting speech cues. As a result, drastic improvements have been observed in speech perception performance of CI users in general (Clark, 2015; Zeng, 2004; also see Figure 12–2). Even in the same users, who, after using a single-channel device, were reimplanted with a multichannel device (for example due to device failure), an immediate improvement in speech perception was observed. In such a CI user, only 3 months after implantation, Spillman and Dillier (1989) observed an improvement in recognition of vowels and sentences, and further, this individual could also make use of second formant information, F2, in addition to F1.

In modern CI devices, despite the improvements over the years in device design, surgical techniques, and speech coding strategies, speech information transmitted via electric stimulation still only partially mimics that of acoustic hearing (Loizou, 1998; also see an example of a sentence processed with an acoustic simulation of a CI, and represented in electrodogram in Figure 12–3). The modern CI processor bandpass filters the acoustic input of broadband speech signal into a number of frequency bands, to be delivered to distinct electrodes for tonotopical stimulation of the auditory nerve. However, the stimulating carrier current is usually a fixed-rate digital pulse sequence. As a result, what is eventually delivered to the nerve is fixed-rate current pulses at each electrode, modulated by the slow-varying amplitude envelopes of the corresponding spectral band. In the signal used for electric stimulation, the slow-varying temporal envelopes and gross spectral details are preserved;

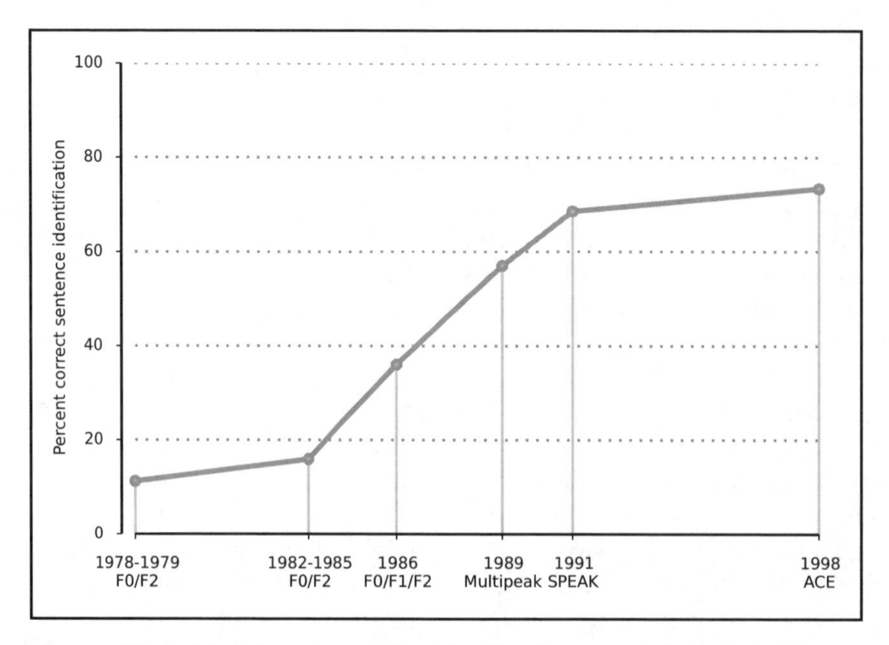

Figure 12–2. Evolution of sentence identification scores through the history of cochlear implants. Adapted from Clark (2015).

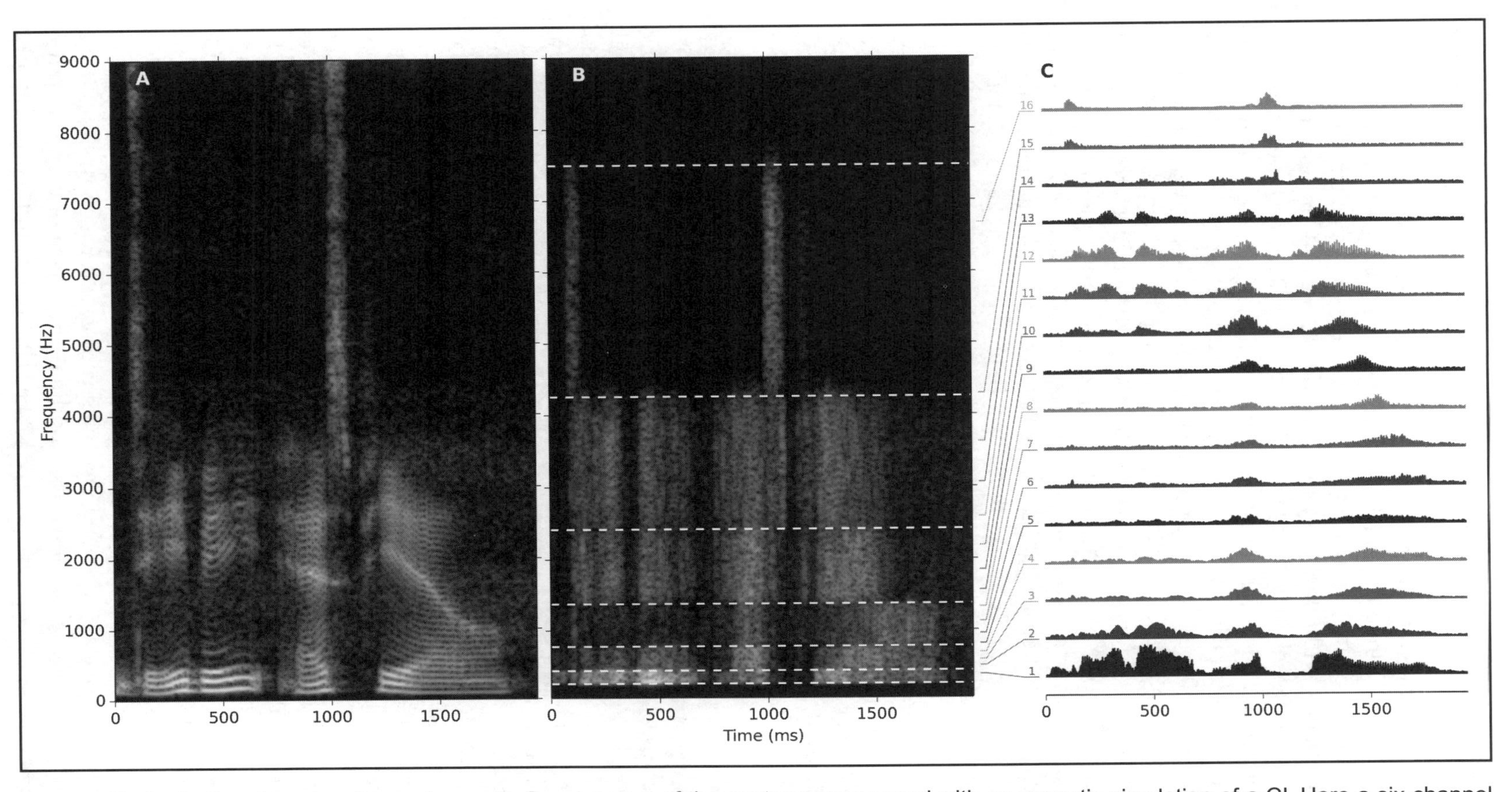

Figure 12–3. A. Spectrogram of a sentence. **B.** Spectrogram of the sentence processed with an acoustic simulation of a CI. Here a six-channel noise-band vocoder is used (based on Shannon et al., 1995). The white dashed lines indicate the band cutoff frequencies of the vocoder bandpass filters. **C.** Electrodogram of the same signal, showing electrical activity in a 16-channel implant using a CIS strategy. Each row corresponds to an electrode, and is connected to the frequency axis of panel B to indicate its center frequency.

however, all spectrotemporal fine structure is lost. While Shannon et al. (1995) have shown early on that even a small number of bands with slow varying envelopes are sufficient for basic level of speech perception, this level of speech detail seems to be insufficient for more advanced levels of speech perception, such as in background noise or talkers (Friesen et al., 2001; Fu & Nogaki, 2005; Stickney et al., 2004), or to achieve other speech-related tasks, such as vocal emotion perception (Chatterjee et al., 2015; Luo, Fu, & Galvin, 2007). Perception (and enjoyment) of more complex, and potentially pleasurable sounds, such as music, seems to be minimally available to CI users (Crew, Galvin, & Fu, 2012; Drennan et al., 2015; Fuller, Free, Maat, & Başkent, 2012; Limb & Rubinstein, 2012).

Perception of Vocal Characteristics

One of the most important speech cues that helps with higher level perception of speech is voice characteristics, namely, voice pitch, directly related to the glottal pulse rate of the speaker, that is, F0, and vocal tract length (VTL), directly related to the size of the speaker (Fitch & Giedd, 1999). While the perception of the former relies both on temporal and place coding of speech (i.e., harmonic structure), the perception of the latter relies mostly on the perception of spectral characteristics, namely, the formant structure (Smith, Patterson, Turner, Kawahara, & Irino, 2005). The way these cues are coded in the acoustic speech signal is shown in Figure 12–4.

A robust perception of voice characteristics is not only important for identifying the speaker characteristics but also plays an important role in everyday life speech perception. In such scenarios, one hears many sounds mixed into one signal where the target speech has to be segregated from the interfering background sounds and the individual audible speech segments be grouped into a meaningful speech stream. For the listener, voice characteristics are powerful acoustic cues that facilitate such segregation. This was shown by the increase in F0 and VTL difference leading to better segregation of concurrent vowels (Vestergaard, Fyson, & Patterson, 2009) and sentences (Darwin, Brungart, & Simpson, 2003). Hence, perception of target speech in a cocktail party seems to heavily rely on effective use of vocal cues.

Even in multichannel CIs the vocal characteristics delivered by the device are considerably weak. In principle, voice pitch (i.e., F0) can be coded through stimulation rate, temporal pattern of stimulation, or place of stimulation, as explained before. However, all have limitations. The coding of voice pitch via stimulation rate is not used in most current processors as they use a relatively high, but fixed, pulse rate. Research has shown that strategies capturing the F0 or the temporal fine structure by timing individual pulses accordingly have some potential to improve speech and music perception (Arnoldner et al., 2007; Laneau, Wouters, & Moonen, 2006). To date, only one clinically used processor is exploiting this form of coding. Another form of pitch coding is through the amplitude modulation pattern of high rate pulse trains. This form of pitch coding, like

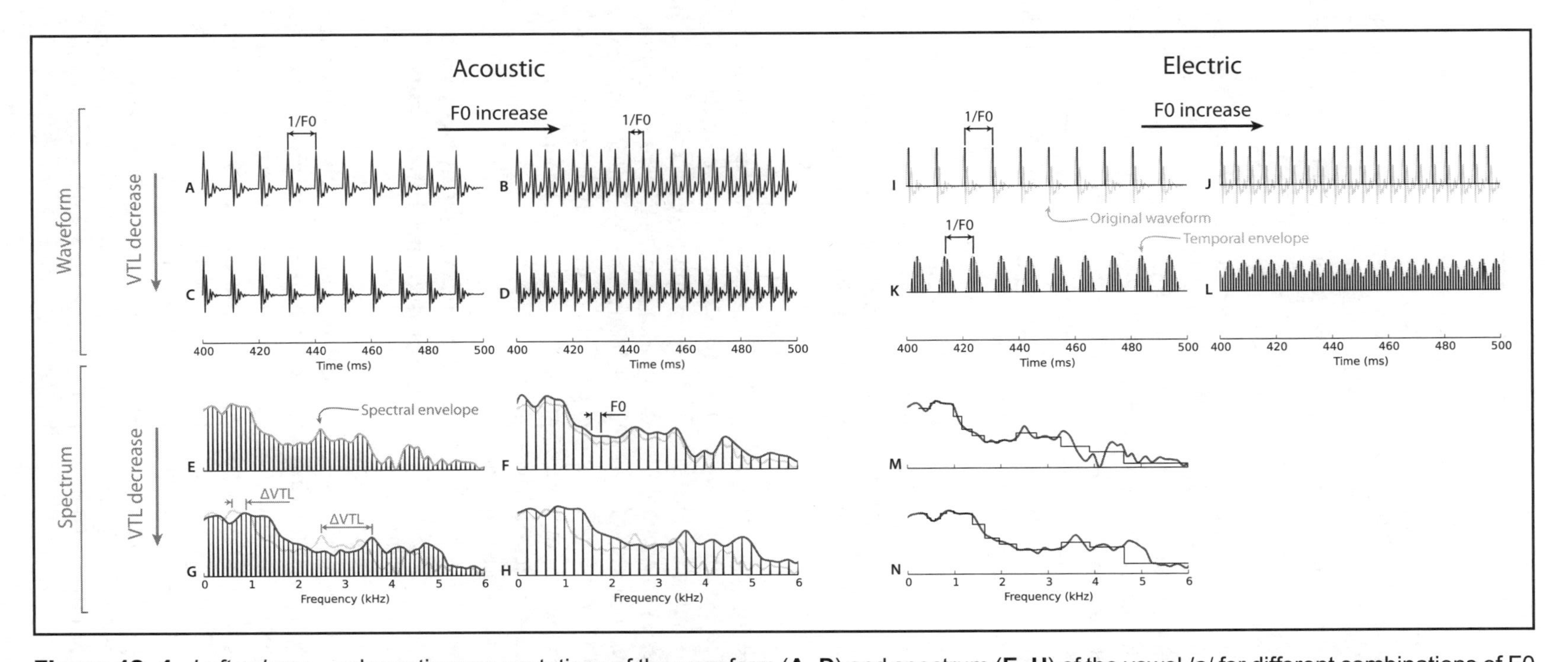

Figure 12–4. *Left column*—schematic representations of the waveform (**A–D**) and spectrum (**E–H**) of the vowel /a/ for different combinations of F0 and VTL. A decrease in VTL results in shrinking the temporal pattern produced by a single glottal pulse (A vs. C), which corresponds to an expansion of the spectral envelope (E vs. G). An increase in F0 results in glottal pulses being more frequent (A vs. B), or to harmonic components to be more spaced while they follow the same spectral envelope (E vs. F). *Right column*—Schematic representation of the coding of F0 (**I–L**) and VTL (**M–N**) in the implant. F0 can be coded by modifying the stimulation pulse rate (I, J) or by using a fixed-rate pulse train with the signal's temporal envelope (K, L). The panels M and N illustrate how spectral quantization affects VTL representation, but do not show the additional effect of spectral smearing that results from current spread.

the previous one, is limited by the rate at which auditory nerve fibers can fire in response to electrical stimulation as well as by the restricted dynamic range offered by electrical stimulation. Consequently, although CI users seem to be able to experience a pitch percept through this mechanism, this percept is consistently reported as being weak (Moore & Carlyon, 2005). To enhance that pitch percept, researchers have developed strategies that code the F0 by explicitly modulating the stimulation pattern rather than relying on natural amplitude modulations of the original stimulus (Milczynski, Wouters, & Van Wieringen, 2009; Vandali & van Hoesel, 2011). Finally, the coding of voice pitch via place of stimulation in CIs is primarily limited by the low spectral resolution available through the implant.

In contrast to large number of studies on F0 perception in CIs, perception of the other dominant voice characteristic, namely VTL, has been only minimally studied. Only recently, Gaudrain and Başkent (2015a) showed that VTL perception is severely impaired in CI users. Using acoustic simulations of CIs, Gaudrain and Başkent (2015b) showed that this limitation was likely due to channel interactions and smeared spectral resolution, similar to the limitation of the place percept of F0. Because VTL is a cue used in many situations by normal-hearing (NH) listeners, this specific impairment of CI users has repercussions on many speech-related tasks. One instance of such consequence that has been demonstrated to date is how speaker gender categorization is impaired in CI users because they can only rely on F0 and cannot access the VTL cue to conduct the task (Fuller et al., 2014).

While impairment in gender categorization per se might not have dramatic consequences in real situations, as other cues are often available to perform this task, its consequences on the ability to hear a specific speaker among other talkers are very real. Brungart (2001) observed that when presenting two competing sentences to NH listeners, intelligibility was much higher if the two sentences were uttered by two speakers of opposite sex, than when the two sentences were uttered by the same speaker, or by speakers of the same gender. When the target and masker had the same intensity level, this talker-gender difference provided an advantage of 52 percentage points. Darwin et al. (2003) later showed that most of this advantage can be explained by F0 and VTL differences. Thus, considering the importance of vocal acoustic cues for the perception of speech on speech, the weakness of their representation in CIs is likely a major limitation for perception of speech in interfering background sounds by CI users.

Speech Perception in Background Interference

The perception of speech in background noise, or in general in presence of any competing sound source, is undoubtedly considered the strongest limitation to CIs, and thus poses the greatest challenge for the research community. Many reports unambiguously show how little robustness speech perception has to competing sound sources in CI users, compared to reference NH listeners (Friesen et al., 2001; Fu & Nogaki, 2004; Stickney et al., 2004).

When the masker is stationary noise, like the one used in most clinical speech tests, vowel and consonant identification are relatively moderately affected by noise level, both in CI and NH listeners. However, word and sentence identification performance quickly becomes challenging for CI users, even at positive signal-to-noise ratios. The nature of the masker also plays an important role. Maskers that are more ecological than stationary noise, such as speech from other speakers, can reveal even more dramatic differences between NH and CI listeners. In the speech-on-speech study by Stickney et al. (2004), where a target sentence was presented simultaneously with a masker sentence, the advantage that NH listeners could derive from voice gender differences reached 49 percentage points (similar to the study by Brungart, 2001). However, in the same conditions, the largest advantage that CI listeners could derive from talker gender differences was only 19 percentage points (Figure 12–5).

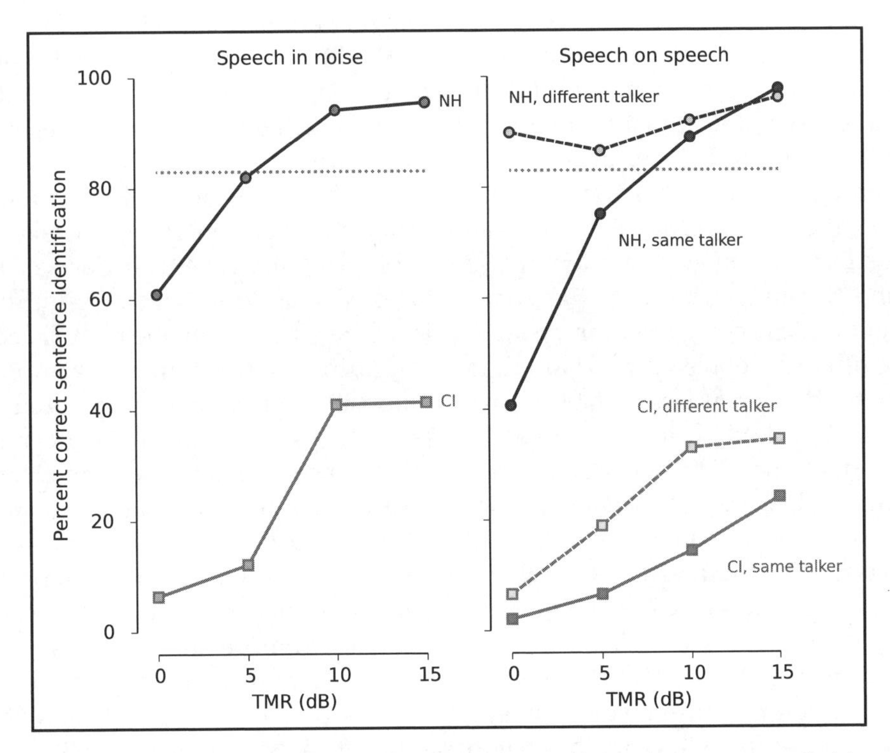

Figure 12–5. *Left panel:* Identification scores for sentences presented in speech-shaped noise for NH and CI listeners, as a function of target-to-masker ratio (TMR). The dotted line shows baseline performance of the CI listeners in quiet. Adapted from Friesen et al. (2001). *Right panel:* Same for speech-on-speech, where the target and masker are either the same talker (*solid lines*) or where the target is a male voice but the masker is a female voice (*dashed lines*). Adapted from Stickney et al. (2004).

Speech-on-speech perception and speech-in-noise perception involve different types of masking, which may explain why CI and NH listeners differ in the way they cope with the two situations. Speech-in-noise perception mostly involves *energetic masking* (i.e., the loss of information representation either in the peripheral auditory system or, in the case of CI listeners, also in the auditory device). Speech-on-speech, on the other hand, involves relatively little energetic masking but instead involves a collection of masking phenomena often gathered under the umbrella term *informational masking*. In fact, the whole phenomenon of perceiving speech in an interferer can be decomposed in order to identify which component mechanisms are affected by the implant limitations.

A component that captures most of the energetic masking is simultaneous segregation (Bregman, 1990) and concerns the perceptual separation of two sound events occurring at the same time. Simultaneous segregation can be studied with the "double vowel" paradigm or with the "concurrent syllables" paradigm. While NH listeners can use F0 (De Cheveigne, 1999) and VTL (Vestergaard et al., 2009), CI listeners do not seem to be able to do so. Luo, Fu, Wu, and Hsu (2009) observed that, similar to competing sentences, CI listeners do not benefit from voice gender differences (which include F0 and VTL differences) in identifying concurrent vowels or syllables. These limitations are thought to be directly related to the poor spectral resolution available through the implant (Qin & Oxenham, 2005).

Another component concerns the way successive speech elements are stringed together to be processed at the linguistic level. In presence of multiple sound sources, successive speech elements must undergo "triage" (i.e., be assigned to the foreground and background auditory streams). This process, known as sequential segregation, can be induced by any perceptual cue that allows discrimination of the two streams (Moore & Gockel, 2002). While CI listeners have been shown to be able to use this mechanism when segregation cues are preserved (Chatterjee et al., 2006; Hong & Turner, 2009), the degradation of the segregation cues that are available in natural speech (like F0 and VTL) also yields weaker stream separation (Gaudrain et al., 2007, 2008).

The literature also described a phenomenon, referred to as *glimpsing*, that is very much related to the segregation mechanisms explained above. Glimpsing is the ability to exploit temporal or spectral sparseness of a masker to "glimpse" unmasked portions of the target signal. While spectral glimpsing is severely limited by the poor spectral resolution of the implant, temporal glimpsing should, in principle, remain possible, just like sequential segregation is in principle less affected by electric hearing than simultaneous segregation. However, using interrupted maskers, researchers have consistently reported that CI listeners seem less able to benefit from glimpses than NH listeners, even when only temporal glimpses are to be used (Gnansia et al., 2010; Nelson & Jin, 2004; Nelson et al., 2003; Qin & Oxenham, 2003).

Simultaneous and sequential segregation, or spectral and temporal glimpsing, are thus all affected by electric hearing, but to different degrees. One could thus venture that different types of maskers have different effects on speech perception in CI than in NH listeners because they involve simultaneous and sequential segregation to different degrees. Kwon, Perry, Wilhelm, and Healy (2012) investigated this hypothesis by creating maskers that either maximized or minimized the need for simultaneous segregation versus sequential segregation. They found that while the sequential condition was easier for NH listeners, CI listeners displayed no such benefit, suggesting that sequential processing is either more degraded in CIs than previously estimated in other studies, or that another component of speech-in-noise perception is also impaired by electrical hearing. Gaudrain and Carlyon (2013) used a more direct approach by developing a purely sequential speech mixture, Zebra-speech, which they compared to a normal speech mixture requiring both simultaneous and sequential segregation (Figure 12–6, Panels A and B). Using noise-band vocoders to simulate some aspects of electrical stimulation, they found sequential segregation was less affected than simultaneous segregation at lower spectral resolutions but also concluded that another mechanism involved in concurrent speech perception must be impaired in CIs.

Indeed, simultaneous and sequential segregation interact together with linguistic processes, whose function it is to transform all accumulated auditory evidence into meaningful semantic content. It has been studied either on its own using interrupted speech (Nelson et al., 2004) or using the phonemic restoration paradigm (Bhargava et al., 2014). This component mechanism strongly hinges on the linguistic and cognitive capacities of the listener (Benard et al., 2014).

Cognitive Factors

In order to get insight into speech processing with CIs, it is important to look beyond the initial sensory input from the device (device- and physiology-related factors described above) and to focus on what the CI users are able to do with this degraded sensory information. Cognitive functions, such as executive functions (e.g., the ability to control and regulate attention, speed of processing, sequential integration) and verbal working memory (the ability to retain and manipulate acoustic signals along the way of the mapping to meaning), are fundamental to speech perception (e.g., Cleary et al., 2000; Nittrouer et al., 2013).

Speech perception requires the ability to adapt the processing of acoustic information toward different speakers and different surrounding acoustics, that is, the allocation of attention to relevant acoustic events and selective inhibition of information that is redundant to a conversation (e.g., Cherry, 1953). Understanding speech also requires the ability to retain acoustic information in memory (Baddeley, 1997) and to integrate these acoustic events with other sources of information, such as grammatical structure, and semantic context from preceding sentential context (e.g., Dahan & Tanenhaus, 2004). To obtain these goals, perception relies on higher-level cognitive processes, which allow top-down processing to enhance the acoustic bottom-up information. Our scientific knowledge of these processes is primarily based on NH populations. However, by now, there is enough evidence that these processes are not

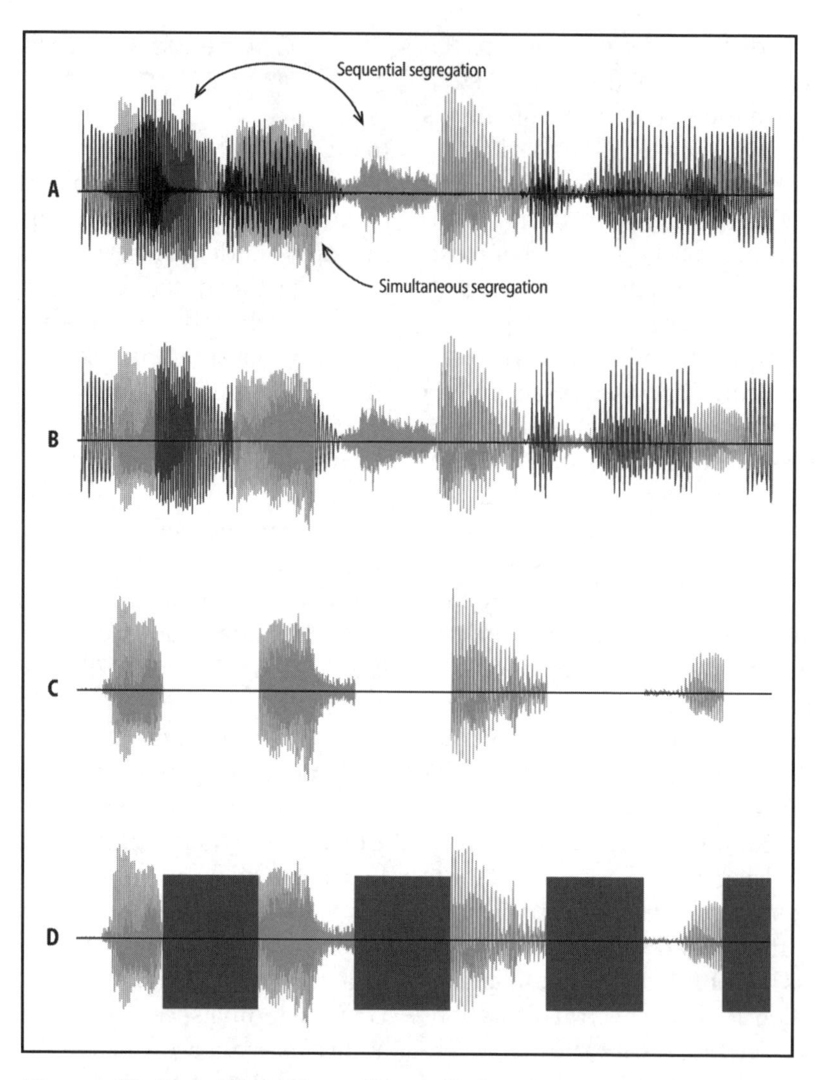

Figure 12–6. A. Waveforms of two simultaneous sentences overlapping. The arrows indicate speech segments whose processing would involve sequential and simultaneous segregation. **B.** Waveform of the Zebra-speech version of the same overlapping sentences. **C.** Waveform of one of the sentences periodically interrupted, similar to the ones used in interrupted speech experiments. **D.** Same as C but where the silent intervals are filled with noise.

equivalent for CI populations, including children and postlingually deafened adults, who in addition show also a great deal of interindividual variation.

The development of cognitive functions in normally developing NH children takes its course in parallel to their speech development (Singer & Bashir, 1999); auditory deprivation during critical developmental phases leads to atypical development of executive functions (Kronenberger et al., 2014).

In NH infants, this parallel development of speech perception and executive functions leads to an optimization of processing of speech in adult listeners. Many prelingually deaf pediatric CI users, however, score significantly below their age-matched peers in a variety of neurocognitive processing measures, including those assessing short-term and working memory, sequential learning, verbal rehearsal, and executive functioning (AuBuchon et al., 2015; Beer et al., 2011; Conway et al., 2011; Harris et al., 2013; Kronenberger et al., 2014; Kronenberger et al., 2013; Pisoni et al., 2011; Van Wieringen & Wouters, 2015). Furthermore, language outcomes for prelingually deaf, pediatric CI users have been at least partially attributed to individual differences in these domains, even when statistically controlling for several potentially confounding variables such as age, duration of deafness, duration of device use, age at onset of deafness, number of active electrodes, and communication mode (e.g., Cleary & Pisoni, 2002; Geers & Sedey, 2011; Marschark et al., 2007; Pisoni & Cleary, 2003; Pisoni & Geers, 2000).

For adult NH populations, understanding native speech is fast, robust, and effortless. This is thanks to a native language specialization of linguistic processes (Cutler, 2012) that are based on and develop in parallel with higher-level cognitive functions. Such perceptual specialization enables listeners to quickly attend to acoustic information that is distinctive and reliable within their native language (Iverson et al., 2003; Wagner et al., 2006). These mechanisms of automatic selection of acoustic cues appear to be limited to speech and do not easily generalize to stimuli that simulate the signals transmitted via CIs (Iverson et al., 2011; Iverson et al., 2016). Furthermore, self-regulatory mechanism of attention shifts between speech signals and masking noise also appears to be guided by the acoustic details that are present in natural speech (Wöstmann et al., 2015). It is hence unclear whether and which of these native perceptual strategies can also be applied by CI listeners when processing spectro-temporally impoverished speech signals. Furthermore, individuals' ability to perceptually adapt to CI signals depends also on listeners' cognitive abilities.

Even among NH young adults, listeners with stronger neurocognitive skills might be better able to understand spoken words, especially in adverse conditions, such as with poor room acoustics, accented or reduced speech, or increased cognitive load (e.g., Francis & Nusbaum, 2009; Tamati et al., 2013), though findings of this link are so far not entirely conclusive (Akeroyd et al., 2014). For postlingually deafened CI users, individual differences in neurocognitive processing mechanisms, for example, after a long period of sensory deprivation, have been found to contribute to speech perception and recognition skills (e.g., Collison et al., 2004; Heydebrand et al., 2007; Holden et al., 2013; Lazard et al., 2010; Lazard et al., 2013).

Taken together, what the CI users are able to do with the information received through a CI is as important for speech and language outcomes as the sensory information itself. Neurocognitive processing skills related to information-processing operations used in the encoding, storage, rehearsal, and retrieval of the phonological and lexical representations of spoken words seem to contribute to the vast amounts of individual differences in the spoken language outcomes of adult and pediatric CI users.

Top-Down Compensation

The fact that CI users can understand speech, given the limitations of the device, as well as the demands that the spectrotemporally impoverished signal set on listeners' cognitive functions, demonstrates a great deal of plasticity of the perceptual system. This plasticity suggests that long-term exposure to speech transmitted via CIs by itself can lead to successful adaptation of the perceptual system, an adjustment of the processing toward the demands of the degraded signal (Svirsky et al., 2001). This implies that compensation mechanisms that help NH listeners to cope with degraded signals could, maybe in adapted ways, also be employed by CI listeners. Among such compensation mechanisms is phonemic restoration, the ability to perceptually complete masked parts of the speech signal by top-down interpretations.

One way to test this idea in the lab is to measure intelligibility performance for *interrupted speech* (see Figure 12–6, panel C). Interrupted, or "gated," sentences are produced by applying a square wave modulation to the original sentence's waveform, which thus periodically turns some segments silent. The parameters are the interruption rate, typically varied from about 1 to 32 Hz, and the duty cycle (i.e., the proportion of remaining speech to the silence). Interruption rate has a very clear effect on interrupted speech perception. Nelson and Jin (2004) observed that, at 2-Hz interruptions, NH listeners' performance drops from nearly 100% to 30% correct, while with 32-Hz interruptions, performance only drops to about 80%. In contrast, CI listeners' performance drops from 80% correct in the uninterrupted case to about 5% and 10% correct for 2- and 32-Hz interruptions, respectively. Bhargava et al. (2016) found similar differences in how CI and NH individuals cope with interruptions at different rates but also found that CI listeners were less able to take advantage of increased duty cycle than NH listeners. In a subsequent experiment, Bhargava et al. selected NH participants whose age was matching individually that of the CI participants, and used noise band vocoders (Shannon, 1995) that were adjusted so that the intelligibility of uninterrupted speech of each NH listener was also matching that of their paired CI listener. They then found that interruptions had a similar or more deleterious effect on performance for these NH listeners. In other words, adding the second degradation of interruptions to an already degraded CI signal amplifies the loss of intelligibility significantly.

Two phenomena happen when speech is interrupted. First, some speech elements are removed. This phenomenon is similar to the one that takes place when listening to speech in noise, where some speech elements can be masked and thus made inaccessible to the listener. But a second phenomenon also takes place; by introducing silences, spurious speech cues are introduced. Indeed, sharp onsets and offsets, and pauses, all carry phonetic and linguistic information. By artificially inserting silences in sentences, speech elements are not only lost but also replaced with potentially misleading speech cues. Dealing with these spurious cues requires more active cognitive processes than dealing with the loss of information alone.

This active top-down restoration can be investigated in the lab using the *phonemic restoration* (PR) paradigm (Warren, 1970; Warren & Sherman, 1974). PR is measured as the difference in intelligibility between interrupted sentences (see Figure 12–6, panel C) and sentences interrupted in the same way except that the silent interruptions are filled with noise bursts (see Figure 12–6, panel D). In NH listeners, adding the noise in the silent interruptions protects against the apparition of spurious cues, allowing a more faithful interpretation of the remaining speech segments, and thus resulting in an increase in intelligibility (Figure 12–7). As such, PR is considered a measure of top-down restoration.

Studies using vocoders to simulate CI processing in NH listeners have led to the conclusion that spectral and temporal degradations such as those occurring in electrical stimulation hinder phonemic restoration (Başkent, 2012; Benard & Başkent, 2014). However, Bhargava et al. (2014) showed that in actual CI users, phonemic restoration can be observed when the duty cycle is made more favorable (see Figure 12–7, left panel).

The discrepancy between the acoustic CI simulations and actual CI listeners seems to come from the type of vocoder—a noise-band vocoder—that was used both by Başkent (2012) and by Bhargava et al. (2014). In order for phonemic restoration to take place, the noise segments must be clearly identified as a potential masker of the speech. When spectral resolution is reduced, and when a noise carrier is used to excite the vocoder, the noisy interruptions become difficult to distinguish from the vocoded speech, which is then itself noisy in nature. Clarke et al. (2016) showed that when the F0 information is restored in the vocoder, phonemic

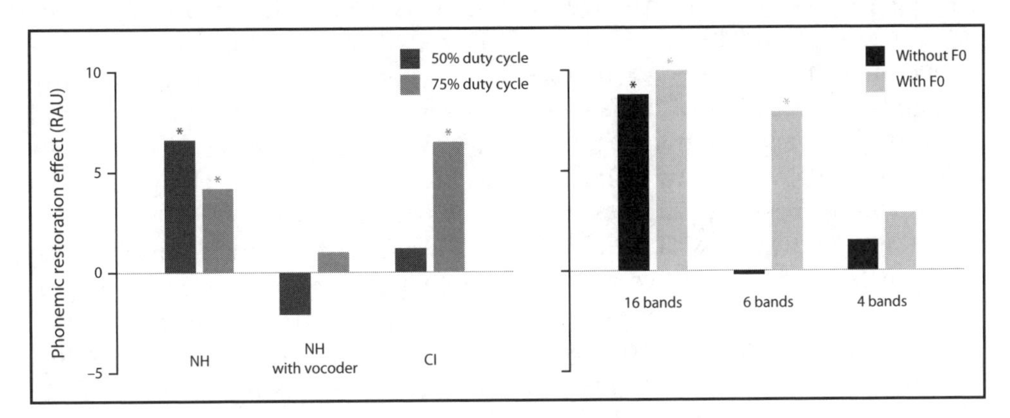

Figure 12–7. *Left panel:* Phonemic restoration effect in rationalized arcsine units (RAU) for NH listeners, NH listeners presented with noise-band vocoded stimuli, and CI listeners. the dark and medium vertical gray bar graphs correspond to two duty cycles (see legend). Adapted from Bhargava et al. (2014). *Right panel:* Phonemic restoration effect (RAU) for NH listeners as a function of number of bands in a vocoder. The black and light gray vertical bar graphs correspond to whether the F0 was preserved or discarded in the vocoder (see legend). Adapted from Clarke, Başkent, and Gaudrain (2016).

restoration is also restored, for intermediate spectral resolutions (see Figure 12–7, right panel). Although F0 information is severely degraded in CI listeners, they generally remain able to distinguish noisy signals from periodic ones, which give them the ability to use top-down restoration.

Speech Perception Mechanisms with Degraded Speech

Adverse conditions, such as hearing impairment or noisy surroundings, may change the functioning of cognitive mechanisms that underlie automatic speech perception in ideal conditions (Mattys et al., 2012). The current audiological assessment methods of speech perception only provide a measure of intelligibility, expressed in a single number, such as percent correct score or speech reception threshold. Hence, they do not fully reveal the underlying mechanisms of speech perception and the potential changes in them as a result of the speech degradation.

One of such changes is an increase in cognitive processing load (listening effort) to decode the degraded speech (Winn, Edwards, & Litovsky, 2015; Pals, Sarampalis & Başkent, 2013; Sarampalis, Kalluri, Edwards, & Hafter, 2009). While an increase in cognitive processing can be a good compensation mechanism to enhance perception of degraded speech, if sustained, it can lead to mental fatigue (Bess & Hornsby, 2014; Wild et al., 2012). Furthermore,, taking up more of the cognitive resources for speech comprehension may lead to fewer cognitive resources left for other mental tasks, such as remembering previously heard message and applying it for predictive processing of successive speech contents (Wagner, Pals, de Blecourt, Sarampalis & Başkent).

While CI users are able to understand the speech signal transmitted via CIs, speech perception is likely a more effortful task for them than NH individuals. Studies with CI simulations show that reducing the spectral resolution of the speech signal increases the effort involved in processing it. This was shown both with pupil dilation, a physiological measure of listening effort (Winn et al., 2015), as well as a dual-task paradigm (Pals et al., 2013), in which response times to a secondary task (either linguistic or nonlinguistic mental manipulation) paired with primary speech perception task reflect listening effort. The rationale behind the latter paradigm is that simultaneous cognitive tasks compete for cognitive resources, which are limited (Kahneman, 1973). Hence, increasing increasing spectral resolution of in the signal reduces the response time needed by listeners to perform the secondary task. This confirms the presence of competition between parallel cognitive processes and shows that processing of degraded speech takes up cognitive resources, making speech perception a more effortful task.

Recently, new studies started investigating changes in speech perception mechanisms, more comprehensively and not limited to listening effort only. An eye-tracking study that combined the measure of the time course of lexical access (i.e., the mapping of the signal to meaning) and of the effort involved in speech perception by NH listeners presented with CI simulations. This study showed that degradation of the signal obscured listeners' use of cues transmitted in the signal, slowed down lexical access, and increased the effort involved in listening (Wagner, Toffanin, & Başkent, 2016). A similarly designed study with CI listeners showed that despite great individual differences among CI users, in general, experienced CI users are able to adapt their use of acoustic cues that are also employed by NH listeners. Durational cues are in particular susceptible to reweighting since they are reliably transmitted through the CI (Wagner, et al., 2016). Similar adaptation toward higher cue weighting of durational cues in experienced CI users relative to NH participants has been reported by Winn et al. (2012). However, Wagner, et al. (2016) additionally found that despite an adaptation toward a stronger reliance on reliably transmitted cues, the process of lexical access was still delayed and prolonged for experienced CI users. In the same vain, Moberly et al. (2014) found evidence for reweighting of the perceptual use of acoustic cues, such as duration and spectral cues. This study, however, also found individual differences in the reweighting of these cues in experienced CI users, and concluded that CI individuals with cue-weighting most similar to NH listeners, who thus relied more on (degraded) spectral cues, showed better performance in word identification.

Potentially, compensation for such phonetic processing could come from a stronger reliance on semantic and contextual information. Effects of such top-down filling in of information based on sources other than the acoustic signal alone were investigated by Bhargava et al. (2014), mentioned above. This study shows that CI users benefit from phonemic restoration, but to a lesser degree than NH listeners, and this benefit is more limited by conditions of testing. Wagner, Pals, et al. (2016) investigated the time course of integration of semantic information from sentential context by means of eye-tracking, using CI simulations. In this study, degradation of the signal reduced listeners' ability to benefit from previously heard information, delayed the integration of contextual information, and these changes came at the cost of a longer and more effortful lexical access. For CI listeners this implies limitations to their ability to enhance the speech signal through top-down information. Whereas NH listeners can use the sentential information preceding the target to anticipate upcoming words, this ability is restricted for CI users. Figure 12–8, right panel, shows the patterns of integration of information in a sentence that differed between the processing of natural and degraded speech. The figure shows the gaze fixation patterns when recognizing a target word, such as *tree,* when it is presented within the sentence, "Since when grows a tree so fast." The verb in this sentence disambiguates the following target as something animate that can grow. In the experimental paradigm, the picture of a tree (target) was presented together with the picture of a child (presenting ambiguity as a child can also grow), and the pictures of two inanimate distractor objects (see Figure 12–8, left panel).

In the figure, the thick solid lines depict the proportion of the gaze fixations toward the target tree, the dashed lines depict the proportion of the gaze fixations toward the competitor child, and the thin dotted lines show the proportion fixations toward the inanimate distractor objects. During the processing

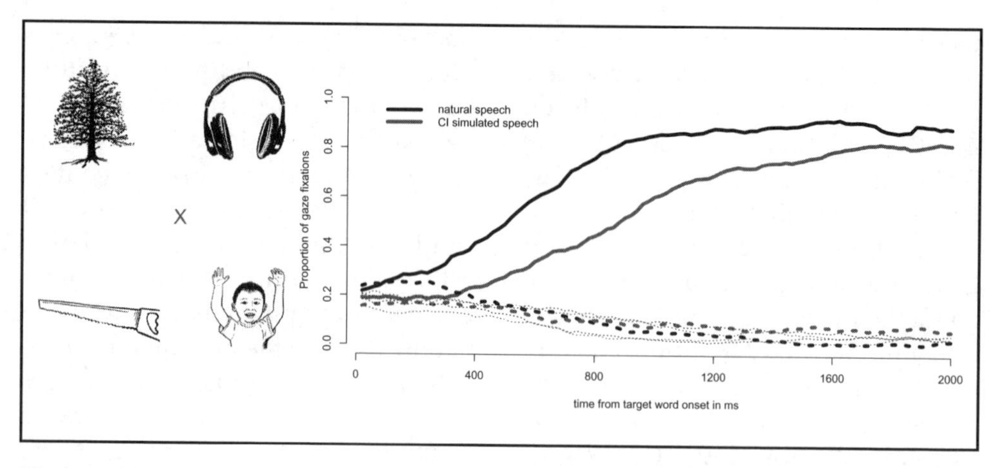

Figure 12–8. *Left panel:* The display presented to participants in Wagner, Pals, et al. (2016). *Right panel:* The time curves of gaze fixations towards the target picture (tree), the semantically viable picture (baby), and the two inanimate distractor pictures (saw and headphone).

of natural speech (black line), listeners integrate the semantic information from the verb, which restricts their gaze fixations toward the two animate objects and discards the possibility that the inanimate objects could be the target word. When listening to CI simulations (gray line), listeners presented with degraded speech take about 300 ms longer to clearly identify the tree as the target, and they tend to fixate the inanimate objects to the same degree as the animate object child. This shows that the integration of semantic information from the sentence context is delayed and cannot be used timely to limit the search for the target from objects that are excluded by the verb (inanimate objects in this example). And note that the experimental setting in this experiment mimicked optimal conditions. Listeners' need to integrate semantic and grammatical sources of information is likely to be even greater in real-life situations.

Real-Life Speech Perception

Speech communication in real life can be very challenging. The adverse conditions of everyday speech communication not only involve the background noise, competing talkers or signals, or poor room acoustics, but also involve natural variability in the speech signal (e.g., Mattys et al., 2012). In normal, everyday environments, a talker makes stylistic changes to their speech depending on the speech environment, speaking rate, or the surrounding phonetic or phonological context. Additionally, individuals or social groups also have diverse speech patterns, reflecting their language and developmental histories (Abercrombie, 1967). Thus, real-life speech is characterized by a great deal of variability in the realization of sounds and words.

In everyday, real-life listening environments, listeners draw upon linguistic knowledge and cognitive resources in order to achieve the perceptually and cognitively demanding task of adapting to these sources of variability in communication. While some speaking styles, talkers, and accents are relatively easy to understand, others may be more difficult, such as unfamiliar regional or foreign accents (e.g., Clopper & Bradlow, 2008; Mason, 1946), fast speech (e.g., Bradlow et al., 1996; Picheny et al., 1989), or reduced speech involving syllable and segment reduction or deletion (e.g., Ernestus et al., 2002; Janse et al., 2007). Furthermore, the presence of multiple talkers and multiple sources of variability can present additional challenges to successful speech recognition (e.g., Mullennix et al., 1989). However, in normal hearing, speech communication in real-life environments is not entirely hindered. NH individuals are able to rapidly adapt to and use variation in the talker's speech characteristics and draw upon prior linguistic knowledge to successfully understand the utterance, while also extracting information about the environment, context, and talker (e.g., Johnson & Mullennix, 1997; Johnsrude et al., 2013; Tamati et al., 2014).

CI users, similar to NH listeners, are also faced with multiple sources of variability in real-life listening environments. However, because CI users must rely on a signal that is less detailed in acoustic-phonetic information than is typically available to NH listeners, their capacity to encode fine context-sensitive episodic information to reliably perceive and use subtle variation may be limited. Additionally, due to periods of auditory deprivation, some CI users may have to rely upon disrupted perceptual and linguistic systems, and atypical neurocognitive skills, both of which seem to be associated with poor speech perception performance, as discussed above. Thus, the adverse effects of speech variability may be exacerbated for CI users who are receiving limited information from a degraded signal, and additionally may have poor spoken language skills and/or limited or atypically developed cognitive functions.

Despite these concerns, in contrast to our growing knowledge of NH perception of real-life speech variability, the speech perception skills of CI populations in these conditions are relatively unknown. A simple approach to studying CI perception of real-life speech variability is to assess their speech recognition with materials that more closely reflect real-life speech. Some recent studies indicate that highly variable speech, more reflective of real-life conditions, presents a significant challenge for CI users. A few recent studies have shown that CI listeners are disproportionately less accurate than NH listeners at recognizing speech produced by multiple talkers from different regions of origin (Faulkner, Tamati, Gilbert, & Pisoni, 2015; Faulkner, Tamati, & Pisoni, 2016), fast speech (Ji et al., 2013), foreign-accented speech (Ji et al., 2014), and reduced speech (Tamati et al., 2015).

Another way to study how CI users perceive, encode, and store robust information about real-life speech variability is to examine the discrimination or identification of different sources of speech variability, and the influence of linguistic information on these nonlinguistic judgments. Studies using this approach have suggested that CI listeners may not be able to make use of detailed talker-specific acoustic-phonetic information to the same extent as normal-hearing (NH) listeners to discriminate or identify different sources of real-life variability. In a recent study, Tamati and Pisoni (2015) used a foreign-accent intelligibility rating task to investigate the perception of foreign-accented speech by prelingually deaf, long-term CI users. The CI users, and age-matched NH listeners, rated the intelligibility of short sentences produced by native and nonnative speakers of American English. Both the CI and NH listeners perceived the foreign-accented sentences as less intelligible than native sentences. However, compared to the NH listeners, the CI listeners perceived a smaller difference in intelligibility for foreign-accented and native speech. These findings suggest that although the CI users are sensitive to some subtle acoustic-phonetic differences between foreign-accented and native speech, they are much less so than NH listeners.

The discrimination of different sources of speech variability has also been found to be closely linked to the perception of the linguistic information in the utterance. Examining talker discrimination abilities of pediatric CI users, Cleary and colleagues (Cleary & Pisoni, 2002; Cleary et al., 2005) found that CI users were able to make more accurate talker discrimination judgments when the linguistic content was constant across two items, compared to when the sentences differed. This suggests that the pediatric CI users were poor at attending to the relevant dimension (i.e., talker voices) and ignoring the irrelevant dimension (i.e., the linguistic content). Furthermore, they found that children who were better on the talker discrimination tasks were also more accurate at recognizing spoken words. Similarly, Tamati and Pisoni (2015) further analyzed the foreign-accent intelligibility ratings by comparing individual differences in accent discrimination scores (i.e., difference in intelligibility ratings for foreign-accented speech compared to native speech). They also found that the accent discrimination scores were related to several measures of sentence recognition abilities. Figure 12–9 displays the relation between the discrimination scores on the foreign-accent rating task and scores on the PRESTO (Perceptually Robust English Sentence Test Open-Set), sentence recognition test for prelingually deaf, long-term CI users. Taken together, CI users' ability to deal with real-life speech variability may reflect the development and use of basic speech and language-processing skills. Thus, CI users may benefit from better basic speech and language skills. In addition, some case studies with postlingually deaf adults suggest that CI users may also be able to use prior linguistic knowledge and experience to improve their perception of some sources of real-life speech (e.g., Tamati et al., 2014).

Despite the challenges of real-life speech, most previous research studies and clinical assessments with CI users have concerned the effects of sources of environmental degradation, such as background noise or competing talkers, and only a few tests used to assess benefits and outcomes of CI implantation in adults contain speech more characteristic of everyday speech environments. As a consequence, we do not have a full

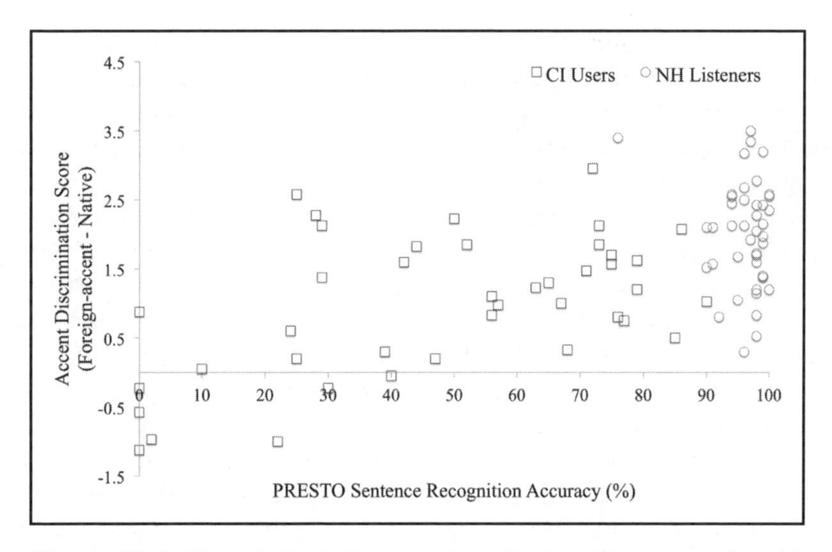

Figure 12–9. The relation between accent discrimination scores (*y*-axis) and PRESTO sentence recognition scores (*x*-axis) for CI users (*open squares*) and NH listeners (*open circles*). Adapted from Tamati and Pisoni (2015).

picture of the potential communicative challenges CI users face in their everyday lives, nor do we have the appropriate tools available to improve their communicative abilities.

New Assessment Techniques

The conventional tests of spoken word recognition commonly used in the clinics have been developed with very simple, familiar materials, slowly and clearly produced by a single talker with no discernable accent. The Hearing in Noise Test (HINT; Nilsson, Soli, & Sullivan, 1994) is an example of a widely used conventional, low-variability sentence recognition test. In U.S. clinics, it is commonly used as a tool to determine CI candidacy and measure outcome and benefit for CI users (e.g., Fabry et al., 2009). The sentences in the HINT are short and syntactically simple, selected or modified to have roughly the same length and the same intelligibility at a fixed noise level, and were produced by a single male talker, with a standard unmarked General American regional dialect. Because listeners benefit from the simple structure of the sentence and the lack of natural talker variability, such sentence tests, like the HINT, likely result in artificially high sentence recognition scores and ceiling effects (e.g., Gifford et al., 2008). Additionally, these simple clinical tests likely do not capture the actual challenges CI users may face in real life and as such fail to identify listeners who may struggle in real-life complex listening environments. Using these materials may also mask potential beneficial outcomes of some new device features or training approaches, leading manufacturers and clinicians to discount them.

With advances in device technology, signal-processing strategies, and the expansion of implant candidacy criteria to listeners with greater residual hearing (Gifford et al., 2010), more sensitive tests are thus needed to assess candidacy, as well as real-life outcomes and benefit from CI use. Recently, several new, more challenging sentence recognition tests have been developed with materials that incorporate more sources of natural speech variability, such as the Az-Bio test (Spahr & Dorman, 2004), the Az-TIMIT test (King et al., 2012), the STARR test (Sentence Test with Adaptive Randomized Roving levels; Boyle, Nunn, O'Connor, & Moore, 2013), and the PRESTO test (Gilbert et al., 2013). These new tests require a listener to adapt to a talker's individual idiolectal speech characteristics. The PRESTO test, for example, was designed to contain a great deal of natural variability to reflect foundational components of real-life listening environments. The test includes sentences produced by unique (i.e., no talker was repeated within a list) male and female talkers from several different dialect regions of the United States. The PRESTO test has been shown to be more challenging for CI users, compared to conventional low-variability materials, better reflecting the basic speech perception and neurocognitive abilities (Faulkner, Tamati, & Pisoni, 2016; Tamati & Pisoni, 2015). Figure 12–10 displays individual scores of prelingually deaf, long-term CI users on HINT and PRESTO. Although listeners who performed well on the HINT sentences also tended to perform well on the PRESTO sentences, group

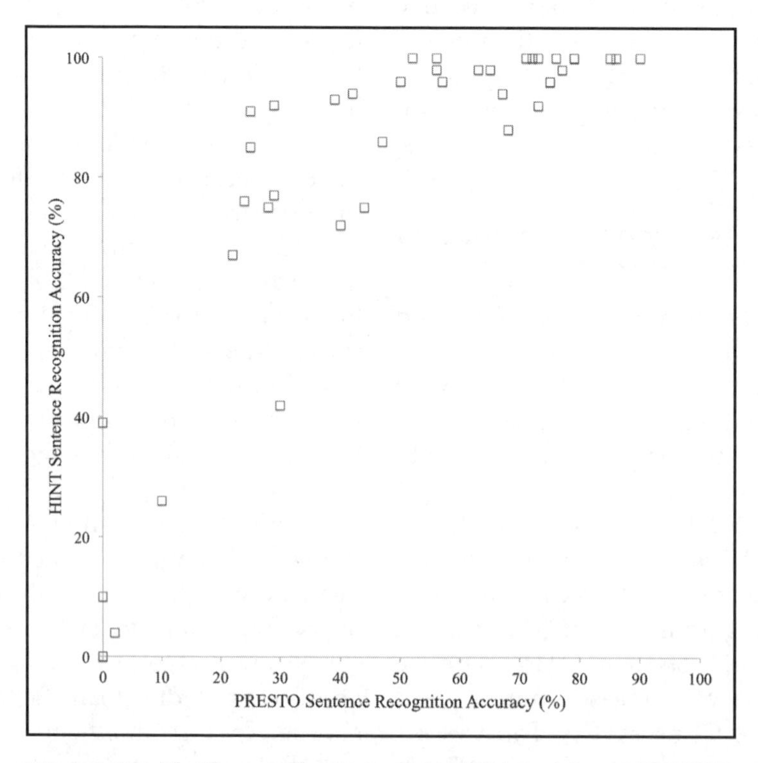

Figure 12–10. The relation between HINT (*y*-axis) and PRESTO (*x*-axis) sentence recognition scores for CI users. Adapted from Tamati and Pisoni (2015) and Faulkner, Tamati, et al. (2016).

and individual patterns of performance differed by test. As can be seen in the figure, the CI users were near ceiling on the HINT sentences in quiet, also demonstrating little listener variability. The CI users were less accurate at recognizing words on the PRESTO sentences, and they also showed a vast amount of individual variability.

Despite development of these new tests that include natural variability more reflective of real-life environments, most research and clinical tests on CI speech perception still focus on a individual's ability to recognize ideal speech. In addition, few high-variability sentence recognition tests have been developed for different languages, and there is a lack of high-variability materials and resources for perceptual training tools in the clinic. As a result, our knowledge of the mechanisms underlying CI speech perception and spoken word recognition remains limited. More widespread use of these tests that incorporate more natural variability and future materials within the clinic and research labs, will thus allow us to better understand CI users' real-life speech perception abilities, as well as the vast amounts of individual variability in these abilities. Furthermore, future improvements on clinical and research materials for CI users will allow for the development of more effective rehabilitation programs to meet the real-life needs of this population.

New Training Approaches

The basic auditory and cognitive processes underlying speech perception performance, as outlined above, have also been a focus of training programs developed for CI users. The basic premise of these programs is that improving CI users' skills in these areas will also lead to improvements in real-life speech communication. Conventional training programs aiming at improving auditory and perceptual skills have focused on challenging areas for CI users, such as fine-grained discrimination of phonemes or pitch, or speech in noise perception. Auditory training using a more bottom-up approach involves training listeners to attend to fine-grained acoustic details to identify a linguistic category, such as a vowel or consonant, or attend to the relevant dimension of a target item, such as a target word in noise. Studies using this approach have been successful at improving vowel and consonant recognition (Dawson & Clark, 1997) and lexical tone recognition in Mandarin, a tonal language (Wu et al., 2007), in CI users. Other studies have demonstrated that speech-in-noise training yielded improvements on speech recognition in noise for CI users (Fu & Galvin, 2008; Ingvalson et al., 2013).

These studies have shown that there is great potential in providing targeted training programs to CI users in improving speech perception and related auditory skills. However, as is often seen in training, most improvement has usually been observed in the skills that are specifically used in the training, and a transfer of learning to other related skills has been limited. Such a transfer of learning has been often seen with musicians, suggesting an overlap between music and speech recognition networks in the auditory system, leading to better use of acoustic cues for speech perception, and perhaps also an improvement in general cognitive skills related to auditory perception that can both be applied to speech and music perception (Besson et al., 2011; Micheyl et al., 2006). As a result, several studies have shown that musically trained individuals perform better not only in music-related tasks but also in perception of speech in noise, compared to nonmusicians (e.g., Başkent & Gaudrain, in press; Parbery-Clark et al., 2009). However, this benefit seems to depend on the specific form of interfering noise (Fuller et al., 2014; Swaminathan et al., 2014). In CI users, music training can improve performance in music-related tasks (Galvin et al., 2007, 2012; Gfeller et al., 2002), but a transfer of learning to speech perception has not yet been shown. Only with CI simulations, recently, Fuller et al. (2014) have shown that the musician advantage can persist with degraded stimuli, but again the benefit seemed to become smaller from music- to speech-related perceptual tasks. On the other hand, music, specifically its perception and appreciation, is another challenging area for CI users (e.g., Fuller et al., 2012; Galvin et al., 2009; Gfeller et al., 2002), on which training with music could also potentially have a positive effect (Looi et al., 2012; Yücel et al., 2009). An additional advantage of training with music could be the fun factor, encouraging the CI users to participate more actively. A pilot study from our group has shown that CI users seem to be more willing to participate in an interactive and fun training program, such as music therapy, than in a scientifically proven but less interactive program, such as computer-based training (Free et al., 2014). Perhaps a combination of two would provide a both fun but also beneficial training option to CI users.

These same principles motivate cognitive training programs with CI users as well. Speech and language outcomes of CI users have been linked to individual differences in underlying neurocognitive processes. In particular, studies on language and cognitive development in CI users, as well as individual differences among CI users, have identified working memory as a key component of speech development and processing. As such, a new direction for cognitive training for CI users is to use training to modify working memory capacity in order to improve speech and language outcomes of CI users. That is, skills that are acquired through the working memory training can be transferred to speech perception skills. Kronenberger and colleagues (2011) used working memory training with nine prelingually deaf children with CIs. They found that the children improved on most of the training exercises and also improved, compared to baseline scores, on measures of verbal and nonverbal working memory, parent reports of the child's memory behavior, and sentence repetition immediately following training. Improvements declined after training for the working memory measures, with no improvements remaining after 6 months, but improvements with sentence repetition were more lasting with improvements being maintained after 6 months. In a larger study, Ingvalson, Young, and Wong (2014) trained the phonological awareness and working memory skills of 10 prelingually deaf children with CIs, with an additional 9 children with CIs serving as controls without training. They found that the children who completed the training showed improvements on oral expressive language and spoken language composite scores, compared to the untrained control listeners. These studies suggest that cognitive training may be a useful new direction for improving speech and language outcomes of CI users.

Taken together, training CI users to make better use of the degraded sensory information they are receiving, in combination with advances in improving the quality of this sensory input, may help CI users achieve greater speech and language outcomes. Further, combining the current and more traditional training methods with new approaches based on new findings on CI speech perception in adverse conditions may offer a path to achieve real-life speech communication benefits for current and future CI users.

Summary

Cochlear implants have provided many deaf individuals the function of hearing, and hence, the ability to communicate. The speech signal transmitted via CIs differs from that of normal hearing and is spectrotemporally degraded. Many CI users seem to adapt to and learn to make use of this degraded speech for communication. However, speech perception in adverse conditions still seems to be a challenge, and there is also a large amount of variability in CI outcome across individuals. Demographics and device- and physiology-related factors have already been identified as contributing to this inter-individual variability in adaptation success and limitations. Recent research shows that, among acoustic factors, the voice-related cues seem to contribute greatly to speech perception, and these are also the cues that are not delivered properly by the device. Further, cognitive processes of speech perception and potential top-down enhancement mechanisms may be altered by the degraded speech input of a CI, and possibly, individual cognitive abilities also contribute to the variability across individuals. Traditionally, the tools used for research on CIs and the rehabilitation for CI users have focused mainly on measures of speech intelligibility, capturing only the end result of speech perception. These, however, do not reveal changes or interactions of speech perception and comprehension mechanisms with degraded speech, especially in more realistic real-life listening environments where the CI users have to deal with different types of background noises, as well as various realizations of speech, such as reduced speech, regional accents, or other types of speaker-induced variability. In this chapter, we present the most contemporary research on these areas. We cover new techniques and methods, which also take into account cognitive factors that can more fully identify the performance and limitations of speech perception and comprehension by CI users, especially in more realistic listening conditions.

References

Abercrombie, D. (1967). *Elements of general phonetics*. Edinburgh, UK: Edinburgh University Press.

Akeroyd, M. (2008). Are individual differences in speech reception related to individual differences in cognitive ability? A survey of twenty experimental studies with normal and hearing-impaired adults. *International Journal of Audiology, 47*(Suppl. 2), S53–S71.

Arnoldner, C., Riss, D., Brunner, M., Durisin, M., Baumgartner, W. D., & Hamzavi, J. S. (2007). Speech and music perception with the new fine structure speech coding strategy: Preliminary results. *Acta Oto-laryngologica, 127*(12), 1298–1303.

AuBuchon, A. M., Pisoni, D. B., & Kronenberger, W. G. (2015). Verbal processing speed and executive functioning in long-term cochlear implant users. *Journal of Speech, Language, and Hearing Research, 58*(1), 151–162.

Baddeley, A. D. (1997). *Human memory: Theory and practice*. Hove, UK: Psychology Press.

Başkent, D. (2012). Effect of speech degradation on top-down repair: Phonemic restoration with simulations of cochlear implants and combined electric–acoustic stimulation. *Journal of the Association for Research in Otolaryngology, 13*(5), 683–692.

Başkent, D., & Gaudrain, E. (in press). Musician advantage for speech-on-speech perception. *The Journal of the Acoustical Society of America.*

Başkent, D., & Shannon, R. V. (2004). Frequency-place compression and expansion in cochlear implant listeners. *The Journal of the Acoustical Society of America, 116*(5), 3130–3140.

Başkent, D., & Shannon, R. V. (2005). Interactions between cochlear implant electrode insertion depth and frequency-place mapping. *The Journal of the Acoustical Society of America, 117*(3, Pt. 1), 1405–1416.

Beer, J., Kronenberger, W. G., & Pisoni, D. B. (2011). Executive function in everyday life: Implications for young cochlear implant users. *Cochlear Implants International, 12*(Suppl. 1), S89–S91.

Benard, M. R., & Başkent, D. (2014). Perceptual learning of temporally interrupted and spectrally degraded speech. *The Journal of the Acoustical Society of America, 136*(3), 1344–1351.

Benard, M. R., Mensink, J. S., & Başkent, D. (2014). Individual differences in top-down restoration of interrupted speech: Links to linguistic and cognitive abilities. *The Journal of the Acoustical Society of America, 135*(2), EL88–EL94.

Berliner, K. I., Tonokawa, L. L., Dye, L. M., & House, W. F. (1989). Open-set speech recognition in children with a single-channel cochlear implant. *Ear and Hearing, 10*(4), 237–242.

Bess, F. H., & Hornsby, B. W. (2014). Commentary: Listening can be exhausting—Fatigue in children and adults with hearing loss. *Ear and Hearing, 35*(6), 592–599.

Besson, M., Chobert, J., & Marie, C. (2011). Transfer of training between music and speech: Common processing, attention, and memory. *Frontiers in Psychology, 2*, 94.

Bhargava, P., Gaudrain, E., & Başkent, D. (2014). Top–down restoration of speech in cochlear-implant users. *Hearing Research, 309*, 113–123.

Bhargava, P., Gaudrain, E., & Başkent, D. (2016). The intelligibility of interrupted speech: Cochlear implant users and normal hearing listeners. *Journal of the Association for Research in Otolaryngology.* Manuscript under revision.

Bierer, J. A., & Faulkner, K. F. (2010). Identifying cochlear implant channels with poor electrode-neuron interface: Partial tripolar, single-channel thresholds and psychophysical tuning curves. *Ear and Hearing, 31*(2), 247–258.

Blamey, P., Artieres, F., Başkent, D., Bergeron, F., Beynon, A., Burke, E., . . . Gallégo, S. (2013). Factors affecting auditory performance of postlinguistically deaf adults using cochlear implants: An update with 2251 patients. *Audiology and Neurotology, 18*(1), 36–47.

Boyle, P. J., Nunn, T. B., O'Connor, A. F., & Moore, B. C. (2013). STARR: A speech test for evaluation of the effectiveness of auditory prostheses under realistic conditions. *Ear and Hearing, 34*(2), 203–212.

Bradlow, A. R., Torretta, G. M., & Pisoni, D. B. (1996). Intelligibility of normal speech I: Global and fine-grained acoustic-phonetic talker characteristics. *Speech Communication, 20*(3), 255–272.

Brungart, D. S. (2001). Informational and energetic masking effects in the perception of two simultaneous talkers. *The Journal of the Acoustical Society of America, 109*(3), 1101–1109.

Chatterjee, M., Sarampalis, A., & Oba, S. I. (2006). Auditory stream segregation with cochlear implants: A preliminary report. *Hearing Research, 222*(1), 100–107.

Chatterjee, M., Zion, D. J., Deroche, M. L., Burianek, B. A., Limb, C. J., Goren, A. P., . . . Christensen, J. A. (2015). Voice emotion recognition by cochlear-implanted children and their normally-hearing peers. *Hearing Research, 322*, 151–162.

Cherry, E. C. (1953). Some experiments on the recognition of speech, with one and with two ears. *The Journal of the Acoustical Society of America, 25*(5), 975–979.

Clark, G. M. (2015). The multi-channel cochlear implant: Multi-disciplinary development of electrical stimulation of the cochlea and the resulting clinical benefit. *Hearing Research, 322*, 4–13.

Clarke, J., Başkent, D., & Gaudrain, E. (2016). Pitch and spectral resolution: A systematic comparison of bottom-up cues for top-down repair of degraded speech. *The Journal of the Acoustical Society of America, 139*, 395–405.

Cleary, M., & Pisoni, D. B. (2002). Talker discrimination by prelingually deaf children with cochlear implants: Preliminary results. *The Annals of Otology, Rhinology & Laryngology Supplement, 189*, 113–118.

Cleary, M., Pisoni, D. B., & Kirk, K. I. (2000). Working memory spans as predictors of spoken word recognition and receptive vocabulary in children with cochlear implants. *The Volta Review, 102*(4), 259–280.

Cleary, M., Pisoni, D. B., & Kirk, K. I. (2005). Influence of voice similarity on talker discrimination in children with normal hearing and children with cochlear implants. *Journal of Speech, Language, and Hearing Research, 48*(1), 204–223.

Clopper, C. G., & Bradlow, A. R. (2008). Perception of dialect variation in noise: Intelligibility and classification. *Language and Speech, 51*(3), 175–198.

Collison, E. A., Munson, B., & Carney, A. E. (2004). Relations among linguistic and cognitive skills and spoken word recognition in adults with cochlear implants. *Journal of Speech, Language, and Hearing Research, 47*(3), 496–508.

Conway, C. M., Karpicke, J., Anaya, E. M., Henning, S. C., Kronenberger, W. G., & Pisoni, D. B. (2011). Nonverbal cognition in deaf children following cochlear implantation: Motor sequencing disturbances mediate language delays. *Developmental Neuropsychology, 36*(2), 237–254.

Crew, J. D., Galvin, J. J., III, & Fu, Q.-J. (2012). Channel interaction limits melodic pitch perception in simulated cochlear implants. *The Journal of the Acoustical Society of America, 132*(5), EL429–EL435.

Cullington, H., & Zeng, F. (2010). Bimodal hearing benefit for speech recognition with competing voice in cochlear implant subject with normal hearing in contralateral ear. *Ear and Hearing, 31*(1), 70–73.

Cutler, A. (2012). *Native listening: Language experience and the recognition of spoken words*. Cambridge, MA: MIT Press.

Dahan, D., & Tanenhaus, M. K. (2004). Continuous mapping from sound to meaning in spoken-language comprehension: Immediate effects of verb-based thematic constraints. *Journal of Experimental Psychology Learning Memory and Cognition, 30*(2), 498–513.

Danley, M., & Fretz, R. (1982). Design and functioning of the single-electrode cochlear implant. *The Annals of Otology, Rhinology & Laryngology Supplement, 91* (2, Pt. 3), 21.

Darwin, C. J., Brungart, D. S., & Simpson, B. D. (2003). Effects of fundamental frequency and vocal-tract length changes on attention to one of two simultaneous talkers. *The Journal of the Acoustical Society of America, 114*(5), 2913–2922.

Dawson, P., & Clark, G. M. (1997). Changes in synthetic and natural vowel perception after specific training for congenitally deafened patients using a multichannel cochlear implant. *Ear and Hearing, 18*(6), 488–501.

De Cheveigné, A. (1999). Waveform interactions and the segregation of concurrent vowels. *The Journal of the Acoustical Society of America, 106*, 2959–2972.

Djourno, A., & Eyriès, C. (1957). Prothèse auditive par excitation électrique à distance du nerf sensoriel à l'aide d'un bobinage inclus a demeure. *Presse médicale, 65*(63), 1417–1417.

Dorman, M., & Spahr, A. (2006). *Speech perception by adults with multichannel cochlear implants.* New York, NY: Thieme Medical.

Dorman, M. F., Loizou, P. C., & Rainey, D. (1997). Simulating the effect of cochlear-implant electrode insertion depth on speech understanding. *The Journal of the Acoustical Society of America, 102*(5), 2993–2996.

Douek, E. E., Fourcin, A. J., Moore, B. C. J., & Clark, G. P. (1977). A new approach to the cochlear implant. *Proceedings of the Royal Society of Medicine, 70*, 379–383.

Drennan, W. R., Oleson, J. J., Gfeller, K., Crosson, J., Driscoll, V. D., Won, J. H., . . . Rubinstein, J. T. (2015). Clinical evaluation of music perception, appraisal and experience in cochlear implant users. *International Journal of Audiology, 54*(2), 114–123.

Ernestus, M., Baayen, H., & Schreuder, R. (2002). The recognition of reduced word forms. *Brain and Language, 81*(1), 162–173.

Fabry, D., Firszt, J., Gifford, R., Holden, L., & Koch, D. (2009). Evaluating speech perception benefit in adult cochlear implant recipients. *Audiology Today, 21*, 36–43.

Faulkner, K. F., Tamati, T. N., Gilbert, J. L., & Pisoni, D. B. (2015). List equivalency of PRESTO for the evaluation of speech recognition. *Journal of the American Academy of Audiology, 26*(6), 582–594.

Faulkner, K. F., Tamati, T. N., & Pisoni, D. B. (2016). *Individual differences in high-variability speech recognition: Some new findings using PRESTO in clinical and non-clinical populations.* Manuscript in preparation.

Finley, C., & Skinner, M. (2008). Role of electrode placement as a contributor to variability in cochlear implant outcomes. *Otology & Neurotology, 29*(7), 920–928.

Fitch, W. T., & Giedd, J. (1999). Morphology and development of the human vocal tract: A study using magnetic resonance imaging. *The Journal of the Acoustical Society of America, 106*(3), 1511–1522.

Francis, A. L., & Nusbaum, H. C. (2009). Effects of intelligibility on working memory demand for speech perception. *Attention, Perception, & Psychophysics, 71*(6), 1360–1374.

Free, R., Fuller, C., Maat, B., & Basskent, D. (2014, February). *The effect of music therapy and training on speech and music perception in cochlear-implant users.* Paper presented at the 37th Annual Mid-winter Meeting of the Association for Research in Otolaryngology, San Diego, CA.

Friesen, L. M., Shannon, R. V., Başkent, D., & Wang, X. (2001). Speech recognition in noise as a function of the number of spectral channels: Comparison of acoustic hearing and cochlear implants. *The Journal of the Acoustical Society of America, 110*(2), 1150–1163.

Friesen, L. M., Shannon, R. V., & Cruz, R. J. (2005). Effects of stimulation rate on speech recognition with cochlear implants. *Audiology and Neurotology, 10*(3), 169–184.

Fu, Q.-J. (2002). Temporal processing and speech recognition in cochlear implant users. *Neuroreport, 13*(13), 1635–1639.

Fu, Q.-J., & Galvin, J. J., III. (2008). Maximizing cochlear implant patients' performance with advanced speech training procedures. *Hearing Research, 242*(1–2), 198–208.

Fu, Q.-J., & Nogaki, G. (2005). Noise susceptibility of cochlear implant users: The role of spectral resolution and smearing. *Journal of the Association for Research in Otolaryngology, 6*(1), 19–27.

Fuller, C., Free, R., Maat, B., & Başkent, D. (2012). Musical background not associated with self-perceived hearing performance or speech perception in postlingual cochlear-implant users. *The Journal of the Acoustical Society of America, 132*(2), 1009–1016.

Fuller, C. D., Galvin, J. J., III, Maat, B., Free, R. H., & Başkent, D. (2014). The musician effect: Does it persist under degraded pitch conditions of cochlear implant simulations? *Frontiers in Neuroscience, 8*, 179.

Fuller, C. D., Gaudrain, E., Clarke, J. N., Galvin, J. J., III, Fu, Q.-J., Free, R. H., & Başkent, D. (2014). Gender categorization is abnormal in cochlear implant users. *Journal of the Association for Research in Otolaryngology, 15*(6), 1037–1048.

Galvin, J. J., III, Fu, Q. J., & Shannon, R. V. (2009). Melodic contour identification and music perception by cochlear implant users. *Annals of the New York Academy of Sciences, 1169*(1), 518–533.

Gantz, B. J., Turner, C., Gfeller, K. E., & Lowder, M. W. (2005). Preservation of hearing in cochlear implant surgery: Advantages of combined electrical and acoustical speech processing. *Laryngoscope, 115*(5 February), 796–802.

Gaudrain, E., & Başkent, D. (2015a, February). *Discrimination of vocal characteristics in cochlear implants.* Paper presented at the 38th Annual Mid-winter Meeting of the Association for Research in Otolaryngology, Baltimore, MD.

Gaudrain, E., & Başkent, D. (2015b). Factors limiting vocal-tract length discrimination in cochlear implant simulations. *The Journal of the Acoustical Society of America, 137*(3), 1298–1308.

Gaudrain, E., & Carlyon, R. P. (2013). Using zebra-speech to study sequential and simultaneous speech segregation in a cochlear-implant simulation. *The Journal of the Acoustical Society of America, 133*(1), 502–518.

Gaudrain, E., Grimault, N., Healy, E. W., & Béra, J. C. (2007). Effect of spectral smearing on the perceptual segregation of vowel sequences. *Hearing Research, 231*, 32–41.

Gaudrain, E., Grimault, N., Healy, E. W., & Béra, J. C. (2008). Streaming of vowel sequences based on fundamental frequency in a cochlear-implant simulation. *The Journal of the Acoustical Society of America, 124*(5), 3076–3087.

Geers, A. E., & Sedey, A. L. (2011). Language and verbal reasoning skills in adolescents with 10 or more years of cochlear implant experience. *Ear and Hearing, 32*(1 Suppl.), 39S–48S.

Gfeller, K., Turner, C., Mehr, M., Woodworth, G., Fearn, R., Knutson, J. F., . . . Stordahl, J. (2002). Recognition of familiar melodies by adult cochlear implant recipients and normal-hearing adults. *Cochlear Implants International, 3*(1), 29–53.

Gifford, R.H., Dorman, M., McKarns, S., & Spahr, A. (2007). Combined electric and contralateral acoustic hearing: Word and sentence recognition with bimodal hearing. *Journal of Speech, Language, and Hearing Research, 50*(4), 835–843.

Gifford, R. H., Shallop, J. K., & Peterson, A. M. (2008). Speech recognition materials and ceiling effects: Considerations for cochlear implant programs. *Audiology and Neurotology, 13*(3), 193–205.

Gilbert, J. L., Tamati, T. N., & Pisoni, D. B. (2013). Development, reliability and validity of PRESTO: A new high-variability sentence recognition test. *Journal of the American Academy of Audiology, 24*(1), 26–36.

Gnansia, D., Pressnitzer, D., Péan, V., Meyer, B., & Lorenzi, C. (2010). Intelligibility of interrupted and interleaved speech for normal-hearing listeners and cochlear implantees. *Hearing Research, 265*(1), 46–53.

Grayden, D. B., & Clark, G. M. (2006). Implant design and development. In H. Cooper & L. Craddock (Eds.), *Cochlear implants: A practical guide* (pp. 1–20). Chichester, UK: Whurr.

Greenwood, D. D. (1990). A cochlear frequency-position function for several species—29 years later. *The Journal of the Acoustical Society of America, 87*, 2592–2605.

Harris, M. S., Kronenberger, W. G., Gao, S., Hoen, H. M., Miyamoto, R. T., & Pisoni, D. B. (2013). Verbal short-term memory development and spoken language outcomes in deaf children with cochlear implants. *Ear and Hearing, 34*(2), 179–192.

Henry, B. A., Turner, C. W., & Behrens, A. (2005). Spectral peak resolution and speech recognition in quiet: Normal hearing, hearing impaired, and cochlear implant listeners. *The Journal of the Acoustical Society of America, 118*(2), 1111–1121.

Heydebrand, G., Hale, S., Potts, L., Gotter, B., & Skinner, M. (2007). Cognitive predictors of improvements in adults' spoken word recognition six months after cochlear implant activation. *Audiology and Neurotology, 12*(4), 254–264.

Hochmair, E. S., & Hochmair-Desoyer, I. J. (1983). Percepts elicited by different speech coding strategies. In C. W. Parkins & S. W. Anderson (Eds.), *Cochlear prostheses—An international symposium* (Vol. 405). New York, NY: Annals of the New York Academy of Sciences.

Hochmair, I., Hochmair, E., Nopp, P., Waller, M., & Jolly, C. (2015). Deep electrode insertion and sound coding in cochlear implants. *Hearing Research, 322*, 14–23.

Holden, L. K., Finley, C. C., Firszt, J. B., Holden, T. A., Brenner, C., Potts, L. G., . . . Skinner, M. W. (2013). Factors affecting open-set word recognition in adults with cochlear implants. *Ear and Hearing, 34*, 342–360.

Hong, R. S., & Turner, C. W. (2009). Sequential stream segregation using temporal periodicity cues in cochlear implant recipients. *The Journal of the Acoustical Society of America, 126*(1), 291–299.

Ingvalson, E. M., Lee, B., Fiebig, P., & Wong, P. C. (2013). The effects of short-term computerized speech-in-noise training on postlingually deafened adult cochlear implant recipients. *Journal of Speech, Language, and Hearing Research, 56*(1), 81–88.

Ingvalson, E. M., Young, N. M., & Wong, P. C. (2014). Auditory–cognitive training improves language performance in prelingually deafened cochlear implant recipients. *International Journal of Pediatric Otorhinolaryngology, 78*(10), 1624–1631.

Iverson, P., Kuhl, P. K., Akahane-Yamada, R., Diesch, E., Tohkura, Y. i., Kettermann, A., & Siebert, C. (2003). A perceptual interference account of acquisition difficulties for non-native phonemes. *Cognition, 87*(1), B47–B57.

Iverson, P., Wagner, A., Pinet, M., & Rosen, S. (2011). Cross-language specialization in phonetic processing: English and Hindi perception of /w/-/v/ speech and nonspeech. *The Journal of the Acoustical Society of America, 130*(5), EL297–EL303.

Iverson, P., Wagner, A., & Rosen, S. (2016). Effects of language experience on pre-categorical perception: Distinguishing general from specialized processes in speech perception. *Journal of the Acoustical Society of America*. Manuscript in revision.

Janse, E., Nooteboom, S. G., & Quené, H. (2007). Coping with gradient forms of /t/-deletion and lexical ambiguity in spoken word recognition. *Language and Cognitive Processes, 22*(2), 161–200.

Ji, C., Galvin, J. J., Chang, Y.-P., Xu, A., & Fu, Q.-J. (2014). Perception of speech produced by native and nonnative talkers by listeners with normal hearing and listeners with cochlear implants. *Journal of Speech, Language, and Hearing Research, 57*(2), 532–554.

Ji, C., Galvin, J. J., III, Xu, A., & Fu, Q.-J. (2013). Effect of speaking rate on recognition of synthetic and natural speech by normal-hearing and cochlear implant listeners. *Ear and Hearing, 34*(3), 313–323.

Johnson, K., & Mullennix, J. W. (1997). *Talker variability in speech processing*. San Francisco, CA: Morgan Kaufmann.

Johnsrude, I. S., Mackey, A., Hakyemez, H., Alexander, E., Trang, H. P., & Carlyon, R. P. (2013). Swinging at a cocktail party voice familiarity aids speech perception in the presence of a competing voice. *Psychological Science, 24* (10), 1995–2004.

Kahneman, D. (1973). *Attention and effort.* Englewood Cliffs, NJ: Prentice-Hall.

Kasturi, K., Loizou, P. C., Dorman, M., & Spahr, T. (2002). The intelligibility of speech with "holes" in the spectrum. *The Journal of the Acoustical Society of America, 112*(3, Pt. 1), 1102–1111.

Khan, A. M., Handzel, O., Burgess, B. J., Damian, D., Eddington, D. K., & Nadol, J. B., Jr. (2005). Is word recognition correlated with the number of surviving spiral ganglion cells and electrode insertion depth in human subjects with cochlear implants? *Laryngoscope, 115*(4), 672–677.

King, S. E., Firszt, J. B., Reeder, R. M., Holden, L. K., & Strube, M. (2012). Evaluation of TIMIT sentence list equivalency with adult cochlear implant recipients. *Journal of the American Academy of Audiology, 23*(5), 313–331.

Kronenberger, W., Beer, J., Castellanos, I., Pisoni, D., & Miyamoto, R. (2014). Neurocognitive risk in children with cochlear implants. *Journal of the American Medical Society Otolaryngology-Head & Neck Surgery, 140,* 608–615.

Kronenberger, W. G., Pisoni, D. B., Henning, S. C., & Colson, B. G. (2013). Executive functioning skills in long-term users of cochlear implants: A case control study. *Journal of Pediatric Psychology, 38,* 902–914.

Kronenberger, W. G., Pisoni, D. B., Henning, S. C., Colson, B. G., & Hazzard, L. M. (2011). Working memory training for children with cochlear implants: A pilot study. *Journal of Speech, Language, and Hearing Research, 54*(4), 1182–1196.

Kwon, B. J., Perry, T. T., Wilhelm, C. L., & Healy, E. W. (2012). Sentence recognition in noise promoting or suppressing masking release by normal-hearing and cochlear-implant listeners. *The Journal of the Acoustical Society of America, 131*(4), 3111–3119.

Laneau, J., Wouters, J., & Moonen, M. (2006). Improved music perception with explicit pitch coding in cochlear implants. *Audiology and Neurotology, 11*(1), 38–52.

Lazard, D., Lee, H., Gaebler, M., Kell, C., Truy, E., & Giraud, A.-L. (2010). Phonological processing in post-lingual deafness and cochlear implant outcome. *Neuroimage, 49*(4), 3443–3451.

Lazard, D. S., Innes-Brown, H., & Barone, P. (2014). Adaptation of the communicative brain to post-lingual deafness. Evidence from functional imaging. *Hearing Research, 307,* 136–143.

Lazard, D. S., Lee, H. J., Truy, E., & Giraud, A. L. (2013). Bilateral reorganization of posterior temporal cortices in post-lingual deafness and its relation to cochlear implant outcome. *Human Brain Mapping, 34*(5), 1208–1219.

Limb, C. J., & Rubinstein, J. T. (2012). Current research on music perception in cochlear implant users. *Otolaryngologic Clinics of North America, 45*(1), 129–140.

Loizou, P. C. (1998, September). Mimicking the human ear: An overview of signal-processing strategies for converting sound into electrical signals in cochlear implants. *IEEE Signal Processing Magazine,* 101–130.

Looi, V., Gfeller, K., & Driscoll, V. (2012). Music appreciation and training for cochlear implant recipients: A review. *Seminars in Hearing, 33*(4), 307–334. doi:10.1055/s-0032-1329222

Luo, X., Fu, Q.-J., & Galvin, J. J. (2007). Vocal emotion recognition by normal-hearing listeners and cochlear implant users. *Trends in Amplification, 11*(4), 301–315.

Luo, X., Fu, Q.-J., Wu, H.-P., & Hsu, C.-J. (2009). Concurrent-vowel and tone recognition by Mandarin-speaking cochlear implant users. *Hearing Research, 256*(1), 75–84. doi:10.1055/s-0032-1329222

Marschark, M., Rhoten, C., & Fabich, M. (2007). Effects of cochlear implants on children's reading and academic achievement. *Journal of Deaf Studies and Deaf Education, 12*(3), 269–282.

Mason, H. M. (1946). Understandability of speech in noise as affected by region of origin of speaker and listener. *Communications Monographs, 13*(2), 54–58.

Mattys, S. L., Davis, M. H., Bradlow, A. R., & Scott, S. K. (2012). Speech recognition in adverse conditions: A review. *Language and Cognitive Processes, 27*(7–8), 953–978.

Merzenich, M. M., Michelson, R. P., Schindler, R. A., Pettit, C. R., & Reid, M. (1973). Neural encoding of sound sensation evoked by electrical stimulation of the acoustic nerve. *Annals of Otology, 82*, 486–503.

Micheyl, C., Delhommeau, K., Perrot, X., & Oxenham, A. J. (2006). Influence of musical and psychoacoustical training on pitch discrimination. *Hearing Research, 219*(1–2), 36–47.

Milczynski, M., Wouters, J., & Van Wieringen, A. (2009). Improved fundamental frequency coding in cochlear implant signal processing. *The Journal of the Acoustical Society of America, 125*, 2260–2271.

Millar, I., Tong, Y., & Clark, G. M. (1984). Speech processing for cochlear implant prostheses. *Journal of Speech, Language, and Hearing Research, 27*(2), 280–296.

Moberly, A. C., Lowenstein, J. H., Tarr, E., Caldwell-Tarr, A., Welling, D. B., Shahin, A. J., & Nittrouer, S. (2014). Do adults with cochlear implants rely on different acoustic cues for phoneme perception than adults with normal hearing? *Journal of Speech, Language, and Hearing Research, 57*(2), 566–582.

Moore, B. C. J., & Carlyon, R. P. (2005). Perception of pitch by people with cochlear hearing loss and by cochlear implant users. In C. J. Plack, A. J. Oxenham, R. R. Fay, & A. N. Popper (Eds.), *Pitch perception* (pp. 234–277). New York, NY: Springer.

Moore, B. C. J., & Gockel, H. (2002). Factors influencing sequential stream segregation. *Acta Acustica United with Acustica, 88*, 320–333.

Mullennix, J. W., Pisoni, D. B., & Martin, C. S. (1989). Some effects of talker variability on spoken word recognition. *The Journal of the Acoustical Society of America, 85*(1), 365–378.

Nelson, P. B., & Jin, S. H. (2004). Factors affecting speech understanding in gated interference: cochlear implant users and normal-hearing listeners. *The Journal of the Acoustical Society of America, 115*(5, Pt. 1), 2286–2294.

Nelson, P. B., Jin, S. H., Carney, A. E., & Nelson, D. A. (2003). Understanding speech in modulated interference: Cochlear implant users and normal-hearing listeners. *The Journal of the Acoustical Society of America, 113*(2), 961–968.

Nilsson, M., Soli, S. D., & Sullivan, J. A. (1994). Development of the Hearing in Noise Test for the measurement of speech reception thresholds in quiet and in noise. *The Journal of the Acoustical Society of America, 95*, 1085–1099.

Nittrouer, S., Caldwell-Tarr, A., & Lowenstein, J. H. (2013). Working memory in children with cochlear implants: Problems are in storage, not processing. *International Journal of Pediatric Otorhinolaryngology, 77*(11), 1886–1898.

Pals, C., Sarampalis, A., & Başkent, D. (2013). Listening effort with cochlear implant simulations. *Journal of Speech, Language, and Hearing Research, 56*, 1075–1084.

Parbery-Clark, A., Skoe, E., Lam, C., & Kraus, N. (2009). Musician enhancement for speech-in-noise. *Ear and Hearing, 30*(6), 653–661.

Picheny, M. A., Durlach, N. I., & Braida, L. D. (1989). Speaking clearly for the hard of hearing: III. An attempt to determine the contribution of speaking rate to differences in intelligibility between clear and conversational speech. *Journal of Speech, Language, and Hearing Research, 32*(3), 600–603.

Pisoni, D., Kronenberger, W., Roman, A., & Geers, A. (2011). Measures of digit span and verbal rehearsal speed in deaf children following more than 10 years of cochlear implantation. *Ear and Hearing, 32*, 60S–74S.

Pisoni, D. B., & Cleary, M. (2003). Measures of working memory span and verbal rehearsal speed in deaf children after cochlear implantation. *Ear and Hearing, 24*(1, Suppl.), 106S–120S.

Pisoni, D. B., & Geers, A. E. (2000). Working memory in deaf children with cochlear implants: Correlations between digit span and measures of spoken language processing. *The Annals of Otology, Rhinology & Laryngology Supplement, 185*, 92–93.

Qin, M. K., & Oxenham, A. J. (2003). Effects of simulated cochlear-implant processing on speech reception in fluctuating maskers. *The Journal of the Acoustical Society of America, 114*, 446–454.

Qin, M. K., & Oxenham, A. J. (2005). Effects of envelope-vocoder processing on f0 discrimination and concurrent-vowel identification. *Ear and Hearing, 26*(5), 451–460.

Rosen, S., Walliker, J., Brimacombe, J. A., & Edgerton, B. J. (1989). Prosodic and segmental aspects of speech perception with the House/3M single-channel implant. *Journal of Speech, Language, and Hearing Research, 32*(1), 93–111.

Rouger, J., Lagleyre, S., Fraysse, B., Deneve, S., Deguine, O., & Barone, P. (2007). Evidence that cochlear-implanted deaf patients are better multisensory integrators. *Proceedings of the National Academy of Sciences, 104*(17), 7295–7300.

Rubinstein, J. T. (2004). How cochlear implants encode speech. *Current Opinion in Otolaryngology & Head and Neck Surgery, 12*(5), 444–448.

Rubinstein, J. T., & Hong, R. (2003). Signal coding in cochlear implants: Exploiting stochastic effects of electrical stimulation. *The Annals of Otology, Rhinology & Laryngology Supplement, 191*, 14–19.

Sarampalis, A., Kalluri, S., Edwards, B., & Hafter, E. (2009). Objective measures of listening effort: effects of background noise and noise reduction. *Journal of Speech, Language, and Hearing Research, 52*(5), 1230–1240.

Schoof, T., & Rosen, S. (2014). The role of auditory and cognitive factors in understanding speech in noise by normal-hearing older listeners. *Frontiers in Aging Neuroscience, 6*, 307. doi:10.3389/fnagi.2014.00307

Sennaroğlu, L. (2010). Cochlear implantation in inner ear malformations—a review article. *Cochlear Implants International, 11*(1), 4–41.

Shannon, R. V. (1983). Multichannel electrical stimulation of the auditory nerve in man: II. Channel interaction. *Hearing Research, 12*(1), 1–16.

Shannon, R. V., Galvin, J. J., III, & Başkent, D. (2002). Holes in hearing. *Journal of the Association for Research in Otolaryngology, 3*(2), 185–199.

Shannon, R. V., Zeng, F.-G., Kamath, V., Wygonski, J., & Ekelid, M. (1995). Speech recognition with primarily temporal cues. *Science, 270*, 303–304.

Siciliano, C., Faulkner, A., Rosen, S., & Mair, K. (2010). Resistance to learning binaurally mismatched frequency-to-place maps: Implications for bilateral stimulation with cochlear implants. *The Journal of the Acoustical Society of America, 127*, 1645–1660.

Singer, B. D., & Bashir, A. S. (1999). What are executive functions and self-regulation and what do they have to do with language-learning disorders? *Language, Speech, and Hearing Services in Schools, 30*(3), 265–273.

Skinner, M. W., Ketten, D. R., Holden, L. K., Harding, G. W., Smith, P. G., Gates, G. A., & Blocker, B. (2002). CT-derived estimation of cochlear morphology and electrode array position in relation to word recognition in Nucleus-22 recipients. *Journal of the Association for Research in Otolaryngology, 3*(3), 332–350.

Smith, D. R., Patterson, R. D., Turner, R., Kawahara, H., & Irino, T. (2005). The processing and perception of size information in speech sounds. *The Journal of the Acoustical Society of America, 117*(1), 305–318.

Spahr, A. J., & Dorman, M. F. (2004). Performance of subjects fit with the Advanced Bionics CII and Nucleus 3G cochlear implant devices. *Archives of Otolaryngology-Head & Neck Surgery, 130*(5), 624–628.

Spillman, T., & Dillier, N. (1989). Comparison of single-channel extracochlear and multichannel intracochlear electrodes in the same patient. *British Journal of Audiology, 23*, 25–31.

Stickney, G. S., Loizou, P. C., Mishra, L. N., Assmann, P. F., Shannon, R. V., & Opie, J. M. (2006). Effects of electrode design and configuration on channel interactions. *Hearing Research, 211*(1–2), 33–45.

Stickney, G. S., Zeng, F. G., Litovsky, R., & Assmann, P. (2004). Cochlear implant speech recognition with speech maskers. *The Journal of the Acoustical Society of America, 116*(2), 1081–1091.

Svirsky, M. A., Silveira, A., Suarez, H., Neuburger, H., Lai, T. T., & Simmons, P. M. (2001). Auditory learning and adaptation after cochlear implantation: A preliminary study of discrimination and labeling of vowel sounds by cochlear implant users. *Acta Oto-Laryngologica, 121*(2), 262–265.

Swaminathan, J., Mason, C., Streeter, T., Best, V., Kidd, G., Jr., & Patel, A. (2014). Musical training, individual differences and the cocktail party problem. *Scientific Reports, 5*, 11628.

Tamati, T. N., & Pisoni, D. B. (2015). The perception of foreign-accented speech by cochlear implant users. In The Scottish Consortium for ICPhS 2015 (Ed.), *Proceedings of the 18th International Congress of Phonetic Sciences*, Glasgow, UK: The University of Glasgow.

Tamati, T. N., Janse, E., & Başkent, D. (2015, April). *The perception of real-life speaking styles under cochlear implant simulation.* Paper presented at the Improving Cochlear Implant Performance Meeting 2015, London, UK.

Tamati, T. N., Gilbert, J. L., & Pisoni, D. B. (2013). Some factors underlying individual differences in speech recognition on PRESTO: A first report. *Journal of the American Academy of Audiology, 24*(7), 616–634.

Tamati, T. N., Gilbert, J. L., & Pisoni, D. B. (2014). Influence of early linguistic experience on regional dialect categorization by an adult cochlear implant user: A case study. *Ear and Hearing, 35*(3), 383–386.

Van Wieringen, A., & Wouters, J. (2015). What can we expect of normally-developing children implanted at a young age with respect to their auditory, linguistic and cognitive skills? *Hearing Research, 322*, 171–179.

Vandali, A. E., & van Hoesel, R. J. (2011). Development of a temporal fundamental frequency coding strategy for cochlear implants. *The Journal of the Acoustical Society of America, 129*(6), 4023–4036.

Vandali, A. E., Whitford, L. A., Plant, K. L., & Clark, G. M. (2000). Speech perception as a function of electrical stimulation rate: Using the Nucleus 24 cochlear implant system. *Ear and Hearing, 21*(6), 608–624.

Venail, F., Mathiolon, C., Menjot de Champfleur, S., Piron, J. P., Sicard, M., Villemus, F., & Uziel, A. (2015). Effects of electrode array length on frequency-place mismatch and speech perception with cochlear implants. *Audiology and Neurotology, 20*(2), 102–111.

Vestergaard, M. D., Fyson, N. R., & Patterson, R. D. (2009). The interaction of vocal characteristics and audibility in the recognition of concurrent syllables. *The Journal of the Acoustical Society of America, 125*(2), 1114–1124.

Wagner, A., Ernestus, M., & Cutler, A. (2006). Formant transitions in fricative identification: The role of native fricative inventory. *The Journal of the Acoustical Society of America, 120*(4), 2267–2277.

Wagner, A., Pals, C., de Blecourt, C., Sarampalis, A., & Başkent, D. (2016). Does signal degradation affect top-down processing of speech? In van Dijk, P., Başkent, D., Gaudrain, E., de Kleine, E., Wagner, A., Lanting, C. (Eds.), *Physiology, Psychoacoustics and Cognition in Normal and Impaired Hearing*. New York, NY: Springer.

Wagner, A., Toffanin, P., & Başkent, D. (2016). The timing and effort of lexical access in natural and degraded speech. *Frontiers in Psychology*, Section Auditory

Cognitive Neuroscience. Manuscript in revision.

Wagner, A., Opie, J., & Başkent, D. (2016). The time-course of pre-lexical and lexical processing in experienced CI users. Manuscript in preparation.

Warren, R. M. (1970). Perceptual restoration of missing speech sounds. *Science, 167*, 392–393.

Warren, R. M., & Sherman, G. L. (1974). Phonemic restorations based on subsequent context. *Attention, Perception, & Psychophysics, 16*(1), 150–156.

Wild, C. D., Yusuf, A., Wilson, D. E., Peelle, J. E., Davis, M. H., Davis, & Johnsrude, I. S. (2012). Effortful listening: The processing of degraded speech depends critically on attention. *The Journal of Neuroscience, 32*(40), 14010–14021.

Wilson, B. S., Finley, C. C., Lawson, D. T., Wolford, R. D., Eddington, D. K., & Rabinowitz, W. M. (1991). Better speech recognition with cochlear implants. *Nature, 352*, 236–238.

Winn, M. B., Chatterjee, M., & Idsardi, W. J. (2012). The use of acoustic cues for phonetic identification: Effects of spectral degradation and electric hearing. *The Journal of the Acoustical Society of America, 131*(2), 1465–1479.

Winn, M. B., Edwards, J. R., & Litovsky, R. Y. (2015). The impact of auditory spectral resolution on listening effort revealed by pupil dilation. *Ear and Hearing, 36*(4), e153–e165.

Wöstmann, M., Schröger, E., & Obleser, J. (2015). Acoustic detail guides attention allocation in a selective listening task. *Journal of Cognitive Neuroscience, 27*(5), 988–1000.

Wu, J.-L., Yang, H.-M., Lin, Y.-H., & Fu, Q.-J. (2007). Effects of computer-assisted speech training on Mandarin-speaking hearing-impaired children. *Audiology and Neurotology, 12*(5), 307–312.

Yücel, E., Sennaroğlu, G., & Belgin, E. (2009). The family oriented musical training for children with cochlear implants: Speech and musical perception results of two year follow-up. *International Journal of Pediatric Otorhinolaryngology, 73*(7), 1043–1052.

Zeng, F.-G. (2004). Trends in cochlear implants. *Trends in Amplification, 8*(1), 1–34.

Zeng, F.-G., Grant, G., Niparko, J., Galvin, J. J., III, Shannon, R., Opie, J., & Segel, P. (2002). Speech dynamic range and its effect on cochlear implant performance. *The Journal of the Acoustical Society of America, 111*(1, Pt. 1), 377–386.

Zwolan, T. A., Kileny, P. R., Ashbaugh, C., & Telian, S. A. (1996). Patient performance with the Cochlear Corporation "20+2" implant: Bipolar versus monopolar activation. *Otology & Neurotology, 17*(5), 717–723.

CHAPTER 13

Theoretical Considerations in Developing an APD Construct: A Neuroscience Perspective

Dennis J. McFarland and Anthony T. Cacace

In this chapter, we address the fundament concern that the central auditory processing disorder (APD or CAPD) lacks a strong theoretical foundation. To improve this area of investigation, we offer two theories, provide a useful definition, and emphasize how this disorder should be best construed for advancing the field.

A central auditory processing disorder, also known as auditory processing disorder, is a modality-specific perceptual dysfunction that is not due to peripheral hearing loss (Cacace & McFarland, 2005). The utility of this definition lies in the fact that it is explicit, straightforward, and simple; there are no ambiguities in terms of "what is" and "what is not" an APD. Additionally, and to clarify, this definition does not imply that people with peripheral hearing loss cannot have APD, only that the disorder is typically construed in the absence of hearing loss. Furthermore and importantly, APD represents a theoretical construct. As others have emphasized and as we have noted previously (Cacace & McFarland, 2013), a theoretical construct is an explanatory variable that is *not* directly observable; it represents "some postulated attribute of people, assumed to be reflected in test performance" (e.g., Cronbach & Meehl, 1955). Of course, theoretical constructs are not limited to hearing science (audiology), to perception in general, or, for that matter, to auditory perception in particular. All scientific disciplines have theoretical constructs and use them to explain various types of phenomena. For example, in physics, atoms, gravity, black holes, and the Higgs particle represent theoretical constructs; in psychology, intelligence, motivation, personality, emotions, and moods are theoretical constructs; in biology, genes, evolution, and taxonomic categories are theoretical constructs, and so on. In contradistinction to the view that APD is a theoretical construct, several national organizations consider APD as constituting observable behaviors. This position is espoused in consensus statements or guidelines or other documents (American Speech-Language-Hearing Association [ASHA], 1996, 2005; Jerger & Musiek, 2000; American Academy of Audiology Guidelines). For example, the ASHA (2005) report on APD states, "Typically, screening questionnaires, checklists, and related measures probe auditory behaviors related to academic achievement, listening skills, and communication" (p. 5). It further states, "The operational definition of (C)APD serves as a guide to the types and categories of auditory skills and behaviors that should be assessed during a central auditory diagnostic evaluation" (p. 6). In the AAA (2010) guidelines, there are over 50 notations referring to APD as a "behavior." Others take a similar position when characterizing the accuracy of central auditory test batteries in individuals with brain lesions. For example, Musiek et al. (2011, p. 357, reference note 1) describe "negative auditory behaviors," which include misinterpretation of acoustical information, frequently asking for speech to be repeated, difficulty hearing in background noise, and so forth as representing symptoms of APD. Taken together, the view that there are "auditory behaviors" suggests a one-to-one correspondence between auditory processing, APDs, and the tests that measure these phenomena. We view this position as being untenable since it creates a quagmire and misrepresents how to conceptualize, identify, and diagnose an APD.

Consider, for example, the behavior of frequently asking for speech to be repeated that Musiek et al. (2011) describe as an "auditory behavior." This specific behavior might have any one of several causes. The individual may have a peripheral hearing loss, a central hearing problem, or a poor grasp of the English language; may have been daydreaming; or may have a tendency to be distracted by objects in his or her peripheral visual field. One could imagine other possibilities as well. The point here is that an APD is only one of several hypotheses that should be considered in this case. This behavior in isolation is not sufficient to diagnose an APD and additional information should be obtained by the examiner in order to rule out alternative hypothesis as to the cause of this behavior.

While this is only one of several pertinent issues, a common theme that we emphasize herein relates to the fact that in over five decades since the APD construct has been introduced into the clinical literature, conspicuously absent in discussions, debates, sponsored technical papers, consensus statements, papers by working groups, guidelines from national and international organizations, etc. is a comprehensive discussion of a coherent theory on this topic. Theoretical considerations are necessary since they serve as a foundation to justify diagnosis based on currently used tests, some of which are almost five decades old but continue to be used without apparent justification.

The Need for Theory in APD

Apart from several criticisms noted above, the discussion we would like to advance in this chapter is the need for theory in order for APD to develop into a useful entity. We think that this represents a good starting point as far as testing is concerned and we can build on this framework as time goes on. In this context, we consider that there are at least two types of theory that might underlie APD testing. These tests might aim to establish (1) site-of-lesion and/or (2) they might have a purely functional basis in dissociating modality specific and supramodal dysfunctions. Site-of-lesion testing needs to be closely linked to existing knowledge of auditory neuroscience, since this is the area where most advances, in our view, will occur.

Site-of-Lesion Testing

Lesion analysis is a good place to start since it has a foundation in neuroscience and/or neuropsychology, has proponents (experts) that are well versed (credible) in the area, and there is already a body of data focusing on areas associated with this topic in the domains of language, memory, and perception (see Damasio & Damasio, 1989). Furthermore, site-of-lesion assessment is not limited to human studies; animal models can also play an important role. However, this approach is not perfect, and to be most effective in terms of localization of function, it is best employed to circumscribed lesions.

There are several reasons why the lesion approach can be useful, and the rationale for this approach, which has been championed by Damasio and Damasio (1989), will be explicated herein. In the most general sense, if one has a preexisting theory about a normal brain operation and how it might mediate performance on some type of experimental task, then a brain lesion can be construed as a "probe" to test the validity of a particular theory. Evaluation procedures can be made on the basis of within subject baseline operations, or for example, by comparing performances across control subjects. While a detailed historical overview is beyond the scope of this exposé, such an approach is well documented in the aphasia literature, with notable pros and cons cited (Cacace & McFarland, 2012). However, the lesion method is only as good as the testing methodology and neuroanatomical localization potential, the sophistication of the theoretical constructs being tested by the lesion probes, how the area in question and the underlying anatomical substrate is conceptualized in current scientific dogma, and of course, the resolution of the paradigms or procedures in question. In this context, single cases or groups can be used; both provide useful information and can advance science (see Rorden & Karnath, 2004).

Tests of central auditory processing were originally designed to identify the site-of-lesion(s) within the central auditory system. For example, Bocca et al. (1954) introduced the filtered speech test as a sensitive measure of lesions of the auditory cortex. Other investigators developed schemes for identifying the site-of-lesion at various levels of the ascending auditory system (e.g., Lynn & Gilroy, 1977; Musiek, 1985). Subsequently, tests of central auditory processing were then applied to individuals *without* known lesions (e.g., Jerger et al., 1988; Welsh et al., 1982). These investigators suggested that the similarity in the pattern of results between cases with known lesions and those with functional deficits indicated disordered central auditory pathways in the latter.

These early attempts to develop test batteries for the evaluation of central auditory processing abilities were mostly based on empirical observations. For example, the criterion for a given test was whether it was sensitive to brain lesions (e.g., Bocca et al., 1954; Musiek, 1985). Perhaps the most popular neural model of central auditory processing was suggested by Sparks et al. (1970) to explain lesion effects on dichotic listening performance (simultaneous presentation of different stimuli to both ears). This model notes that there are both ipsilateral and contralateral pathways from each ear to each hemisphere. Given dichotic presentation of messages, the ipsilateral pathways are assumed to be suppressed so that the message travels to the contralateral hemisphere. In the context of verbal material, assuming that final processing of the speech message occurs in the left hemisphere, this model predicts both a contralateral "ear effect" (i.e., a lesion will affect recall of the message to the ear on the opposite side) and a "right ear superiority" for processing verbal material. Note that this model could be characterized as a "pathway" model since it emphasizes transmission of a message rather than the computations that are performed by neural circuits. As will be discussed later, more recent views of the ascending auditory pathways conceptualize these as part of a hierarchy of processing stages (e.g., Plack et al., 2014) rather than merely passive conduits for information. This is in contrast to the simple transmission model of Sparks et al. (1970). The Sparks et al. (1970) model has been commonly used to interpret dichotic listening performance. However, there are many more factors that might affect dichotic listening task than identified by this model. For example, the typical dichotic listening task requires that the individual responds by reproducing the stimuli verbally. Comparing verbal reproduction or a recognition-based response selection task, Lawfield et al. (2011) found that whether or not a right ear advantage was observed depended on the type of response selection required. Their results suggest that asymmetries in verbal response mechanisms may produce the right ear effect. This interpretation differs markedly from models of dichotic speech perception that attributes the right ear advantage to structural properties of afferent auditory pathways.

While some APD models continue to emphasize site-of-lesion (e.g., Katz & Smith, 1991), the heterogeneity in the profiles of APD test results has more recently been dealt with by systems of test classification based on the process that these tests are thought to measure. The system suggested by an ASHA committee of experts (ASHA, 2005) is an example. This system considers five types of behavioral tests: (1) auditory discrimination, (2) auditory temporal processing and patterning, (3) dichotic speech tests, (4) monaural low-redundancy speech tests, and (5) binaural interaction tests. It is notable that this system was developed by a consensus panel, and empirical evidence was not offered in support of the system in that document (ASHA, 2005). Other experts have suggested other schemes. For example, Keith (1981) earlier suggested that auditory abilities might include discrimination, localization, auditory attention, auditory figure ground, auditory closure, auditory blending, auditory analysis, auditory association, and auditory memory. The British Society of Audiology (2011) disagreed with the ASHA model and considered attention to be a key element in auditory processing.

Several investigators have tried to identify a system of auditory processing skills by use of factor analysis. For example, Schow et al. (2000) included a number of tests that were described by terminology closely following ASHA (1996) recommendations. They reported that a covariance structural model that included monaural separation/closure, binaural integration, auditory patterning/temporal ordering, and binaural separation provided adequate fit of the data. However, McFarland and Cacace (2002) showed these data could just as readily be accounted for by a model with a single general factor and four specific factors on which each pair of left- and right-ear presentations were included. This result shows that there are often several models that might account for the same data. In fact, if the Schow et al. (2000) model were valid, they would need to validate the four measures they identified. So far, the authors have been silent in this validation process.

More recently, Ahmmed et al. (2014) used principal components analysis with tests of auditory processing and measures of nonauditory and supramodal factors. They found a general auditory factor in addition to two other nonauditory cognitive factors. Interestingly, their two auditory and two visual attention tasks loaded on the same factor. This result suggests that the auditory versions of these attention tasks do not primarily assess a modality-specific process, as the visual versions were closely matched in terms of task requirements. Unfortunately, these researchers did not include matched visual versions of their other auditory tasks so that the modality specificity of their general auditory factor is uncertain. In any case, it is important to keep in mind that factor analysis does *not* identify underlying dimensions of individual differences. Rather, it identifies dimensions along which tests are similar (see McFarland, 2014, for a detailed discussion). As such, it is highly dependent on the sampling of tests that are included in the analysis.

If a functional classification of auditory abilities is to be based on the results of factor analysis, then the sampling of tests is critical. In order to establish that the constructs are truly auditory abilities and not of a supramodal nature, it is necessary to include matched tests in another modality, such as vision. We have already seen that the visual and auditory tests of attention did not segregate on separate factors in the Ahmmed et al. (2014) study.

Demonstrating modality specificity with matched visual and auditory tasks that segregate on separate factors (or having loadings of opposite sign) should be a requirement for all constructs identified as auditory factors. In addition, if tests are to be grouped together as representatives of a single construct, such as auditory discrimination or auditory temporal processing, then several members of each group should be included in the analysis. For example, the ASHA (2005) technical report states that auditory temporal processing tests include "sequencing and patterns, gap detection, fusion discrimination, integration, forward and backward masking" (p. 12). If auditory temporal processing is to be considered a distinct ability, then performance on these tests should be correlated. In addition, consistently poor performance of all of these tests should be expected in individuals purported to have auditory temporal processing deficits.

The authors are not aware of a thorough analysis of auditory temporal processing tasks used in APD batteries. However, some evidence does exist. For example, in a recent study, Tomlin et al. (2015) report that none of the correlations between pitch patterns, masking level difference, and detection of gaps in noise were significant. In contrast, correlations between pitch patterns and dichotic digits were significant. This is not the pattern of results to be expected if pitch patterns, masking level difference, and gaps in noise are all measures of a singular auditory temporal processing construct. Examining correlations between these measures in a large population of children is exactly what is needed to evaluate constructs such as auditory temporal processing. What is missing here is inclusion of psychometrically matched tests in the visual modality. We have previously reviewed some of the auditory temporal processing literature (McFarland & Cacace, 2009) and concluded that it cannot be taken for granted that different temporal processing tasks listed by the ASHA (2005) working group all measure some unitary factor. If factor analysis is to serve as the basis of auditory processing constructs, then a sample of multiple representatives of each along with corresponding visual analogues needs to be examined in a large sample.

We have previously suggested that a functional approach might be practical, given the uncertainty in the site-of-lesions responsible for disorders of auditory perception (McFarland & Cacace, 1995). However, there have been considerable advances in neuroscience over the past 20 years. While understanding of the central auditory system is far from complete, some aspects of its nature are beginning to emerge. These include a realization that the ascending auditory pathways are part of a hierarchy of processing stages and that different perceptual features are processed in distinct neural networks. These concepts are distinct from the pathway model (Sparks et al., 1970) discussed earlier. Thus, while auditory neuroscience is not yet mature enough to serve as the basis for task selection, it could serve as a basis for further research.

Sound perception involves the activity of neurons in dozens of subcortical nuclei and cortical areas (Hackett, 2006). The subcortical pathways include five major nuclear groups (i.e., the cochlear nuclei, superior olivary complex, lateral lemniscus, inferior colliculus, and medial geniculate complex) that mediate ascending and descending projections of multiple parallel pathways. Cant and Benson (2003) describe a variety of cell types in dorsal and ventral cochlear nuclei that process incoming signals from the acoustic nerve differently and project to diverse rostral sites. Recent studies have shown that the brainstem pathways are involved in active processing of the auditory signal and not just passive conduits for the transmission of auditory information. For example, Spencer et al. (2015)

showed that fast feedforward inhibition in the ventral nucleus of the lateral lemniscus suppresses spectral splatter that results from rapid onset of stimuli. Likewise, the dorsal nucleus of the lateral lemniscus acts as a filter to suppress spurious localization cues (Meffin & Grothe, 2009). Brainstem nuclei are also involved in spatial release from masking (Lane & Delgutte, 2005). Findings such as these show that brainstem mechanisms play an active role in acoustic signal processing. Again, this is in contrast to a simple pathway model. Of course, shaping the sensory input vis-à-vis descending activity from the olivocochlear efferent system to the periphery (cochlear hair cells) is an important consideration. A comprehensive review of this area is found in Chapter 10 of this book.

Atencin et al. (2012) report that the dimensionality of receptor fields increases as one ascends the auditory nervous system. Neuronal elements at the level of the inferior colliculus tend to respond to a single stimulus feature while those in auditory cortex tend to be driven by multiple auditory features. Thus, the nature of the information processing becomes more complex as one goes from the midbrain to the cortex. The auditory cortex can be characterized as having core, belt, and parabelt areas (Hackett, 2010), which appear to be specialized for processing different kinds of auditory information. Several lines of evidence suggest that there are two major auditory pathways, one devoted to the recognition of auditory objects and the other devoted to auditory spatial features (Ahveninen et al., 2013; Clarke et al., 2000; Lomber & Malhotra, 2008; Rauschecker & Scott, 2009) and perhaps another devoted to action (Bizley & Cohen, 2013).

Recent developments in neural imaging have led to new conceptualizations of cortical organization. Analysis of resting state and task-related covariance in slow fluctuations of blood oxygen-level dependent (BOLD) signals has resulted in the identification of functionally coupled brain networks. An auditory network is generally identified in more thorough analyses (e.g., He et al., 2009; Mesmoudi et al., 2013). A similar modular architecture that includes an auditory/language area is obtained from an analysis of the covariation in cortical thickness across individual subjects (Chen et al., 2008). This latter technique is particularly interesting since it is based on individual differences in brain anatomy. As such, it might be one correlate of individual differences in auditory abilities.

These new imaging methods have the potential to identify brain networks associated with purported tests of central auditory abilities. For example, Wack et al. (2014) found activations in auditory cortex and inferior colliculus associated with the performance of a masking level difference task. Schmithorst et al. (2013) reported functional activation in frontal eye fields during dichotic listening and concluded that this task was not specific to auditory processing. These studies show how recent advances in neuroimaging can be used to relate auditory tasks to brain function. Indeed, technical considerations, a review of imaging physics, and relevant psychoacoustic studies are thoroughly covered in Chapter 14 of this book.

It would be impossible to adequately review the neuroscience of auditory processing in this brief chapter. However, several trends are beginning to emerge from the literature. Sensory pathways are no longer viewed as simple transmission lines. Information processing occurs at each synapse so as to form a convergent and divergent cascade of feature-extracting recursive filters (Grossberg & Kazerounian, 2011). Similar to the visual system (Goodale & Milner, 1992), there are separate "what" and "where" pathways in the central auditory nervous system. The classification of auditory tests might better follow this scheme than some of the others we have discussed. Information processing is more frequently attributed to brain networks, rather than centers as has been done in the past. At the same time, these networks appear to have a modular structure.

As this brief review illustrates, there have been considerable advances in neuroscience since the time when most of the currently used tests of central auditory processing were developed. In addition, theory has also advanced. Modern neuroscience is beginning to characterize the nature of auditory information processing. This is likely to suggest new methods for evaluating central auditory processing that were not conceivable to the early pioneers (e.g., Bocca et al., 1954) over 60 years ago when most of the current tests were formulated. No doubt, both methods and theory will continue to evolve. It is important that the field of central auditory processing disorders evolve along with the rest of neuroscience.

References

Ahmmed, A. U., Ahmmed, A. A., Bath, J. R., Ferguson, M. A., Plank, C. J., & Moore, D. R. (2014). Assessment of children with suspected auditory processing disorder: A factor analysis study. *Ear and Hearing, 35,* 295–305.

Ahveninen, J., Huang, S., Nummenmaa, A., Belliveau, J. W., Hung, A., Jaaskelainen, I. P., . . . Raij, T., (2013). Evidence for distinct auditory cortex regions for sound location versus identity processing. *Nature Communications, 4,* 2585.

American Speech-Language-Hearing Association. (1996). Central auditory processing: Current status of research and implications for clinical practice. *American Journal of Audiology, 5,* 41–53.

American Speech-Language-Hearing Association. (2005). *(Central) auditory processing disorders.* Retrieved from http://www.asha.org/members/deskref-journals/deskref/default.AAA

Atencio, C. A., Sharpee, T. O., & Schreiner, C. E. (2012). Receptive field dimensionality increases from auditory midbrain to cortex. *Journal of Neurophysiology, 107,* 2594–2603.

Bizley, J. K., & Cohen, Y. E. (2013). The what, where and how of auditory-object perception. *Nature Review Neuroscience, 14,* 693–707.

Bocca, E., Calearo, C., & Cassinari, V. (1954). A new method for testing hearing in temporal lobe tumors. *Acta Oto-Laryngologica, 44,* 219–224.

British Society of Audiology. (2011). *Position statement: Auditory processing disorder (APD).* Berkshire, UK: British Society of Audiology.

Cacace, A. T., & McFarland, D. J. (2005). The importance of modality specificity in diagnosing central auditory processing disorder. *American Journal of Audiology, 14,* 112–123.

Cacace, A. T., & McFarland, D. J. (2012). Single and double dissociations as a frame of reference: Application to auditory processing disorders (APDs). In R. Goldfard (Ed.), *Translational speech-language pathology and audiology* (pp. 179–184). San Diego, CA: Plural.

Cacace, A. T., & McFarland, D. J. (2013). Factors influencing tests of auditory processing: A perspective on current issues and relevant concerns. *Journal of the American Academy of Audiology, 24,* 1–18.

Cant, N. B., & Benson, C. G. (2003). Parallel auditory pathways: Projection patterns of the different neuronal populations in the dorsal and ventral cochlear nuclei. *Brain Research Bulletin, 60,* 457–474.

Chen, Z. J., He, Y., Rosa-Neto, P., Germann, J., & Evans, A. C. (2008). Revealing modular architecture of human brain structural networks by using cortical thickness from MRI. *Cerebral Cortex, 18,* 2374–2381.

Clarke, S., Bellmann, A., Meuli, R. A., Assal, G. & Steck, A. J. (2000). Auditory agnosia and auditory spatial deficits following left hemisphere lesions: Evidence for distinct processing pathways. *Neuropsychologia, 38,* 797–807.

Cronbach, L. J., & Meehl, P. E. (1955). Construct validity in psychological tests. *Psychological Bulletin, 52,* 281–302.

Damasio, H., & Damasio, A. R. (1989). *Lesion analysis in neuropsychology.* New York, NY: Oxford University Press.

Goodale, M. A., & Milner, A. D. (1992). Separate visual pathways for perception and action. *Trends in Neuroscience, 15,* 20–25.

Grossberg, S., & Kazerounian, S. (2011). Laminar cortical dynamics of conscious speech perception: Neural model of phonemic restoration using subsequent context in noise. *Journal of the Acoustical Society of America, 130,* 440–460.

Hackett, T. A. (2009). Organization of the central auditory pathways in nonhuman primates and humans. In A. T. Cacace & D. J. McFarland (Eds.), *Controversies in central auditory processing disorder* (pp. 15–45). San Diego, CA: Plural.

He, Y., Wang, J., Wang, L., Chen, Z. J., Yan, C., Yang, H., . . . Evans, A. C., (2009). Uncovering intrinsic modular organization of spontaneous brain activity in humans. *PLoS ONE, 4,* e5226.

Jerger, S., Johnson, K., & Loiselle, L. (1998). Pediatric central auditory dysfunction: Comparison of children with confirmed lesions versus suspected processing disorders. *American Journal of Otology, 9*(Suppl.), 63–71.

Katz, J., & Smith, P. S. (1991). The staggered spondaic word test: A ten minute look at the central nervous system through the ears. *Annals of the New York Academy of Sciences, 620,* 233–251.

Keith, R. W. (1981). Tests of central auditory function. In R. J. Roeser & M. P. Downs (Eds.), *Auditory disorders in school children* (pp. 159–173). New York, NY: Thieme-Stratton.

Lane, C. C., & Delgutte, B. (2005). Neural correlates and mechanisms of spatial release from masking: Single-unit and population responses in the inferior colliculus. *Journal of Neurophysiology, 94,* 1180–1198.

Lawfield, A., McFarland, D. J., & Cacace, A. T. (2011). Dichotic and dichoptic digit perception in normal adults. *Journal of the American Academy of Audiology, 22,* 332–341.

Lomber, S. G., & Malhotra, S. (2008). Double dissociation of "what" and "where" processing in auditory cortex. *Nature Neuroscience, 11,* 609–616.

Lynn, G. E., & Gilroy, J. (1977). Evaluation of central auditory dysfunction in patients with neurological disorders. In R. W. Keith (Ed.), *Central auditory dysfunction* (pp. 177–221). New York, NY: Grune & Stratton.

McFarland, D. J. (2014). Simulating the effects of common and specific abilities on test performance: An evaluation of factor analysis. *Journal of Speech, Language, and Hearing Research, 57,* 1919–1928.

McFarland, D. J., & Cacace, A. T. (1995). Modality specificity as a criterion for diagnosing central auditory processing disorder. *American Journal of Audiology, 4,* 36–48.

McFarland, D. J., & Cacace, A. T. (2002). Factor analysis in CAPD and the "unimodal" test battery: Do we have a model that will satisfy? *American Journal of Audiology, 11,* 7–9.

McFarland, D. J., & Cacace, A. T. (2009). Models of central auditory processing ability. In A.T. Cacace & D. J. McFarland (Eds.), *Controversies in central audi-*

tory processing disorder (pp. 93–107). San Diego, CA: Plural.

Meffin, H., & Grothe, B. (2009). Selective filtering to spurious localization cues in the mammalian auditory brainstem. *Journal of the Acoustical Society of America, 126,* 2437–2454.

Mesmoudi, S., Perlbarg, V., Rudrauf, D., Messe, A., Pinsard, B., Hasboun, D., . . . Burnod, Y., (2013). Resting state networks corticotopy: The dual intertwined rings architecture. *PLoS ONE, 8,* e67444.

Musiek, F. E. (1985). Application of central auditory tests: an overview. In J. Katz, W. L. Gabbay, D. S. Ungerleider, & L. Wilde (Eds.), *Handbook of clinical audiology* (3rd ed., pp. 321–336). Baltimore, MD: Williams & Wilkins.

Musiek, F. E., Chermak, G. D., Weihing, J., Zappulla, M., & Nagle, S. (2011). Diagnostic accuracy of established central auditory processing test batteries in patients with documented brain lesions. *Journal of the American Academy of Audiology, 22,* 342–358.

Plack, C. J., Barker, D., & Hall, D. A. (2014). Pitch coding and pitch processing in the human brain. *Hearing Research, 307,* 53–64.

Rauschecker, J., & Scott, S. (2009). Maps and streams in the auditory cortex: Nonhuman primates illuminate human speech processing. *Nature Neuroscience, 12,* 718–724.

Rorden, C., & Karnath, H-O. (2004) Using Human brain lesions to infer function: A relic from a past era in the fMRI age? *Nature Reviews Neuroscience, 5,* 813–819.

Schmithorst, V. J., Farah, R., & Keith, R. W. (2013). Left ear advantage in speech-related dichotic listening is not specific to auditory processing disorder in children: A machine-learning fMRI and DTI study. *NeuroImage: Clinical, 3,* 8–17.

Schow, R. L., Seikel, J. A., Chermak, G. D., & Berent, M. (2000). Central auditory processes and test measures: ASHA 1996 revisited. *American Journal of Audiology, 9,* 63–68.

Sparks, R., Goodglass, H., & Nickel, B. (1970). Ipsilateral versus contralateral extinction in dichotic listening resulting from hemisphere lesions. *Cortex, 6,* 249–260.

Spencer, M. J., Nayagam, D. A. X., Clarey, J. C., Paolini, A. G., Meffin, H., Burkitt, A. N., & Grayden, D. B. (2015). Broadband onset inhibition can suppress spectral splatter in the auditory brainstem. *PLoS ONE, 10,* e0126500.

Tomlin, D., Dillion, H., Sharma, M., & Rance, G. (2015). The impact of auditory processing and cognitive abilities in children. *Ear and Hearing, 36,* 527–542.

Wack, D. S., Polak, P., Furuyama, J., & Burkard, R. F. (2014). Masking level differences: A diffusion tensor imaging and functional MRI study. *PLoS ONE, 9,* e88466.

Welsh, L. W., Welsh, J. J., Healy, M., & Cooper, B. (1982). Cortical, subcortical, and brainstem dysfunction: A correlation in dyslexic children. *Annals of Otology, Rhinology & Laryngology, 91,* 310–315.

CHAPTER 14

Normal Sound Processing: fMRI

Stefan Uppenkamp and Roy D. Patterson

Neuroimaging is only one of many experimental methods for investigating auditory perception. But with carefully designed stimulus paradigms, neuroimaging can provide unique contributions to our understanding of how acoustic stimuli are transformed into the neural code that provides the basis of our mental representation of the auditory world. This chapter focuses on two versions of magnetic resonance imaging commonly used to investigate the auditory system. Magnetic resonance imaging (MRI) uses the effect of nuclear magnetic resonance and takes advantage of differences in the relaxation processes for different types of brain tissue to generate structural image contrasts. A secondary metabolic response is exploited in functional MRI (fMRI) to provide activity contrasts. The combination of structural and functional MR images provides a unique direct link from recorded maps of brain activation to the respective anatomical structures involved. Substantial progress has been achieved in recent years to make this method suitable for the auditory modality.

Figure 14–1 from Patterson, Uppenkamp, Johnsrude, and Griffiths (2002) illustrates the concepts of auditory functional MRI. The figure shows several small sections of three slices through anatomical images of the human brain. The colored layers typically used (represented here in gray scale) show what is usually referred to as "activation maps" of auditory cortex obtained with fMRI. They are superimposed on gray-scale structural images obtained with MRI. The intensity of the gray-scale is essentially presenting a tissue contrast between gray matter, white matter, and cerebrospinal fluid in the brain. The structures are slightly blurred because these images present average structural scans across a group of nine participants. The exact shape and location of the gyri and sulci in individual brains are different, but the gross anatomy is fairly similar across adults.

Superimposed on the anatomic scans are four different gray scale activation

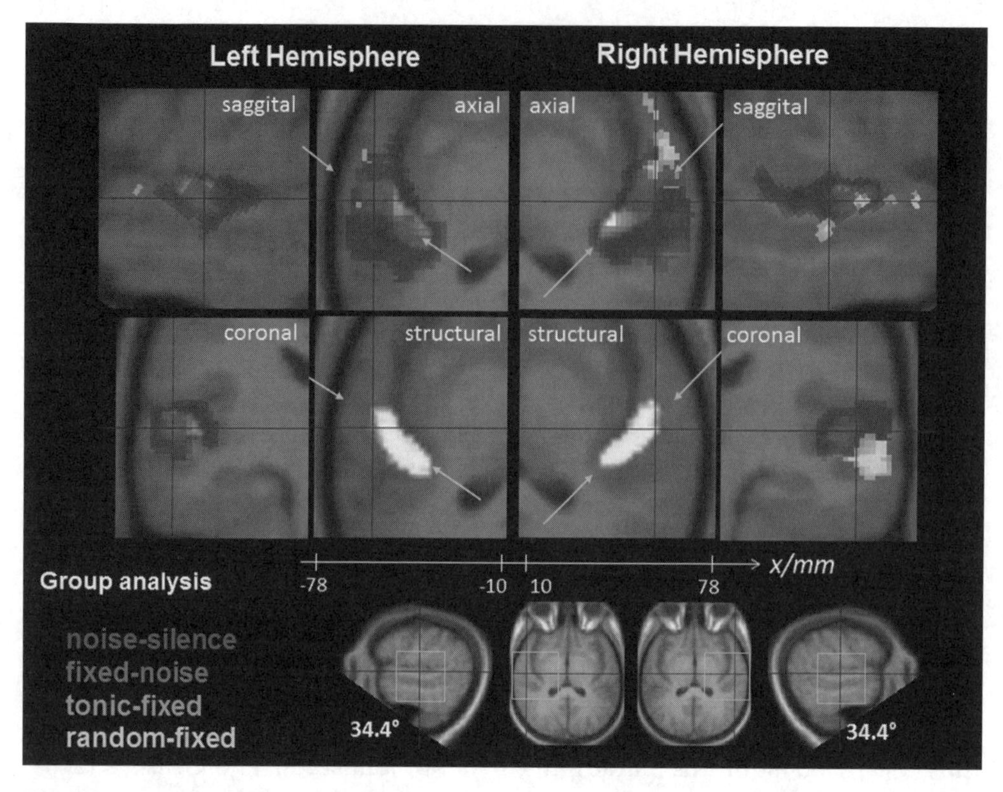

Figure 14–1. Example result from an auditory fMRI study. Reproduced from Patterson et al. (2002), with permission of Elsevier.

maps. This part of the figure has also been created using data gained from magnetic resonance imaging, but in a fairly indirect way. If we think of this image of the brain as composed of a large number of small cubic volume elements, referred to as "voxels," then each of the activation maps shows the results when a statistical test is performed voxel by voxel to assess whether there is a significant difference in the strength of the responses measured in the scanner to two different experimental conditions. The response itself is referred to as the "blood-oxygen level dependent" response or simply the BOLD response. The different experimental conditions in this particular study contrasted the perception of noise with two buzzy sounds, one having a fixed pitch and one having a pitch that varied to produce a melody. The details are presented in the second part of this chapter. The figure reveals that different regions of auditory cortex deal with different sound features and variation of these features. The results suggest that there is a hierarchy of pitch processing, with a comparatively small, bilateral, pitch region in the anterolateral section of the first transverse temporal gyrus (*Heschl's gyrus*), and several melody-specific regions outside the primary auditory regions in the superior temporal gyrus and sulcus, and at the pole of the temporal lobes. The question that the chapter attempts to answer is how we can use MRI and fMRI to obtain these activation maps for groups of participants.

The chapter has two parts. The first provides a technical introduction to MRI in general and auditory fMRI in particular. The second describes imaging experiments designed to test theories about how the periodicity and intensity of sounds are transformed into their perceptual correlates of pitch and loudness, and where the processing takes place.

MRI Physics

The principle of MRI is radiofrequency excitation and relaxation of nuclear magnetic moments that have been aligned with a strong external magnetic field. Functional MRI is based on the narrow magnetic resonance of proton spins. The spatial resolution and signal-to-noise ratio are directly linked to the strength of the external magnetic field. For imaging human subjects, magnetic flux densities from 1.5 tesla to 7 tesla are used. With the currently available techniques, auditory fMRI provides a spatial resolution of a few millimeters and a temporal resolution on the order of a few seconds.

Nuclear Magnetic Moments

Magnetic resonance imaging is based on the physical effect of nuclear magnetic resonance (NMR). Atomic nuclei are composed of protons and neutrons, together called nucleons. Each nucleon has an intrinsic angular momentum, the nuclear spin, which is reflected by a corresponding magnetic moment. There are two possible states for the spin of each nucleon, "spin-up" and "spin-down," with different potential energies. The magnetic moments of all nucleons add up to a total nuclear magnetic moment. The overall spin for nuclei with an even number of both protons and neutrons is zero (pairwise cancellation of spin-up and spin-down). Nuclei with an odd number of either protons or neutrons have a net, nonzero spin, which in case of a single proton, or Hydrogen nucleus ^1H, is either spin-up or spin-down. The hydrogen nuclear spin is the one which is used in most applications of magnetic resonance imaging.

The potential energy of the possible spin states in an external magnetic field of flux density $\vec{B}$ is given by the scalar multiplication of the nuclear magnetic moment $\vec{m}$ and the external field:

$$E = -\vec{m} \cdot \vec{B}. \qquad (1)$$

The two possible states for the spin of ^1H are given by magnetic moments $\pm\frac{1}{2}$ $\gamma\hbar$ with γ denoting the gyromagnetic ratio, a constant characteristic for each type of nucleus, and $\hbar$, the Planck constant h divided by 2π ($h = 6.626 \cdot 10^{-34}\,\mathrm{J \cdot s}$). The gyromagnetic ratio γ determines the precession frequency of a magnetic moment about the axis of an external magnetic field. Its value is $\gamma = 42.58$ MHz/tesla for protons.

With no external magnetic field, the two possible spin states will be evenly distributed in any large ensemble of protons that are embedded in different chemical environments in the body, so that there is no macroscopic magnetization. If, however, the observed object is placed in a strong magnetic field, $\vec{B}$, the ratio of spin-up and spin-down nuclei will be shifted. In the thermodynamic equilibrium, this ratio is given by the Boltzmann distribution:

$$\frac{N^-}{N^+} = e^{\Delta E / k_B \cdot T}, \quad (2)$$

Where $\Delta E = \gamma \hbar B$ denotes the energy difference between the two possible states, $k_B = 1.381 \cdot 10^{-23} J/K$ is Boltzmann's constant, and T = 310 K (≡ 37°C) that is, human body temperature. For protons in one cubic millimeter of water, and for an external magnetic field at 1 tesla, this results in a net magnetization $M_0 = 3 \cdot 10^{-3}$ A/m pointing in the direction of the external field.

The NMR experiment is a sophisticated manipulation and observation of this macroscopic magnetization, the basis of which is a quantum mechanical effect. Luckily, no further quantum mechanics is needed to understand the principle of MRI.

Radiofrequency Excitation and the NMR Signal

As stated above, there is a fixed energy difference between the two possible states of proton spin. If we add energy from the outside, we can manipulate the distribution of these two states within the spin ensemble, as long as we preserve this energy difference. This is the nature of the NMR effect. The energy is fed in by an electromagnetic wave, generated by a radiofrequency coil. The required excitation frequency is the precession frequency of the nuclear magnetic moment, called the Larmor frequency:

$$\nu_0 = \gamma \cdot B_z, \quad (3)$$

where γ denotes the gyromagnetic ratio from above, and B_z the strength of the static magnetic field. We have now introduced a coordinate system with a z-axis parallel to the direction of the static magnetic field, so $\vec{B} = (0,0, B_z)$. Equation (3) is the central equation of MRI, and the key to spatial encoding of the NMR signal. If B_z becomes a function of space, the Larmor frequency will also be a function of space, so we can transform spatial information into frequency information (see next section). For a typical field strength of 3 Tesla and for proton spins, the required frequency is 127.74 MHz, which is in the radiofrequency range. Thus, MRI machines need to include radiofrequency shielding.

Figure 14–2 illustrates the steps performed during the NMR experiment. A strong static magnetic field $\vec{B}_0 = (0,0, B_z)$ creates a net magnetization $\vec{M}_0$ in the z-direction (Figure 14–2A). Using a radiofrequency coil, an electromagnetic wave at the NMR frequency is applied for a short period of time. This moves the direction of the magnetization vector $\vec{M}$ away from the z-direction by a certain angle (Figure 14–2B). This angle, called the flip angle α, depends on the duration of the radio frequency (RF) excitation. The result is a transversal component M_T of the magnetization, which then performs a precession movement about the z-axis caused by the effective angular momentum of $\vec{B}_0$ acting on $\vec{M}$. Now the RF coil that was initially used to apply the RF excitation acts as a receiver (Figure 14–2C). Due to the rotation of the transversal component M_T of the vector $\vec{M}$, the magnetic flux through the coil is changing with time. This induces a voltage in the receiver coil. The result is an alternating cur-

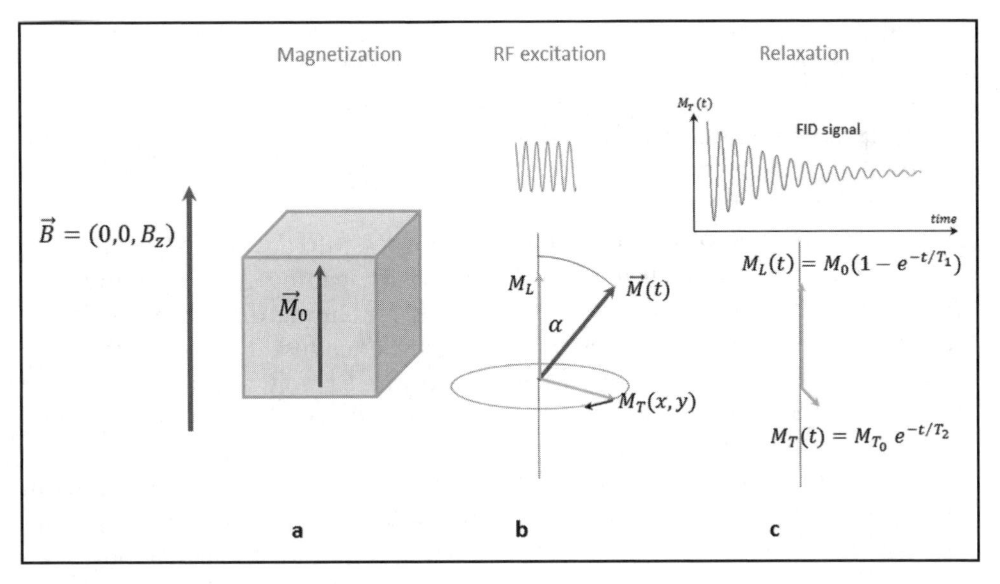

Figure 14–2. Principle of magnetic resonance imaging.

rent through this coil, at the Larmor frequency, which can be detected as the NMR signal or *free induction decay* (FID) signal. The strength of this signal is proportional to M_T. Finally, when the RF excitation is switched off, M_T will decline exponentially with a time constant T_2 (transversal relaxation), caused by dephasing due to nearest-neighbor interactions (spin-spin interactions). At the same time, the initial longitudinal magnetization given by the thermodynamic equilibrium (see Equation 2) is restored with a time constant T_1 (longitudinal relaxation, or spin-lattice interaction).

Note that the relaxation times T_1 and T_2 are related to two different physical mechanisms. T_1 (in the order of seconds) is always bigger than T_2 (in the order of 100 ms). Different chemical environments and therefore different tissue types vary in their proton densities and their relaxation time constants. The image contrast in MRI reflects these differences. The advantage of MRI is that image contrast can be adjusted using two simple external parameters set in the MRI scanner, namely the time of repeat (TR), which is the time span between two successive RF excitations, and the time of echo (TE), which is linked to the time between RF excitation and readout of the induced NMR signal. This allows the collection of both anatomic images with high spatial resolution (typically T_1 weighted images, where image contrast is based on differences in longitudinal relaxation time) and functional images (typically T_2 weighted images) gathered with a very fast acquisition rates (tens of milliseconds for a single slice) and a corresponding reduction in spatial resolution. The combination of these two image types is the basis of human brain mapping in neuroscience. The MRI scanner is a unique tool that allows the collection of both image types by the same machine.

Spatial Encoding of the NMR Signal

Tomographic imaging involves the creation of a stack of two-dimensional images, or "slices," through the tissue covering the volume of interest. Spatial encoding in MRI is achieved using additional gradient coils that restrict RF excitation to predefined slices. The coding strategy has to include three steps, one for each dimension in space.

The first step is the introduction of a stationary gradient in $\vec{B}_0 = (0,0, B_z)$ along one dimension during the measurement in, say, the z-direction, that is, $\vec{B} = \vec{B}_0 + \frac{\partial B_z}{\partial z} \cdot \vec{e}_z$. The magnetic field gradient $G_z = \frac{\partial B_z}{\partial z}$ reduces the initial three-dimensional (3D) problem to two dimensions, since the resonance condition (Equation 3 above) is only fulfilled in a single slice. The steeper the gradient G_z, the thinner will be the respective slice. There is, however, a limit to this, as the slice must be thick enough to generate a measurable signal relative to the noise background. Typical values for the slice thickness are on the order of 1 mm for anatomic images and on the order of a few millimeters for functional images.

The task now is to measure the transversal magnetization as a function of the two other spatial dimensions, y and x. This is done with two separate techniques, phase encoding and frequency encoding. For phase encoding, along the y-direction, a gradient $G_y = \frac{\partial B_z}{\partial y}$ is repeatedly switched on for a short time T_y between successive steps of RF excitation and the readout of the signal. During this time, the precession frequency will vary with y, imprinting a phase difference ω_P between adjacent spin ensembles as a function of y: $\omega_P = -\gamma \cdot G_y \cdot y \cdot y$. For each of the phase encoding steps, the y-component of the initial transversal magnetization will then become

$$M_T(y,T_y) = M_{T_0}(y) \cdot e^{-i\gamma G_y \cdot y \cdot T_y}$$

During the MRI acquisition, a whole set of phase encoding steps is recorded.

The remaining dimension, x, is frequency encoded. During each readout of the RF coil, a gradient $G_x = \frac{\partial B_z}{\partial x}$ is switched on. The FID signal is now a mix of several frequencies corresponding to the different positions in the x-direction, rather than a signal at the initial Larmor frequency only. The x-component of M_{T_0} will then become

$$M_T(x,t) = M_{T_0}(x) \cdot e^{-i\gamma G_x \cdot x \cdot t}$$

In essence, for each slice selected by a particular gradient G_z, the recorded signal in the RF coil is given by the two-dimensional integral along x- and y-axes across the complete slice

$$S(t,T_y) = \iint M_{T_0}(x, y) \cdot e^{-i\gamma G_x \cdot x \cdot t} \cdot e^{-i\gamma G_y \cdot y \cdot T_y} dxdy \quad (4)$$

With the substitutions $k_x = \gamma \cdot G_x \cdot t$ and $k_y = \gamma \cdot G_y \cdot T_y$, this expression becomes

$$S(k_x, k_y) = \iint M_{T_0}(x, y) \cdot e^{-ik_x \cdot x} \cdot e^{-ik_y \cdot y} dxdy \quad (5)$$

Thus, the signal recorded with the RF coil is the two-dimensional spatial Fourier transform of the desired transversal magnetization $M_{T_0}(x,y)$, where k_x and k_y are spatial frequencies. The MRI scanner records data in the Fourier space, or k-space. The images need to be computed from the recorded data by inverse Fourier transform.

The BOLD Response and Functional MRI

Functional MRI involves a secondary, metabolic response rather than primary neural activation. Active neurons consume oxygen, which is replaced by oxygen from arterial blood. The oxygen is carried by hemoglobin and oxygenated hemoglobin is diamagnetic, while deoxygenated hemoglobin is paramagnetic. Thus, an increase in the flow of oxygenated blood, following an increase in neuronal activity, causes a local change in the strength of the magnetic field and a subsequent change in the FID signal. This is the famous BOLD response (blood oxygen-level dependent response). The fMRI signal is strongest in the venous drainage from the active brain region, so there is a small mismatch between the location of the active neurons and their position in the fMRI data. The BOLD response also has a comparatively long latency relative to stimulus onset (about 6 seconds in auditory cortex) because it is mediated by a metabolic response to neural activity rather than the neural activity itself.

Activated regions appear lighter than regions at the baseline metabolic rate. This is the basis for an image contrast in fMRI between activated regions and rest (overview in Buxton, 2002). However, since the observed NMR signal changes in activated regions of the brain are very small (on the order of 1% of the baseline or less), the analysis of fMRI data is heavily dependent on repeated measures and statistical models that test for significant differences between conditions (Worsley, Evans, Marrett, & Neelin, 1992). One common approach involves statistical parametric mapping (SPM), based on the theory of Gaussian random fields (Frackowiak et al., 2004; http://www.fil.ion.ucl.ac.uk/spm).

Specific Techniques for Auditory fMRI

There are several problems with fMRI as a technique for studying the auditory system, the most prominent of which is the noise produced by the scanner itself (Hall et al., 2000; Scarff, Dort, Eggermont, & Goodyear, 2004). The magnetic field gradients required for spatial encoding (see section on spatial encoding above) are created by coils mounted in the bore of the scanner. Abrupt switching of the gradient coils in the presence of the static magnetic field results in the generation of considerable background sounds due to mechanical vibrations caused by large Lorentz forces acting on the coils. The principle is very similar to a dynamic loudspeaker. Acoustically, an MRI machine seems like a very expensive way of generating loud impact noise. Popular imaging sequences often result in noise pulses with sound pressure levels in excess of 100 dB. Accordingly, scanning precautions are required to avoid damaging the subject's ears and to ensure that the unavoidable auditory response to the scanner noise does not contaminate the auditory response to the experimental test sounds. One common solution to this latter problem is referred to as "sparse temporal sampling" (Hall et al., 1999). It is based on clustered volume acquisition (Edmister, Talavage, Ledden, & Weisskoff, 1999) and the fact that the fMRI signal is a delayed metabolic response to the underlying neuronal activity.

The traditional "continuous imaging" paradigm is illustrated in the upper panel of Figure 14–3; sparse temporal imaging is illustrated in the lower panel. During continuous imaging (upper panel), the response to the test sound (gray) is mixed with the auditory response to the continuous scanner noise (black). Given the complexity of the processes in the brain, it seems unlikely that the two BOLD signals will simply superimpose linearly, so it is not clear that the mixture can be decomposed. In sparse temporal sampling (lower panel), the test sounds are presented during silent periods when the scanner is off. The BOLD response to the test sound reaches its maximum 6 to 8 seconds after stimulus onset, at which point the test sound ends and the response decays back to baseline 10 to 12 seconds after onset. The repetition time (TR) of the scanner is set to this comparatively long 10 to 12 seconds, and a full volume of echo-planar imaging (EPI) slices is gathered in a short period (2–3 seconds, which is the shortest duration dictated by the data acquisition time for one slice and the number of slices in the volume). The BOLD response to the scanner noise dies away as the response to the next test sound is building up. In this paradigm, the NMR signals picked up in the scans are dominated by test sound responses with the minimum interference from scanner noise. The rate of data acquisition is slower in sparse imaging, but the quality of the data in terms of signal-to-noise ratio often makes it the paradigm of choice

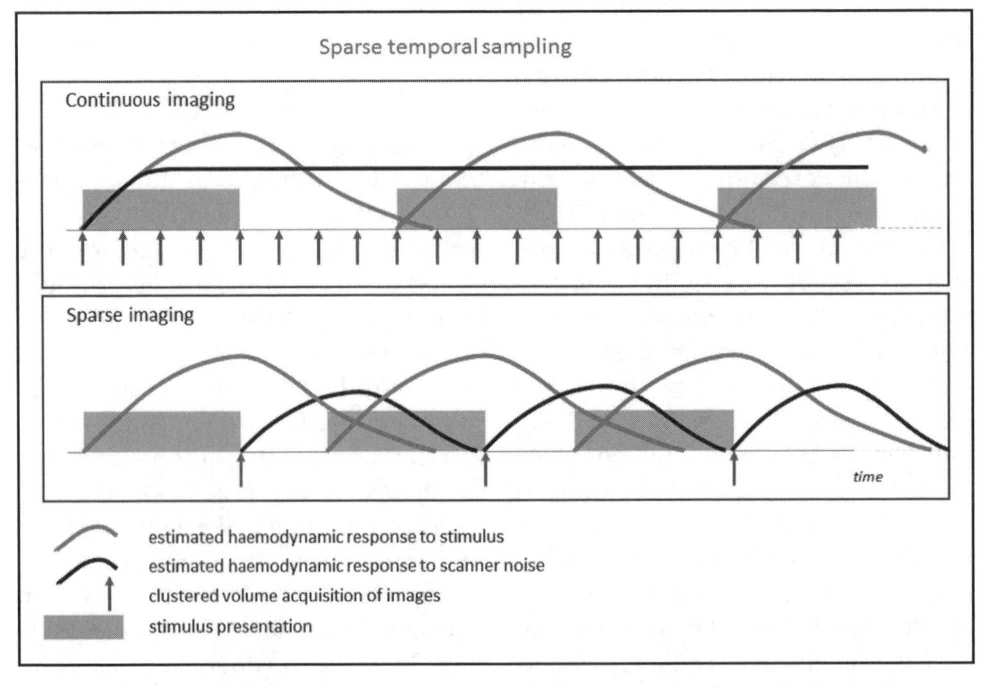

Figure 14–3. Experimental paradigm of *sparse temporal sampling* to minimize interference between acoustic stimuli of interest and scanner noise during an auditory fMRI measurement.

(e.g., Gaab, Gabrieli, & Glover, 2007a, 2007b; Gaab, Gaser, Zaehle, Jancke, & Schlaug, 2003; Griffiths, Uppenkamp, Johnsrude, Josephs, & Patterson, 2001; Langers, Backes, & van Dijk, 2003; Schwarzbauer, Davis, Rodd, & Johnsrude, 2006).

There are also other methods dealing with scanner noise. Gradient echo sequences with long gradient-ramp times (Brechmann, Baumgart, & Scheich, 2002; Thaerig et al., 2008) produce less scanner noise to begin with. High-fidelity headphones that are MRI compatible are available in conventional ear defenders, and they attenuate external noise by 15 to 40 dB. Sound-absorbing material can be fitted around the subject's head and along the inside walls of the scanner bore. In addition to these passive measures, active noise cancellation (ANC) has been implemented to attenuate scanner noise in the subject's ear canals under ear defenders (Chambers, Akeroyd, Summerfield, & Palmer, 2001; Chen, Chiueh, & Chen, 1999; Goldman, Gossman, & Friedlander, 1989). This can provide an additional reduction of 10 to 15 dB in the scanner noise at the eardrum. However, it is strongly dependent on the frequency content of the gradient coil noise. It is not very effective against the short, high-frequency tone pulses typical of echo planar imaging sequences.

Results From Normal-Hearing Listeners With Respect to Basic Sound Features

Broadly speaking, the goal of auditory neuroimaging is to understand how the initial sensory representation of sound is transformed into our perception of auditory events—events that provide the basis for speech and music. Two perceptual features, pitch and loudness, play a major role in the perception of speech and music and this part of the chapter is concerned with attempts to image the auditory processing associated with pitch and loudness. The acoustic correlates of pitch and loudness are stimulus periodicity and stimulus intensity, respectively, and over the years, physiologic investigations have assembled a comprehensive characterization of the sensory coding of periodicity and intensity in the periphery of the auditory system, including the cochlea, auditory nerve, and auditory brainstem (Pickles, 2012). But the transformation of sensation into perception at the cortical level is much less well understood, partly because a host of nonauditory factors involving cognition, context, and even personality contribute to the way we listen to and interpret speech and music. Auditory functional MRI reveals the cortical component of auditory activation, which can help us identify the neural correlates of auditory perception and the influence of nonauditory factors in perception.

Periodicity and Pitch Perception

The voiced parts of speech and the tones of music are effectively periodic sounds and the pitch we hear when presented with such sounds is the perceptual correlate of the repetition rate of the corresponding acoustic wave. The pitch of a vowel or musical tone can be determined by matching it to the pitch produced by a temporally regular click train. The correspondence between the repetition rate of a sound and its pitch is so close that we commonly use the physical variable, cycles per second, or hertz (= sec^{-1}) to specify not only the repetition rate of the sound but also the pitch we hear when presented with the sound, although the actual neural code for pitch has yet to be determined. The notes of music have repetition rates (and thus pitches) that range from about 30 Hz to 4000 Hz.

In the laboratory, it is possible to make stimuli where the degree of temporal regularity in the acoustic wave varies in a regular way—a manipulation that varies the salience of the pitch in the resulting perception. These regular interval sounds (RIS) have been used in auditory imaging experiments (e.g., Griffiths et al., 2001; Patterson et al., 2002) to study pitch processing in auditory cortex because they facilitate fMRI contrasts that cannot be achieved with natural vowels or musical tones. In these experiments, the RIS is generated from broadband noise by repeatedly delaying a sample of the noise and adding it back into the original noise (Yost, 1996a, 1996b). The resulting sound has some of the hiss of the original noise, but it also has a pitch corresponding to the inverse of the delay. The pitch becomes stronger, relative to the hiss, as the delay-and-add process is repeated. When the pitch is less than about 125 Hz (corresponding to delays in excess of 8 ms) and the stimuli are high-pass filtered at about 500 Hz, the RIS excites the frequency channels in the cochlea in much the same way as random noise. It is hypothesized (Krumbholz, Patterson, Nobbe, & Fastl, 2003) that the time intervals in the peripheral neural representation of the sound are used to extract the pitch information. An RIS generated with a delay of 8 ms has an excess of 8-ms time intervals and the model of auditory processing detects the overrepresentation of 8-ms time intervals relative to others in that time-interval range.

Example: Auditory fMRI Experiment With RIS Pitch

Griffiths et al. (2001) and Patterson et al. (2002) investigated the processing of temporal regularity as an RIS ascends the auditory pathway from the cochlear nucleus to auditory cortex using fMRI. These studies illustrate the precautions required to isolate the neural response to sound in the auditory pathway. The sounds were presented as sequences of 32 notes at a rate of 4 notes per second. The primary contrasts in the experiment were between the activity produced by sequences of random noise bursts and matched sequences of RIS having either fixed pitch or a pitch that varied between notes to produce a novel melody. The pitch range for the melodies was 50 to 110 Hz. The pitch in the fixed-pitch sequences was varied between sequences to cover the same range as the melodies over the course of the experiment. The stimuli were bandpass filtered between 500 Hz and 4000 Hz and presented to the subjects at a level of 75 dB SPL via an MR-compatible, electrostatic headphone system mounted in conventional ear defenders. There was also a silence control condition. Each condition was repeated 48 times during the experiment and the order of conditions was randomized over the complete set.

Sparse temporal sampling was used to separate the scanner noise and the experimental sounds in time. BOLD contrast-image volumes were acquired every 12 seconds using a 2-T MRI scanner with gradient EPI. Forty-eight axial slices were acquired covering the brain from cortex down to the small structures in the brainstem, including the cochlear nucleus and the inferior colliculus. These nuclei are not fixed to the skull and their momentary position is affected by pulsing in a large artery running alongside the brainstem. Without intervention, the motion would reduce the signal recorded from these nuclei significantly. Accordingly, the procedure included cardiac gating in which data acquisition is synchronized to the individual's heartbeat using a continuous signal derived from a pulse oximeter on the finger. There were nine normal-hearing participants in the experiment.

A summary of the main results at the level of auditory cortex is presented in Figure 14–1. It shows the average data from all listeners in a statistical *fixed-effects* model, which makes the assumption that each participant represents the group in some way. If we contrast any of the sound conditions with the silence condition, most of the surface of the temporal lobe, including primary and secondary auditory cortex along Heschl's gyrus (HG), and the auditory association areas in planum temporale (PT) just behind HG, exhibits a significant BOLD signal (the darker gray region in Figure 14–1). This is the traditional expression of neural activation in auditory cortex in response to sound. It is bilateral and typically symmetric across the hemispheres for diotic stimulus presentation, as in the current experiments. Most listeners also show bilateral, pitch-specific activation toward the lateral edge of HG outside primary auditory cortex (medium gray in Figure 14–1). Activation specific to melody with changing pitch is observed outside of HG in adjacent cortical areas, mainly in superior temporal gyrus and sulcus (lighter gray tones in Figure 14–1). This melody-specific processing appears to be asymmetric across the hemispheres, with somewhat more activation in the right hemisphere for most listeners.

The variability across participants in this supports the general assumptions made about the processing of pitch and melody information in the auditory pathway. The processing up to and including HG is symmetric across hemispheres given diotic sounds with fixed pitch, and this pitch-processing activity appears to mainly reflect the final stages of sensory coding. In contrast, the hemispheric asymmetry of melody-specific processing and the increase in participant variability suggest process higher up in the hierarchy, involving perceptual processing and even cognitive processing associated with the way the participant is currently listening to the melodies. These latter stages of the hierarchy are where we might also expect to observe differences in musical aptitude.

The observation that pitch-specific processing is observed in the lateral part of Heschl's gyrus was confirmed in a series of subsequent studies involving fMRI (Penagos, Melcher, & Oxenham, 2004), MEG imaging (Chait, Poeppel, & Simon, 2006; Gutschalk, Patterson, Scherg, Uppenkamp, & Rupp, 2004; Hertrich, Mathiak, Menning, Lutzenberger, & Ackermann, 2005), and electrophysiologic recordings from both macaque monkeys (Bendor & Wang, 2005) and human listeners (Schönwiesner & Zatorre, 2008). This interpretation of a general pitch processing center was recently challenged in a series of fMRI studies by Hall, Garcia, and Plack, employing sinusoidal pitch and an exotic binaural "Huggins pitch" (Garcia, Hall, & Plack, 2010; Hall & Plack, 2007a, 2007b, 2009), produced by presenting a noise to both ears that has a specific interaural phase difference in one ear in a narrow frequency band. The pitch is quite weak and it is only heard when listening to the dichotic signal with both ears through headphones. The results indicated that not all pitch phenomena were reliably accompanied by neural activity in lateral HG, and Hall et al. suggested that the activity observed with RIS might have more to do with the spectro-temporal fluctuations in their specific neural activity patterns, rather than the perception of pitch itself. A subsequent fMRI study by Puschmann, Uppenkamp, Kollmeier, and Thiel (2010) demonstrated that lateral HG is actually activated by Huggins pitch and by several other binaural pitch stimuli, including a dichotic $N_{\pi}S_0$ stimulus. In this study, melody-specific activation was restricted to planum temporale and planum polare as before, so the results appear to support the view that pitch processing is hierarchically organized in the auditory pathway (Patterson et al., 2002), for humans at least. It is not currently clear why the data of Puschmann et al. and the data of Hall et al. are not more compatible. Attention is known to modulate cortical activity, so perhaps the differences are arise from the task differences in the different fMRI experiments.

Sound Intensity and Loudness Perception

The physical intensity of a sound is readily measured with a sound-level meter, and it is normally expressed on a logarithmic ratio scale in dB. Loudness is the perceptual correlate of sound intensity and is usually measured psychophysically with a scaling procedure. However, the relationship between intensity and loudness is complicated by the fact that intensity is not the only physical variable that affects loudness. It is also affected by the bandwidth of the sound, its duration, any modulation, and possibly more physical variables (see Jestaedt & Leibold, 2011, for details). Moreover, there are nonacoustic factors that influence how listeners answer the simple question, "How loud is this sound?" These include the precise details of the procedure employed to gather the loudness judgments (Marks & Florentine, 2011), context effects (Arieh & Marks, 2011), the individual's hearing status (e.g., Smeds & Leijon, 2011), and nonauditory factors like, for example, personality (Ellermeier, Eigenstetter, & Zimmer, 2001; Stephens, 1970).

The factors that affect loudness have been thoroughly studied in psychoacoustic experiments with a huge variety of tasks (for a concise review, see Marks & Florentine, 2011). Collectively, the results suggest the primary determinants of the properties of the signal, as expected, and, possibly, individual hearing status (e.g., Chalupper & Fastl, 2002; Moore & Glasberg, 1996; Zwicker & Scharf, 1965). By comparison, the number of studies on the relationship between brain activation and either intensity or loudness is very small. Some of the main results are summarized below.

Loudness-Related Brain Activation

Hall et al. (2001) used functional MRI in 10 normal-hearing listeners to produce activation maps in response to low-frequency sinusoids (300 Hz) at six levels between 66 and 91 dB SPL, as well as a broadband harmonic complex tone at two different levels. The data were analyzed in relation to sound intensity (in dB SPL) as well as loudness level (in phons) as estimated with the loudness model of Moore and Glasberg (1996). Combining the results for tones and complex tones, Hall et al. found a significant positive correlation between loudness and two fMRI measures, the extent and the magnitude of the cortical activation. The corresponding correlation with intensity was not significant. Although the correlation with loudness was weak ($r^2 = 0.12–0.13$), the result does suggest the transformation of the physical variable intensity into the perceptual variable loudness at the cortical level.

Subsequently, an fMRI study with a fairly homogeneous group of 45 normal-hearing listeners (Röhl & Uppenkamp, 2012) set out to determine whether the neural activity in auditory cortex is more closely related to the physical variable intensity or the perceptual variable loudness when you use individual loudness ratings. The functional dependence of activation size and activation magnitude, as measured by the number of activated voxels and voxel intensity, was examined for a broadband stimulus whose level ranged from close to threshold up to a value just below the individual categorical rating of "very loud." Individual loudness sensitivity was assessed by a categorical loudness scaling procedure (Heller, 1985) while participants were inside the scanner. The results reveal strong nonlinear growth of activation, both with increasing categorical loudness judgments and sound intensity. In contrast, the BOLD signal grows almost linearly with sound intensity and linearly with categorical loudness in auditory cortex over the whole range of presentation levels. At the cortical level, BOLD signal strength discriminates between different categories of loudness even at a fixed sound pressure level. This relationship is not observed at the level of the IC, or at the level of the MGB in auditory thalamus. Thus, the neural activity in auditory cortex appears to be a direct linear reflection of subjective loudness, rather than a display of physical sound pressure level, and the transformation of sensation into perception is not completed at the brainstem level. These findings with regard to loudness coding in auditory cortex are largely in line with the results found in a quite different study by Langers, van Dijk, Schoemaker, and Backes (2007).

Spectral Loudness Summation

Spectral loudness summation refers to the fact that, over a certain range of bandwidths, the loudness of a sound increases with increasing bandwidth, even when the sound pressure level remains constant. This was originally explained within the concept of critical bands (Fletcher, 1940; Scharf, 1961; Zwicker, Flottorp, & Stephens, 1957).

The question arises at what stage in the auditory pathway beyond the cochlea does the final summation of loudness across the initially independent channels take place. With the background of the results from the study of Röhl and Uppenkamp (2012), it is hypothesized that the effect of spectral loudness summation is not complete prior to primary auditory cortex (PAC). Loudness judgements and brain activation maps were analyzed in detail in a second fMRI study with a group of 22 normal-hearing listeners (Röhl, Kollmeier, & Uppenkamp, 2011). Stimuli were pink noises at a fixed SPL of 70 dB, with varying bandwidths from 50 Hz to 8 kHz. Loudness follows a nonlinear trend that appears to be related to spectral loudness summation for larger bandwidths and to peak listening for smaller bandwidths where amplitude fluctuation becomes pronounced. While the nonlinear trend of loudness was reflected in corresponding neural activation in PAC, the linear trend of the physical measure of bandwidth was reflected in corresponding neural activation in brainstem. Thus, the link between loudness and neural activation is not observed until the level of PAC, in line with the initial hypothesis.

Comodulation Masking and Audibility

Psychoacoustical masking experiments have long been used to examine both peripheral and central processing in the auditory system. Nevertheless, it is still not clear how the main physical parameters of sound level, temporal structure, and spectral structure are transformed into perceptual measures like loudness or audibility (the perceptual correlate of signal-to-noise ratio, SNR). There are two fMRI studies that investigated the representation of changes in overall level and the representation of signal-to-noise ratio in cortical activation maps (Ernst, Uppenkamp, & Verhey, 2010; Ernst, Verhey, & Uppenkamp, 2008). They are described below and, broadly speaking, they demonstrate a spatial dissociation in auditory cortex for the representation of level changes and audibility changes.

In the first study, sinusoids with frequencies between 440 and 587 Hz were presented as short melodies in the presence of a broadband, unmodulated noise at SNRs from −18 dB to +24 dB (Ernst et al., 2008). For small values (−18, −12, −6 dB), the overall level of the signal is nearly constant, effectively determined by the level of the noise. The audibility of the sinusoid increases with SNR, as they become detectable at the threshold SNR of −18 dB. At 0 dB SNR and above, the sinusoid is always audible and the perceived change is mainly an increase in the overall loudness, which is now dominated by the level of the sinusoid. This perceptual dissociation between changes in SNR and overall loudness is also manifest in fMRI activation maps gathered at the level of auditory cortex. Regions that are specifically involved in representing changes of the overall level (mainly in PT) and regions showing a parametric change in activation with the audibility of a tone in the presence of a masking noise (mainly in the lateral HG) are largely separate, with few overlapping image voxels (Ernst et al., 2008). These results provide further evidence that cortex employs different coding strategies for overall sound level and the audibility of sinusoidal signals.

In the second study, the masking noise was amplitude modulated to investigate the neural correlates of the perceptual phenomenon referred to as "comodulation masking release" (CMR) wherein detection of a sinusoid presented in a narrow band of modulated noise is rendered more detectable by the addition of comodulated flanking bands (Ernst et al., 2010). The results of the previous study suggest that comparison of modulated and unmodulated masker might reveal (1) regions in auditory cortex that will show an increase in activation as overall level increases, independent of the masker type, and (2) separate regions that show an increase in activation with audibility. These latter regions should be more sensitive to sinusoids embedded in modulated noise than in uncorrelated noise. Thus the study provides another test of the hypothesis that the coding of audibility and overall loudness takes place in separate regions, as least to some degree. The comparison of fMRI activation maps for signals presented in modulated and unmodulated noise revealed that those regions in lateral HG previously shown to represent audibility rather than overall level exhibit stronger activation in modulated, as opposed to unmodulated maskers. This might be interpreted as a physiologic correlate of comodulation masking release at the level of auditory cortex, although it does not imply that CMR is purely cortical.

One Example of a Nonauditory Factor

There have been several reports of relationships between personality traits, serotonergic modulation, and the dependence of auditory evoked potentials on loudness (Juckel, Schmidt, Rommelspacher, & Hegerl, 1995; Pogarell et al., 2008). These relationships have also been explored as they appear in fMRI activation maps (Röhl & Uppenkamp, 2010). Using Cloninger's tridimensional personality questionnaire (TPQ; Cloninger, 1987), the personality trait "impulsivity" was identified as one factor associated with the cortical volume of sound-induced activation. Listeners with high impulsivity scores on the TPQ scale showed approximately twice the volume of activation when compared with persons having low impulsivity scores. This apparent link between evoked brain activation and personality is one example of nonauditory factors that might contribute to the variability across listeners in otherwise objective measures of auditory function.

Summary

Recent progress in the techniques of auditory brain imaging means that it is now possible to use fMRI to investigate the neural correlates of auditory perception—properties like pitch, loudness, and audibility—despite the disruptive noise of the scanning process. Recent fMRI studies with normal-hearing listeners have shown:

- A comparatively small region in lateral Heschl's gyrus is involved in pitch-specific processing, and this activity is largely robust to changes in the form of sound used to create the pitch percept. The processing is bilateral and largely symmetric.

- The formation of the overall loudness percept, including spectral summation, is not completed prior to auditory cortex. Nonauditory factors like context and personality can modulate loudness in specific individuals, which is reflected in the associated cortical activation.
- The processing that produces the psychologic correlates of periodicity and intensity, that is, pitch and loudness, takes place in spatially separate regions of auditory cortex.
- The detection of a sinusoid in a noise is based on the audibility of its pitch in the hiss of the noise. Systematic changes in the audibility of the pitch are reflected as systematic change of the corresponding brain activation.

The challenge now is to develop existing functional imaging measures to serve as diagnostics for hearing disorders.

References

Arieh, Y., & Marks, L. E. (2011). Measurement of loudness: Part II. Context effects. In M. Florentine, A. N. Popper, & R. R. Fay (Eds.), *Loudness. Springer handbook of auditory research* (Vol. 37, pp. 57–87). New York, NY: Springer.

Bendor, D., & Wang, X. (2005). The neuronal representation of pitch in primate auditory cortex. *Nature, 436*, 1161–1165.

Brechmann, A., Baumgart, F., & Scheich, H. (2002). Sound-level-dependent representation of frequency modulations in human auditory cortex: A low-noise fMRI study. *Journal of Neurophysiology, 87*, 423–433.

Buxton, R. B. (2002). *Introduction to functional magnetic resonance imaging*. Cambridge, UK: Cambridge University Press.

Chait, M., Poeppel, D., & Simon, J. Z. (2006). Neural response correlates of detection of monaurally and binaurally created pitches in humans. *Cerebral Cortex, 16*, 835–848.

Chalupper, J., & Fastl, H. (2002). Dynamic loudness model (DLM) for normal and hearing-impaired listeners. *Acta Acustica United With Acustica, 88*, 378–386.

Chambers, J., Akeroyd, M. A., Summerfield, A. Q., & Palmer, A. R. (2001). Active control of the volume acquisition noise in functional magnetic resonance imaging: Method and psychoacoustical evaluation. *Journal of the Acoustical Society of America, 110*, 3041–3054.

Chen, C. K., Chiueh, T. D., & Chen, J. H. (1999). Active cancellation system of acoustic noise in MR imaging. *IEEE Transactions on Biomedical Engineering, 46*, 186–191.

Cloninger, C. R. (1987). A systematic method for clinical description and classification of personality variants—a proposal. *Archives of General Psychiatry, 44*, 573–588.

Edmister, W. B., Talavage, T. M., Ledden, P. J., & Weisskoff, R. M. (1999). Improved auditory cortex imaging using clustered volume acquisitions. *Human Brain Mapping, 7*, 89–97.

Ellermeier, W., Eigenstetter, M., & Zimmer, K. (2001). Psychoacoustic correlates of individual noise sensitivity. *Journal of the Acoustical Society of America, 109*, 1464–1473.

Ernst, S. M. A., Verhey, J. L., & Uppenkamp, S. (2008). Spatial dissociation of changes of level and signal-to-noise ratio in auditory cortex for tones in noise. *Neuroimage, 43*, 321–328.

Ernst, S. M. A., Uppenkamp. S., & Verhey, J. L. (2010). Cortical representation of release from auditory masking. *Neuroimage, 49*, 835–842.

Fletcher, H. (1940). Auditory patterns. *Review of Modern Physics, 12*, 47–65.

Frackowiak, R. S. J., Friston, K. J., Frith, C. D., Dolan, R. J., Price, C. J., Zeki, S., . . .

Penny, W. (2004). *Human brain function* (2nd ed.). London, UK: Elsevier Academic Press.

Gaab, N., Gabrieli, J. D. E., & Glover, G.H. (2007a). Assessing the influence of scanner background noise on auditory processing: I. An fMRI study comparing three experimental designs with varying degrees of scanner noise. *Human Brain Mapping, 28*, 703–720.

Gaab, N., Gabrieli, J. D. E., & Glover, G.H. (2007b). Assessing the influence of scanner background noise on auditory processing: II. An fMRI study comparing auditory processing in the absence and presence of recorded scanner noise using a sparse design. *Human Brain Mapping, 28*, 721–732.

Gaab, N., Gaser, C., Zaehle, T., Jancke, L., & Schlaug, G. (2003). Functional anatomy of pitch memory—an fMRI study with sparse temporal sampling. *Neuroimage, 19*, 1417–1426.

Garcia, D., Hall, D. A., & Plack, C. J. (2010). The effect of stimulus context on pitch representations in the human auditory cortex. *Neuroimage, 51*, 808–816.

Goldman, A. M., Gossman, W. E., & Friedlander, P. C. (1989). Reduction of sound levels with antinoise in MR imaging. *Radiology, 173*, 549–550.

Griffiths, T. D., Uppenkamp, S., Johnsrude, I., Josephs, O., & Patterson, R. D. (2001). Encoding of the temporal regularity of sound in the human brainstem. *Nature Neuroscience, 4*, 633–637.

Gutschalk, A., Patterson, R. D., Scherg, M., Uppenkamp, S., & Rupp, A. (2004). Temporal dynamics of pitch in human auditory cortex. *Neuroimage, 22*, 755–766.

Hall, D. A., Haggard, M. P., Akeroyd, M. A., Palmer, A. R., Summerfield, A. Q., Elliott, M. R., Gurney, E. M., & Bowtell, R. W. (1999). "Sparse" temporal sampling in auditory fMRI. *Human Brain Mapping, 7*, 213–223.

Hall, D. A., Haggard, M. P., Akeroyd, M. A., Palmer, A. R., Summerfield, A. Q., & Bowtell, R. W. (2001). Functional magnetic resonance imaging measurements of sound-level encoding in the absence of background scanner noise. *Journal of the Acoustical Society of America, 109*, 1559–1570.

Hall, D. A., & Plack, C. J. (2007a). Searching for a pitch centre in human auditory cortex. In B. Kollmeier, G. Klump, V. Hohmann, U. Langemann, M. Mauermann, S. Uppenkamp, & J. Verhey (Eds.), *Hearing—From sensory processing to perception* (pp. 83–93). Berlin, Germany: Springer.

Hall, D. A., & Plack, C. J. (2007b). The human "pitch center" responds differently to iterated noise and Huggins pitch. *Neuroreport, 18*, 323–327.

Hall, D. A., & Plack, C. J. (2009). Pitch processing sites in the human auditory brain. *Cerebral Cortex, 19*, 576–585.

Hall, D. A., Summerfield, A. Q., Goncalves, M. S., Foster, J. R., Palmer, A. R., & Bowtell, R. W. (2000). Time-course of the auditory BOLD response to scanner noise. *Magnetic Resonance in Medicine, 43*, 601–606.

Heller, O. (1985). Hörfeldaudiometrie mit dem Verfahren der Kategorienunterteilung. *Psychologische Beiträge, 27*, 478–493.

Hertrich, I., Mathiak, K., Menning, H., Lutzenberger, W., & Ackermann, H. (2005). MEG responses to rippled noise and Huggins pitch reveal similar cortical representations. *Neuroreport, 16*, 193–196.

Jestaedt, W., & Leibold, L. J. (2011). Loudness in the laboratory: Part I. Steady-state sound. In M. Florentine, A. N. Popper, & R. R. Fay (Eds.), *Loudness. Springer handbook of auditory research* (Vol. 37, pp. 109–144). New York, NY: Springer.

Juckel, G., Schmidt, L. G., Rommelspacher, H., & Hegerl, U. (1995). The tridimensional personality questionnaire and the intensity dependence of auditory evoked dipole source activity. *Biological Psychiatry, 37*, 311–317.

Krumbholz, K., Patterson, R. D., Nobbe. A., & Fastl, H. (2003). Microsecond temporal resolution in monaural hearing without spectral cues? *Journal of the Acoustical Society of America, 113*, 2790–2800.

Langers, D. R. M., Backes, W., & van Dijk, P. (2003). Spectrotemporal features of the auditory cortex: The activation in response to dynamic ripples. *Neuroimage, 20,* 265–275.

Langers, D. R. M., van Dijk, P., Schoemaker, E. S., & Backes, W. H., (2007). fMRI activation in relation to sound intensity and loudness. *Neuroimage, 35,* 709–718.

Marks, L. E., & Florentine, M. (2011). Measurement of loudness: Part I. Methods, problems, and pitfalls. In M. Florentine, A. N. Popper, R. R. Fay (Eds.), *Loudness. Springer handbook of auditory research* (Vol. 37, pp. 17–56). New York, NY: Springer.

Moore, B. C. J., & Glasberg, B. R. (1996). A revision of Zwicker's loudness model. *Acustica United With Acta Acustica, 82,* 335–345.

Patterson, R. D., Uppenkamp, S., Johnsrude, I. S., & Griffiths, T. D. (2002). The processing of temporal pitch and melody information in auditory cortex. *Neuron, 36,* 767–776.

Penagos, H., Melcher, J. R., & Oxenham, A. J. (2004). A neural representation of pitch salience in nonprimary human auditory cortex revealed with functional magnetic resonance imaging. *Journal of Neuroscience, 24,* 6810–6815.

Pickles, J. O. (2012). *An introduction to the physiology of hearing* (4th ed.). London, UK: Emerald.

Pogarell, O., Koch, W., Schaaff, N., Pöpperl, G., Mulert, C., Juckel, G., . . . Tatsch, K. (2008). ADAM brainstem binding correlates with the loudness dependence of auditory evoked potentials. *European Archives of Psychiatry and Clinical Neuroscience, 258,* 40–47.

Puschmann, S., Uppenkamp, S., Kollmeier, B., & Thiel, C. M. (2010). Dichotic pitch activates pitch processing centre in Heschl's gyrus. *Neuroimage, 49,* 1641–1649.

Röhl, M., & Uppenkamp, S. (2010). An auditory fMRI correlate of impulsivity. *Psychiatry Research: Neuroimaging, 181,* 145–150.

Röhl, M., & Uppenkamp, S. (2012). Neural coding of sound intensity and loudness in the human auditory system. *Journal of the Association of Research in Otolaryngology, 13,* 369–379.

Röhl, M., Kollmeier, B., & Uppenkamp, S. (2011). Spectral loudness summation takes place in the primary auditory cortex. *Human Brain Mapping, 32,* 1483–1496.

Scarff, C. J., Dort, J. C., Eggermont, J. J., & Goodyear, B. G. (2004). The effect of MR scanner noise on auditory cortex activity using fMRI. *Human Brain Mapping, 22,* 341–349.

Scharf, B. (1961). Complex sounds and critical bands. *Psychological Bulletin, 58,* 205–217.

Schönwiesner, M., & Zatorre, R. J. (2008). Depth electrode recordings show double dissociation between pitch processing in lateral Heschl's gyrus and sound onset processing in medial Heschl's gyrus. *Experimental Brain Research, 187,* 97–105.

Schwarzbauer, C., Davis, M. H., Rodd, J. M., & Johnsrude, I. (2006). Interleaved silent steady state (ISSS) imaging: A new sparse imaging method applied to auditory fMRI. *Neuroimage, 29,* 774–782.

Smeds, K., & Leijon, A. (2011), Loudness and hearing loss. In M. Florentine, A. N. Popper, R. R. Fay (Eds.), *Loudness. Springer handbook of auditory research* (Vol. 37, pp. 223–259). New York, NY: Springer.

Stephens, S. D. G. (1970). Personality and slope of loudness function. *Quarterly Journal of Experimental Psychology, 22,* 9–13.

Thaerig, S., Behne, N., Schadow, J., Lenz, D., Scheich, H., Brechmann, A., & Herrmann, C. S. (2008). Sound level dependence of auditory evoked potentials: Simultaneous EEG recording and low-noise fMRI. *International Journal of Psychophysiology, 67,* 235–241.

Worsley, K. J., Evans, A. C., Marrett, S., & Neelin, P. (1992). A three-dimensional statistical analyis for CBF activation studies in human brain. *Journal of Cerebral Blood Flow and Metabolism, 12,* 900–918.

Yost, W. A. (1996a). Pitch of iterated rippled noise. *Journal of the Acoustical Society of America, 100,* 511–518.

Yost, W. A. (1996b). Pitch strength of iterated rippled noise. *Journal of the Acoustical Society of America, 100,* 3329–3335.

Zwicker, E., Flottorp, G., & Stevens, S. S. (1957). Critical bandwidths in loudness summation. *Journal of the Acoustical Society of America, 29,* 548–557.

Zwicker, E., & Scharf, B. (1965). A model of loudness summation. *Psychological Review, 72,* 3–26.

CHAPTER 15

Tinnitus Neurophysiology According to Structural and Functional Magnetic Resonance Imaging

Dave R. M. Langers and Emile de Kleine

Tinnitus is a prevalent and debilitating hearing disorder for which no generally effective treatment currently exists (Baguley et al., 2013; Langguth et al., 2013). Although the pathophysiology of tinnitus is not well understood, it is known that central neural mechanisms play a key role. It is thought that the onset of tinnitus is usually triggered by some form of peripheral hearing loss; yet, its intrusiveness and persistence are determined by maladaptive plastic changes in the central auditory pathways.

Several hypotheses that involve a number of key brain structures (Figure 15–1) have been forwarded to explain tinnitus.

- In the lower auditory pathway in the brainstem, the central gain hypothesis (Noreña, 2011) states that hearing loss–induced sensory deprivation of auditory neurons is counteracted by a homeostatic upregulation of their excitability. This increases the likelihood that neurons respond to random input that is not stimulus related but spontaneously generated, which is perceived as tinnitus.
- At the level of the thalamus, an inhibitory asymmetry may occur between neural assemblies that result in resonant coherence between low- and high-frequency oscillations in the brain. Such coherent neural activity putatively leads to the generation of a conscious percept (Llinás et al., 1999).
- At the level of the cortex, and perhaps below, changes may occur in the tonotopic organization. Animal studies suggest that the representation of tinnitus frequencies expands

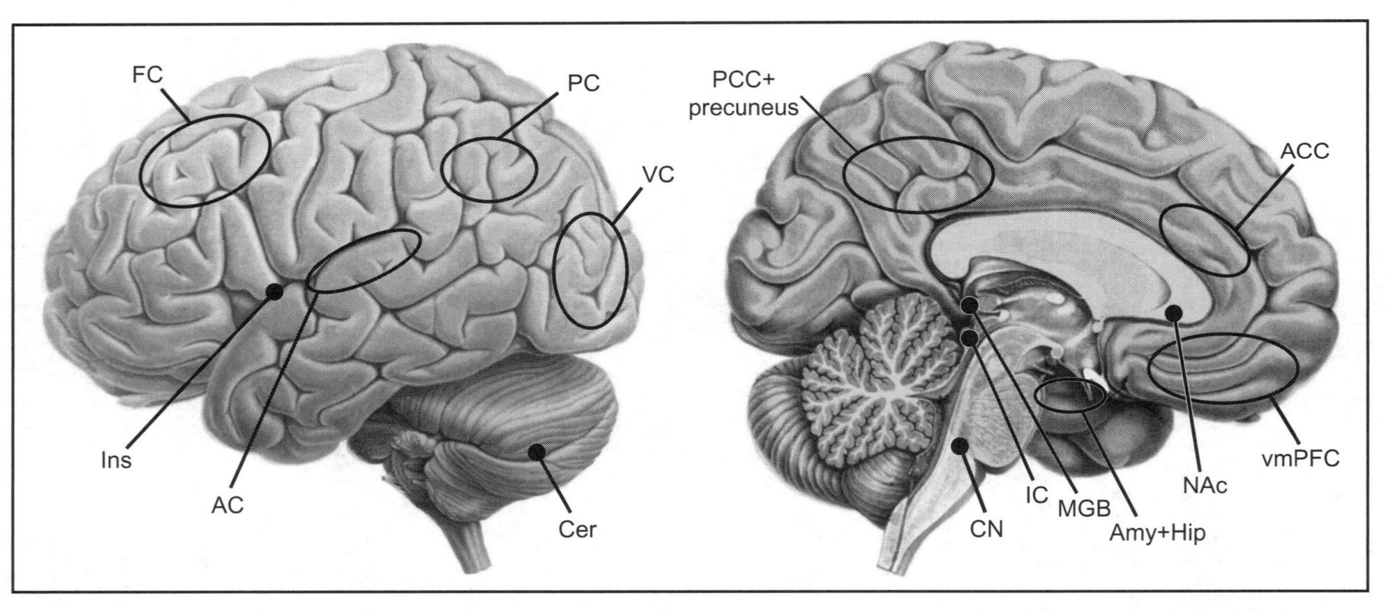

Figure 15–1. Lateral (*left panel*) and medial (*right panel*) views of the human brain, labeling various brain structures with particular significance in tinnitus pathophysiology. According to some models, neural gain is upregulated in the auditory pathway following hearing loss, most likely in auditory brainstem nuclei like the cochlear nucleus (CN) and inferior colliculus (IC). Other models propose that tinnitus is caused by an inhibitory/excitatory imbalance in the auditory thalamus, comprising the medial geniculate body (MGB). Alternatively, maladaptive tonotopic reorganization may occur in the auditory cortex (AC), resulting in an overrepresentation of certain sound frequencies associated with the hearing loss and the tinnitus percept. Outside of the classical auditory system, the mediotemporal limbic lobe that contains the amygdala (Amy) and (para)hippocampus (Hip) assigns affective value to incoming sound; the nucleus accumbens (NAc) of the basal ganglia and the adjacent ventromedial prefrontal cortex (vmPFC) may use this evaluation to block irrelevant sounds at the level of the MGB. When this feedback loop is compromised, unwanted neural activity may be conducted to the AC, giving rise to consciously perceived tinnitus. Finally, additional roles have been proposed for the salience network that consists of the insula (Ins) and anterior cingulate cortex (ACC), the dorsal attention network comprising frontal (FC) and parietal cortex (PC), the default mode network that includes the posterior cingulate cortex (PCC) and precuneus, the visual cortex (VC), and the cerebellum (Cer), although it remains unclear whether abnormal activity in these networks is a cause or a side effect of the chronic presence of tinnitus.

on the cortical surface (Stolzberg et al., 2011). Thus, abnormally large populations of neurons may synchronize their activity, which the brain interprets as a sound.

- Another model posits that structures in the prefrontal cortex and basal ganglia assign excessive relevance and salience to the perceived sound based on an affective evaluation by the amygdala (Rauschecker et al., 2010). This subsequently results in a failure to inhibit the phantom activity at a thalamic level in the auditory pathway.
- In more recent years, an involvement of various other brain networks has been suggested (De Ridder et al., 2014). Apart from auditory and limbic structures, this includes systems related to vision, attention, memory, and introspection.

A number of techniques have been employed to study tinnitus. In humans, objective techniques are mainly limited to noninvasive neuroimaging, that is, electro- and magnetoencephalography (E/MEG), positron emission tomography (PET), and magnetic resonance imaging (MRI). This chapter aims to describe how MRI-based techniques have contributed to our understanding of tinnitus.

MR-Neuroimaging Approaches

Functional MRI studies can be categorized into studies that (1) employ sound stimulation, (2) involve some somatic manipulation that affects the tinnitus percept, or (3) focus on spontaneous brain activity. Besides that, (4) structural changes in macro- and microscopic brain structure have been assessed.

Sound-Evoked fMRI

A number of studies have been performed in which brain activity evoked by external sounds was measured in patients with tinnitus. Although this kind of evoked brain activity does not directly relate to the tinnitus percept, it may give information about sound processing in the auditory brain, which is probably disturbed in patients with tinnitus. The studies performed, as described below, in general compare the MR signal during sound presentations with the signal during silent periods. The difference between the two signals is then called activation—due to the sound stimulus. An important issue with auditory fMRI is that the scanner itself produces an intense sound of over 100 dB (SPL) when the MR signal is read out. In the late 1990s, a method was devised that solves this problem to a great extent, called "sparse temporal sampling" (Hall et al., 1999; see also Chapter 14).

Melcher et al. (2000) were the first to explore a sound-evoked paradigm in tinnitus, reasoning that modulating brain activity with external sounds was likely to reveal tinnitus-related abnormalities. Furthermore, they selected subjects with lateralized tinnitus, hypothesizing that in these patients fMRI activation would exhibit abnormal asymmetries. Findings from these subjects were compared with (1) patients with nonlateralized tinnitus and

(2) normal-hearing participants. Importantly, hearing levels between groups were chosen to be comparable, so only individuals with normal or near-normal thresholds were selected. In patients with unilateral tinnitus, it was observed that the activation of the inferior colliculi (IC) was abnormally asymmetric: The activation of the IC contralateral to the tinnitus percept was unusually low. No differences were found between the nonlateralized tinnitus subjects and controls. The asymmetric activation was interpreted to result from elevated activity in the contralateral IC during rest. This abnormally high activity in the IC would then correspond to the tinnitus percept.

Kovacs et al. (2006) and Smits et al. (2007) expanded on these findings by investigating the central auditory system not only at the level of the IC, but also in the medial geniculate body (MGB) and the auditory cortex (AC). Activation was found to be symmetrical in patients with bilateral tinnitus, and asymmetric in patients with unilateral tinnitus. However, in contrast to Melcher et al. (2000), the activation was lateralized toward the tinnitus side. In these two studies, the patient group was very heterogeneous, and particularly included patients with all degrees of hearing loss. Also, neither these two studies nor Melcher et al. (2000) made use of a sparse sampling paradigm, so the scanner noise may well have interfered with the sound stimuli.

All studies from the year 2008 onward employed sparse sampling. Lanting et al. (2008) and Melcher et al. (2009) focused on unilateral tinnitus, in most cases with normal or near-normal hearing. Both studies, however, could not confirm the lateralized activation in IC, MGB, or AC, neither in patients with bilateral nor unilateral tinnitus. Melcher et al. (2009) performed extensive analyses trying to understand this. They concluded that methodologic details may underlie the disparate findings in these studies. In particular, the presence of background noise caused by the helium pump of the MR scanner possibly interfered with the results. Notably, both studies reported elevated responses to sound in the IC of patients with tinnitus, compared to controls. Lanting et al. (2008) used monaural stimuli, whereas Melcher et al. (2009) used binaural stimuli, so these findings appear not to depend on the laterality of stimulation.

Hyperacusis was put forward as an important confound in the imaging of tinnitus, as reflected by the increased responses to external sound in the latter two studies (Lanting et al., 2008; Melcher et al., 2009). Hyperacusis is characterized by a decreased tolerance to sound (Baguley et al., 2013). There is a large overlap with tinnitus: 40% to 60% of the patients with tinnitus as primary complaint also have hyperacusis. Assessment of hyperacusis is troublesome; audiologic measures such as the loudness discomfort level and hyperacusis questionnaires often do not correlate (Meeus et al., 2010). The pathophysiology is unknown, although recent models try to understand hyperacusis as a gain enhancement in the central auditory system in response to peripheral hearing loss (Auerbach et al., 2014; Knipper et al., 2013). Gu et al. (2010) performed a sound-evoked fMRI study on subjects with clinically normal-hearing thresholds. They compared brain responses of four groups of subjects: with or without tinnitus and with or without hyperacusis. Subjects with hyperacusis showed increased activation in IC, MGB, and AC compared to subjects without hyperacusis. Conversely, subjects with tinnitus showed elevated activation in the AC, but not in the subcortical centers. These findings underline that tinnitus and hyperacusis show associations, and should be studied together.

Leaver et al. (2011) and Seydell-Greenwald et al. (2012) focused on brain activation caused by binaural sounds that were matched in frequency to the tinnitus of their specific patients. To each patient, one control subject was linked, receiving the same set of stimuli. However, Leaver et al. (2011) did not match their two subject groups regarding age and hearing levels, although they performed correlation analyses with these parameters. They found a small cluster near the right nucleus accumbens (NAc) that showed enhanced activation in tinnitus subjects. In the work of Seydell-Greenwald et al. (2012), 20 patients with tinnitus and 20 controls were matched on age, sex, and mostly also on hearing. The enhanced activation in the NAc could not be confirmed, but they did find reduced activation of the right ventromedial prefrontal cortex (vmPFC). Furthermore, enhanced activation of a cluster in the right superior temporal gyrus was shown, which the authors assumed to be a tinnitus-related effect of auditory attention, rather than a correlate of the tinnitus percept itself. Given that the NAc and the vmPFC are important constituents of their previously formulated tinnitus model (Rauschecker et al., 2010), the findings in these two studies were interpreted to support their ideas. The model describes thalamic gating that can suppress the tinnitus signal. Because the gating mechanism involves input from the vmPFC, tinnitus may be caused by a problem at the level of the vmPFC.

Tonotopic reorganization in patients with tinnitus was evaluated in a study by Langers et al. (2012). Binaural tone pips at six octave frequencies were used as stimuli in a study with 20 patients with tinnitus and in 20 controls, all having normal-hearing thresholds. Tonotopic gradients could be distinguished robustly, and were in line with earlier findings on cortical tonotopy in humans. Yet, no significant differences between the cortical maps of the two groups could be identified (Langers, 2014). One difference in activation magnitude was found in the left lateral Heschl's gyrus, which showed hyperactivation in patients. Interestingly, this area is tuned to low frequencies, whereas almost all patients perceived high-pitched tinnitus.

Two studies from one group used the same experimental paradigm, but compared different subject groups (Boyen et al., 2014; Lanting et al., 2014). Broadband sound stimuli were monaurally presented at various levels. Boyen et al. (2014) selected hearing-impaired subjects with and without tinnitus. All subjects had mild to moderate hearing loss, and groups were matched with respect to age, gender, handedness, and hearing loss. Voxel-wise analyses did not show any group differences, but region-of-interest (ROI) analyses did: Enhanced activation was observed in the right cochlear nucleus (CN) and left MGB of patients with tinnitus (although in most cases tinnitus was not lateralized). Hyperacusis was assessed by a questionnaire, but proved to correlate with tinnitus handicap, so it was not possible to disentangle their effects. Lanting et al. (2014) compared subjects with unilateral tinnitus and subjects without tinnitus. All subjects had no or mild hearing loss, where the patients' hearing thresholds were slightly worse. In this case, groups showed no difference in activation, except for one cluster of voxels, located in the vermis of the cerebellum, that showed higher activation in the patients with unilateral tinnitus.

The studies that have been performed until now show very diverse results, possibly due to the fact that subject selection and employed paradigms differed substantially. Mostly, differences between tinnitus and nontinnitus subjects were small in magnitude and extent. Several brain divisions have been identified that may be involved in tinnitus: (1) the auditory system, (2) the limbic system, (3) the frontal lobe, and (4) the cerebellum. Studies from different labs have hardly been able to confirm each other's findings.

Somatic Modulations

Traditional imaging with fMRI or $H_2{}^{15}O$ PET makes use of scanning the brain in two or more different conditions, which can be contrasted during the analysis to get brain activation data (see Chapter 14). In tinnitus, however, the problem with this method is that in most patients, tinnitus is continuously present. Therefore, one cannot normally contrast scans "in the presence of tinnitus" to those "in the absence of tinnitus" to detect tinnitus-related brain activity. In the first attempts to image tinnitus-related brain activity, a special group of patients with tinnitus was considered, that is, patients able to modulate the loudness of their tinnitus by somatic maneuvers like eye gaze or jaw movement. In this way, louder and softer tinnitus could be contrasted, yielding images of the activation corresponding to this difference in tinnitus loudness (Cacace et al., 1996; Giraud et al., 1999).

Since the beginning of the 1990s, growing evidence suggests a relationship between tinnitus and the somatosensory system. In several studies, it was shown that tinnitus loudness and pitch could be altered by movement of head, neck, and jaw, but also by external pressure applied to these regions (Shore et al., 2007; Simmons et al., 2008; Won et al., 2013). Such manipulations may also evoke tinnitus in healthy people. Systematic examination of patients using these maneuvers has been termed somatic testing and is nowadays used in the clinical practice. The prevalence of tinnitus modulation by contraction of head and neck muscles ranges from 50% to 80% of the tinnitus population, so it appears to be a rather common phenomenon (Levine et al., 2007; Won et al., 2013). In general, neck maneuvers decrease tinnitus loudness and jaw maneuvers increase loudness (Won et al., 2013).

Findings from animal studies have provided evidence that the dorsal cochlear nucleus (DCN) in the brainstem may be responsible for these interactions between the somatosensory and auditory central systems (Dehmel et al., 2008; Shore, 2011). Plastic changes in the DCN may result when input from either the auditory or somatosensory system is altered. The observation that somatosensory input can alter the sound-evoked responses of DCN neurons may explain the ability of patients to modify their tinnitus by somatic maneuvers (Shore, 2011).

A special group of patients with tinnitus consists of patients in whom a vestibular schwannoma was surgically removed. During this procedure, the auditory nerve is typically sectioned. These patients often perceive tinnitus in the deafferented ear, sometimes already before surgery (55%), probably due to the tumor pressing on the nerve. After surgery, tinnitus most often persists (83%), while in a minority it decreases (17%). In the group without preoperative tinnitus, it developed in 34% of patients (Baguley et al., 2006). (Moreover, 50% of these patients became extra sensitive to noise after the operation.) This knowledge was important for our current understanding that tinnitus is a central, rather than a peripheral, auditory problem. Almost exclusively in this group, a subset of patients can modulate their tinnitus by gaze (19%; Baguley et al., 2006) or cutaneous stimulation. The pathophysiology of this phenomenon is not clear, but it must presumably be understood in terms of cross-modal interactions in the brainstem that follow sectioning of the VIIIth nerve (van Gendt et al., 2012).

Around the year 2000, the first four papers appeared that used somatic modulations of tinnitus to investigate brain activation (Cacace et al., 1999; Giraud et al., 1999; Lockwood et al., 1998, 2001). Three of these papers use PET, and one used fMRI, which was a rather new technique at the time (Cacace et al., 1999). Rather small numbers of patients were included, ranging from $n = 2$ to $n = 8$. Patients were compared to themselves, and in two papers also to a control group (Lockwood et al., 1998, 2001). A common finding from these studies is that increased (respectively: decreased) tinnitus loudness was found to correspond to increased (respectively: decreased) activation of certain brain areas. The areas in which activation correlated with tinnitus loudness differed, however, between the various studies. Two studies reported the auditory association cortex, two the primary auditory cortex, one the thalamus, one the lower brainstem, and one the cerebellum. The dissimilarity of these findings may have to do with the limited patient numbers, the limited spatial resolution of the scanning, or the heterogeneity of the subjects.

In two more recent studies, fMRI was used to study tinnitus modulated by jaw protrusion (Lanting et al., 2010) and eye gaze (van Gendt et al., 2012). Lanting et al. (2010) recruited 13 patients with tinnitus that could modulate their tinnitus by jaw protrusion and 20 control subjects with normal hearing. In most patients, tinnitus was louder during protrusion. When brain activity during jaw protrusion was contrasted with rest, activation of the entire central auditory system was found. Remarkably, this was also the case for the normal-hearing control subjects. Small differences between patients and controls were found in the brainstem (CN and IC): Activation levels due to jaw protrusion were higher in patients than in controls. For the patients, this activation was interpreted to be associated with the tinnitus modulation. For the controls, however, the interpretation of this finding is not straightforward. The differences between patients and controls may be due to the tinnitus, but also to the difference in hearing thresholds (controls had better hearing). At the same time, these findings are in agreement with animal work that showed that peripheral hearing loss enhances somatosensory input to the DCN (Shore et al., 2008).

The most recent paper on somatic modulation of tinnitus is an fMRI study on gaze-evoked tinnitus (van Gendt et al., 2012). The authors recruited 18 patients who could modulate their tinnitus by sustained gaze deviating from the central axis and 9 control subjects. All patients had a history of removal of a vestibular schwannoma, resulting in one deaf ear. Brain activity was measured using fMRI by contrasting peripheral with central gaze. In the normal-hearing control subjects, peripheral gaze resulted in inhibition of the AC and no response in the MGB and IC. In contrast, in the patients with tinnitus, peripheral gaze reduced the inhibition in the AC, inhibited the MGB, and activated the IC. Moreover, increased tinnitus loudness correlated with increased activation of the CN and IC, and reduced inhibition of the AC. These increased responses in the brainstem are probably due to plastic reorganization after removal of the vestibular schwannoma. Changes in the thalamus and auditory cortex are consistent with a model that explains tinnitus as thalamocortical dysrhythmia (Llinás et al., 1999). Although the experiment was performed with a very special group of patients with tinnitus, these findings may help our understanding of the pathophysiology of tinnitus in general.

Somatic modulation of tinnitus is potentially a powerful tool in the imaging of tinnitus in humans. First, it allows for within-subject comparisons, so there is no direct need for matched control subjects. Second, since the tinnitus can be modulated, the difference in brain activity is more likely related to the tinnitus percept itself, unlike most other methods. Surprisingly, up to now, only a few papers have appeared. This may be caused by the fact that only a minority of patients is able to modulate their tinnitus. Also potential technical difficulties like movement artifacts may play a role. Findings from the existing studies generally suggest that increasing loudness of the tinnitus is associated with increasing activity in the central auditory system. Since the number of studies is low, there is great need for replication of these findings.

Functional Connectivity and Resting State Fluncational Magnetic Resonance Imaging (rs-fMRI)

All results so far were derived from conventional fMRI that requires multiple conditions to be contrasted, either in the form of stimulus presentations or task manipulations. However, tinnitus intrinsically is a phenomenon that is unceasing and uncontrollable by the patient. The measured sound- or task-evoked effects do not necessarily capture the neural activity patterns that directly underlie the tinnitus percept itself. Other methods of fMRI data acquisition and analysis may be able to measure more direct neural correlates of tinnitus (Husain & Schmidt, 2014).

Even during a state of comparative "rest," slow spontaneous variations occur in the distribution of neural activity across the brain. Such resting-state activity remains coherent across similar collections of brain regions that are found in particular task paradigms (Fox & Raichle, 2007). The involved time scales of neural fluctuations in resting-state fMRI exceed those in E/MEG by several orders of magnitude and range from 10 to 100 seconds (Fransson, 2005). Whereas stimulus- or task-evoked activation can be detected by correlating the signal in a particular brain area with the experimental paradigm, functional connectivity is derived from correlations between the signals from multiple areas. Importantly, this does not provide any information about the directed causal relationships that link these regions, nor does it suggest the existence of direct anatomical connections. However, it does allow the brain to be parcellated into networks, many of which play a role in sound processing and prove important in tinnitus (Langers & Melcher, 2011). These include the auditory and visual systems involved in unimodal sensory processing, the salience network involved in the detection of sensory events, the dorsal attention network involved in allocation of attention, the affective and mnemonic limbic networks involved in emotion and memory, and the default mode network involved in introspective mental processing. Because signal fluctuations can be induced by neural and nonneural sources alike, physiologic confounds like head motion, respiration, and heartbeat should be kept in mind when interpreting functional connectivity maps (Beall & Lowe, 2010). Furthermore, acoustic scanner noise affects not only the auditory system but also networks related to vision, attention, emotion, memory, and introspection (Langers & van Dijk, 2011).

Tinnitus-related abnormalities in functional connectivity in the auditory pathways have been reported in a couple of studies (Boyen et al., 2014; Lanting et al., 2014). These studies did not employ a true resting-state design, but included monaural sound presentations at varying levels, such that the observed functional relationships are likely to be driven by sound-evoked processing (Arbabshirani et al., 2013). Nevertheless, functional connectivity between subcortical nuclei and cortical regions was significantly decreased in patients with tinnitus compared to controls with matched hearing loss. The authors attributed this dissociation between subcortical and cortical sound-evoked activity to a change in function of the MGB, which relays sound information from the brainstem nuclei to the cortex. This is compatible with a key role of the auditory thalamus as hypothesized in some pathophysiologic models (Llinás et al., 1999; Rauschecker et al., 2010).

Several studies have investigated the functional interactions between auditory and nonauditory regions of the brain. In one pilot study (Kim et al., 2012), the amygdala, which is part of the affective limbic system and receives input from both the auditory cortex and auditory thalamus, was bilaterally more strongly involved in patients than in controls. In addition, functional connectivity was reduced between left and right auditory cortex, suggesting an imbalance of the normal excitatory and inhibitory interhemispheric connections. Although another study could not reproduce these findings (Davies et al., 2014), this mirrors identical findings concerning auditory hallucinations (Gavrilescu et al., 2010).

In a pair of reports, an elaborate network of nonauditory regions that was anticorrelated with auditory cortex in controls proved to be uncorrelated in patients with tinnitus (Maudoux et al., 2012a, 2012b). This included most notably the parahippocampus, but also the amygdala, regions in the frontal and parietal lobes, and parts of the basal ganglia, thalamus, and brainstem. Some of these differences covaried with the severity of the tinnitus. Consistent with Kim et al. (2012), there was a trend for left and right auditory cortices to show less similar signals in patients. Finally, visual regions in the posterior occipital lobe and posterior cingulate gyrus were more strongly anticorrelated to the auditory network in patients with tinnitus compared to controls.

Another study included a third group besides patients with tinnitus and normal-hearing controls that consisted of individuals with similar hearing loss as the patients but no tinnitus (Schmidt et al., 2013). Functional connectivity in the superior temporal plane proved less extensive in the presence of hearing loss, likely due to deafferentation. In the tinnitus group, parahippocampal regions were functionally more strongly connected to parts of the auditory and dorsal attention networks. The default mode network appeared less strongly connected to other brain regions. However, most significant group differences comprised small isolated foci with limited significance.

Finally, Burton et al. (2012) compared functional connectivity in patients with bothersome tinnitus to that in controls and observed that auditory cortex was extensively connected to the superior temporal plane, to adjoining areas in the insula and lateral orbitofrontal cortex, and to anterior cingulate cortex. These areas tended to be more interconnected in patients than in controls. In addition, striking anticorrelations were observed between auditory and visual areas in patients but not in controls. Interestingly, the same lab applied their approach to patients with nonbothersome tinnitus (Wineland et al., 2012) as well as patients who could change their tinnitus using orofacial manipulations (Lee et al., 2012), but found no significant differences in those cases. The findings are furthermore reminiscent of similar changes in functional connectivity between networks that occur following sad mood induction (Harrison et al., 2008). Taken together, this suggests that the reported differences in cortical connectivity are conditional on the tinnitus being intrusive.

In summary, the above exposition shows that the fMRI-based functional connectivity literature suggests extensive changes related to tinnitus that include brain networks that are not nominally involved in sound processing (De Ridder et al., 2014; Georgiewa et al., 2006; Jastreboff, 1990; Salviati et al., 2014). Given the lack of a guiding experimental paradigm, the retrospective interpretation of the reported effects is not straightforward. This task is further complicated by the fact that an increase in functional connectivity can refer to a weak positive correlation becoming stronger, or to a negative correlation becoming weaker or flipping sign, for instance. Still, evidence suggests that in patients with tinnitus, the activity in auditory cortex is disproportionately correlated with the salience network and the affective limbic system, anticorrelated with the visual system, and decorrelated with the subcortical auditory nuclei, the homologous auditory cortex in the contralateral hemisphere, and the default mode network. Additional decorrelations with the dorsal attention network and the mnemonic limbic system may exist, although these effects are more ambiguous.

In our view, the picture that emerges from these investigations is that the auditory cortex couples more tightly to the regions in the brain that serve to evaluate the valence of incoming stimuli and to draw attention to them in a bottom-up fashion, while at the same time it disengages from networks related to top-down allocation of attention to competing nonauditory events, whether extrinsic (e.g., vision) or intrinsic (e.g., introspection and memory). This interpretation appears to be consistent with the inability of patients with chronic bothersome tinnitus to divert their attention away from their tinnitus percept. At the same time, the coherence within the central auditory system itself is decreased. These dissociations putatively arise from signals of abnormal origin being conducted along the auditory pathways, and may be more closely related to the emergence of the tinnitus percept itself.

Structural MRI

In addition to functional MRI that measures correlates of neural activity, it has been suggested that tinnitus may be accompanied by structural abnormalities in the brain. This link is plausible because all brain functions are fundamentally embodied in structural characteristics of synapses, neurons, and networks. Conversely, sustained changes in neural activity patterns may themselves induce structural brain changes. Structural abnormalities have for instance been shown in relation to hearing loss and deafness (Shibata, 2007) and changes in brain shape were also reported in relation to musical training and aptitude (Hyde et al., 2007). In this section, we briefly explain how such features are measured *in vivo* in humans using noninvasive MRI, and discuss the current evidence for tinnitus-related abnormalities (see also Adjamian et al., 2014).

Large-scale structural features of the brain are most commonly measured using two types of techniques that are referred to as surface- and voxel-based morphometry, respectively. Both methods aim to quantify morphologic measures of brain tissues or subdivisions (i.e., location, dimension, volume, shape, etc.). Surface-based morphometry is a conceptually simple approach that comprises the (automated) delineation of two 2-dimensional curved surfaces that bound the cortical gray matter, one corresponding with the outer pial boundary of the brain and another corresponding with the underlying boundary between gray and white matter. From these surfaces, the thickness, area, volume, and curvature of the cerebral cortex can be derived. Voxel-based morphometry is an alternative approach that operates on the basis of segmentations of 3-dimensional brain volumes. Image intensities in the individual anatomic images are combined with prior knowledge regarding the location of gray and white matter structures in order to assign a likelihood for each brain voxel to contain gray or white matter. This approach allows local differences in tissue composition to be assessed across individuals.

Perhaps the most obvious place to look for tinnitus-related abnormalities in brain morphology is the central auditory system. Schneider et al. (2009) reported a reduction by approximately one third in the thickness of the cortex in Heschl's gyrus that hosts primary auditory cortex, even after accounting for the effects of other confounds like age, gender, or body size. Interestingly, changes appeared to be related to tinnitus laterality: gray matter shrinkage occurred ipsilaterally to the affected ear in unilateral tinnitus, or on both sides in bilateral tinnitus. Although smaller on average, reductions in thickness of auditory cortex were similarly reported by Aldhafeeri et al. (2012). Schecklmann et al. (2013) argued that these effects were correlated particularly with the amount of tinnitus distress. Additional effects in the central auditory system have incidentally been reported in the form of a gray matter increase for the MGB (Mühlau et al., 2006) and a gray matter decrease for the IC (Landgrebe et al., 2009).

Given the substantial size of the reported effects in auditory cortex, it is surprising that numerous other studies neither revealed comparable tinnitus-related reductions nor confirmed a relation with tinnitus laterality. This may partly be related to the large variability in anatomy in the region of the auditory cortex (Rademacher et al., 2001). In an effort to understand cortical changes in more detail, two studies tried to disentangle the effects of tinnitus and hearing loss by including three subject groups: individuals with hearing loss with or without tinnitus and normal-hearing controls (Boyen et al., 2013; Husain et al., 2011). Although their results diverged in some other respects, both studies failed to reveal evidence for tinnitus-related decreases in cortical gray matter. Instead, any observed reductions were attributed to hearing loss—a comorbid factor that was poorly controlled for in previous studies. Tinnitus was even argued to have a preservative effect: Auditory cortex tends to shrink following a reduction of sensory input from the ear, but this decline may be averted if the input is replaced by "phantom" signals (i.e., tinnitus).

More specific tinnitus-related effects may possibly be found outside the central auditory system. Mühlau et al. (2006) published the earliest study that revealed morphologic changes in the NAc, an area underneath the anterior part of the corpus callosum. Further evidence in favor of the involvement of this region and nearby vmPFC was provided by the same lab (Leaver et al., 2011, 2012). Aldhafeeri et al. (2012) also confirmed the existence of prefrontal effects, but in less spatial detail. Altogether, significantly reduced gray matter was observed, which could not be explained by the measured hearing thresholds. These findings led to the formulation of the model by Rauschecker et al. (2010) that is based on a dysfunctional feedback loop comprising the limbic system, auditory and prefrontal cortices, and several thalamic formations, including the MGB.

Despite this evidence, the potential involvement of subcallosal and prefrontal regions, as well as their precise role, remains debated. Two independent studies were set up specifically to reproduce the original findings by Mühlau et al. (2006), but failed to do so even under relaxed statistical thresholds (Landgrebe et al., 2009; Melcher et al., 2013). These authors suggested that the discrepancies between outcomes might be due to differences in the particular subject characteristics, like tinnitus laterality or severity, or the presence of hearing loss at high frequencies. Therefore, independent confirmation of the findings that underlie the mentioned tinnitus model remains desirable to unequivocally settle the matter.

Finally, other nonauditory regions of the brain have been implicated in tinnitus based on observed anatomic brain differences. These include the cingulate cortex and insula (which together form the salience network), the hippocampus (involved in auditory memory and habituation), and the occipitoparietal cortex and cerebellum (involved in sensorimotor integration). Despite that many of the mentioned divisions occur in several functional as well as structural studies and are therefore assumed to play some role in tinnitus (De Ridder et al., 2014), most such reported structural results proved to be poorly reproducible so far.

Whereas MRI currently does not provide the ability to resolve tissue characteristics at the cellular level directly, certain imaging sequences are nevertheless sensitive to microscopic features. Most prominently, diffusion-weighted imaging (DWI) allows images to be constructed that depend on the diffusion of water molecules (Le Bihan et al., 2001). The ability to detect water diffusion is of interest because intra- and extracellular fluids cannot diffuse freely: Particularly in white matter tissue, diffusion is severely hindered by the presence of cell membranes, microtubules, and myelin sheaths that densely align along the principal direction of axonal tracts. Two quantities of interest that are commonly reported are the mean diffusivity (MD; measured in mm^2/s, and sometimes called the apparent diffusion coefficient, ADC) that quantifies the observed overall amount of diffusion averaged over all directions in space, and the fractional anisotropy (FA; measured in dimensionless units, and ranging from 0 to 1) that summarizes how much the diffusion is found to vary along different directions. Healthy white matter has low MD and high FA compared to unmyelinated gray matter or cerebrospinal fluid; increased MD or decreased FA therefore serve as surrogate markers of compromised white matter integrity.

Very few studies have applied DWI to assess the auditory tracks. Crippa et al. (2010) specifically targeted the lemniscal connections from the IC to the AC as part of the classical auditory pathway, as well as extralemniscal connections with the amygdala along the nonclassical pathway (Møller, 2007). Patients with tinnitus showed stronger connections compared to a control group as quantified by an increased anisotropy along the reconstructed fiber tracts, especially between the auditory cortex and amygdala. Their results were corroborated and expanded on by Seydell-Greenwald et al. (2014), who showed lower diffusivity and higher anisotropy in tinnitus patients than in controls in the vicinity of the acoustic radiations beneath the AC and IC. These results are suggestive of denser white matter tissue and stronger myelination, which can be interpreted to indicate more potent neural connections in patients with tinnitus.

A larger number of studies considered the major fiber bundles that connect the various lobes of the brain. Initially, widespread reductions in white matter integrity were shown (Aldhafeeri et al., 2012; Lee et al., 2007), including the anterior thalamic radiation and prefrontal areas (connecting the thalamus to the frontal lobe, or vice versa), the longitudinal fasciculi (connecting the brain lobes from front to back, v.v.), and the corpus callosum (connecting the hemispheres from left to right, v.v.). However, when controlling for hearing loss, the precise opposite was found: Anisotropy was elevated in patients with tinnitus, including in the (left) anterior thalamic radiation and longitudinal fasciculi (Benson et al., 2014). This agrees with other findings that indicate that white matter integrity may be decreased by hearing loss but is actually more likely to be increased by the presence of tinnitus (Husain et al., 2011; Seydell-Greenwald et al., 2014).

In summary, structural abnormalities have been found in both the gray and white matter of the central auditory system, but these are more likely attributable to the effects of peripheral hearing loss rather than to tinnitus itself. Tinnitus may even provide a substitute source of stimulation that prevents atrophy-like brain deteriorations in auditory cortex and distributed white matter degeneration from occurring. There is some evidence for gray matter decline in a subcallosal region comprising the NAc and vmPFC. However, these effects seem too small to be of clinical significance and remain to be reproduced more robustly. All other reported morphologic abnormalities appear to be spurious results that should be interpreted with caution until rigorously confirmed by other studies.

Discussion

In the past two decades, it has become clear that central neural mechanisms play a key role in tinnitus, generally combined with a peripheral hearing loss. From animal models and basic neuroscience, several possible mechanisms have been put forward in explaining tinnitus. These models mostly describe a brain response to a disturbed auditory input, caused by hearing loss. An increasing number of studies have tried to get a better understanding of these mechanisms in humans using functional and structural MRI. Taking all studies together, findings are diverse and show few clear outcomes. It seems that not only the central auditory system is involved, but also the limbic system, and possibly the frontal lobe and the cerebellum. Also, resting-state fluctuations in the whole brain seem to be stronger in patients with tinnitus. Furthermore, there are indications that functional connections within the central auditory system are reduced in patients with tinnitus.

For future studies, it has become clear that patient matching—especially with respect to hearing loss and hyperacusis—is very important when comparing subject groups with and without tinnitus. Moreover, more standardization of diverging paradigms may allow for an easier comparison of findings from different research groups, which in turn will increase our insight in tinnitus pathophysiology.

References

Adjamian, P., Hall, D. A., Palmer, A. R., Allan, T. W., & Langers, D. R. M. (2014). Neuroanatomical abnormalities in chronic tinnitus in the human brain. *Neuroscience and Biobehavioral Reviews, 45C*, 119–133.

Aldhafeeri, F. M., Mackenzie, I., Kay, T., Alghamdi, J., & Sluming, V. (2012). Neuroanatomical correlates of tinnitus revealed by cortical thickness analysis and diffusion tensor imaging. *Neuroradiology, 54*, 883–892.

Arbabshirani, M. R., Havlicek, M., Kiehl, K. A., Pearlson, G. D., & Calhoun, V. D. (2013). Functional network connectivity during rest and task conditions: A comparative study. *Human Brain Mapping, 34*, 2959–2971.

Auerbach, B. D., Rodrigues, P. V., & Salvi, R. J. (2014). Central gain control in tinnitus and hyperacusis. *Frontiers in Neurology, 5*, 206.

Baguley, D., Andersson, G., McFerran, D., & McKenna, L. (2013). *Tinnitus: A multidisciplinary approach.* Chichester, UK: Wiley-Blackwell.

Baguley, D. M., Phillips, J., Humphriss, R. L., Jones, S., Axon, P. R., & Moffat, D. A. (2006). The prevalence and onset of gaze modulation of tinnitus and increased sensitivity to noise after translabyrinthine vestibular schwannoma excision. *Otology & Neurotology, 27*, 220–224.

Beall, E. B., & Lowe, M. J. (2010). The nonseparability of physiologic noise in functional connectivity MRI with spatial ICA at 3T. *Journal of Neuroscience Methods, 191*, 263–276.

Benson, R. R., Gattu, R., & Cacace, A. T. (2014). Left hemisphere fractional anisotropy increase in noise-induced tinnitus: A diffusion tensor imaging (DTI) study of white matter tracts in the brain. *Hearing Research, 309*, 8–16.

Boyen, K., de Kleine, E., van Dijk, P., & Langers, D. R. M. (2014). Tinnitus-related dissociation between cortical and subcortical neural activity in humans with mild to moderate sensorineural hearing loss. *Hearing Research, 312*, 48–59.

Boyen, K., Langers, D. R. M., de Kleine, E., & van Dijk, P. (2013). Gray matter in the brain: Differences associated with tinnitus and hearing loss. *Hearing Research, 295*, 67–78.

Burton, H., Wineland, A., Bhattacharya, M., Nicklaus, J., Garcia, K. S., & Piccirillo, J. F. (2012). Altered networks in bothersome tinnitus: A functional connectivity study. *BMC Neuroscience, 13*, 3.

Cacace, A. T., Cousins, J. P., Moonen, C. T. W., van Gelderen, P., Miller, D., Parnes, S. M., & Lovely, T. J. (1996). *In-vivo* localization of phantom auditory perceptions during functional magnetic resonance imaging of the human brain. In *Proceedings of the Fifth International Tinnitus Seminar* (pp. 397–401). Portland, OR: American Tinnitus Association.

Cacace, A. T., Cousins, J. P., Parnes, S. M., Semenoff, D., Holmes, T., McFarland, D. J., Davenport, C., Stegbauer, K., & Lovely, T. J. (1999). Cutaneous-evoked tinnitus. I. Phenomenology, psychophysics and functional imaging. *Audiology & Neuro-Otology, 4*, 247–257.

Crippa, A., Lanting, C. P., van Dijk, P., & Roerdink, J. B. T. M. (2010). A diffusion tensor imaging study on the auditory system and tinnitus. *The Open Neuroimaging Journal, 4*, 16–25.

Davies, J., Gander, P. E., Andrews, M., & Hall, D. A. (2014). Auditory network connectivity in tinnitus patients: A resting-state fMRI study. *International Journal of Audiology, 53*, 192–198.

De Ridder, D., Vanneste, S., Weisz, N., Londero, A., Schlee, W., Elgoyhen, A. B., & Langguth, B. (2014). An integrative model of auditory phantom perception: Tinnitus as a unified percept of interacting separable subnetworks. *Neuroscience and Biobehavioral Reviews, 44C*, 16–32.

Dehmel, S., Cui, Y. L., & Shore, S. E. (2008). Cross-modal interactions of auditory and somatic inputs in the brainstem and midbrain and their imbalance in tinnitus and deafness. *American Journal of Audiology, 17*, S193–S209.

Fox, M. D., & Raichle, M. E. (2007). Spontaneous fluctuations in brain activity observed with functional magnetic resonance imaging. *Nature Reviews Neuroscience, 8*, 700–711.

Fransson, P. (2005). Spontaneous low-frequency BOLD signal fluctuations: An fMRI investigation of the resting-state default mode of brain function hypothesis. *Human Brain Mapping, 26*, 15–29.

Gavrilescu, M., Rossell, S., Stuart, G. W., Shea, T. L., Innes-Brown, H., Henshall, K., . . . Egan, G. F. (2010). Reduced connectivity of the auditory cortex in patients with auditory hallucinations: A resting state functional magnetic resonance imaging study. *Psychological Medicine, 40,* 1149–1158.

Georgiewa, P., Klapp, B. F., Fischer, F., Reisshauer, A., Juckel, G., Frommer, J., & Mazurek, B. (2006). An integrative model of developing tinnitus based on recent neurobiological findings. *Medical Hypotheses, 66,* 592–600.

Giraud, A. L., Chéry-Croze, S., Fischer, G., Fischer, C., Vighetto, A., Grégoire, M. C., . . . Collet, L. (1999). A selective imaging of tinnitus. *Neuroreport, 10,* 1–5.

Gu, J. W., Halpin, C. F., Nam, E.-C., Levine, R. A., & Melcher, J. R. (2010). Tinnitus, diminished sound-level tolerance, and elevated auditory activity in humans with clinically normal hearing sensitivity. *Journal of Neurophysiology, 104,* 3361–3370.

Hall, D. A., Haggard, M. P., Akeroyd, M. A., Palmer, A. R., Summerfield, A. Q., Elliott, M.R., . . . Bowtell, R. W. (1999). "Sparse" temporal sampling in auditory fMRI. *Human Brain Mapping, 7,* 213–223.

Harrison, B. J., Pujol, J., Ortiz, H., Fornito, A., Pantelis, C., & Yücel, M. (2008). Modulation of brain resting-state networks by sad mood induction. *PLoS One, 3,* e1794.

Husain, F. T., Medina, R. E., Davis, C. W., Szymko-Bennett, Y., Simonyan, K., Pajor, N. M., & Horwitz, B. (2011). Neuroanatomical changes due to hearing loss and chronic tinnitus: A combined VBM and DTI study. *Brain Research, 1369,* 74–88.

Husain, F. T., & Schmidt, S. A. (2014). Using resting state functional connectivity to unravel networks of tinnitus. *Hearing Research, 307,* 153–162.

Hyde, K. L., Lerch, J. P., Zatorre, R. J., Griffiths, T. D., Evans, A. C., & Peretz, I. (2007). Cortical thickness in congenital amusia: When less is better than more. *The Journal of Neuroscience, 27,* 13028–13032.

Jastreboff, P. J. (1990). Phantom auditory perception (tinnitus): Mechanisms of generation and perception. *Neuroscience Research, 8,* 221–254.

Kim, J., Kim, Y., Lee, S., Seo, J.-H., Song, H.-J., Cho, J. H., & Chang, Y. (2012). Alteration of functional connectivity in tinnitus brain revealed by resting-state fMRI? A pilot study. *International Journal of Audiology, 51,* 413–417.

Knipper, M., Van Dijk, P., Nunes, I., Rüttiger, L., & Zimmermann, U. (2013). Advances in the neurobiology of hearing disorders: Recent developments regarding the basis of tinnitus and hyperacusis. *Progress in Neurobiology, 111,* 17–33.

Kovacs, S., Peeters, R., Smits, M., De Ridder, D., Van Hecke, P., & Sunaert, S. (2006). Activation of cortical and subcortical auditory structures at 3 T by means of a functional magnetic resonance imaging paradigm suitable for clinical use. *Investigative Radiology, 41,* 87–96.

Landgrebe, M., Langguth, B., Rosengarth, K., Braun, S., Koch, A., Kleinjung, T., . . . Hajak, G. (2009). Structural brain changes in tinnitus: grey matter decrease in auditory and non-auditory brain areas. *NeuroImage, 46,* 213–218.

Langers, D. R. M. (2014). Assessment of tonotopically organised subdivisions in human auditory cortex using volumetric and surface-based cortical alignments. *Human Brain Mapping, 35,* 1544–1561.

Langers, D. R. M., de Kleine, E., & van Dijk, P. (2012). Tinnitus does not require macroscopic tonotopic map reorganization. *Frontiers in Systems Neuroscience, 6,* 2.

Langers, D. R. M., & Melcher, J.R. (2011). Hearing without listening: Functional connectivity reveals the engagement of multiple nonauditory networks during basic sound processing. *Brain Connectivity, 1,* 233–244.

Langers, D. R. M., & van Dijk, P. (2011). Robustness of intrinsic connectivity networks in the human brain to the presence of acoustic scanner noise. *NeuroImage, 55*, 1617–1632.

Langguth, B., Kreuzer, P. M., Kleinjung, T., & De Ridder, D. (2013). Tinnitus: Causes and clinical management. *Lancet Neurology, 12*, 920–930.

Lanting, C. P., De Kleine, E., Bartels, H., & Van Dijk, P. (2008). Functional imaging of unilateral tinnitus using fMRI. *Acta Oto-Laryngologica, 128*, 415–421.

Lanting, C. P., de Kleine, E., Eppinga, R. N., & van Dijk, P. (2010). Neural correlates of human somatosensory integration in tinnitus. *Hearing Research, 267*, 78–88.

Lanting, C. P., de Kleine, E., Langers, D. R. M., & van Dijk, P. (2014). Unilateral tinnitus: Changes in connectivity and response lateralization measured with fMRI. *PLoS One, 9*, e110704.

Leaver, A. M., Renier, L., Chevillet, M. A., Morgan, S., Kim, H. J., & Rauschecker, J. P. (2011). Dysregulation of limbic and auditory networks in tinnitus. *Neuron, 69*, 33–43.

Leaver, A. M., Seydell-Greenwald, A., Turesky, T. K., Morgan, S., Kim, H. J., & Rauschecker, J. P. (2012). Cortico-limbic morphology separates tinnitus from tinnitus distress. *Frontiers in Systems Neuroscience, 6*, 21.

Le Bihan, D., Mangin, J. F., Poupon, C., Clark, C. A., Pappata, S., Molko, N., & Chabriat, H. (2001). Diffusion tensor imaging: Concepts and applications. *Journal of Magnetic Resonance Imaging: JMRI, 13*, 534–546.

Lee, M. H., Solowski, N., Wineland, A., Okuyemi, O., Nicklaus, J., Kallogjeri, D., . . . Burton, H. (2012). Functional connectivity during modulation of tinnitus with orofacial maneuvers. *Otolaryngology-Head and Neck Surgery, 147*, 757–762.

Lee, Y.-J., Bae, S.-J., Lee, S.-H., Lee, J.-J., Lee, K.-Y., Kim, M.-N., . . . Chang, Y. (2007). Evaluation of white matter structures in patients with tinnitus using diffusion tensor imaging. *Journal of Clinical Neuroscience, 14*, 515–519.

Levine, R. A., Nam, E. C., Oron, Y., & Melcher, J. R. (2007). Evidence for a tinnitus subgroup responsive to somatosensory-based treatment modalities. *Progress in Brain Research, 166*, 195–207.

Llinás, R. R., Ribary, U., Jeanmonod, D., Kronberg, E., & Mitra, P. P. (1999). Thalamocortical dysrhythmia: A neurological and neuropsychiatric syndrome characterized by magnetoencephalography. *Proceedings of the National Academy of Sciences of the United States of America, 96*, 15222–15227.

Lockwood, A. H., Salvi, R. J., Coad, M. L., Towsley, M. L., Wack, D. S., & Murphy, B. W. (1998). The functional neuroanatomy of tinnitus: Evidence for limbic system links and neural plasticity. *Neurology, 50*, 114–120.

Lockwood, A. H., Wack, D. S., Burkard, R. F., Coad, M. L., Reyes, S. A., Arnold, S. A., & Salvi, R.J. (2001). The functional anatomy of gaze-evoked tinnitus and sustained lateral gaze. *Neurology, 56*, 472–480.

Maudoux, A., Lefebvre, P., Cabay, J.-E., Demertzi, A., Vanhaudenhuyse, A., Laureys, S., & Soddu, A. (2012a). Connectivity graph analysis of the auditory resting state network in tinnitus. *Brain Research, 1485*, 10–21.

Maudoux, A., Lefebvre, P., Cabay, J.-E., Demertzi, A., Vanhaudenhuyse, A., Laureys, S., & Soddu, A. (2012b). Auditory resting-state network connectivity in tinnitus: A functional MRI study. *PLoS One, 7*, e36222.

Meeus, O. M., Spaepen, M., Ridder, D. D., & de Heyning, P. H. V. (2010). Correlation between hyperacusis measurements in daily ENT practice. *International Journal of Audiology, 49*, 7–13.

Melcher, J. R., Knudson, I. M., & Levine, R. A. (2013). Subcallosal brain structure: Correlation with hearing threshold at supra-clinical frequencies (>8 kHz), but not with tinnitus. *Hearing Research, 295*, 79–86.

Melcher, J. R., Levine, R. A., Bergevin, C., & Norris, B. (2009). The auditory midbrain of people with tinnitus: Abnormal sound-evoked activity revisited. *Hearing Research, 257*, 63–74.

Melcher, J. R., Sigalovsky, I. S., Guinan, J. J., & Levine, R. A. (2000). Lateralized tinnitus studied with functional magnetic resonance imaging: Abnormal inferior colliculus activation. *Journal of Neurophysiology, 83*, 1058–1072.

Møller, A. R. (2007). The role of neural plasticity in tinnitus. *Progress in Brain Research, 166*, 37–45.

Mühlau, M., Rauschecker, J. P., Oestreicher, E., Gaser, C., Röttinger, M., Wohlschläger, A. M., . . . Sander, D. (2006). Structural brain changes in tinnitus. *Cerebral Cortex, 16*, 1283–1288.

Noreña, A. J. (2011). An integrative model of tinnitus based on a central gain controlling neural sensitivity. *Neuroscience and Biobehavioral Reviews, 35*, 1089–1109.

Rademacher, J., Morosan, P., Schormann, T., Schleicher, A., Werner, C., Freund, H. J., & Zilles, K. (2001). Probabilistic mapping and volume measurement of human primary auditory cortex. *NeuroImage, 13*, 669–683.

Rauschecker, J. P., Leaver, A. M., & Mühlau, M. (2010). Tuning out the noise: Limbic-auditory interactions in tinnitus. *Neuron, 66*, 819–826.

Salviati, M., Bersani, F. S., Valeriani, G., Minichino, A., Panico, R., Romano, G. F., . . . Cianfrone, G. (2014). A brain centred view of psychiatric comorbidity in tinnitus: From otology to hodology. *Neural Plasticity, 2014*, 817852.

Schecklmann, M., Lehner, A., Poeppl, T. B., Kreuzer, P. M., Rupprecht, R., Rackl, J., . . . Landgrebe, M. (2013). Auditory cortex is implicated in tinnitus distress: A voxel-based morphometry study. *Brain Structure & Function, 218*, 1061–1070.

Schmidt, S. A., Akrofi, K., Carpenter-Thompson, J. R., & Husain, F. T. (2013). Default mode, dorsal attention and auditory resting state networks exhibit dif-ferential functional connectivity in tinnitus and hearing loss. *PLoS One, 8*, e76488.

Schneider, P., Andermann, M., Wengenroth, M., Goebel, R., Flor, H., Rupp, A., & Diesch, E. (2009). Reduced volume of Heschl's gyrus in tinnitus. *NeuroImage, 45*, 927–939.

Seydell-Greenwald, A., Leaver, A. M., Turesky, T. K., Morgan, S., Kim, H. J., & Rauschecker, J. P. (2012). Functional MRI evidence for a role of ventral prefrontal cortex in tinnitus. *Brain Research, 1485*, 22–39.

Seydell-Greenwald, A., Raven, E. P., Leaver, A. M., Turesky, T. K., & Rauschecker, J. P. (2014). Diffusion imaging of auditory and auditory-limbic connectivity in tinnitus: Preliminary evidence and methodological challenges. *Neural Plasticity, 2014*, 145943.

Shibata, D. K. (2007). Differences in brain structure in deaf persons on MR imaging studied with voxel-based morphometry. *American Journal of Neuroradiology, 28*, 243–249.

Shore, S. E. (2011). Plasticity of somatosensory inputs to the cochlear nucleus—implications for tinnitus. *Hearing Research, 281*, 38–46.

Shore, S., Zhou, J., & Koehler, S. (2007). Neural mechanisms underlying somatic tinnitus. *Progress in Brain Research, 166*, 107–123.

Shore, S. E., Koehler, S., Oldakowski, M., Hughes, L. F., & Syed, S. (2008). Dorsal cochlear nucleus responses to somatosensory stimulation are enhanced after noise-induced hearing loss. *European Journal of Neuroscience, 27*, 155–168.

Simmons, R., Dambra, C., Lobarinas, E., Stocking, C., & Salvi, R. (2008). Head, neck, and eye movements that modulate tinnitus. *Seminars in Hearing, 29*, 361–370.

Smits, M., Kovacs, S., de Ridder, D., Peeters, R. R., van Hecke, P., & Sunaert, S. (2007). Lateralization of functional magnetic resonance imaging (fMRI) activation in the auditory pathway of patients with lateralized tinnitus. *Neuroradiology, 49*, 669–679.

Stolzberg, D., Chen, G.-D., Allman, B. L., & Salvi, R. J. (2011). Salicylate-induced peripheral auditory changes and tonotopic reorganization of auditory cortex. *Neuroscience, 180*, 157–164.

van Gendt, M. J., Boyen, K., de Kleine, E., Langers, D. R. M., & van Dijk, P. (2012). The relation between perception and brain activity in gaze-evoked tinnitus. *The Journal of Neuroscience, 32*, 17528–17539.

Wineland, A. M., Burton, H., & Piccirillo, J. (2012). Functional connectivity networks in nonbothersome tinnitus. *Otolaryngology-Head and Neck Surgery, 147*, 900–906.

Won, J. Y., Yoo, S., Lee, S. K., Choi, H. K., Yakunina, N., Le, Q., & Nam, E.-C. (2013). Prevalence and factors associated with neck and jaw muscle modulation of tinnitus. *Audiology & Neuro-Otology, 18*, 261–273.

Index

Note: Page numbers in **bold** reference non-text material.

A

D

E

F

G

H

N

O

P

Q

R

S

U

V

W